CRC SERIES IN AGRICULTURE

Editor-in-Chief

Angus A. Hanson, Ph.D.
Vice President
and
Director of Research
W-L Research, Inc.
Highland, Maryland

HANDBOOK OF SOILS AND CLIMATE IN AGRICULTURE

Editor
Victor J. Kilmer (Deceased)
Chief
Soils and Fertilizer Research Branch
National Fertilizer Development Center
Tennessee Valley Authority
Muscle Shoals, Alabama

HANDBOOK OF PLANT SCIENCE IN AGRICULTURE

Editor
B. R. Christie, Ph.D.
Professor
Department of Crop Science
Ontario Agricultural College
University of Guelph
Guelph, Ontario, Canada

HANDBOOK OF PEST MANAGEMENT IN AGRICULTURE

Editor
David Pimentel, Ph.D.
Professor
Department of Entomology
New York College of Agricultural
and Life Sciences
Cornell University
Ithaca, New York

HANDBOOK OF ENGINEERING IN AGRICULTURE

Editor
R. H. Brown, Ph.D.
Chairman
Division of Agricultural Engineering
Agricultural Engineering Center
University of Georgia
Athens, Georgia

HANDBOOK OF TRANSPORTATION AND MARKETING IN AGRICULTURE

Editor
Essex E. Finney, Jr., Ph.D.
Assistant Center Director
Agricultural Research Center
U.S. Department of Agriculture
Beltsville, Maryland

HANDBOOK OF PROCESSING AND UTILIZATION IN AGRICULTURE

Editor
Ivan A. Wolff, Ph.D. (Retired)
Director
Eastern Regional Research Center
Science and Education Administration
U.S. Department of Agriculture
Philadelphia, Pennsylvania

HANDBOOK OF ANIMAL SCIENCE
IN AGRICULTURE

Handbook of Soils and Climate in Agriculture

Editor

Victor J. Kilmer
Chief, Soils and Fertilizer Research Branch (Deceased)
Division of Agricultural Development
Tennessee Valley Authority
Muscle Shoals, Alabama

CRC Series in Agriculture

Editor-in-Chief

A. A. Hanson
Vice President and
Director of Research
W-L Research, Inc.
Highland, Maryland

CRC Press, Inc.
Boca Raton, Florida

Library of Congress Cataloging in Publication Data
Main entry under title:

Handbook of soils and climate in agriculture

(CRC series in agriculture)
Bibliography.
Includes index.
1. Soils and climate. I. Kilmer, Victor
James, 1913- . II. Series.
S596.3.H36 631.4 81-9960
ISBN 0-8493-3811-7 AACR2

This book represents information obtained from authentic and highly regarded sources. Reprinted material is quoted with permission, and sources are indicated. A wide variety of references are listed. Every reasonable effort has been made to give reliable data and information, but the author and the publisher cannot assume responsibility for the validity of all materials or for the consequences of their use.

Direct all inquiries to CRC Press, Inc., 2000 Corporate Blvd., N.W., Boca Raton, Florida, 33431.

International Standard Book Number 0-8493-3811-7

Library of Congress Card Number 81-9960
Printed in the United States

PREFACE

CRC SERIES IN AGRICULTURE

Agriculture, because of its pivotal role in the development of civilized societies, contributed much to the development of various scientific disciplines. Thus, agricultural pursuits led to the practical application of chemistry, and gave rise to such major disciplines as economics and statistics. The expansion of scientific frontiers, and the concomitant specilization within disciplines, has proceeded to the point where agricultural scientists classify themselves in an array of disciplines and subdisciplines, i.e., nematologist, geneticist, physicist, virologist, and so forth. Nevertheless, within the framework of these various disciplines and mission oriented agricultural research, information of primary interest and concern in the solution of agriculturally oriented problems is generated. Although some of the basic information finds its way into the plethora of reference books available within most disciplines, no attempt has been made to develop a comprehensive handbook series for the agricultural sciences.

It is recognized that there are serious difficulties in developing a meaningful handbook series in agriculture because of the range and complexity of agricultural enterprises. In fact, the single common denominator that applies to all agricultural scientists is their universal concern with at least some aspect of the production and utilization of farm products. The disciplines and resources that are called for in a specific investigation are either the same or similar to those utilized in any area of biological research, or in any one of several fields of scientific endeavor.

The sections in this handbook series reflect the input of different editors and advisory boards, and as a consequence, there is considerable variation in both the depth and coverage offered within a given area. However, an attempt has been made throughout to bring together pertinent information that will serve the needs of nonspecialists, provide a quick reference to material that might otherwise be difficult to locate, and furnish a starting point for further study.

The project was undertaken with the realization that the initial volumes in the series could have some obvious deficiencies that will necessitate subsequent revisions. In the meantime, it is felt that the primary objectives of the Section Editors and their Advisory Boards has been met in this first edition.

A. A. Hanson
Editor-in-Chief

PREFACE

This volume is one of a series within the CRC Series in Agriculture. The aim throughout has been to present in condensed form reliable information on soil science and climate as it relates to crop production. Convenience in form and the possibility of wide use and distribution have been kept uppermost in mind during the preparation of this volume. The tables and figures have been compiled especially for this Handbook by highly qualified and competent scientists who are recognized authorities in their respective fields of specialization.

An attempt has been made to include material pertaining to all major branches of soil science. However, a decision was made to exclude types of data of use only in highly specialized lines of work. The final product is not intended as a guide to agricultural practices, because specific recommendations are constantly changing on the basis of new research findings.

Soils are systems of extreme complexity and variability. Consequently, soils tend to defy accurate characterization on a broad scale by means of absolute constants and fixed values. Soil systems invariably exhibit ranges in particular properties and characteristics. The difficulties involved in attempting to quantify long range climatic trends are well known to both professional meterologists and laymen. In spite of these obstacles, it is hoped that the Handbook will be of significant value to all who seek and use information concerning soils and climate.

Lastly, it is appropriate that this volume be dedicated to Dr. Gilbert L. Terman. Dr. Terman's long and valuable contributions to soil science and agronomy included a service as a coeditor of this volume. Characteristically, he performed these duties in a highly competent and efficient manner to the very end of his life, which occurred on January 31, 1979.

THE EDITOR-IN-CHIEF

Angus A. Hanson, Ph.D., is Vice President and Director of Research, W-L Research, Inc., Highland, Maryland, and has had broad experience in agricultural research and development. He is a graduate of the University of British Columbia, Vancouver, McGill University, Quebec, and received the Ph.D. degree from the Pennsylvania State University, University Park, in 1951.

An employee of the U.S. Department of Agriculture from 1949 to 1979, Dr. Hanson worked as a Research Geneticist at University Park, 1949 to 1952, and at Beltsville, Maryland, serving successively as Research Leader for Grass and Turf Investigations, 1953 to 1965, Chief of the Forage and Range Research Branch, 1965 to 1972, and Director of the Beltsville Agricultural Research Center, 1972 to 1979. He has been appointed to a number of national and regional task forces charged with assessing research needs and priorities, and has participated in reviewing agricultural needs and research programs in various foreign countries. As Director at Beltsville, he was directly responsible for programs that included most dimensions of agricultural research.

In his personal research, he has emphasized the improvement of forage crops, breeding and management of turfgrasses, and the breeding of alfalfa for multiple pest resistance, persistence, quality, and sustained yield. He is the author of over 100 technical and popular articles on forage crops and turfgrasses, and has served as Editor of *Crop Science* and the *Journal of Environmental Quality.*

THE EDITOR

Victor J. Kilmer was a native of Wisconsin and majored in soil science at the University of Wisconsin. Following completion of the M.S. Degree in 1941, he accepted a position with the U.S. Department of Agriculture in the Upper Mississippi Valley Soil Conservation Experiment Station at LaCrosse, Wisconsin. Following 4 years of military service during World War II, he returned to the University of Wisconsin on special assignment, and then spent a year as a soil surveyor in central Wisconsin. In 1947 he joined the U.S. Department of Agriculture at the Plant Industry Station, Beltsville, Maryland as a soil scientist. He became a staff member on the Tennessee Valley Authority National Fertilizer Development Center, Muscle Shoals, Alabama, in 1961, where he served as Chief of the Soils and Fertilizer Research Branch from 1963 until retirement in 1978.

In addition to his duties at the Center, he was a consultant to other government agencies and universities both in the U.S. and in foreign countries.

He was a fellow of the American Society of Agronomy and the Soil Science Society of America, and president of the latter organization in 1977. He was a member of the Council of Agricultural Science and Technology, and the International Society of Soil Science.

He contributed to technical journals, books and encyclopedias as an author and as an editor, and participated in numerous national and international symposia.

His primary research interest was in the areas of soil classification, plant nutrient losses from soils, soil fertility and methods of soil analysis.

ADVISORY BOARD AND CONTRIBUTORS

Donald R. Christenson, Ph.D.
Professor
Crop and Soil Sciences Department
Michigan State University
East Lansing, Michigan

Frank B. Dazzo, Ph.D.
Assistant Professor
Department of Microbiology and Public Health
Michigan State University
East Lansing, Michigan

William A. Jury, Ph.D.
Associate Professor
Department of Soil and Environmental Sciences
University of California
Riverside, California

Richard H. Loeppert, Jr., Ph.D.
Assistant Professor
Soil and Crop Sciences Department
Texas A & M University
College Station, Texas

Jerry V. Mannering, Ph.D.
Professor
Department of Agronomy
Purdue University
West Lafayette, Indiana

Robert D. Munson, Ph.D.
Midwest Director
Potash and Phosphate Institute
St. Paul, Minnesota

Robert H. Shaw, Ph.D.
Distinguished Professor
Department of Agronomy
Iowa State University
Ames, Iowa

Roy W. Simonson, Ph.D.
Director (Ret.)
Soil Classification and Correlation
Soil Conservation Service
U.S. Department of Agriculture
Washington, District of Columbia

Edward L. Skidmore, Ph.D.
Soil Scientist
Wind Erosion Research Unit
U.S. Department of Agriculture
Agronomy Department
Kansas State University
Manhattan, Kansas

Lewis H. Stolzy, Ph.D.
Professor
Department of Soil and Environmental Sciences
University of California
Riverside, California

Grant W. Thomas, Ph.D.
Professor
Agronomy Department
University of Kentucky
Lexington, Kentucky

James M. Tiedje, Ph.D.
Professor
Department of Crop and Soil Sciences
Mighigan State University
East Lansing, Michigan

Bob G. Volk, Ph.D.
Associate Professor
Soil Science Department
University of Florida
Gainesville, Florida

TABLE OF CONTENTS

CLIMATE OF THE UNITED STATES

R. H. Shaw

Climate is a term used to describe the average weather conditions over several years. *Weather* is the day-to-day fluctuation in temperature, rainfall, and associated conditions. To better understand weather and climate, a brief look at the factors which control the climate is necessary. These nine principal factors are the sun, land and water surfaces, altitude, mountain barriers, wind and pressure patterns, ocean currents, storms, and human beings. These nine factors interact to determine weather and climate.

The sun is the major controller of the world's climate. Solar energy is received in different intensities and durations. Maximum solar energy on a clear day is received when the sun is overhead. Day length varies with latitude. At the equator (0° latitude) the day is always about 12 hr long. At the poles it varies from 24 to 0 hr. In the middle latitudes it varies from 15 to 16 hr in midsummer to 8 to 9 hr in midwinter.

Land surfaces tend to warm and cool faster than water surfaces. This causes the weather in the middle of large land masses to have large temperature fluctuations, both daily and annually, compared with islands and areas close to water. Crops growing in the land-dominated areas are subjected to greater temperature fluctuations.

Where mountains are present, temperatures decrease with increasing altitude at a rate near 3.3°F/1000 ft (2°C/300 m). The growing season also shortens with elevation. These same mountains may also act as barriers to the wind flow. Rainfall may be high on the ocean side of the mountains, but low on the downwind side, i.e., causing the dry Great Plains area of the U.S.

Air masses determine the type of air brought into a region. Those that originate over oceans may bring significant precipitation to land areas, while those that originate over land tend to be dry. Winds and pressure patterns also help determine the kind of air present or moving into a region. Areas dominated by high pressure tend to be dry. Those dominated by low pressure areas tend to be wet.

Ocean currents have a particularly important effect on coastal areas, but may affect the climate many miles inland if winds are onshore and no mountain barriers are present. Storms are an important feature of the climate because much of the summer rainfall is received from thunderstorms. These same storms cause hail and wind damage to crops. Without these storms a drought may occur.

Human actions are now becoming important factors affecting climate because of the air pollution created. Climatologists do not agree upon what the overall effect is, but it may be changing the climate in some areas.

The U.S. has a wide range of climates, but in terms of modern climate classification, only three of six climate zones of the world have extensive areas in the lower 48 states. These are favorable climates for agriculture, particularly grain production. Trewartha and Horn[1] have presented a classification which is useful for such a comparison (Table 1, Figure 1). Only the tip of Florida has a tropical (A) climate. The subtropical climate (C) is located in the southeastern U.S. (humid type) and in California (dry summer type). A very extensive area of temperate climate (D) is present. The warmest subgroup covers the "Corn Belt", with the cooler subgroup area growing shorter-season grains. A marine temperate climate is present on the northern west coast and extends into southern Alaska. The Great Plains area has a dry semiarid steppe climate with desert regions in the southwest U.S. Hawaii has both A and B climates. Alaska has mostly E and F climates.

Table 1
CLIMATE GROUPS

Group	Description of climate
A	Tropical — frost free; may be rainy all year (r) or winter dry (w)
B	Dry — potential evaporation exceeds rainfall; may be steppe or desert
C	Subtropical — 8 to 12 months 50°F (10°C) or above; (f) has no dry season, (s) has dry summer
D	Temperate — 4 to 7 months 50°F (10°C) or above; (c) is continental, (o) is oceanic, (a) has hot summer, (b) has cool summer
E	Boreal — 1 to 3 months 50°F (10°C) or above
F	Polar — all months below 50°F; (t) is tundra

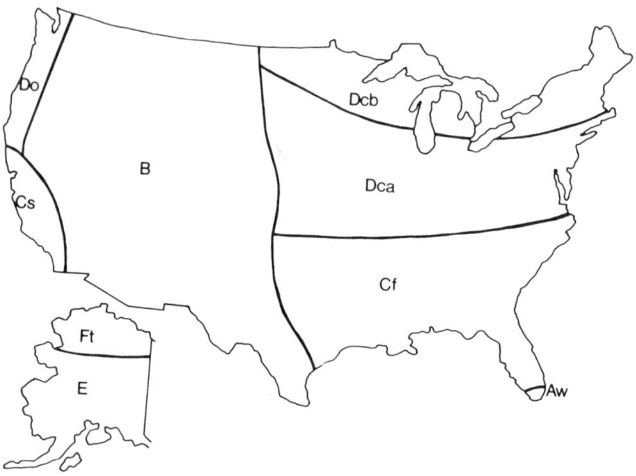

FIGURE 1. Climatic zones of the United States according to the Trewartha climate system. The Hawaiian Islands (not shown) would generally have A climates on the windward sides with some B climates on the leeward sides. C types may be found in the mountains.

PRECIPITATION

Precipitation is the water equivalent of all forms of moisture which fall — rain, drizzle, snow, hail, ice pellets, and glaze ice. The normal (30-year average) precipitation for selected stations is given in Table 2 for each month and for the annual total. The monthly and annual snowfall amounts are given in Table 3.

TEMPERATURE

Temperature is recorded for each day as the highest (maximum) and lowest (minimum) reading. The normal maximum temperatures are given in Table 4, the normal minimum temperatures in Table 5. The average for each month can be obtained by

$$\frac{\text{maximum} + \text{minimum}}{2} = \text{average}$$

The highest temperature of record for each month is given in Table 6, the lowest in Table 7.

The mean date of the last occurrence of 32°F in the spring, the first occurrence in the fall, and the length of the freeze-free period are summarized in Table 8.

Growing degree days for warm temperature crops have been calculated by use of the following equation:

$$\frac{\text{maximum + minimum}}{2} \quad 50 = \text{growing degree days}$$

In this procedure any minimum below 50°F is designated as 50°F, any maximum above 86°F is designated as 86°F, and growing degree days cannot be less than zero. These values are summarized in Table 9.

OTHER FACTORS

Relative Humidity

Maximum relative humidity values usually occur in the morning, with minimum values occurring in the afternoon. The average values for selected morning and afternoon values are given in Table 10.

Sunshine and Radiation

Sunshine is recorded as the percent of possible sunshine for each day. This gives the percent of the time the sun is shining above a selected level, but does not tell how much above this base. Data for selected stations are presented in Table 11. Some stations record the intensity of solar radiation (g cal/cm^2/day = langleys/day). Data for these stations are given in Table 12.

Evaporation

The class A evaporation pan is used as a standard for evaporation. This is a 4-ft diameter pan. Evaporation values give a measure of the evaporation power of the atmosphere. Data for selected stations are given in Table 13.

Table 2
AVERAGE INCHES OF PRECIPITATION IN THE U.S. AND CERTAIN PACIFIC ISLANDS (NORMALS 1941-1970)[2]

Location	Years	Jan.	Feb.	Mar.	April	May	June	July	Aug.	Sept.	Oct.	Nov.	Dec.	Annual
Alabama														
Birmingham	30	4.84	5.28	6.17	4.62	3.62	4.00	5.22	4.31	3.64	2.58	3.72	5.23	53.23
Huntsville	30	5.13	5.16	5.79	4.79	3.86	3.97	4.88	3.46	3.29	2.57	3.88	5.38	52.15
Mobile	30	4.71	4.76	7.07	5.59	4.52	6.09	8.86	6.93	6.59	2.55	3.39	5.92	66.98
Montgomery	30	4.02	4.30	6.02	4.45	3.47	4.03	5.09	3.47	4.41	2.24	3.43	4.93	49.86
Alaska														
Anchorage	30	0.84	0.84	0.56	0.56	0.59	1.07	2.07	2.32	2.37	1.43	1.02	1.07	14.74
Annette	30	10.42	9.27	9.15	8.77	6.20	5.05	5.42	7.15	10.01	17.85	12.87	12.14	114.30
Barrow	30	0.23	0.20	0.19	0.21	0.17	0.35	0.88	1.04	0.58	0.55	0.30	0.19	4.89
Barter Island	30	0.55	0.33	0.26	0.23	0.31	0.53	1.12	1.28	0.89	0.81	0.45	0.29	7.05
Bethel	30	0.54	0.74	0.79	0.43	0.83	1.24	1.98	3.97	2.42	1.32	0.96	0.62	15.84
Bettles	30	0.72	0.77	0.82	0.63	0.62	1.22	1.79	2.77	1.78	1.23	0.81	0.82	14.18
Big Delta	30	0.36	0.27	0.33	0.31	0.94	2.20	2.49	1.92	1.23	0.56	0.41	0.42	11.44
Cold Bay	30	2.42	2.59	1.93	1.54	2.19	1.84	2.22	3.89	3.95	4.31	3.90	2.45	33.23
Fairbanks	30	0.60	0.53	0.48	0.33	0.65	1.42	1.90	2.19	1.08	0.73	0.66	0.65	11.22
Gulkana	30	0.58	0.47	0.34	0.22	0.63	1.34	1.84	1.58	1.72	0.88	0.75	0.76	11.11
Homer	30	1.70	1.54	1.22	1.09	0.91	1.06	1.70	2.56	2.85	3.38	2.76	2.29	23.08
Juneau	30	3.94	3.44	3.57	2.99	3.31	2.93	4.69	5.00	6.90	7.85	5.53	4.52	54.67
King Salmon	30	0.94	0.99	1.16	0.90	1.13	1.44	2.18	3.46	3.07	2.00	1.43	1.05	19.75
Kodiak	30	5.01	4.89	3.85	3.81	4.35	4.12	3.54	4.30	6.11	6.29	5.41	5.03	56.71
Kotzebue	30	0.29	0.30	0.33	0.33	0.40	0.52	1.55	2.26	1.43	0.61	0.41	0.33	8.78
McGrath	30	0.85	0.90	0.86	0.66	0.80	1.70	2.28	3.28	2.14	1.22	1.03	1.02	16.74
Nome	30	0.90	0.84	0.79	0.73	0.70	0.95	2.42	3.57	2.40	1.42	0.98	0.74	16.44
St. Paul Island	30	1.84	1.37	1.26	1.09	1.28	1.25	2.23	3.44	3.02	3.15	2.58	2.03	24.54
Talkeetna	30	1.63	1.79	1.54	1.12	1.46	2.17	3.48	4.89	4.52	2.54	1.79	1.71	28.54
Unalakleet	30	0.50	0.53	0.60	0.49	0.71	0.92	2.32	3.89	2.20	1.03	0.56	0.41	14.16
Valdez	30	5.06	5.30	4.33	3.06	3.20	2.70	4.31	5.80	7.74	6.75	5.67	5.39	59.31
Yakutat	30	10.36	9.28	9.57	7.65	8.02	5.68	8.46	10.81	15.45	19.52	14.80	12.86	132.46
Arizona														
Flagstaff	30	1.89	1.47	1.83	1.33	0.56	0.57	2.48	2.65	1.68	1.37	1.33	2.15	19.31
Phoenix	30	0.71	0.60	0.76	0.32	0.14	0.12	0.75	1.22	0.69	0.46	0.46	0.82	7.05

Tucson	30	0.77	0.70	0.64	0.35	0.14	0.20	2.38	2.34	1.37	0.66	0.56	0.94	11.05
Winslow	30	0.42	0.38	0.44	0.36	0.28	0.28	1.18	1.50	0.85	0.69	0.36	0.59	7.33
Yuma	30	0.38	0.27	0.21	0.11	0.03	0.00	0.18	0.44	0.22	0.27	0.22	0.34	2.67
Arkansas														
Fort Smith	30	2.38	3.20	3.64	4.74	5.48	3.93	3.24	2.91	3.31	3.47	3.08	2.89	42.27
Little Rock	30	4.24	4.42	4.93	5.25	5.30	3.50	3.38	3.01	3.55	2.99	3.86	4.09	48.52
California														
Bakersfield	30	0.96	1.03	0.83	0.85	0.19	0.06	0.02	0.01	0.08	0.26	0.69	0.74	5.72
Bishop	30	1.20	1.06	0.43	0.41	0.27	0.09	0.17	0.10	0.10	0.26	0.58	1.05	5.72
Blue Canyon	30	13.66	9.42	8.55	5.47	3.14	0.99	0.10	0.27	0.52	4.14	9.04	12.28	67.58
Eureka	30	7.42	5.15	4.83	2.95	2.11	0.66	0.14	0.27	0.65	3.23	5.77	6.58	39.76
Fresno	30	1.84	1.72	1.62	1.24	0.32	0.06	0.00	0.02	0.07	0.42	1.22	1.71	10.24
Long Beach	30	2.26	2.16	1.50	0.89	0.07	0.04	0.00	0.02	0.09	0.19	1.38	1.65	10.25
Los Angeles														
International Airport	30	2.52	2.32	1.71	1.10	0.08	0.03	0.01	0.02	0.07	0.22	1.76	1.75	11.59
City	30	3.00	2.77	2.19	1.27	0.13	0.03	0.00	0.04	0.17	0.27	2.02	2.16	14.05
Mount Shasta	30	6.65	5.61	4.03	3.05	1.87	1.08	0.32	0.31	0.69	2.54	5.09	6.25	37.49
Oakland	30	4.03	2.83	2.32	1.58	0.55	0.14	0.01	0.03	0.18	1.08	2.37	3.57	18.69
Red Bluff	30	4.48	3.17	2.51	1.79	0.98	0.47	0.04	0.18	0.31	1.17	3.05	3.91	22.06
Sacramento	30	3.73	2.68	2.17	1.54	0.51	0.10	0.01	0.05	0.19	0.99	2.13	3.12	17.22
Sandberg	30	2.15	2.45	1.33	1.12	0.25	0.03	0.03	0.06	0.21	0.34	2.05	1.95	11.97
San Diego	30	1.88	1.48	1.55	0.81	0.15	0.05	0.01	0.07	0.13	0.34	1.25	1.73	9.45
San Francisco														
International Airport	30	4.37	3.04	2.54	1.59	0.41	0.13	0.01	0.03	0.16	0.98	2.29	3.98	19.53
City	30	4.51	2.97	2.77	1.63	0.54	0.17	0.01	0.05	0.17	1.06	2.60	4.18	20.66
Santa Maria	30	2.25	2.40	1.98	1.31	0.19	0.04	0.03	0.02	0.10	0.52	1.36	2.05	12.25
Stockton	30	2.91	2.11	1.96	1.37	0.42	0.07	0.01	0.03	0.17	0.72	1.72	2.68	14.17
Colorado														
Alamosa	30	0.25	0.26	0.35	0.63	0.62	0.52	1.17	1.15	0.71	0.69	0.24	0.35	6.94
Colorado Springs	30	0.31	0.34	0.77	1.45	2.12	2.31	3.10	2.58	1.11	0.92	0.45	0.27	15.73
Denver	30	0.61	0.67	1.21	1.93	2.64	1.93	1.78	1.29	1.13	1.13	0.76	0.43	15.51
Grand Junction	30	0.64	0.61	0.75	0.79	0.63	0.55	0.46	1.05	0.84	0.93	0.61	0.55	8.41
Pueblo	30	0.32	0.32	0.68	1.29	1.65	1.36	1.87	1.96	0.79	0.96	0.42	0.29	11.91
Connecticut														
Bridgeport	30	2.71	2.71	3.49	3.39	3.57	2.56	3.44	3.80	2.88	2.79	3.83	3.44	38.61
Hartford	30	3.28	3.17	3.82	3.75	3.50	3.53	3.41	3.94	3.55	3.03	4.33	4.06	43.37

Table 2 (continued)
AVERAGE INCHES OF PRECIPITATION IN THE U.S. AND CERTAIN PACIFIC ISLANDS (NORMALS 1941-1970)[2]

Location	Years	Jan.	Feb.	Mar.	April	May	June	July	Aug.	Sept.	Oct.	Nov.	Dec.	Annual
Delaware														
Wilmington	30	2.85	2.75	3.74	3.20	3.35	3.24	4.31	3.98	3.42	2.60	3.49	3.32	40.25
Washington, D.C.														
Dulles Airport	30	2.84	2.61	3.48	2.96	3.68	3.61	4.12	4.25	3.29	2.74	3.06	3.47	40.11
National Airport	30	2.62	2.45	3.33	2.86	3.68	3.48	4.12	4.67	3.08	2.66	2.90	3.04	38.89
Florida														
Apalachicola	30	3.07	3.78	4.70	3.61	2.78	5.30	8.02	8.07	9.00	2.88	2.68	3.32	57.21
Daytona Beach	30	2.05	2.92	3.37	2.39	2.65	6.60	6.69	6.84	7.10	5.52	2.13	1.96	50.22
Fort Myers	30	1.64	2.03	3.06	2.03	3.99	8.89	8.90	7.72	8.71	4.37	1.31	1.30	53.95
Jacksonville	30	2.78	3.58	3.56	3.07	3.22	6.27	7.35	7.89	7.83	4.54	1.79	2.59	54.47
Key West	30	1.67	1.85	1.56	2.17	2.51	4.55	4.11	4.47	7.34	5.57	2.67	1.52	39.99
Lakeland	30	2.32	2.52	4.02	2.57	3.44	6.70	8.09	7.18	6.06	2.84	1.60	2.09	49.43
Miami	30	2.15	1.95	2.07	3.60	6.12	9.00	6.91	6.72	8.74	8.18	2.72	1.64	59.80
Orlando	30	2.28	2.95	3.46	2.72	2.94	7.11	8.29	6.73	7.20	4.07	1.56	1.90	51.21
Pensacola	30	4.37	4.69	6.31	4.99	4.25	6.30	7.33	6.67	8.15	3.13	3.37	4.66	64.22
Tallahassee	30	3.74	4.77	5.93	4.07	4.04	6.62	8.92	6.89	6.64	2.93	2.81	4.22	61.58
Tampa	30	2.33	2.86	3.89	2.10	2.41	6.49	8.43	8.00	6.35	2.54	1.79	2.19	49.38
West Palm Beach	30	2.60	2.60	3.32	3.51	5.17	8.14	6.52	6.91	9.85	8.75	2.48	2.21	62.06
Georgia														
Athens	30	4.76	4.55	5.70	4.37	4.02	4.27	5.23	3.56	3.70	2.73	3.38	4.33	50.60
Atlanta	30	4.34	4.41	5.84	4.61	3.71	3.67	4.90	3.54	3.15	2.50	3.43	4.24	48.34
Augusta	30	3.44	3.75	4.67	3.37	3.38	3.66	5.09	4.21	3.26	2.17	2.21	3.42	42.63
Columbus	30	4.34	4.40	6.03	4.50	4.05	4.12	5.75	4.22	3.67	1.97	2.96	4.93	50.94
Macon	30	3.67	4.38	5.15	3.55	3.57	3.86	4.52	3.65	3.10	2.31	2.39	4.31	44.46
Rome	30	4.99	5.19	6.17	4.77	3.93	3.74	4.74	3.49	3.92	2.90	3.76	4.99	52.59
Savannah	30	2.92	2.86	4.41	2.93	4.20	5.89	7.87	6.47	5.57	2.81	1.94	3.28	51.15
Hawaii														
Hilo	30	9.07	12.90	13.69	12.88	10.07	6.61	9.54	10.88	7.44	10.96	13.77	15.76	133.57
Honolulu	30	4.40	2.46	3.18	1.36	0.96	0.32	0.60	0.76	0.67	1.51	2.99	3.69	22.90
Kahului	30	3.59	2.64	2.87	1.25	0.62	0.21	0.39	0.34	0.27	1.08	2.33	2.84	18.43
Lihue	30	6.24	4.28	4.67	3.25	2.43	1.57	1.87	2.21	1.85	3.84	5.63	6.34	44.18

Idaho														
Boise	30	1.47	1.16	1.01	1.14	1.32	1.06	0.15	0.30	0.41	0.80	1.32	1.36	11.50
Lewiston	30	1.27	0.85	0.96	1.13	1.58	1.83	0.53	0.60	0.86	1.08	1.25	1.27	13.21
Pocatello	30	1.05	0.80	0.94	1.06	1.29	1.28	0.36	0.62	0.61	0.75	1.05	0.99	10.80
Illinois														
Cairo	30	3.99	3.77	4.74	4.37	5.18	4.39	3.24	3.50	3.47	2.68	3.85	3.94	47.12
Chicago														
O'Hare Airport	30	1.70	1.30	2.52	3.38	3.41	4.15	3.46	2.73	3.01	2.32	2.10	1.64	31.72
Midway Airport	30	1.85	1.59	2.73	3.75	3.41	3.95	4.09	3.14	3.00	2.62	2.20	2.11	34.44
Moline	30	1.66	1.29	2.57	3.82	3.92	4.42	4.58	3.37	3.84	2.69	1.87	1.77	35.80
Peoria	30	1.82	1.50	2.80	4.36	3.87	3.91	3.76	3.07	3.55	2.51	2.02	1.89	35.06
Rockford	30	1.79	1.29	2.65	3.85	3.86	4.42	4.27	3.66	4.00	2.85	2.37	1.71	36.72
Springfield	30	1.75	1.77	2.70	4.14	3.54	4.16	3.81	2.74	3.27	3.08	2.14	1.92	35.02
Indiana														
Evansville	30	3.40	3.27	4.69	4.06	4.38	3.57	3.77	2.95	2.80	2.52	3.17	3.30	41.88
Fort Wayne	30	2.49	2.05	2.90	3.56	3.85	3.91	3.89	2.90	2.56	2.84	2.65	2.20	35.80
Indianapolis	30	2.86	2.36	3.75	3.87	4.08	4.16	3.67	2.80	2.87	2.51	3.10	2.71	38.74
South Bend	30	2.37	1.94	2.75	4.01	3.17	3.72	3.67	3.26	3.07	3.06	2.71	2.47	36.20
Iowa														
Burlington	30	1.58	1.25	2.65	3.79	3.58	4.71	3.76	3.36	3.74	3.02	1.60	1.59	34.63
Des Moines	30	1.14	1.05	2.31	2.94	4.21	4.90	3.28	3.30	3.07	2.14	1.42	1.09	30.85
Dubuque	30	1.73	1.26	2.97	4.15	4.67	5.29	4.31	4.02	4.64	2.83	2.48	1.92	40.27
Sioux City	30	0.65	0.94	1.45	2.19	3.54	4.59	3.30	2.95	2.84	1.63	0.91	0.75	25.74
Waterloo	30	1.00	0.92	2.25	3.43	4.18	5.01	4.75	3.50	3.63	2.38	1.50	1.19	33.74
Kansas														
Concordia	30	0.66	0.93	1.55	2.26	4.23	5.00	3.31	3.15	3.13	1.82	0.78	0.78	27.60
Dodge City	30	0.50	0.63	1.13	1.71	3.13	3.34	3.08	2.64	1.67	1.65	0.59	0.51	20.58
Goodland	30	0.37	0.39	0.88	1.42	2.54	2.92	2.71	2.14	1.35	0.99	0.53	0.41	16.65
Topeka	30	0.97	0.98	2.17	3.62	4.01	5.80	4.21	4.18	3.28	2.65	1.26	1.53	34.66
Wichita	30	0.85	0.98	1.78	2.95	3.60	4.49	4.35	3.10	3.69	2.50	1.17	1.12	30.58
Kentucky														
Covington-Cincinnati, Ohio Airport	30	3.34	3.04	4.09	3.64	3.74	3.81	4.12	2.62	2.55	2.15	3.08	2.86	39.04
Lexington	30	3.95	3.42	4.80	3.87	4.16	4.31	4.83	3.40	2.65	2.12	3.36	3.62	44.49
Louisville	30	3.53	3.47	5.05	4.10	4.20	4.05	3.76	2.99	2.94	2.35	3.33	3.34	43.11
Louisiana														
Baton Rouge	30	4.40	4.76	5.14	5.10	4.39	3.77	6.51	4.67	3.79	2.65	3.84	5.03	54.05
Lake Charles	30	4.04	4.47	3.84	4.33	5.06	5.04	6.55	4.75	4.13	3.48	4.08	5.70	55.47

Table 2 (continued)
AVERAGE INCHES OF PRECIPITATION IN THE U.S. AND CERTAIN PACIFIC ISLANDS (NORMALS 1941-1970)[2]

Location	Years	Jan.	Feb.	Mar.	April	May	June	July	Aug.	Sept.	Oct.	Nov.	Dec.	Annual
New Orleans	30	4.53	4.82	5.49	4.15	4.20	4.74	6.72	5.27	5.58	2.26	3.88	5.13	56.77
Shreveport	30	4.04	3.71	4.10	5.19	5.04	3.34	2.89	2.68	3.07	2.90	3.57	4.19	44.72
Maine														
Caribou	30	2.04	2.11	2.20	2.42	2.96	3.41	3.98	3.78	3.49	3.31	3.50	2.62	35.82
Portland	30	3.38	3.52	3.60	3.34	3.33	3.10	2.61	2.60	3.09	3.31	4.86	4.06	40.80
Maryland														
Baltimore	30	2.91	2.81	3.69	3.07	3.61	3.77	4.07	4.21	3.12	2.81	3.13	3.26	40.46
Massachusetts														
Blue Hill Observatory	30	4.12	3.97	4.51	3.64	3.62	3.15	2.95	3.83	3.65	3.62	5.06	4.70	46.82
Boston	30	3.69	3.54	4.01	3.49	3.47	3.19	2.74	3.46	3.16	3.02	4.51	4.24	42.52
Worcester	30	3.35	3.18	3.85	3.83	3.97	3.60	3.62	4.19	3.52	3.54	4.66	3.93	45.24
Michigan														
Alpena	30	1.66	1.35	1.87	2.43	2.70	2.87	2.50	2.69	3.25	1.97	2.45	1.85	27.59
Detroit														
City Airport	30	1.93	1.80	2.33	3.08	3.43	3.04	2.99	3.04	2.30	2.52	2.31	2.19	30.98
Metro Airport	30	1.91	1.75	2.47	3.22	3.31	3.42	3.10	3.28	2.16	2.48	2.32	2.27	31.89
Flint	30	1.70	1.58	2.11	2.86	3.12	3.43	2.99	3.25	2.52	2.24	2.29	1.68	29.77
Grand Rapids	30	1.94	1.50	2.50	3.39	3.19	3.42	3.09	2.54	3.30	2.56	2.80	2.16	32.39
Houghton Lake	30	1.46	1.20	1.72	2.37	2.78	3.34	3.07	2.35	3.17	2.58	2.53	1.82	28.39
Lansing	30	1.91	1.62	2.36	2.90	3.32	3.47	2.82	2.79	2.63	2.31	2.26	2.00	30.39
Marquette	30	1.54	1.52	1.88	2.60	2.93	3.44	3.07	3.01	3.45	2.44	2.99	1.97	30.84
Muskegon	30	2.25	1.76	2.39	3.16	2.71	2.64	2.51	2.57	3.36	2.70	3.04	2.44	31.53
Sault Ste. Marie	30	1.92	1.48	1.74	2.22	3.01	3.31	2.60	3.10	3.85	2.85	3.26	2.36	31.70
Minnesota														
Duluth	30	1.16	0.85	1.76	2.55	3.41	4.44	3.73	3.79	3.06	2.30	1.73	1.40	30.18
International Falls	30	0.85	0.71	1.10	1.67	2.75	3.91	3.98	3.39	3.32	1.69	1.30	0.98	25.65
Minneapolis — St. Paul	30	0.73	0.84	1.68	2.04	3.37	3.94	3.69	3.05	2.73	1.78	1.20	0.89	25.94
Rochester	30	0.65	0.65	1.70	2.36	3.46	4.63	3.74	3.59	3.08	1.82	1.02	0.77	27.47
Saint Cloud	30	0.76	0.78	1.33	2.30	3.60	4.64	3.23	3.89	2.65	1.69	1.12	0.85	26.84
Mississippi														
Jackson	30	4.53	4.62	5.63	4.65	4.38	3.40	4.27	3.59	2.99	2.22	3.87	5.04	49.19

Meridian	30	4.33	4.86	6.21	5.10	3.84	3.68	5.12	3.89	3.29	2.18	3.51	5.57	51.58
Missouri														
Columbia	30	1.57	1.72	2.58	3.83	4.68	4.59	3.89	3.19	4.39	3.38	1.79	1.78	37.39
Kansas City	30	1.25	1.25	2.55	3.50	4.28	5.55	4.37	3.81	4.21	3.24	1.47	1.52	37.00
Saint Joseph	30	1.09	1.03	2.40	3.21	4.65	6.50	3.82	4.06	3.83	2.47	1.24	1.35	35.65
St. Louis	30	1.85	2.06	3.03	3.92	3.86	4.42	3.69	2.87	2.89	2.79	2.47	2.04	35.89
Springfield	30	1.67	2.22	2.99	4.27	4.93	4.72	3.62	2.94	4.11	3.44	2.34	2.45	39.70
Montana														
Billings	30	0.70	0.64	1.01	1.56	2.08	2.61	0.87	1.00	1.39	0.88	0.73	0.68	14.15
Glasgow	30	0.39	0.32	0.37	0.71	1.31	2.72	1.43	1.51	0.85	0.56	0.39	0.31	10.87
Great Falls	30	0.88	0.75	0.97	1.18	2.37	3.11	1.27	1.09	1.17	0.68	0.81	0.71	14.99
Havre	30	0.52	0.40	0.49	1.02	1.48	2.55	1.38	1.05	1.11	0.67	0.46	0.42	11.55
Helena	30	0.55	0.38	0.69	0.93	1.76	2.38	0.96	0.98	0.97	0.59	0.61	0.58	11.38
Kalispell	30	1.52	0.97	0.86	1.01	1.80	2.56	1.03	1.30	1.13	1.21	1.39	1.46	16.24
Miles City	30	0.49	0.51	0.65	1.26	2.06	3.32	1.55	1.20	1.19	0.71	0.51	0.48	13.93
Missoula	30	1.17	0.74	0.71	1.01	1.68	2.12	0.92	0.92	1.06	0.92	0.96	1.13	13.34
Nebraska														
Grand Island	30	0.52	0.76	1.18	2.47	3.78	4.40	3.00	2.54	2.51	1.08	0.61	0.56	23.41
Lincoln	30	0.73	1.14	1.54	2.59	3.87	5.16	3.24	3.63	3.25	1.69	0.98	0.79	28.61
Norfolk	30	0.62	0.78	1.37	2.15	3.69	4.88	3.18	2.66	2.41	1.33	0.62	0.63	24.32
North Platte	30	0.45	0.52	0.99	1.93	3.26	3.77	2.98	2.07	2.01	0.99	0.52	0.41	19.90
Omaha	30	0.76	0.98	1.59	2.97	4.11	4.94	3.71	3.97	3.27	1.93	1.11	0.84	30.18
Omaha (North)	30	0.69	0.96	1.62	2.82	3.99	4.93	3.71	4.01	3.16	1.76	1.01	0.77	29.43
Scottsbluff	30	0.32	0.33	0.76	1.52	2.81	3.36	1.76	0.95	1.15	0.82	0.39	0.40	14.57
Valentine	30	0.31	0.53	0.76	1.77	2.80	3.60	2.50	2.38	1.48	0.92	0.45	0.30	17.80
Nevada														
Elko	30	1.16	0.77	0.83	0.82	1.03	1.01	0.41	0.61	0.34	0.66	1.01	1.13	9.78
Ely	30	0.64	0.60	0.85	1.00	0.93	0.93	0.61	0.56	0.61	0.60	0.66	0.71	8.70
Las Vegas	30	0.45	0.30	0.33	0.27	0.10	0.09	0.44	0.49	0.27	0.22	0.43	0.37	3.76
Reno	30	1.21	0.86	0.70	0.47	0.66	0.40	0.26	0.22	0.23	0.42	0.68	1.09	7.20
Winnemucca		0.97	0.81	0.71	0.73	0.91	1.01	0.23	0.26	0.28	0.65	0.97	0.94	8.47
New Hampshire														
Concord	30	2.67	2.45	2.77	2.92	3.02	3.35	3.14	2.89	3.06	2.68	3.96	3.26	36.17
Mt. Washington	30	5.12	6.51	5.60	5.46	5.84	6.50	6.77	7.19	6.36	6.12	7.67	7.03	76.17
New Jersey														
Atlantic City	30	3.56	3.37	4.31	3.37	3.54	3.38	4.36	4.90	2.99	3.46	4.21	4.01	45.46
Newark	30	2.91	2.95	3.93	3.44	3.60	2.99	4.03	4.27	3.44	2.82	3.61	3.46	41.45
Trenton	30	2.76	2.70	3.81	3.15	3.40	3.21	4.74	4.17	3.17	2.53	3.25	3.28	40.17

Table 2 (continued)
AVERAGE INCHES OF PRECIPITATION IN THE U.S. AND CERTAIN PACIFIC ISLANDS (NORMALS 1941-1970)[2]

Location	Years	Jan.	Feb.	Mar.	April	May	June	July	Aug.	Sept.	Oct.	Nov.	Dec.	Annual
New Mexico														
Albuquerque	30	0.30	0.39	0.47	0.48	0.53	0.50	1.39	1.34	0.77	0.79	0.29	0.52	7.77
Clayton	30	0.28	0.36	0.67	1.27	2.64	1.75	2.76	2.67	1.77	1.11	0.33	0.30	15.91
Roswell	30	0.40	0.37	0.47	0.49	1.00	1.24	1.71	1.48	1.47	1.22	0.29	0.47	10.61
New York														
Albany	30	2.20	2.11	2.58	2.70	3.26	3.00	3.12	2.87	3.12	2.63	2.84	2.93	33.36
Binghamton	30	2.32	2.25	2.87	3.18	3.83	3.59	3.83	3.61	3.02	3.00	3.10	2.75	37.35
Buffalo	30	2.90	2.55	2.85	3.15	2.97	2.23	2.93	3.53	3.25	3.01	3.74	3.00	36.11
Central Park	30	2.71	2.92	3.73	3.30	3.47	2.96	3.68	4.01	3.27	2.85	3.76	3.53	40.19
JFK Airport	30	2.69	3.05	3.77	3.59	3.54	2.98	4.04	4.30	3.31	2.76	3.90	3.60	41.53
La Guardia Airport	30	2.88	3.10	3.98	3.56	3.36	2.89	3.85	4.48	3.15	2.96	3.77	3.63	41.61
Rochester	30	2.25	2.42	2.57	2.74	2.80	2.54	2.89	2.97	2.35	2.62	2.83	2.35	31.33
Syracuse	30	2.68	2.79	3.03	3.08	3.02	3.09	3.08	3.50	2.71	3.09	3.25	3.09	36.41
North Carolina														
Asheville	30	3.39	3.60	4.66	3.53	3.31	3.97	4.87	4.50	3.57	3.25	2.94	3.59	45.18
Cape Hatteras	30	4.26	4.15	3.84	3.07	3.28	4.83	5.90	6.75	5.76	4.79	4.45	4.54	55.62
Charlotte	30	3.51	3.83	4.52	3.40	2.90	3.70	4.57	3.96	3.46	2.69	2.74	3.44	42.72
Greensboro	30	3.22	3.37	3.72	3.15	3.04	3.91	4.39	4.30	3.55	2.94	2.62	3.15	41.36
Raleigh	30	3.22	3.32	3.44	3.07	3.32	3.67	5.08	4.93	3.78	2.81	2.82	3.08	42.54
Wilmington	30	3.18	3.39	4.05	2.94	3.97	5.58	8.34	6.82	5.64	3.31	3.01	3.36	53.59
North Dakota														
Bismarck	30	0.51	0.44	0.73	1.44	2.17	3.58	2.20	1.96	1.32	0.80	0.56	0.45	16.16
Fargo	30	0.50	0.44	0.83	2.08	2.29	3.20	3.16	2.85	1.84	1.09	0.72	0.62	19.62
Williston	30	0.59	0.53	0.63	1.24	1.62	3.25	2.04	1.56	1.21	0.64	0.53	0.49	14.33
Ohio														
Akron	30	2.69	2.16	3.15	3.32	3.87	3.50	3.80	2.77	2.60	2.37	2.50	2.40	35.13
Cincinnati	30	3.40	2.95	4.13	3.85	3.96	3.92	3.96	3.02	2.69	2.20	3.08	2.87	40.03
Cleveland	30	2.56	2.18	3.05	3.49	3.49	3.28	3.45	3.00	2.80	2.57	2.76	2.36	34.99
Columbus	30	2.87	2.32	3.44	3.71	4.10	4.13	4.21	2.86	2.41	1.89	2.68	2.39	37.01
Dayton	30	2.76	2.24	3.21	3.34	3.76	3.88	3.54	2.55	2.28	1.94	2.57	2.29	34.36
Mansfield	30	2.35	1.99	2.83	3.33	3.78	3.49	3.69	2.71	2.85	1.97	2.52	2.16	33.67

Toledo	30	2.08	1.75	2.52	2.95	3.33	3.38	3.23	3.07	2.40	2.24	2.32	2.24	31.51
Youngstown	30	2.94	2.42	3.24	3.67	3.90	3.59	3.90	3.23	2.64	2.94	2.97	2.55	37.99
Oklahoma														
Oklahoma City	30	1.11	1.32	2.05	3.47	5.20	4.22	2.66	2.56	3.55	2.57	1.40	1.26	31.37
Tulsa	30	1.43	1.72	2.52	4.17	5.11	4.69	3.51	2.95	4.07	3.22	1.87	1.64	36.90
Oregon														
Astoria	30	9.73	7.82	6.62	4.61	2.72	2.45	0.96	1.46	2.83	6.80	9.78	10.57	66.34
Burns	30	1.76	1.18	0.92	0.70	1.03	0.97	0.32	0.44	0.46	0.89	1.43	1.73	11.83
Eugene	30	7.54	4.67	4.43	2.31	2.06	1.28	0.26	0.58	1.26	4.00	6.53	7.64	42.56
Medford	30	3.54	2.15	1.64	1.00	1.44	0.89	0.25	0.33	0.56	2.05	3.10	3.69	20.64
Pendleton	30	1.60	1.07	1.00	1.01	1.24	1.01	0.26	0.34	0.64	1.11	1.50	1.53	12.31
Portland	30	5.88	4.06	3.64	2.22	2.09	1.59	0.47	0.82	1.60	3.59	5.61	6.04	37.61
Salem	30	6.90	4.79	4.33	2.29	2.09	1.39	0.35	0.57	1.46	3.98	6.08	6.85	41.08
Sexton Summit	30	6.54	4.05	3.43	2.06	2.22	1.26	0.33	0.44	1.03	3.63	5.61	6.19	36.79
Pacific														
Guam	30	5.16	4.26	2.94	4.03	4.49	5.19	9.59	12.16	14.08	14.40	8.51	5.85	90.66
Johnston Island	30	2.59	1.95	2.85	2.53	2.22	1.03	1.09	2.58	2.05	3.48	2.30	3.39	28.06
Koror Island	30	10.79	7.43	7.79	8.69	14.53	13.75	16.19	15.40	12.99	12.59	10.67	12.87	143.69
Kwajalein Island	30	3.74	2.35	6.21	5.91	9.71	9.93	9.70	10.20	10.87	11.97	11.46	9.23	101.28
Majuro, Marshall Is.	30	8.96	6.98	9.25	10.31	12.15	12.17	13.64	11.78	14.39	16.00	15.38	11.88	142.89
Pago Pago, American Samoa	30	12.27	14.16	12.07	12.26	10.69	9.41	6.94	7.40	6.08	12.10	10.31	14.85	128.54
Ponape Island	30	11.75	11.44	15.04	19.58	19.43	15.66	17.82	16.70	16.76	16.42	17.17	15.95	193.72
Truk, Caroline Islands	30	8.59	6.40	7.98	13.17	15.74	12.59	15.41	13.44	13.07	13.39	11.99	13.92	145.69
Wake Island	30	1.08	1.21	1.60	2.22	2.25	2.35	4.45	6.35	5.69	5.45	3.02	1.73	37.40
Yap Island	30	8.44	5.41	5.58	5.98	10.03	10.90	14.64	15.14	13.09	12.42	10.03	9.88	121.54
Pennsylvania														
Allentown	30	3.02	2.78	3.61	3.79	3.78	3.47	4.36	4.18	3.59	2.73	3.59	3.59	42.49
Erie	30	2.47	2.12	2.75	3.55	3.63	3.50	3.52	3.35	3.56	3.24	3.70	2.81	38.20
Harrisburg	30	2.57	2.42	3.22	2.98	3.76	3.11	3.70	3.22	2.66	2.57	3.19	3.07	36.47
Philadelphia	30	2.81	2.62	3.69	3.29	3.35	3.70	4.09	4.11	3.03	2.53	3.39	3.32	39.93
Pittsburgh														
International Airport	30	2.79	2.35	3.60	3.40	3.63	3.48	3.84	3.15	2.52	2.52	2.47	2.48	36.23
City	30	2.61	2.29	3.58	3.44	3.59	3.74	3.78	3.18	2.53	2.47	2.49	2.52	36.22
Avoca	30	2.04	1.96	2.50	3.06	3.50	3.40	4.09	3.21	2.82	2.71	3.01	2.51	34.81
Williamsport	30	2.52	2.58	3.53	3.42	3.99	3.25	4.19	3.44	3.03	3.20	3.74	3.10	39.99
Puerto Rico														
San Juan	30	3.73	2.50	2.04	3.40	6.54	5.64	6.41	6.98	6.07	5.64	5.49	4.71	59.15

Table 2 (continued)
AVERAGE INCHES OF PRECIPITATION IN THE U.S. AND CERTAIN PACIFIC ISLANDS (NORMALS 1941-1970)[2]

Location	Years	Jan.	Feb.	Mar.	April	May	June	July	Aug.	Sept.	Oct.	Nov.	Dec.	Annual
Rhode Island														
Block Island	30	3.41	3.32	3.88	3.51	3.25	2.20	2.74	3.86	3.00	2.88	4.35	4.11	40.51
Providence	30	3.52	3.45	3.99	3.72	3.49	2.65	2.85	3.90	3.26	3.27	4.52	4.13	42.75
South Carolina														
Charleston	30	2.90	3.27	4.75	2.95	3.81	6.30	8.21	6.44	5.17	3.05	2.13	3.14	52.12
Columbia	30	3.44	3.67	4.67	3.51	3.35	3.82	5.65	5.63	4.32	2.58	2.34	3.38	46.36
Greenville — Spartanburg	30	4.07	4.43	5.33	4.31	2.95	4.05	4.18	4.06	3.79	3.18	3.10	4.09	47.54
South Dakota														
Aberdeen	30	0.53	0.63	0.94	1.96	2.57	3.63	2.74	2.10	1.71	1.22	0.62	0.45	19.10
Huron	30	0.43	0.75	1.07	1.96	2.75	3.76	2.23	1.98	1.78	1.53	0.67	0.53	19.44
Rapid City	30	0.47	0.57	0.99	2.09	2.81	3.67	2.10	1.47	1.22	0.86	0.48	0.39	17.12
Sioux Falls	30	0.57	1.04	1.40	2.30	3.37	4.32	2.94	2.84	2.85	1.50	0.85	0.74	24.72
Tennessee														
Bristol — Johnson City	30	3.62	3.71	3.91	3.30	3.29	3.52	4.98	3.72	2.88	2.25	2.78	3.51	41.47
Chattanooga	30	5.38	5.19	5.63	4.42	3.43	3.68	5.14	3.22	3.69	2.95	3.94	5.25	51.92
Knoxville	30	4.67	4.71	4.86	3.61	3.28	3.63	4.70	3.24	2.78	2.67	3.56	4.47	46.18
Memphis	30	4.93	4.73	5.10	5.42	4.39	3.46	3.53	3.33	3.01	2.58	3.92	4.70	49.10
Nashville	30	4.75	4.43	5.00	4.11	4.10	3.38	3.83	3.24	3.09	2.16	3.46	4.45	46.00
Oak Ridge	30	5.25	5.24	5.45	4.21	3.52	3.94	5.67	3.85	3.34	2.72	4.05	5.36	52.60
Texas														
Abilene	30	1.02	0.97	0.98	2.47	3.86	2.82	2.34	2.05	2.26	2.60	1.20	1.02	23.59
Amarillo	30	0.54	0.56	0.77	1.23	2.83	3.45	2.95	2.93	1.93	1.83	0.53	0.73	20.28
Austin	30	1.88	3.09	1.89	3.49	3.97	3.13	1.88	2.20	3.68	3.02	2.04	2.22	32.49
Brownsville	30	1.35	1.48	0.69	1.28	2.51	2.80	1.19	2.66	5.23	3.32	1.34	1.24	25.09
Corpus Christi	30	1.58	1.95	1.10	2.15	3.17	2.67	1.88	3.20	4.90	2.77	1.63	1.53	28.53
Dallas — Fort Worth	30	1.80	2.36	2.54	4.30	4.47	3.05	1.84	2.26	3.15	2.68	2.03	1.82	32.30
Del Rio	30	0.60	1.02	0.72	1.57	2.41	2.03	1.02	1.22	3.05	2.07	0.65	0.52	16.88
El Paso	30	0.39	0.42	0.39	0.24	0.32	0.60	1.53	1.12	1.16	0.78	0.32	0.50	7.77
Galveston	30	3.02	2.67	2.60	2.63	3.16	4.05	4.41	4.40	5.60	2.83	3.16	3.67	42.20
Houston	30	3.57	3.54	2.68	3.54	5.10	4.52	4.12	4.35	4.65	4.05	4.03	4.04	48.19
Lubbock	30	0.55	0.50	0.89	1.08	3.17	2.78	2.23	1.87	2.19	2.05	0.49	0.61	18.41

Midland — Odessa	30	0.59	0.56	0.59	0.85	2.16	1.49	1.82	1.52	1.54	1.38	0.49	0.52	13.51
Port Arthur	30	4.06	4.24	3.05	4.19	4.94	4.81	5.89	5.69	5.34	3.71	4.26	4.89	55.07
San Angelo	30	0.81	0.79	0.87	1.66	2.70	1.88	1.23	1.44	2.74	1.86	0.85	0.70	17.53
San Antonio	30	1.66	2.06	1.54	2.54	3.07	2.79	1.69	2.41	3.71	2.84	1.77	1.46	27.54
Victoria	30	1.76	2.20	1.89	2.65	3.96	3.31	2.79	3.15	4.61	3.63	2.31	1.95	34.29
Waco	30	1.87	2.38	2.36	4.02	4.60	2.73	1.47	1.81	3.19	2.55	2.27	2.01	31.26
Wichita Falls	30	1.07	1.16	1.62	3.16	4.58	3.39	2.16	1.77	3.00	2.68	1.35	1.28	27.22
Utah														
Milford	30	0.61	0.70	1.04	0.90	0.61	0.56	0.51	0.68	0.61	0.78	0.67	0.73	8.40
Salt Lake City	30	1.27	1.19	1.63	2.12	1.49	1.30	0.70	0.93	0.68	1.16	1.31	1.39	15.17
Vermont														
Burlington	30	1.74	1.68	1.93	2.62	3.01	3.46	3.54	3.72	3.05	2.74	2.86	2.19	32.54
Virginia														
Lynchburg	30	2.77	2.79	3.46	2.73	3.22	3.43	4.05	4.05	3.30	2.60	2.66	3.21	38.27
Norfolk	30	3.35	3.31	3.42	2.71	3.34	3.62	5.70	5.92	4.20	3.06	2.94	3.11	44.68
Richmond	30	2.86	3.01	3.38	2.77	3.42	3.52	5.63	5.06	3.58	2.94	3.20	3.22	42.59
Roanoke	30	2.74	3.09	3.33	2.80	3.47	3.51	3.74	4.15	3.42	3.19	2.48	3.11	39.03
Washington														
Olympia	30	7.93	5.97	4.81	3.14	1.88	1.57	0.70	1.17	2.12	5.28	7.98	8.19	50.74
Quillayute	30	14.60	11.95	10.79	8.15	4.71	3.50	2.36	2.75	5.16	11.58	13.84	15.60	104.99
Seattle														
Urban Site	30	5.17	3.93	3.24	2.41	1.71	1.57	0.87	0.88	1.75	3.43	5.34	5.35	35.65
International Airport	30	5.79	4.19	3.61	2.46	1.70	1.53	0.71	1.08	1.99	3.91	5.88	5.94	38.79
Spokane	30	2.47	1.68	1.53	1.12	1.46	1.36	0.40	0.58	0.83	1.42	2.20	2.37	17.42
Stampede Pass	30	12.89	10.29	8.93	6.63	4.17	3.96	1.54	2.35	4.64	8.89	12.46	14.31	91.06
Walla Walla	30	2.07	1.40	1.37	1.43	1.58	1.18	0.33	0.45	0.85	1.49	1.89	1.97	16.01
Yakima	30	1.33	0.78	0.58	0.51	0.55	0.73	0.16	0.25	0.31	0.58	1.07	1.15	8.00
West Virginia														
Beckley	30	3.46	3.33	4.23	3.31	3.78	4.22	4.38	3.77	3.40	2.49	2.92	3.33	42.62
Charleston	30	3.39	3.11	4.03	3.33	3.48	3.31	5.04	3.68	2.94	2.45	2.81	3.18	40.75
Elkins	30	3.29	2.92	3.93	3.62	3.88	4.76	4.94	4.02	3.18	2.71	2.72	3.25	43.22
Huntington	30	3.15	2.90	4.07	3.26	3.82	3.37	4.19	3.34	2.86	2.09	2.86	2.97	38.88
Parkersburg	30	3.08	2.77	3.75	3.45	3.56	4.01	4.28	3.34	2.80	2.11	2.52	2.77	38.44
Wisconsin														
Green Bay	30	1.09	1.01	1.68	2.69	3.10	3.41	3.09	2.62	3.24	1.93	1.88	1.27	27.01
La Crosse	30	0.96	0.87	2.02	2.63	3.70	4.44	3.52	3.02	3.38	2.05	1.45	1.04	29.08
Madison	30	1.25	0.95	1.93	2.66	3.41	4.33	3.81	3.05	3.36	2.16	1.87	1.47	30.25
Milwaukee	30	1.63	1.13	2.24	2.76	2.88	3.58	3.41	2.68	3.02	1.98	2.01	1.75	29.07

Table 2 (continued)
AVERAGE INCHES OF PRECIPITATION IN THE U.S. AND CERTAIN PACIFIC ISLANDS
(NORMALS 1941-1970)[2]

Location	Years	Jan.	Feb.	Mar.	April	May	June	July	Aug.	Sept.	Oct.	Nov.	Dec.	Annual
Wyoming														
Casper	30	0.50	0.50	0.91	1.45	1.94	1.44	0.95	0.57	0.87	0.92	0.68	0.49	11.22
Cheyenne	30	0.46	0.46	1.05	1.57	2.52	2.41	1.82	1.45	1.03	0.95	0.58	0.35	14.65
Lander	30	0.48	0.66	1.18	2.36	2.59	1.93	0.61	0.42	1.05	1.24	0.87	0.45	13.84
Sheridan	30	0.69	0.77	1.21	2.12	2.45	2.99	1.07	0.95	1.28	1.02	0.92	0.69	16.16

Table 3
AVERAGE INCHES OF SNOWFALL IN THE U.S. AND CERTAIN PACIFIC ISLANDS
(DATA THROUGH 1977)[2]

Location	Years	Jan.	Feb.	Mar.	April	May	June	July	Aug.	Sept.	Oct.	Nov.	Dec.	Annual
Alabama														
Birmingham	34	0.5	0.2	T	T	0.0	0.0	0.0	0.0	0.0	T	T	0.5	1.2
Huntsville	10	1.3	0.7	0.2	T	0.0	0.0	0.0	0.0	0.0	0.0	0.1	0.2	2.5
Mobile	36	0.2	0.2	T	0.0	0.0	0.0	0.0	0.0	0.0	0.0	T	0.1	0.5
Montgomery	33	0.3	0.1	T	0.0	0.0	0.0	0.0	0.0	0.0	0.0	T	T	0.4
Alaska														
Anchorage	34	10.8	11.9	9.3	5.7	0.6	0.0	0.0	0.0	0.3	6.6	10.1	14.9	70.2
Annette	30	15.6	12.7	10.8	2.9	0.2	0.0	0.0	0.0	0.0	0.3	3.5	13.1	59.1
Barrow	57	2.2	2.2	1.9	2.2	1.9	0.5	0.6	0.7	3.1	7.0	3.7	2.6	28.6
Barter Island	29	5.2	2.9	2.8	2.6	3.3	1.7	0.6	1.4	5.7	9.2	5.0	3.3	43.7
Bethel	19	5.3	6.2	7.6	6.9	2.5	0.2	T	0.0	0.2	4.5	7.8	8.2	49.4
Bettles	26	11.7	9.0	9.5	7.6	1.3	T	0.0	0.1	1.3	11.6	12.1	12.4	76.6
Big Delta	33	5.9	4.3	5.0	3.0	0.6	T	0.0	T	1.4	8.1	6.4	5.8	40.5
Cold Bay	27	8.8	10.5	9.5	5.9	2.1	T	T	0.0	T	3.3	6.6	9.6	56.3
Fairbanks	26	10.3	9.8	7.5	3.7	0.7	T	0.0	T	1.0	10.7	12.8	12.3	68.8
Gulkana	34	7.0	5.9	4.8	2.1	0.6	T	0.0	0.1	1.0	7.0	8.5	9.6	46.6
Homer	34	10.4	12.3	10.7	3.9	0.6	T	0.0	0.0	T	2.2	7.2	12.8	60.1
Juneau	34	25.6	21.2	19.0	4.8	0.1	T	0.0	0.0	T	1.3	12.1	24.1	108.2

King Salmon	28	7.2	7.2	7.8	5.1	1.0	T	0.0	0.0	T	3.1	5.8	8.9	46.1
Kodiak	15	18.9	20.8	16.9	8.5	1.1	T	0.0	0.0	T	3.3	8.1	14.7	92.3
Kotzebue	34	6.0	5.4	5.8	5.1	1.8	0.2	T	T	1.1	6.7	8.2	6.9	47.2
McGrath	35	13.9	12.5	12.1	8.2	0.7	T	0.0	T	0.8	9.7	15.2	16.4	89.5
Nome	31	9.2	6.0	7.6	6.5	2.1	0.1	0.0	T	0.4	4.8	9.9	7.9	54.5
St. Paul Island	54	12.5	10.2	9.4	5.9	2.3	0.1	0.0	0.0	0.1	2.7	5.9	9.7	58.8
Talkeetna	47	18.0	18.2	18.1	8.1	0.7	T	0.0	T	0.1	9.8	14.8	18.6	106.4
Unalakleet	24	5.1	6.0	5.9	3.9	1.0	T	0.0	0.0	0.9	4.3	7.3	5.5	39.9
Valdez	6	58.3	47.9	46.8	34.5	0.8	0.0	0.0	0.0	0.0	13.0	33.1	63.9	298.3
Yakutat	29	42.2	44.5	45.2	20.4	2.1	0.0	0.0	0.0	T	6.6	24.4	44.3	229.7
Arizona														
Flagstaff	28	15.6	14.4	19.7	10.6	2.3	T	0.0	0.0	0.1	2.6	8.4	14.9	88.6
Phoenix	40	T	T	T	T	0.0	0.0	0.0	0.0	0.0	0.0	0.0	T	—
Tucson	37	0.2	0.2	0.4	0.1	0.0	0.0	0.0	0.0	0.0	T	0.2	0.3	1.4
Winslow	45	2.5	2.0	2.0	0.5	T	0.0	0.0	0.0	T	0.3	0.7	3.1	11.1
Yuma	27	0.0	0.0	0.0	0.0	0.0	0.0	0.0	0.0	0.0	0.0	0.0	T	T
Arkansas														
Fort Smith	32	2.0	1.5	0.6	T	0.0	0.0	0.0	0.0	0.0	0.0	0.7	0.9	5.7
Little Rock	35	2.2	1.2	0.6	T	0.0	0.0	0.0	0.0	0.0	0.0	0.2	0.9	5.1
California														
Bakersfield	40	T	T	T	0.0	0.0	0.0	0.0	0.0	0.0	0.0	0.0	T	T
Bishop	30	3.8	2.0	0.8	0.5	0.1	0.0	0.0	0.0	T	T	0.2	1.2	8.6
Blue Canyon	37	46.3	41.0	48.6	28.2	7.8	0.5	0.0	0.0	0.4	2.1	22.0	39.9	236.8
Eureka	67	0.2	T	T	T	0.0	0.0	0.0	0.0	0.0	0.0	T	T	0.2
Fresno	28	0.1	T	T	0.0	0.0	0.0	0.0	0.0	0.0	T	0.0	T	0.1
Long Beach	34	T	0.0	0.0	0.0	0.0	0.0	0.0	0.0	0.0	0.0	0.0	0.0	T
Los Angeles														
International Airport	42	T	T	0.0	0.0	0.0	0.0	0.0	0.0	0.0	0.0	0.0	T	T
City	37	T	T	0.0	0.0	0.0	0.0	0.0	0.0	0.0	0.0	0.0	T	T
Mount Shasta	35	32.9	17.2	17.4	9.4	1.0	T	0.0	0.0	0.0	0.4	9.6	22.2	110.1
Oakland	49	T	T	T	0.0	0.0	0.0	0.0	0.0	0.0	0.0	0.0	T	T
Red Bluff	33	1.3	0.1	0.2	T	0.0	0.0	0.0	0.0	0.0	T	0.1	0.7	2.4
Sacramento	29	T	0.1	0.0	0.0	0.0	0.0	0.0	0.0	0.0	0.0	0.0	T	0.1
Sandberg	44	9.0	4.5	4.8	3.2	0.4	0.0	0.0	0.0	T	0.1	1.4	5.1	28.5
San Diego	37	T	0.0	0.0	0.0	0.0	0.0	0.0	0.0	0.0	0.0	0.0	T	T
San Francisco														
International Airport	50	T	T	T	0.0	0.0	0.0	0.0	0.0	0.0	0.0	0.0	T	T

Table 3 (continued)
AVERAGE INCHES OF SNOWFALL IN THE U.S. AND CERTAIN PACIFIC ISLANDS
(DATA THROUGH 1977)[2]

Location	Years	Jan.	Feb.	Mar.	April	May	June	July	Aug.	Sept.	Oct.	Nov.	Dec.	Annual
City	36	T	T	T	0.0	0.0	0.0	0.0	0.0	0.0	0.0	0.0	T	T
Santa Maria	35	T	T	0.0	0.0	0.0	0.0	0.0	0.0	0.0	0.0	T	0.0	T
Stockton	23	0.0	T	T	T	0.0	0.0	0.0	0.0	0.0	0.0	0.0	T	T
Colorado														
Alamosa	32	4.4	4.8	6.4	5.0	1.4	T	0.0	0.0	0.3	3.5	4.0	5.6	35.4
Colorado Springs	29	4.4	4.4	8.7	7.1	1.1	T	0.0	0.0	1.5	3.2	4.8	4.1	39.3
Denver	43	8.0	7.8	12.7	9.5	1.5	T	0.0	0.0	1.8	3.8	7.7	6.2	59.0
Grand Junction	31	7.5	4.2	4.3	1.2	T	0.0	0.0	0.0	0.1	0.6	3.0	5.4	26.3
Pueblo	37	5.3	4.4	7.2	3.2	0.5	0.0	0.0	0.0	0.6	1.1	4.2	4.4	30.9
Connecticut														
Bridgeport	29	7.3	8.1	5.2	0.4	T	0.0	0.0	0.0	0.0	T	0.5	5.3	26.8
Hartford	23	11.9	12.7	12.2	1.6	0.1	0.0	0.0	0.0	0.0	T	1.7	12.8	53.0
Delaware														
Wilmington	30	5.7	5.5	3.7	0.1	T	0.0	0.0	0.0	0.0	T	1.0	3.9	19.9
District of Columbia														
Washington														
Dulles Airport	15	5.5	6.5	2.8	0.2	T	0.0	0.0	0.0	0.0	T	1.5	5.3	21.8
National Airport	34	4.7	4.9	2.3	T	T	0.0	0.0	0.0	0.0	T	0.7	3.7	16.3
Florida														
Apalachicola	48	T	T	T	0.0	0.0	0.0	0.0	0.0	0.0	0.0	0.0	T	T
Daytona Beach	34	T	T	0.0	0.0	0.0	0.0	0.0	0.0	0.0	0.0	0.0	T	T
Fort Myers	38	0.0	0.0	0.0	0.0	0.0	0.0	0.0	0.0	0.0	0.0	0.0	0.0	0.0
Jacksonville	36	T	T	T	0.0	0.0	0.0	0.0	0.0	0.0	0.0	0.0	T	T
Key West	29	0.0	0.0	0.0	0.0	0.0	0.0	0.0	0.0	0.0	0.0	0.0	0.0	0.0
Lakeland	37	T	T	0.0	0.0	0.0	0.0	0.0	0.0	0.0	0.0	0.0	0.0	T
Miami	35	0.0	0.0	0.0	0.0	0.0	0.0	0.0	0.0	0.0	0.0	0.0	0.0	0.0
Orlando	35	T	0.0	0.0	0.0	0.0	0.0	0.0	0.0	0.0	0.0	0.0	0.0	T
Pensacola	14	0.2	0.1	0.0	0.0	0.0	0.0	0.0	0.0	0.0	0.0	0.0	0.0	0.3
Tallahassee	16	T	T	T	0.0	0.0	0.0	0.0	0.0	0.0	0.0	0.0	T	T
Tampa	31	T	T	0.0	0.0	0.0	0.0	0.0	0.0	0.0	0.0	0.0	0.0	T
West Palm Beach	32	T	0.0	0.0	0.0	0.0	0.0	0.0	0.0	0.0	0.0	0.0	0.0	T

Georgia														
Athens	34	0.7	0.5	0.2	0.0	0.0	0.0	0.0	0.0	0.0	0.0	0.1	0.2	1.7
Atlanta	43	0.7	0.4	0.2	T	0.0	0.0	0.0	0.0	0.0	0.0	T	0.2	1.5
Augusta	27	0.2	0.7	T	0.0	0.0	0.0	0.0	0.0	0.0	0.0	T	T	0.9
Columbus	32	T	0.4	T	T	0.0	0.0	0.0	0.0	0.0	0.0	T	T	0.4
Macon	29	0.3	0.7	T	0.0	0.0	0.0	0.0	0.0	0.0	0.0	T	T	1.0
Rome	31	0.7	0.5	0.4	T	0.0	0.0	0.0	0.0	0.0	T	T	0.4	2.0
Savannah	27	0.1	0.3	T	0.0	0.0	0.0	0.0	0.0	0.0	0.0	0.0	T	0.4
Hawaii														
Hilo	35	0.0	0.0	0.0	0.0	0.0	0.0	0.0	0.0	0.0	0.0	0.0	0.0	0.0
Honolulu	31	0.0	0.0	0.0	0.0	0.0	0.0	0.0	0.0	0.0	0.0	0.0	0.0	0.0
Kahului	23	0.0	0.0	0.0	0.0	0.0	0.0	0.0	0.0	0.0	0.0	0.0	0.0	0.0
Lihue	28	0.0	0.0	0.0	0.0	0.0	0.0	0.0	0.0	0.0	0.0	0.0	0.0	0.0
Idaho														
Boise	38	7.5	3.8	2.0	0.8	0.1	T	T	0.0	0.0	0.1	1.8	5.4	21.5
Lewiston	31	7.1	2.6	1.7	0.1	T	0.0	0.0	0.0	0.0	0.1	1.7	4.6	17.9
Pocatello	28	9.9	5.7	5.4	4.9	0.4	T	0.0	0.0	0.1	2.0	3.7	7.9	40.0
Illinois														
Cairo	35	3.0	2.1	2.4	T	0.0	0.0	0.0	0.0	0.0	0.0	0.4	1.5	9.4
Chicago														
O'Hare Airport	19	9.5	7.5	7.6	1.8	0.1	0.0	0.0	0.0	T	0.5	1.8	8.6	37.4
Midway Airport	35	9.9	8.3	7.5	1.4	T	0.0	0.0	0.0	0.0	0.3	2.9	10.8	41.1
Moline	45	7.7	5.7	6.6	0.8	T	0.0	0.0	0.0	T	0.2	2.4	6.9	30.3
Peoria	34	6.0	4.7	4.6	0.6	T	0.0	0.0	0.0	0.0	0.1	2.4	5.9	24.3
Rockford	27	7.9	6.1	7.3	1.3	T	0.0	0.0	0.0	0.0	0.1	2.9	8.5	34.1
Springfield	30	5.5	5.9	4.1	0.6	T	0.0	0.0	0.0	0.0	T	2.2	4.8	23.1
Indiana														
Evansville	37	3.7	3.2	2.9	0.3	0.0	0.0	0.0	0.0	0.0	T	0.8	2.5	13.4
Fort Wayne	31	6.8	6.7	5.9	1.6	T	0.0	0.0	0.0	0.0	0.1	3.5	6.9	31.5
Indianapolis	46	5.2	5.0	3.7	0.5	T	0.0	0.0	0.0	0.0	T	2.2	5.0	21.6
South Bend	38	16.5	13.6	9.6	2.1	T	0.0	0.0	0.0	T	0.7	9.0	17.0	68.5
Iowa														
Burlington	80	7.2	5.5	4.6	1.1	T	0.0	0.0	0.0	T	0.3	1.5	5.5	25.7
Des Moines	38	8.5	7.0	6.8	1.6	T	0.0	0.0	0.0	T	0.1	2.7	6.4	33.1
Dubuque	27	8.3	7.6	10.4	2.4	0.2	0.0	0.0	0.0	0.0	0.2	3.2	10.3	42.6
Sioux City	37	6.3	5.7	7.8	1.2	0.1	0.0	0.0	0.0	T	0.4	3.0	6.1	30.5
Waterloo	23	6.4	6.4	6.5	1.5	T	0.0	0.0	0.0	0.0	T	3.0	7.4	31.2

Table 3 (continued)
AVERAGE INCHES OF SNOWFALL IN THE U.S. AND CERTAIN PACIFIC ISLANDS (DATA THROUGH 1977)[2]

Location	Years	Jan.	Feb.	Mar.	April	May	June	July	Aug.	Sept.	Oct.	Nov.	Dec.	Annual
Kansas														
Concordia	15	5.0	5.8	3.5	0.7	T	0.0	0.0	0.0	0.0	0.3	2.0	4.4	21.7
Dodge City	35	4.0	3.2	4.7	0.7	T	0.0	0.0	0.0	T	0.2	2.1	3.3	18.2
Goodland	38	5.4	4.5	8.6	3.7	0.3	0.0	0.0	0.0	0.1	1.4	4.1	5.5	33.6
Topeka	31	5.7	4.2	4.3	0.6	0.0	0.0	0.0	0.0	0.0	T	1.3	4.7	20.8
Wichita	24	4.2	3.8	3.0	0.2	0.0	0.0	0.0	0.0	0.0	T	1.0	2.9	15.1
Kentucky														
Covington (Cincinnati Airport)	30	6.8	5.0	4.8	0.5	T	0.0	0.0	0.0	0.0	0.1	2.9	3.8	23.9
Lexington	33	5.6	4.6	3.0	0.2	0.0	0.0	0.0	0.0	0.0	T	0.7	1.8	15.9
Louisville	30	5.4	4.2	3.8	0.2	0.0	0.0	0.0	0.0	0.0	T	1.6	2.4	17.6
Louisiana														
Baton Rouge	27	T	0.2	T	0.0	0.0	0.0	0.0	0.0	0.0	0.0	T	T	0.2
Lake Charles	16	0.3	T	T	0.0	0.0	0.0	0.0	0.0	0.0	0.0	T	T	0.3
New Orleans	31	T	0.1	T	0.0	0.0	0.0	0.0	0.0	0.0	0.0	T	0.1	0.2
Shreveport	25	0.7	0.3	0.2	0.0	0.0	0.0	0.0	0.0	0.0	0.0	T	0.2	1.4
Maine														
Caribou	38	23.0	23.0	19.9	8.2	0.9	T	0.0	0.0	T	2.2	12.4	23.3	112.9
Portland	37	18.1	19.6	13.9	3.1	0.2	0.0	0.0	0.0	T	0.3	3.3	16.0	74.5
Maryland														
Baltimore	27	5.1	5.9	4.6	0.1	T	0.0	0.0	0.0	0.0	T	1.1	4.4	21.2
Massachusetts														
Blue Hill Observatory	92	15.4	16.3	11.7	3.1	0.1	0.0	0.0	0.0	0.0	0.2	2.6	10.9	60.3
Boston	42	12.2	11.8	8.1	0.7	T	0.0	0.0	0.0	0.0	T	1.2	8.1	42.1
Worcester	22	16.4	18.8	16.2	3.7	0.6	0.0	0.0	0.0	0.0	0.5	3.4	14.6	74.2
Michigan														
Alpena	18	20.8	15.2	14.1	3.4	0.3	0.0	0.0	0.0	T	0.7	9.0	21.4	84.9
Detroit														
City Airport	38	7.9	7.7	5.4	1.1	T	0.0	0.0	0.0	0.0	T	2.5	7.1	31.7
Metro Airport	19	8.6	8.3	6.8	1.6	T	0.0	0.0	0.0	0.0	T	3.6	11.0	39.9
Flint	36	11.1	9.8	8.0	2.5	T	0.0	0.0	0.0	T	0.1	4.2	9.6	45.3
Grand Rapids	14	20.9	11.3	13.1	3.8	T	0.0	0.0	0.0	T	0.8	9.1	17.6	76.6

Houghton Lake	13	18.9	14.7	14.2	4.5	0.2	0.0	T	0.0	T	0.6	12.0	18.1	83.2
Lansing	73	11.6	9.8	8.7	2.8	0.2	0.0	0.0	0.0	T	0.4	4.7	10.5	48.7
Marquette	40	21.4	19.9	16.7	7.6	1.3	T	0.0	0.0	0.2	2.5	14.8	22.9	107.3
Muskegon	37	30.2	16.9	12.5	2.7	T	0.0	0.0	0.0	T	0.5	9.8	23.3	95.9
Sault Ste. Marie	36	26.7	19.6	15.1	4.8	0.7	T	0.0	0.0	0.2	2.3	15.1	26.3	110.8
Minnesota														
Duluth	34	17.1	11.8	14.1	6.8	0.9	T	0.0	T	T	1.2	10.2	15.7	77.8
International Falls	38	11.5	8.5	10.1	6.1	1.0	T	0.0	0.0	0.1	1.5	10.5	10.8	60.1
Minneapolis-St. Paul	39	9.2	8.3	10.7	2.5	0.2	0.0	0.0	0.0	T	0.4	5.8	9.0	46.1
Rochester	17	8.8	7.6	9.9	3.0	0.1	0.0	0.0	0.0	T	0.4	3.4	11.2	44.4
Saint Cloud	38	8.6	7.6	9.8	2.5	0.1	0.0	0.0	0.0	T	0.5	6.1	7.9	43.1
Mississippi														
Jackson	14	0.5	0.3	0.4	T	0.0	0.0	0.0	0.0	0.0	0.0	T	T	1.2
Meridian	32	0.5	0.3	T	0.0	0.0	0.0	0.0	0.0	0.0	0.0	T	0.6	1.4
Missouri														
Columbia	27	5.0	5.2	4.5	0.7	T	0.0	0.0	0.0	0.0	T	2.1	4.1	21.6
Kansas City	43	5.9	3.8	3.7	0.7	T	0.0	0.0	0.0	0.0	T	1.1	4.3	19.5
Saint Joseph	22	5.8	3.1	4.6	0.6	0.0	0.0	0.0	0.0	0.0	T	0.5	4.6	19.2
St. Louis	41	4.6	4.1	4.4	0.2	T	0.0	0.0	0.0	0.0	T	1.5	3.7	18.5
Springfield	32	3.1	3.6	3.9	0.5	0.0	0.0	0.0	0.0	0.0	T	2.0	2.4	15.5
Montana														
Billings	43	9.4	8.1	10.1	7.9	1.1	T	0.0	0.0	0.9	3.2	6.5	8.8	56.0
Glasgow	22	7.0	3.8	3.4	2.9	0.5	0.0	0.0	0.0	0.1	1.4	2.9	4.9	26.9
Great Falls	40	10.0	8.3	9.8	7.5	1.4	0.4	T	0.0	1.3	3.0	7.3	8.8	57.8
Havre	17	9.5	4.6	7.4	8.3	0.2	T	0.0	0.0	0.3	1.0	4.1	8.2	43.6
Helena	37	9.1	6.0	7.4	5.3	1.5	0.1	T	0.0	1.4	2.0	6.7	8.6	48.1
Kalispell	28	19.1	10.7	6.7	2.3	1.2	T	0.0	0.0	0.1	1.5	8.7	16.7	67.2
Miles City	37	6.4	5.2	5.3	3.3	0.5	0.1	0.0	0.0	0.3	1.5	4.0	5.7	32.3
Missoula	33	13.2	8.0	6.3	2.5	0.8	T	0.0	0.0	T	0.9	6.3	11.2	49.2
Nebraska														
Grand Island	39	5.7	5.8	5.7	1.9	0.2	0.0	0.0	0.0	T	0.2	3.0	6.5	29.0
Lincoln	22	6.8	6.2	5.9	0.9	0.1	0.0	0.0	0.0	0.0	0.3	2.7	5.5	28.4
Norfolk	32	5.7	5.6	6.8	1.3	0.1	0.0	0.0	0.0	T	0.2	2.9	6.2	28.8
North Platte	25	5.4	4.7	7.4	3.0	0.2	0.0	0.0	0.0	T	1.5	3.0	4.7	29.9
Omaha	42	8.5	7.2	6.7	0.9	0.1	0.0	0.0	0.0	T	0.3	2.5	5.8	32.0
Omaha (North)	23	1.4	6.7	7.2	0.9	T	0.0	0.0	0.0	T	0.2	2.9	5.2	24.5
Scottsbluff	34	6.1	4.6	8.7	4.7	0.8	T	0.0	0.0	0.3	2.4	4.4	6.0	38.0

Table 3 (continued)
AVERAGE INCHES OF SNOWFALL IN THE U.S. AND CERTAIN PACIFIC ISLANDS (DATA THROUGH 1977)[2]

Location	Years	Jan.	Feb.	Mar.	April	May	June	July	Aug.	Sept.	Oct.	Nov.	Dec.	Annual
Valentine	22	4.5	4.7	7.5	3.6	T	0.0	0.0	0.0	T	1.0	3.6	3.5	28.4
Nevada														
Elko	46	10.0	6.4	5.6	2.7	0.9	T	0.0	0.0	T	0.8	4.2	8.3	38.9
Ely	39	8.5	7.0	9.3	6.3	2.2	0.2	0.0	0.0	0.1	1.9	4.5	7.6	47.6
Las Vegas	29	1.1	T	T	T	0.0	0.0	0.0	0.0	0.0	T	0.2	0.1	1.4
Reno	35	6.3	5.4	5.3	1.6	1.1	T	0.0	0.0	T	0.3	1.6	4.9	26.5
Winnemucca	28	5.7	3.7	4.9	2.9	0.7	T	0.0	0.0	T	0.3	2.0	5.2	25.4
New Hampshire														
Concord	36	17.2	15.3	11.6	2.2	0.2	0.0	0.0	0.0	0.0	0.1	3.9	14.3	64.8
Mt. Washington	45	36.6	41.7	41.1	28.8	11.2	1.1	T	0.2	1.6	11.6	30.5	42.0	246.4
New Jersey														
Atlantic City	33	4.9	4.8	3.1	0.2	T	0.0	0.0	0.0	0.0	T	0.4	2.4	15.8
Newark	36	6.9	7.9	4.9	0.5	T	0.0	0.0	0.0	0.0	T	0.4	6.7	27.3
Trenton	45	5.7	6.9	4.2	0.4	T	0.0	0.0	0.0	0.0	0.1	0.9	4.5	22.7
New Mexico														
Albuquerque	38	2.3	1.8	2.1	0.5	T	0.0	0.0	0.0	T	T	1.2	2.6	10.5
Clayton	33	3.2	3.0	4.9	1.7	T	0.0	0.0	0.0	0.1	0.5	2.8	2.8	19.0
Roswell	30	2.6	2.5	1.5	0.1	0.0	0.0	0.0	0.0	0.0	0.2	1.2	2.5	10.6
New York														
Albany	31	15.0	15.3	12.0	2.5	0.1	0.0	0.0	0.0	0.0	0.1	4.5	16.2	65.7
Binghamton	26	19.0	19.2	15.4	5.0	0.4	0.0	0.0	0.0	T	0.4	8.2	19.3	86.9
Buffalo	34	23.0	18.1	12.2	3.2	0.1	0.0	0.0	0.0	T	0.3	13.2	22.8	92.9
New York														
Central Park	109	7.6	8.6	5.2	0.9	T	0.0	0.0	0.0	0.0	T	0.9	5.8	29.0
JFK Airport	19	6.7	8.1	4.1	0.2	T	0.0	0.0	0.0	0.0	T	0.2	4.2	23.5
LaGuardia Airport	33	6.4	8.3	4.7	0.6	T	0.0	0.0	0.0	0.0	T	0.4	5.8	26.2
Rochester	37	21.5	21.9	14.6	2.8	0.1	0.0	0.0	T	T	0.2	6.9	18.9	86.9
Syracuse	28	26.3	26.9	17.5	3.2	0.1	0.0	0.0	0.0	0.0	0.7	10.0	26.0	110.7
North Carolina														
Asheville	13	4.4	4.9	3.1	T	0.0	0.0	0.0	0.0	0.0	T	1.6	3.4	17.4
Cape Hatteras	20	0.4	0.7	0.4	0.0	0.0	0.0	0.0	0.0	0.0	0.0	T	0.1	1.6

Charlotte	38	2.1	1.3	1.1	T	0.0	0.0	0.0	0.0	0.0	0.0	0.1	0.7	5.3
Greensboro	49	3.3	1.9	1.8	T	0.0	0.0	0.0	0.0	0.0	0.0	0.2	1.5	8.7
Raleigh	33	2.6	1.8	1.3	T	0.0	0.0	0.0	0.0	0.0	0.0	0.2	0.9	6.8
North Dakota														
Wilmington	26	0.4	0.8	0.3	0.0	0.0	0.0	0.0	0.0	0.0	0.0	T	0.4	1.9
Bismarck	38	7.0	5.9	7.9	3.7	1.0	T	0.0	0.0	0.2	1.3	5.0	6.7	38.7
Fargo	35	7.3	5.5	6.7	3.7	0.1	0.0	0.0	0.0	T	0.7	4.7	6.8	35.5
Williston	16	7.0	4.8	6.7	5.3	0.2	0.0	0.0	0.0	0.4	1.4	4.2	7.2	37.2
Ohio														
Akron	29	10.4	9.4	9.3	2.2	0.1	0.0	0.0	0.0	T	0.6	5.4	10.4	47.8
Cincinnati	57	5.1	4.3	3.4	0.5	T	0.0	0.0	0.0	0.0	0.1	1.6	3.9	18.9
Cleveland	36	10.8	11.0	10.0	2.1	0.1	0.0	0.0	0.0	T	0.8	5.8	11.6	52.2
Columbus	30	7.5	5.8	4.9	0.7	T	0.0	0.0	0.0	T	0.1	2.8	5.9	27.7
Dayton	34	7.0	5.6	5.7	0.7	T	0.0	0.0	0.0	0.0	0.1	2.6	6.1	27.8
Mansfield	18	9.7	10.8	7.3	1.3	0.1	0.0	0.0	0.0	T	T	2.7	9.3	41.2
Toledo	22	9.0	8.1	6.7	1.7	T	0.0	0.0	0.0	T	T	3.6	9.8	38.9
Youngstown	34	12.7	11.3	10.9	2.4	0.2	0.0	0.0	0.0	T	0.5	6.7	12.9	57.6
Oklahoma														
Oklahoma City	38	2.8	2.2	1.7	T	0.0	0.0	0.0	0.0	0.0	T	0.5	1.6	8.8
Tulsa	39	3.1	2.3	1.7	0.1	0.0	0.0	0.0	0.0	0.0	T	0.5	1.4	9.1
Oregon														
Astoria	24	3.1	0.3	0.9	0.1	T	0.0	0.0	0.0	T	T	0.1	1.7	6.2
Burns	27	14.2	6.4	6.7	1.8	0.3	0.1	0.0	T	T	0.6	4.9	12.3	47.3
Eugene	35	5.0	0.4	0.7	T	T	0.0	0.0	0.0	T	T	0.3	1.2	7.6
Medford	48	4.0	1.5	0.8	0.3	T	0.0	0.0	0.0	0.0	T	0.6	1.5	8.7
Pendleton	42	7.7	3.6	0.9	0.2	T	0.0	0.0	0.0	0.0	0.1	1.6	3.6	17.7
Portland	37	4.2	0.7	0.6	T	T	0.0	0.0	0.0	T	T	0.4	1.5	7.4
Salem	40	3.7	0.8	0.8	T	T	0.0	0.0	0.0	0.0	T	0.3	1.4	7.0
Sexton Summit	36	24.9	16.1	21.0	8.8	2.1	0.2	0.0	T	0.1	1.8	9.8	21.6	106.4
Pacific														
Guam	21	0.0	0.0	0.0	0.0	0.0	0.0	0.0	0.0	0.0	0.0	0.0	0.0	0.0
Johnston Island	21	0.0	0.0	0.0	0.0	0.0	0.0	0.0	0.0	0.0	0.0	0.0	0.0	0.0
Koror Island	28	0.0	0.0	0.0	0.0	0.0	0.0	0.0	0.0	0.0	0.0	0.0	0.0	0.0
Kwajalein Island	33	0.0	0.0	0.0	0.0	0.0	0.0	0.0	0.0	0.0	0.0	0.0	0.0	0.0
Majuro, Marshall Islands	23	0.0	0.0	0.0	0.0	0.0	0.0	0.0	0.0	0.0	0.0	0.0	0.0	0.0
Pago Pago, American Samoa	18	0.0	0.0	0.0	0.0	0.0	0.0	0.0	0.0	0.0	0.0	0.0	0.0	0.0
Ponape Island	27	0.0	0.0	0.0	0.0	0.0	0.0	0.0	0.0	0.0	0.0	0.0	0.0	0.0

Table 3 (continued)
AVERAGE INCHES OF SNOWFALL IN THE U.S. AND CERTAIN PACIFIC ISLANDS (DATA THROUGH 1977)[2]

Location	Years	Jan.	Feb.	Mar.	April	May	June	July	Aug.	Sept.	Oct.	Nov.	Dec.	Annual
Truk, Caroline Islands	28	0.0	0.0	0.0	0.0	0.0	0.0	0.0	0.0	0.0	0.0	0.0	0.0	0.0
Wake Island	30	0.0	0.0	0.0	0.0	0.0	0.0	0.0	0.0	0.0	0.0	0.0	0.0	0.0
Yap Island	29	0.0	0.0	0.0	0.0	0.0	0.0	0.0	0.0	0.0	0.0	0.0	0.0	0.0
Pennsylvania														
Allentown	34	7.8	8.6	6.0	0.4	T	0.0	0.0	0.0	0.0	T	1.4	7.3	31.5
Erie	23	20.1	14.7	11.1	2.7	T	0.0	0.0	0.0	0.0	0.5	12.6	21.6	83.3
Harrisburg	39	9.0	9.3	6.4	0.2	T	0.0	0.0	0.0	0.0	0.1	2.0	7.5	34.5
Philadelphia	35	5.7	5.9	3.7	0.2	T	0.0	0.0	0.0	0.0	T	0.7	4.0	20.2
Pittsburgh														
International Airport	25	11.2	10.3	9.1	1.5	0.2	0.0	0.0	0.0	0.0	0.2	4.0	8.8	45.3
City	43	7.6	6.6	5.9	0.8	T	0.0	0.0	0.0	0.0	0.1	2.6	6.4	30.0
Avoca	22	11.0	11.6	10.9	2.7	0.2	0.0	0.0	0.0	T	0.3	3.9	10.0	50.6
Williamsport	33	9.8	11.0	8.9	1.1	T	0.0	0.0	0.0	0.0	0.1	3.3	9.6	43.8
Puerto Rico														
San Juan	22	0.0	0.0	0.0	0.0	0.0	0.0	0.0	0.0	0.0	0.0	0.0	0.0	0.0
Rhode Island														
Block Island	27	5.1	6.3	5.9	0.3	0.0	0.0	0.0	0.0	0.0	T	0.2	2.9	20.7
Providence	24	9.6	10.2	9.0	0.7	0.3	0.0	0.0	0.0	0.0	0.1	0.5	7.6	38.0
South Carolina														
Charleston	35	0.1	0.3	0.1	0.0	0.0	0.0	0.0	0.0	0.0	0.0	T	0.1	0.6
Columbia	30	0.4	0.8	0.2	0.0	0.0	0.0	0.0	0.0	0.0	0.0	T	0.3	1.7
Greenville-Spartanburg	15	2.1	1.6	0.8	0.0	0.0	0.0	0.0	0.0	0.0	0.0	0.1	1.1	5.7
South Dakota														
Aberdeen	48	7.2	6.9	6.6	3.7	0.2	0.0	0.0	0.0	T	0.7	4.6	6.5	36.4
Huron	38	6.8	9.4	9.2	2.2	0.2	0.0	0.0	0.0	T	0.8	4.6	6.3	39.5
Rapid City	35	5.4	6.3	9.6	6.2	0.8	0.1	0.0	0.0	0.1	1.7	4.3	4.8	39.3
Sioux Falls	32	6.1	9.2	10.2	1.6	T	0.0	0.0	0.0	T	0.5	4.0	7.5	39.1
Tennessee														
Bristol-Johnson City	40	4.8	4.1	2.5	T	T	0.0	0.0	0.0	0.0	T	1.3	2.9	15.6
Chattanooga	47	1.7	1.0	0.5	T	T	0.0	0.0	0.0	0.0	T	0.1	0.7	4.0
Knoxville	35	3.8	3.4	1.9	0.2	T	0.0	0.0	0.0	0.0	T	0.9	2.0	12.2

Memphis	27	2.1	1.2	1.2	T	0.0	0.0	0.0	0.0	0.0	0.0	0.1	0.9	5.5
Nashville	36	4.0	2.5	1.7	0.1	0.0	0.0	0.0	0.0	0.0	0.0	0.6	2.0	10.9
Oak Ridge	29	3.2	2.7	1.4	T	0.0	0.0	0.0	0.0	0.0	T	0.4	2.0	9.7
Texas														
Abilene	38	1.8	1.0	0.7	T	0.0	0.0	0.0	0.0	0.0	T	0.5	0.5	4.5
Amarillo	36	3.4	3.6	2.6	0.5	T	0.0	0.0	0.0	T	0.2	1.7	2.3	14.3
Austin	36	0.5	0.4	0.1	0.0	0.0	0.0	0.0	0.0	0.0	0.0	T	T	1.0
Brownsville	38	T	T	T	0.0	0.0	0.0	0.0	0.0	0.0	0.0	T	T	T
Corpus Christi	38	0.1	T	T	0.0	0.0	0.0	0.0	0.0	0.0	0.0	T	0.0	0.1
Dallas-Fort Worth	24	1.5	0.7	0.3	0.0	0.0	0.0	0.0	0.0	0.0	0.0	0.2	0.2	2.9
Del Rio	14	T	0.4	0.2	0.0	0.0	0.0	0.0	0.0	0.0	T	T	T	0.6
El Paso	38	1.4	0.8	0.4	T	0.0	0.0	0.0	0.0	0.0	T	1.1	1.0	4.7
Galveston	107	0.1	0.2	T	0.0	0.0	0.0	0.0	0.0	0.0	0.0	0.0	T	0.3
Houston	43	0.2	0.2	T	0.0	0.0	0.0	0.0	0.0	0.0	0.0	T	T	0.4
Lubbock	29	1.8	2.9	2.0	T	0.0	0.0	0.0	0.0	0.0	0.3	0.8	1.5	9.3
Midland-Odessa	29	1.4	0.9	0.4	T	0.0	0.0	0.0	0.0	0.0	T	0.4	0.3	3.4
Port Arthur	24	0.1	0.3	0.1	0.0	0.0	0.0	0.0	0.0	0.0	0.0	T	T	0.5
San Angelo	30	1.1	0.9	0.3	0.0	0.0	0.0	0.0	0.0	0.0	0.0	0.6	0.1	3.0
San Antonio	35	0.2	0.3	T	0.0	0.0	0.0	0.0	0.0	0.0	0.0	T	T	0.5
Victoria	16	0.1	0.1	0.0	0.0	0.0	0.0	0.0	0.0	0.0	0.0	T	T	0.2
Waco	34	0.9	0.4	0.1	0.0	0.0	0.0	0.0	0.0	0.0	0.0	T	0.1	1.5
Wichita Falls	34	1.7	1.6	0.9	T	0.0	0.0	0.0	0.0	0.0	T	0.4	0.9	5.5
Utah														
Milford	29	8.4	7.2	8.5	5.4	1.2	0.0	0.0	0.0	0.3	1.8	4.0	7.0	43.8
Salt Lake City	49	13.2	9.7	10.4	5.0	0.6	T	0.0	0.0	0.1	1.1	6.2	12.0	58.3
Vermont														
Burlington	34	17.9	18.0	12.4	3.6	0.2	0.0	0.0	0.0	T	0.2	7.4	19.6	79.3
Virginia														
Lynchburg	33	5.2	4.7	3.8	0.3	0.0	0.0	0.0	0.0	0.0	T	0.9	3.2	18.1
Norfolk	29	3.1	1.8	0.9	T	0.0	0.0	0.0	0.0	0.0	0.0	T	1.2	7.0
Richmond	40	5.2	3.3	2.8	0.1	0.0	0.0	0.0	0.0	0.0	T	0.4	2.1	13.9
Roanoke	30	6.4	6.7	4.0	0.3	T	0.0	0.0	0.0	T	T	2.0	4.7	24.1
Washington														
Olympia	36	9.1	2.5	2.2	0.1	T	T	0.0	0.0	T	T	1.3	4.0	19.2
Quillayute	11	9.0	3.9	2.8	0.7		T	0.0	0.0	T	T	0.8	3.4	20.6
Seattle														
Urban Site	27	3.9	0.8	0.8	T	T	0.0	0.0	0.0	0.0	T	0.5	2.3	8.3
International Airport	33	6.8	1.6	1.7	0.1	T	0.0	0.0	0.0	T	0.1	1.1	3.2	14.6

Table 3 (continued)
AVERAGE INCHES OF SNOWFALL IN THE U.S. AND CERTAIN PACIFIC ISLANDS (DATA THROUGH 1977)[2]

Location	Years	Jan.	Feb.	Mar.	April	May	June	July	Aug.	Sept.	Oct.	Nov.	Dec.	Annual
Spokane	30	18.3	7.6	4.8	0.6	0.1	T	0.0	0.0	0.0	0.6	5.8	15.5	53.3
Stampede Pass	34	82.5	71.9	69.4	43.1	12.9	1.5	0.3	T	1.5	18.1	63.0	82.4	446.6
Walla Walla	63	7.8	4.0	1.2	0.2	T	0.0	0.0	0.0	0.0	0.2	1.8	4.8	20.0
Yakima	31	9.9	3.2	1.8	T	T	0.0	0.0	0.0	0.0	0.1	1.7	7.8	24.5
West Virginia														
Beckley	14	14.1	14.8	8.0	1.6	T	0.0	0.0	0.0	T	0.4	5.5	11.4	55.8
Charleston	30	8.9	7.7	4.3	0.4	T	0.0	0.0	0.0	0.0	0.2	2.9	5.2	29.6
Elkins	15	14.8	17.5	9.7	2.3	T	0.0	0.0	0.0	0.0	0.4	7.6	15.0	67.3
Huntington	16	8.2	5.6	4.0	0.1	T	0.0	0.0	0.0	0.0	T	1.8	4.4	24.1
Parkersburg	89	6.9	6.0	4.0	0.7	T	0.0	0.0	0.0	0.0	0.1	1.8	4.4	23.9
Wisconsin														
Green Bay	28	10.0	8.6	9.3	2.1	0.1	0.0	0.0	0.0	T	0.1	4.0	10.4	44.6
La Crosse	27	9.5	8.2	10.5	2.0	T	0.0	0.0	0.0	T	0.1	3.5	9.1	42.9
Madison	29	8.8	7.0	9.4	1.8	T	0.0	0.0	0.0	T	0.1	3.1	10.0	40.2
Milwaukee	37	12.2	9.0	9.5	1.8	T	0.0	0.0	0.0	T	0.1	3.1	10.2	45.9
Wyoming														
Casper	27	9.1	9.9	15.0	13.3	2.4	0.3	0.0	T	0.9	4.8	9.1	9.1	73.9
Cheyenne	42	5.8	5.5	11.9	9.3	3.1	0.3	0.0	0.0	0.7	3.4	6.0	5.2	51.2
Lander	31	8.7	12.0	19.2	22.5	6.2	1.6	0.0	0.0	2.1	10.3	11.8	9.7	104.1
Sheridan	37	10.8	10.8	13.1	10.0	1.6	0.2	0.0	0.0	0.9	3.8	8.7	10.2	70.1

Table 4
AVERAGE DAILY MAXIMUM TEMPERATURE (°F) IN THE U.S. AND CERTAIN PACIFIC ISLANDS[2]
(NORMALS 1941-1970)

Location	Years	Jan.	Feb.	Mar.	April	May	June	July	Aug.	Sept.	Oct.	Nov.	Dec.	Annual
Alabama														
Birmingham	30	54.3	57.7	64.8	75.3	82.5	88.4	90.3	89.7	84.7	75.8	64.0	55.5	73.6
Huntsville	30	50.5	54.0	61.7	73.1	80.9	87.9	90.2	90.2	84.2	74.5	61.6	52.1	71.7
Mobile	30	61.1	64.1	69.5	78.0	85.0	89.8	90.5	90.6	86.5	79.7	69.5	63.0	77.3
Montgomery	30	57.9	61.4	67.7	76.8	83.6	89.2	90.5	90.7	86.5	78.0	67.2	59.3	75.7
Anchorage	30	20.0	26.6	32.8	43.8	55.2	62.9	65.6	63.8	55.7	41.8	28.3	20.6	43.1
Alaska														
Annette	30	38.0	41.5	43.7	48.8	56.2	61.0	64.0	64.6	59.8	51.7	44.3	40.1	51.1
Barrow	30	−8.0	−12.6	−8.6	6.5	24.2	37.1	44.3	42.0	33.4	20.2	5.4	−6.4	14.8
Barter Island	30	−8.5	−13.1	−7.5	8.2	26.5	38.2	45.5	43.5	35.0	21.5	6.3	−6.4	15.8
Bethel	30	12.3	15.8	20.3	32.7	48.5	60.2	61.9	58.5	51.8	36.1	23.6	11.3	36.1
Bettles	30	−5.3	1.0	12.9	31.7	51.6	67.2	68.2	60.9	48.2	26.6	5.8	−4.3	30.4
Big Delta	30	2.5	12.4	23.4	39.3	56.2	67.1	68.8	64.1	51.9	32.1	14.1	3.2	36.3
Cold Bay	30	32.6	32.8	33.7	37.6	44.2	49.8	54.4	55.3	51.6	44.0	38.6	33.4	42.3
Fairbanks	30	−2.2	9.3	23.3	40.4	58.8	70.7	71.8	65.8	54.4	33.5	11.7	−1.5	36.3
Gulkana	30	1.9	14.7	28.1	41.8	55.1	66.0	68.1	64.3	53.7	35.7	13.8	3.0	37.2
Homer	30	28.0	31.8	35.0	42.3	50.3	56.7	60.1	60.1	54.8	44.4	34.5	27.6	43.8
Juneau	30	29.1	33.9	38.2	46.5	55.4	62.0	63.6	62.3	56.1	47.2	37.3	32.0	47.0
King Salmon	30	21.0	24.6	28.6	39.1	51.3	59.7	62.5	60.9	54.8	40.8	29.0	19.6	41.0
Kodiak	30	34.5	35.7	36.9	41.6	47.9	54.6	59.1	60.1	54.9	45.6	39.0	34.3	45.4
Kotzebue	30	3.2	3.3	8.1	22.3	37.8	49.6	58.7	55.9	46.5	28.7	13.4	2.6	27.5
Mc Grath	30	.5	10.9	22.0	37.7	54.5	66.1	67.7	62.1	52.3	32.4	12.9	−θ.8	34.8
Nome	30	13.5	13.7	16.6	27.0	41.4	52.2	55.8	54.6	48.2	34.4	22.1	11.7	32.6
St. Paul Island	30	30.1	27.4	28.6	32.5	38.8	45.0	49.4	50.8	48.2	41.7	36.9	31.7	38.4
Talkeetna	30	19.1	25.8	32.8	44.0	56.1	65.7	67.5	64.1	55.6	40.6	26.1	18.0	42.9
Unalakleet	30	10.3	12.0	18.0	30.1	45.5	55.3	60.5	58.3	50.5	33.3	19.3	8.2	33.4
Valdez	30	25.0	29.8	35.3	44.1	52.5	59.6	61.1	60.0	54.2	44.0	32.2	25.9	43.6
Yakutat	30	31.2	35.0	37.9	43.6	50.8	56.4	59.3	59.6	55.4	47.3	38.2	32.7	45.6
Arizona														
Flagstaff	30	41.4	44.0	47.9	56.9	66.6	76.0	80.8	77.9	73.7	62.9	50.9	43.2	60.2
Phoenix	30	64.8	69.3	74.5	83.6	92.9	101.5	104.8	102.2	98.4	87.6	74.7	66.4	85.1

Table 4 (continued)
AVERAGE DAILY MAXIMUM TEMPERATURE (°F) IN THE U.S. AND CERTAIN PACIFIC ISLANDS[2]
(NORMALS 1941-1970)

Location	Years	Jan.	Feb.	Mar.	April	May	June	July	Aug.	Sept.	Oct.	Nov.	Dec.	Annual
Tucson	30	63.5	67.0	71.5	80.7	89.6	97.9	98.3	95.3	93.1	83.8	72.2	64.8	81.5
Winslow	30	45.6	53.3	60.2	70.1	79.9	89.8	93.6	90.6	85.4	73.2	58.2	46.7	70.6
Yuma	30	67.4	72.6	77.6	85.6	93.4	100.8	106.0	104.4	100.3	89.8	76.5	68.2	86.9
Arkansas														
Fort Smith	30	49.9	54.6	62.1	74.2	81.3	89.0	93.8	93.5	86.3	76.5	62.7	52.2	73.0
Little Rock	30	50.1	53.8	61.8	73.5	81.4	89.3	92.6	92.6	85.8	76.0	62.4	52.1	72.6
California														
Bakersfield	30	57.5	63.3	68.6	75.5	83.6	91.5	99.1	96.5	91.1	80.5	67.8	57.4	77.7
Bishop	30	53.6	57.9	63.7	71.9	80.7	89.7	97.5	95.3	88.7	77.2	63.8	55.4	74.6
Blue Canyon	30	42.9	44.2	45.0	51.7	59.8	67.8	77.5	76.4	72.7	62.6	51.7	45.6	58.2
Eureka	30	53.5	54.4	54.1	54.9	57.2	59.6	60.4	61.2	61.9	60.4	57.9	54.6	57.5
Fresno	30	54.8	60.8	66.6	74.3	82.9	90.3	98.2	96.0	91.0	79.8	66.1	54.6	76.3
Long Beach	30	65.2	66.2	67.7	70.5	73.8	76.8	82.8	84.0	83.3	78.3	72.7	66.9	74.0
Los Angeles														
International Airport	30	63.5	64.1	64.3	65.9	68.4	70.3	74.8	75.8	75.7	72.9	69.6	66.5	69.2
City	30	66.5	67.6	68.6	70.5	73.2	76.5	82.9	83.7	82.5	78.0	73.2	68.0	74.3
Mount Shasta	30	41.8	47.3	51.0	58.5	66.6	74.0	84.9	83.4	77.8	64.9	51.6	43.7	62.1
Oakland	30	54.5	58.0	60.2	62.8	65.4	68.5	69.7	70.2	72.3	68.7	62.0	55.5	64.0
Red Bluff	30	53.6	59.5	63.8	71.6	80.6	89.3	98.0	95.7	90.6	78.3	64.0	54.7	75.0
Sacramento	30	53.0	59.1	64.1	71.3	78.8	86.4	92.9	91.3	87.7	77.1	63.6	53.3	73.2
Sandberg	30	46.4	49.1	51.5	58.3	65.8	75.0	85.1	84.2	79.2	67.8	55.4	48.0	63.8
San Diego	30	64.6	65.6	66.0	67.6	69.4	71.1	75.3	77.3	76.5	73.8	70.1	66.1	70.3
San Francisco														
International Airport	30	55.3	58.6	61.0	63.5	66.6	70.2	70.9	71.6	73.6	70.3	63.3	56.5	65.1
City	30	56.0	58.9	60.1	61.2	62.5	64.5	63.8	64.8	68.8	68.2	63.2	56.8	62.4
Santa Maria	30	62.7	63.6	64.3	66.0	67.4	69.7	71.8	72.1	73.9	73.3	69.2	64.2	68.2
Stockton	30	52.8	59.0	64.8	72.4	80.3	88.1	94.7	92.8	88.8	78.1	64.2	53.3	74.1
Colorado														
Alamosa	30	35.0	40.1	47.3	57.9	68.0	77.6	82.0	79.6	74.2	63.1	48.0	37.3	59.2
Colorado Springs	30	41.0	43.6	47.7	59.2	68.4	78.1	84.4	82.4	74.9	64.2	49.8	43.1	61.4
Denver	30	43.5	46.2	50.1	61.0	70.3	80.1	87.4	85.8	77.7	66.8	53.3	46.2	64.0

Grand Junction	30	36.7	44.0	52.8	64.6	75.8	85.9	93.1	89.1	81.3	67.9	50.9	39.4	65.1
Pueblo	30	45.5	49.8	54.9	66.4	75.5	85.8	91.1	88.8	81.5	70.7	56.5	48.2	67.9
Connecticut														
Bridgeport	30	36.9	37.9	45.0	56.5	66.7	76.0	81.5	80.4	74.5	65.1	53.3	40.5	59.5
Hartford	30	33.4	35.7	44.6	58.9	70.3	79.5	84.1	81.9	74.5	64.3	50.6	36.8	59.6
Delaware														
Wilmington	30	40.2	42.2	51.1	63.0	73.1	81.6	85.5	83.9	78.2	67.8	55.2	43.0	63.7
District of Columbia														
Washington														
Dulles Airport	30	41.2	43.4	52.7	65.0	74.5	82.7	86.4	85.0	78.7	68.2	55.6	43.3	64.7
National Airport	30	43.5	46.0	55.0	67.1	76.6	84.6	88.2	86.6	80.2	69.8	57.2	45.2	66.7
Florida														
Apalachicola	30	61.1	63.1	67.6	75.0	81.8	86.4	87.5	87.7	84.7	78.3	69.0	62.7	75.4
Daytona Beach	30	69.1	70.4	74.5	80.2	85.2	88.4	89.6	89.4	87.0	81.4	75.1	70.2	80.0
Fort Myers	30	74.7	76.0	79.7	84.8	89.0	90.5	91.1	91.5	89.8	85.3	79.9	75.9	84.0
Jacksonville	30	64.6	66.9	72.2	79.0	84.6	88.3	90.0	89.7	86.0	79.2	71.4	65.6	78.1
Key West	30	75.6	76.6	79.4	82.5	85.3	87.9	89.2	89.5	87.8	84.0	79.6	76.4	82.8
Lakeland	30	70.5	72.0	76.2	81.8	86.8	89.7	90.4	90.4	88.0	82.4	76.0	71.5	81.3
Miami	30	75.6	76.6	79.5	82.7	85.3	88.0	89.1	89.9	88.3	84.6	79.9	76.6	83.0
Orlando	30	70.5	71.8	76.0	81.5	86.7	89.3	89.8	90.0	87.9	82.5	76.2	71.5	81.1
Pensacola	30	61.2	64.1	68.9	76.8	84.2	89.0	89.7	90.0	86.3	79.9	69.7	63.2	76.9
Tallahassee	30	64.2	66.5	72.1	80.1	86.7	90.4	90.6	90.5	87.4	80.6	71.4	65.1	78.8
Tampa	30	70.6	71.9	76.1	82.4	87.5	89.9	90.1	90.4	89.0	83.9	77.1	72.0	81.7
West Palm Beach	30	75.0	76.0	79.3	82.9	86.1	88.3	89.6	90.2	88.3	84.3	79.5	76.1	83.0
Georgia														
Athens	30	53.3	56.2	63.2	73.8	81.8	87.8	89.5	88.9	82.9	74.0	63.1	54.2	72.4
Atlanta	30	51.4	54.5	61.1	71.4	79.0	84.6	86.5	86.4	81.2	72.5	61.9	52.7	70.3
Augusta	30	57.6	60.5	67.1	76.9	84.2	89.6	90.8	90.2	85.2	77.0	67.1	58.7	75.4
Columbus	30	57.8	60.9	67.4	77.3	84.7	89.9	90.8	90.8	85.8	77.4	66.8	59.0	75.7
Macon	30	58.7	61.8	68.4	78.3	86.0	90.9	92.1	91.8	86.6	78.2	67.9	59.6	76.7
Rome	30	52.2	55.9	63.4	74.5	82.2	88.2	90.3	90.1	84.6	74.9	62.8	53.4	72.7
Savannah	30	61.1	63.6	69.5	77.8	84.8	89.3	90.8	90.3	85.4	78.2	69.3	62.1	76.8
Hawaii														
Hilo	30	79.6	79.4	78.8	79.8	81.3	82.7	83.0	83.5	83.6	83.2	81.3	79.4	81.3
Honolulu	30	79.3	79.2	79.7	81.4	83.6	85.6	86.8	87.4	87.4	85.8	83.2	80.3	83.3
Kahului	30	79.5	79.3	80.3	81.8	83.8	85.6	86.3	86.7	87.0	85.6	82.8	79.6	83.2
Lihue	30	77.9	77.9	77.9	79.2	81.4	83.3	84.0	84.6	84.8	83.3	80.8	78.2	81.1

Table 4 (continued)
AVERAGE DAILY MAXIMUM TEMPERATURE (°F) IN THE U.S. AND CERTAIN PACIFIC ISLANDS[2]
(NORMALS 1941-1970)

Location	Years	Jan.	Feb.	Mar.	April	May	June	July	Aug.	Sept.	Oct.	Nov.	Dec.	Annual
Idaho														
Boise	30	36.5	43.8	51.6	61.4	70.6	78.3	90.5	87.6	77.6	64.7	48.9	39.1	62.6
Lewiston	30	37.9	46.0	52.9	62.0	70.6	77.9	89.2	87.1	77.6	63.2	48.0	41.0	62.8
Pocatello	30	32.4	38.6	45.8	57.7	68.1	76.5	88.8	86.4	75.7	63.0	45.9	35.5	59.5
Illinois														
Cairo	30	43.7	47.6	56.6	69.5	78.7	87.0	89.8	88.3	81.4	71.5	56.7	46.1	68.1
Chicago														
O'Hare Airport	30	31.1	34.3	44.6	59.4	69.7	79.1	83.1	82.3	75.4	65.5	48.3	35.0	59.0
Midway Airport	30	31.5	34.6	44.6	59.3	70.3	80.6	84.4	83.3	75.8	65.1	48.1	35.3	59.4
Moline	30	30.0	34.3	45.0	61.3	72.0	81.4	85.2	83.8	76.0	66.0	48.1	34.6	59.8
Peoria	30	31.9	36.0	46.5	61.7	72.3	81.7	85.5	84.0	76.4	65.9	48.7	35.7	60.5
Rockford	30	28.9	32.6	43.1	59.1	70.3	79.9	84.2	82.8	74.7	64.3	46.5	33.1	58.3
Springfield	30	34.8	38.9	48.7	63.6	74.1	83.3	86.6	85.0	78.7	68.1	51.0	38.2	62.6
Indiana														
Evansville	30	41.5	45.4	54.6	67.9	77.0	86.0	88.9	88.0	81.4	71.2	55.2	44.0	66.8
Fort Wayne	30	32.6	35.5	45.1	59.5	70.2	80.1	83.6	82.2	75.9	64.6	48.3	35.7	59.4
Indianapolis	30	36.0	39.3	49.0	62.8	72.9	82.3	85.4	84.0	77.7	67.0	50.5	38.7	62.2
South Bend	30	31.5	34.1	43.9	58.4	69.4	79.4	82.8	81.8	74.7	63.7	47.2	35.1	58.5
Iowa														
Burlington	30	31.7	36.2	46.6	62.1	72.7	81.9	86.1	84.6	76.7	66.7	49.2	35.9	60.9
Des Moines	30	27.5	32.5	42.5	59.7	70.9	79.8	84.9	83.2	74.6	64.9	46.4	32.8	58.3
Dubuque	30	26.1	30.5	41.1	57.8	68.5	77.6	82.1	80.6	72.2	62.3	44.3	30.7	56.1
Sioux City	30	28.2	33.3	42.9	61.3	72.5	81.3	86.7	84.8	75.3	65.8	47.0	33.3	59.4
Waterloo	30	25.7	30.2	40.6	58.5	70.0	79.4	83.6	82.1	73.3	63.5	44.6	30.6	56.8
Kansas														
Concordia	30	36.9	42.5	50.8	64.9	74.3	83.9	89.7	89.0	79.8	69.6	52.8	40.4	64.6
Dodge City	30	42.6	47.1	53.9	66.9	76.2	86.0	91.4	90.4	81.4	70.7	55.2	44.6	67.2
Goodland	30	41.5	45.1	50.3	63.3	72.9	83.5	90.5	88.7	79.6	68.3	52.6	43.4	65.0
Topeka	30	38.3	44.1	52.6	66.3	75.8	84.0	89.2	88.5	80.4	70.3	54.3	41.8	65.5
Wichita	30	41.4	47.1	55.0	68.1	77.1	86.5	91.7	91.0	81.9	71.3	55.8	44.3	67.6

Kentucky														
Covington (Cincinnati Airport)	30	39.7	42.7	51.8	65.0	74.4	83.2	86.5	85.8	79.7	68.5	53.2	42.0	64.4
Lexington	30	41.3	44.3	53.4	66.0	75.5	83.5	86.4	85.5	79.6	68.8	53.9	43.7	65.2
Louisville	30	42.0	45.0	54.0	66.9	75.6	83.7	87.3	86.8	80.5	70.3	54.9	44.1	65.9
Louisiana														
Baton Rouge	30	61.5	64.5	70.6	79.0	85.2	90.3	91.2	91.1	87.2	80.4	70.3	63.7	77.9
Lake Charles	30	61.6	64.6	70.0	78.2	84.3	89.6	91.2	91.2	87.9	81.6	71.1	64.2	78.0
New Orleans	30	62.3	65.1	70.4	78.4	84.9	89.6	90.4	90.6	86.6	79.9	70.3	64.2	77.7
Shreveport	30	56.6	60.4	67.3	76.9	83.6	90.1	93.5	93.8	87.9	79.3	67.2	58.9	76.3
Maine														
Caribou	30	19.8	23.0	32.8	45.5	60.3	70.3	75.6	73.1	64.7	52.8	38.2	23.9	48.3
Portland	30	31.2	33.3	40.8	52.8	63.6	73.2	79.1	77.6	69.9	60.2	47.5	34.9	55.3
Maryland														
Baltimore	30	41.9	43.9	53.0	65.2	74.8	83.2	86.7	85.1	79.0	68.3	56.1	43.9	65.1
Massachusetts														
Blue Hill Observatory	30	33.6	35.5	43.2	55.7	66.8	75.4	80.5	78.6	71.3	62.0	49.8	37.0	57.5
Boston	30	35.9	37.5	44.6	56.3	67.1	76.6	81.4	79.3	72.2	63.2	51.7	39.3	58.7
Worcester	30	31.0	33.1	41.2	54.9	66.1	75.0	79.4	77.2	70.1	60.6	47.3	34.3	55.8
Michigan														
Alpena	30	26.9	28.7	36.8	51.6	63.6	74.3	79.0	77.3	68.4	58.9	42.9	31.0	53.3
Detroit														
City Airport	30	31.7	33.7	43.1	57.6	68.5	79.1	83.1	81.6	74.2	63.4	47.7	35.4	58.3
Metro Airport	30	31.9	34.3	43.8	58.1	69.1	79.4	83.4	82.0	74.8	64.1	47.8	35.4	58.7
Flint	30	29.9	32.0	41.4	56.5	67.0	77.0	81.2	79.7	72.2	61.9	45.9	33.8	56.5
Grand Rapids	30	30.3	32.6	42.0	57.3	68.8	79.1	83.3	81.9	73.9	63.1	46.2	33.9	57.7
Houghton	30	25.8	28.0	36.9	53.2	65.2	75.2	78.8	77.0	68.2	58.0	41.7	29.6	53.1
Lansing	30	29.9	32.0	41.7	57.0	68.1	78.4	82.6	81.1	73.0	62.1	46.0	33.9	57.2
Marquette	30	24.8	26.5	34.3	47.9	58.9	69.6	75.3	73.9	65.3	56.3	39.8	29.0	50.1
Muskegon	30	30.1	31.5	40.5	54.9	65.9	76.2	80.4	79.2	71.4	60.7	45.7	34.3	55.9
Sault Ste. Marie	30	22.0	23.7	32.5	47.2	59.4	70.0	75.1	73.4	64.5	54.8	39.0	26.8	49.0
Minnesota														
Duluth	30	17.6	22.1	32.6	47.8	60.0	69.7	76.4	74.4	64.0	54.3	35.3	22.5	48.1
International Falls	30	12.8	19.4	32.3	49.1	62.5	72.4	78.2	75.5	64.2	54.0	32.5	18.1	47.6
Minneapolis—St. Paul	30	21.2	25.9	36.9	55.5	67.9	77.1	82.4	80.8	70.7	60.7	40.6	26.6	53.8
Rochester	30	21.9	26.4	36.5	54.9	67.1	76.4	80.7	79.4	70.2	60.4	40.9	26.9	53.5
Saint Cloud	30	19.2	24.4	35.8	53.8	66.6	76.0	81.8	80.1	69.2	59.0	38.6	24.8	52.4
Mississippi														

Table 4 (continued)
AVERAGE DAILY MAXIMUM TEMPERATURE (°F) IN THE U.S. AND CERTAIN PACIFIC ISLANDS[2] (NORMALS 1941-1970)

Location	Years	Jan.	Feb.	Mar.	April	May	June	July	Aug.	Sept.	Oct.	Nov.	Dec.	Annual
Jackson	30	58.4	61.7	68.7	78.2	85.0	91.0	92.7	92.6	88.0	80.1	68.5	60.5	77.1
Meridian	30	58.4	61.8	68.7	78.4	85.2	91.3	92.5	92.2	86.9	78.8	67.7	59.8	76.8
Missouri														
Columbia	30	38.0	42.7	51.3	65.3	74.5	82.7	87.4	86.4	79.4	69.2	53.6	41.1	64.3
Kansas City	30	35.7	41.4	50.7	64.7	74.2	82.8	88.0	87.2	78.8	68.2	51.4	39.3	63.5
Saint Joseph	30	35.9	41.8	50.9	65.8	75.9	84.2	89.0	87.5	79.5	70.1	53.1	40.0	64.5
St. Louis	30	39.9	44.2	53.0	67.0	76.0	84.9	88.4	87.2	80.1	69.8	54.1	42.7	65.6
Springfield	30	43.2	47.5	55.1	68.0	76.1	84.2	89.0	88.9	81.2	71.1	56.4	45.7	67.2
Montana														
Billings	30	31.2	37.1	42.1	55.8	65.7	73.7	85.6	83.8	71.3	61.0	45.0	35.8	57.3
Glasgow	30	18.9	25.4	35.6	55.0	66.7	73.9	84.3	83.0	70.4	59.2	39.1	26.5	53.2
Great Falls	30	29.3	35.9	40.4	54.5	65.0	72.1	83.7	81.8	70.0	59.4	43.4	34.7	55.9
Havre	30	22.1	28.5	36.9	55.7	67.7	74.2	85.0	83.4	71.4	60.7	41.7	28.9	54.7
Helena	30	28.3	35.8	41.7	55.2	64.8	71.6	83.7	82.0	70.0	58.7	42.6	33.2	55.6
Kalispell	30	27.1	34.1	40.7	53.5	63.6	70.0	80.8	79.1	68.0	54.4	38.7	30.8	53.4
Miles City	30	26.1	32.8	41.6	57.8	68.9	77.4	88.9	87.2	73.5	62.1	43.1	32.2	57.6
Missoula	30	28.6	35.8	43.8	56.6	65.9	72.5	84.3	82.5	70.8	56.8	40.8	31.9	55.9
Nebraska														
Grand Island	30	33.3	38.7	46.7	62.2	72.7	82.4	88.7	87.4	77.2	67.3	50.3	37.8	62.1
Lincoln	30	32.8	38.3	47.0	63.4	73.4	83.1	88.9	87.0	77.5	67.6	50.3	37.7	62.2
Norfolk	30	29.6	34.4	42.9	60.3	71.4	80.9	87.0	85.3	75.2	65.5	47.6	34.6	59.6
North Platte	30	36.6	40.9	47.2	61.4	71.4	80.6	87.6	86.7	77.0	67.0	50.5	39.8	62.2
Omaha	30	32.7	38.5	47.7	64.4	74.4	83.1	88.6	87.2	78.6	69.1	50.9	37.8	62.8
Omaha (North)	30	29.1	34.8	44.1	61.0	71.4	80.2	85.4	84.0	75.2	65.9	47.4	34.3	59.4
Scottsbluff	30	38.5	43.1	48.0	60.4	70.2	79.7	88.8	86.9	77.1	66.0	50.4	40.8	62.5
Valentine	30	33.2	37.2	43.5	59.0	70.1	79.7	88.4	87.0	76.2	65.4	48.0	36.9	60.4
Nevada														
Elko	30	36.0	41.6	48.4	58.8	68.5	77.5	90.4	88.2	78.8	65.8	49.0	38.2	61.8
Ely	30	38.0	41.4	46.6	56.3	66.4	75.7	86.3	84.2	76.1	63.6	49.1	40.4	60.4
Las Vegas	30	55.7	61.3	67.8	77.5	87.5	97.2	103.9	101.5	94.8	81.0	65.7	56.7	79.2
Reno	30	45.4	51.1	56.0	64.0	72.2	80.4	91.1	89.0	81.8	70.0	56.3	46.4	67.0

Winnemucca	30	40.6	46.7	51.9	61.0	70.3	78.8	91.1	88.7	79.9	67.3	52.3	42.9	64.3
New Hampshire														
Concord	30	31.3	33.8	42.4	56.7	68.6	77.7	82.6	80.1	72.4	62.3	47.9	34.6	57.5
Mt. Washington	30	14.1	13.5	19.1	29.3	40.7	50.9	54.6	52.5	46.9	37.5	27.1	16.6	33.6
New Jersey														
Atlantic City	30	41.4	42.9	50.7	62.3	72.4	80.8	84.7	83.0	77.3	67.5	55.9	44.2	63.6
Newark	30	38.5	40.2	48.8	61.2	71.6	81.1	85.6	83.7	77.0	66.9	54.2	41.5	62.5
Trenton	30	38.8	40.6	49.2	61.8	72.0	80.9	84.9	82.8	76.2	66.2	53.9	41.5	62.4
New Mexico														
Albuquerque	30	46.9	52.6	59.2	70.1	79.9	89.5	92.2	89.7	83.4	71.7	57.1	47.5	70.0
Clayton	30	47.5	50.2	55.3	65.6	74.4	83.4	87.3	85.7	78.9	69.3	56.9	49.3	67.0
Roswell	30	55.4	60.9	67.7	78.2	86.4	94.2	94.7	93.4	86.5	77.0	64.8	56.8	76.3
New York														
Albany	30	30.4	32.7	42.6	58.0	69.7	79.4	83.9	81.4	73.7	62.8	48.1	34.1	58.1
Binghamton	30	28.7	30.1	38.9	54.0	65.0	74.5	78.5	76.8	69.4	59.2	44.5	31.3	54.3
Buffalo	30	29.8	31.0	39.0	53.3	64.3	75.1	79.5	77.6	70.9	60.2	46.1	33.6	55.0
New York														
Central Park	30	38.5	40.2	48.4	60.7	71.4	80.5	85.2	83.4	76.8	66.8	54.0	41.4	62.3
JFK Airport	30	38.0	39.1	46.5	58.1	68.4	78.0	83.2	81.7	75.4	65.8	53.7	41.3	60.8
La Guardia Airport	30	37.7	39.2	47.1	59.3	69.8	79.4	84.1	82.1	75.2	65.1	53.2	41.0	61.1
Rochester	30	31.3	32.6	41.1	56.0	67.2	78.0	82.2	80.1	73.1	62.4	47.9	34.9	57.2
Syracuse	30	31.4	32.7	41.5	56.5	67.6	77.7	82.0	80.2	73.3	62.4	48.3	35.0	57.4
North Carolina														
Asheville	30	48.4	50.6	58.3	69.4	76.8	82.5	84.3	83.8	78.0	69.1	58.2	49.3	67.4
Cape Hatteras	30	52.3	53.1	57.9	66.3	73.8	80.5	83.8	83.4	79.5	71.3	63.1	54.8	68.3
Charlotte	30	52.1	54.9	62.2	72.7	80.2	86.4	88.3	87.4	82.0	73.1	62.4	52.5	71.2
Greensboro	30	48.8	51.4	59.4	70.9	78.9	85.4	87.5	86.2	80.5	71.2	60.1	49.9	69.2
Raleigh	30	51.0	53.2	61.0	72.2	79.4	85.6	87.7	86.8	81.5	72.4	62.1	51.9	70.4
Wilmington	30	56.6	58.7	64.9	74.3	81.4	86.8	88.8	88.0	83.5	75.4	66.8	58.2	73.6
North Dakota														
Bismarck	30	19.1	24.5	35.4	54.8	67.1	75.8	84.3	83.5	71.3	60.3	39.4	26.0	53.5
Fargo	30	15.4	20.6	33.5	52.6	66.8	75.9	82.8	81.6	69.6	58.4	37.2	21.9	51.4
Williston	30	19.3	25.7	35.6	54.9	67.0	75.0	84.0	82.8	70.3	58.9	38.3	25.8	53.1
Ohio														
Akron	30	33.9	36.0	45.4	59.3	69.8	79.4	82.6	81.3	74.7	63.7	48.6	36.5	59.3
Cincinnati	30	39.8	42.9	52.2	65.5	75.2	83.6	86.6	86.0	79.8	68.8	53.0	41.8	64.6
Cleveland	30	33.4	35.0	44.1	58.0	68.4	78.2	81.6	80.4	74.2	63.6	48.8	36.4	58.5

Table 4 (continued)
AVERAGE DAILY MAXIMUM TEMPERATURE (°F) IN THE U.S. AND CERTAIN PACIFIC ISLANDS[2]
(NORMALS 1941-1970)

Location	Years	Jan.	Feb.	Mar.	April	May	June	July	Aug.	Sept.	Oct.	Nov.	Dec.	Annual
Columbus	30	36.4	39.2	49.3	62.8	72.9	81.9	84.8	83.7	77.6	66.4	50.9	38.7	62.1
Dayton	30	35.8	38.6	48.1	61.7	72.0	81.6	84.7	83.4	77.0	66.0	50.1	38.3	61.5
Mansfield	30	35.3	37.6	46.9	60.1	70.5	80.1	83.7	82.4	76.0	65.1	49.7	37.6	60.4
Toledo	30	32.4	35.2	45.0	59.5	70.5	80.5	83.8	82.4	75.7	64.8	48.0	35.4	59.4
Youngstown	30	33.0	34.7	44.3	58.4	68.9	78.3	81.8	80.4	73.9	62.9	47.9	35.6	58.4
Oklahoma														
Oklahoma City	30	47.6	52.6	59.8	71.6	78.7	87.0	92.6	92.5	84.7	74.2	60.9	50.7	71.1
Tulsa	30	47.0	52.2	59.7	71.8	79.2	87.3	92.8	92.7	84.8	75.0	60.8	50.1	71.1
Oregon														
Astoria	30	46.5	50.6	52.1	55.6	60.3	63.8	67.7	68.3	67.6	61.0	53.4	48.6	58.0
Burns	30	35.4	41.5	47.9	58.1	66.5	73.9	85.6	83.4	75.4	62.6	47.2	37.5	59.6
Eugene	30	45.6	51.7	55.2	61.2	67.8	74.1	82.6	81.3	76.5	64.0	53.1	47.4	63.4
Medford	30	44.2	51.8	56.7	63.8	71.7	79.4	89.5	87.8	82.1	67.4	52.7	44.2	66.0
Pendleton	30	38.6	46.5	53.2	61.9	70.4	78.3	88.2	85.5	76.9	63.4	48.9	41.8	62.8
Portland	30	43.6	50.1	54.3	60.3	67.0	72.1	79.0	78.1	73.9	62.9	52.1	46.0	61.6
Salem	30	45.3	51.4	54.9	61.0	68.1	74.0	82.4	81.3	76.5	64.1	53.0	47.1	63.3
Sexton Summit	30	39.8	43.1	44.9	51.9	59.6	66.3	75.7	74.7	70.2	58.3	47.1	41.7	56.1
Pacific														
Guam	30	83.2	83.3	84.5	85.6	86.4	86.7	86.2	86.1	85.8	85.6	85.3	84.1	85.1
Johnston Island	30	80.7	80.5	80.6	81.5	82.6	84.0	84.6	85.0	84.9	84.5	83.1	81.5	82.8
Koror Island	30	86.4	86.5	87.1	88.1	88.0	87.7	87.1	87.0	87.5	87.9	88.0	87.2	87.3
Kwajalein Island	30	85.0	85.4	85.9	85.8	86.0	86.2	86.7	87.3	87.3	87.2	86.0	85.5	86.2
Majuro, Marshall Islands	30	84.7	85.1	85.3	85.3	85.6	85.4	85.5	86.0	86.1	86.0	85.6	85.2	85.4
Pago Pago, American Samoa	30	87.0	86.9	86.9	87.0	85.3	84.5	83.6	83.3	84.6	85.1	86.1	86.1	85.5
Ponape Island	30	85.8	85.8	86.3	86.5	86.8	87.1	87.2	87.6	87.7	87.7	87.3	86.6	86.9
Truk, Caroline Islands	30	85.2	85.3	85.5	85.9	86.4	86.8	86.5	86.8	86.9	86.9	86.7	85.8	86.2
Wake Island	30	81.6	81.7	82.5	83.1	84.6	86.3	87.1	87.8	87.6	86.6	84.9	83.1	84.7
Yap Island	30	85.3	85.5	86.2	87.2	87.5	87.6	87.2	87.0	87.4	87.6	87.3	86.1	86.8
Pennsylvania														
Allentown	30	35.7	37.9	47.7	61.3	71.7	81.0	85.4	62.8	75.9	65.6	51.7	38.7	61.3
Erie	30	31.7	32.5	40.4	53.5	63.7	73.5	77.4	76.0	70.2	60.2	46.5	34.8	55.0

Harrisburg	30	37.7	40.5	50.7	64.1	74.5	83.0	86.8	84.6	78.0	66.9	52.9	40.1	63.3
Philadelphia	30	40.1	42.2	51.2	63.5	74.1	83.0	86.8	84.8	78.4	67.9	55.5	43.2	64.2
Pittsburgh														
International Airport	30	35.3	37.3	47.2	60.9	70.8	79.5	82.5	80.9	74.9	63.9	49.3	37.3	60.0
City	30	37.4	39.5	48.9	62.5	72.3	81.0	84.0	82.6	76.7	66.1	51.6	39.8	61.9
Avoca	30	33.5	35.3	44.7	58.9	70.0	79.0	83.0	80.7	73.6	63.0	48.8	36.1	58.9
Williamsport	30	35.0	37.1	47.0	61.2	72.0	80.6	84.4	82.4	75.2	64.5	50.0	37.8	60.6
Puerto Rico														
San Juan	30	81.9	82.1	83.6	84.4	85.6	87.0	87.0	87.5	87.6	87.4	85.0	83.1	85.2
Rhode Island														
Block Island	30	37.4	37.2	42.7	51.7	60.4	69.7	75.6	75.3	69.8	61.4	51.8	41.2	56.2
Providence	30	36.2	37.6	44.7	56.7	66.8	76.3	81.1	79.8	73.1	63.9	52.0	39.6	59.0
South Carolina														
Charleston	30	59.8	61.9	67.8	76.2	83.1	87.7	89.1	88.6	84.5	77.1	68.4	60.8	75.4
Columbia	30	56.9	59.7	66.5	76.9	84.5	90.3	92.0	91.0	85.4	77.1	66.9	57.9	75.4
Greenville—Spartanburg	30	51.6	54.1	61.6	72.0	79.9	85.9	87.6	86.8	81.0	72.4	61.8	52.4	70.6
South Dakota														
Aberdeen	30	20.5	26.0	37.7	56.7	68.6	77.4	85.1	84.7	73.2	61.9	41.4	26.9	55.0
Huron	30	23.4	28.7	39.2	57.9	69.7	79.0	86.8	85.5	74.1	63.0	43.4	29.4	56.7
Rapid City	30	34.2	37.6	42.7	57.2	67.4	76.3	86.3	85.9	74.7	63.6	47.5	38.0	59.3
Sioux Falls	30	24.6	29.7	39.7	57.8	69.7	78.9	85.1	83.8	73.0	62.7	43.5	29.6	56.5
Tennessee														
Bristol—Johnson City	30	46.0	48.9	57.1	68.3	77.1	84.0	85.9	85.3	80.4	70.4	56.9	47.3	67.3
Chattanooga	30	49.9	53.4	61.2	72.9	81.0	87.5	89.5	89.0	83.4	73.5	60.7	50.9	71.1
Knoxville	30	48.9	52.0	60.4	72.0	79.8	86.1	88.0	87.3	82.0	71.8	58.9	49.8	69.8
Memphis	30	49.4	53.1	60.8	72.7	81.2	88.7	91.6	90.6	84.3	74.9	61.5	51.7	71.7
Nashville	30	47.6	50.9	59.2	71.3	79.8	87.5	90.2	89.2	83.5	73.2	59.0	49.6	70.1
Oak Ridge	30	47.1	50.3	58.7	70.5	78.9	85.5	87.5	86.8	81.5	71.1	57.9	48.3	68.6
Texas														
Abilene	30	55.7	59.9	67.3	77.7	83.9	91.6	95.3	95.3	87.5	78.0	66.2	58.2	76.4
Amarillo	30	49.4	53.0	60.0	70.9	79.2	88.0	91.4	90.4	82.9	72.9	60.0	51.5	70.8
Austin	30	60.0	63.8	70.7	79.0	85.2	91.7	95.4	95.9	89.4	81.3	70.2	63.0	78.8
Brownsville	30	69.5	72.7	76.6	83.1	87.1	90.6	92.8	93.0	89.9	84.7	77.5	72.3	82.5
Corpus Christi	30	66.5	69.8	75.5	82.1	86.6	91.2	94.4	94.8	90.0	84.1	75.2	69.3	81.6
Dallas—Fort Worth	30	55.7	59.8	66.6	76.3	82.8	90.8	95.5	96.1	88.5	79.2	67.5	58.7	76.5
Del Rio	30	63.4	68.6	76.4	85.1	90.2	96.1	99.2	98.5	91.9	83.1	72.2	65.0	82.5
El Paso	30	57.0	62.5	68.9	78.5	87.2	94.9	94.6	92.8	87.4	78.5	66.1	57.8	77.2

Table 4 (continued)
AVERAGE DAILY MAXIMUM TEMPERATURE (°F) IN THE U.S. AND CERTAIN PACIFIC ISLANDS[2] (NORMALS 1941-1970)

Location	Years	Jan.	Feb.	Mar.	April	May	June	July	Aug.	Sept.	Oct.	Nov.	Dec.	Annual
Galveston	30	59.4	61.5	66.0	73.3	80.0	85.2	87.4	87.6	84.6	78.0	68.8	62.7	74.5
Houston	30	62.6	66.0	71.8	79.4	85.9	91.3	93.8	94.3	90.1	83.5	73.0	65.8	79.8
Lubbock	30	53.4	57.0	63.8	74.8	82.5	90.6	92.4	91.3	83.8	74.7	63.1	55.2	73.6
Midland—Odessa	30	57.8	62.1	69.4	79.1	86.5	92.8	95.0	94.4	87.9	79.2	67.5	60.1	77.7
Port Arthur	30	61.5	65.0	70.5	78.3	84.3	89.9	92.0	92.6	88.6	81.3	70.9	64.2	78.3
San Angelo	30	59.1	63.2	70.7	80.4	86.5	93.4	96.9	96.9	88.4	79.6	68.5	61.4	78.7
San Antonio	30	61.6	65.6	72.5	80.3	86.2	92.4	95.6	95.9	89.8	81.8	71.1	64.6	79.8
Victoria	30	63.4	66.5	72.4	79.8	85.3	90.5	93.4	93.9	89.1	82.7	73.2	66.5	79.7
Waco	30	57.4	61.5	68.4	77.8	84.4	91.9	96.2	96.7	89.5	80.4	68.7	60.5	77.8
Wichita Falls	30	53.5	58.1	65.8	77.4	84.7	93.9	99.2	99.4	90.3	79.2	66.0	56.2	77.0
Utah														
Milford	30	38.4	44.2	52.4	62.9	73.7	83.5	92.8	90.3	81.8	68.0	52.0	41.3	65.1
Salt Lake City	30	37.4	43.4	50.8	61.8	72.4	81.3	92.8	90.2	80.3	66.4	50.0	39.0	63.8
Vermont														
Burlington	30	25.9	28.2	38.0	53.3	66.1	76.5	81.0	78.3	70.0	58.7	44.3	30.3	54.2
Virginia														
Lynchburg	30	45.8	47.8	56.2	68.1	76.6	83.5	86.1	84.5	78.7	69.0	57.3	46.9	66.7
Norfolk	30	48.8	50.0	57.3	67.7	76.2	83.5	86.6	84.9	79.6	70.1	60.5	50.6	68.0
Richmond	30	47.4	49.9	58.2	70.3	78.4	85.4	88.2	86.6	80.9	71.2	60.6	49.1	68.8
Roanoke	30	45.6	47.9	56.3	67.9	76.1	83.0	85.9	84.9	79.5	69.9	57.6	46.6	66.8
Washington														
Olympia	30	44.0	49.6	53.6	59.9	67.2	71.9	78.4	77.2	72.1	61.2	51.3	45.8	61.0
Quillayute	30	44.0	47.8	49.4	53.7	59.7	63.0	67.7	67.5	66.3	58.7	50.6	45.7	56.2
Seattle														
Urban Site	30	44.7	50.0	53.0	59.0	65.8	70.4	75.8	74.5	69.3	61.8	51.0	46.6	60.2
International Airport	30	43.4	48.5	51.5	57.0	64.1	69.0	75.1	73.8	68.7	59.4	50.4	45.4	58.8
Spokane	30	31.1	39.0	46.2	57.0	66.5	73.6	84.3	81.9	72.5	58.1	41.8	33.9	57.2
Stampede Pass	30	27.3	32.1	35.1	41.5	50.3	57.0	65.5	64.4	59.4	47.6	35.0	29.8	45.4
Walla Walla	30	39.3	46.9	54.1	62.6	70.9	78.7	88.9	86.0	77.4	63.9	49.2	42.5	63.4
Yakima	30	36.4	46.1	54.8	64.1	73.1	79.7	88.1	85.9	78.3	64.7	48.5	39.1	63.2

West Virginia														
Beckley	30	39.9	41.7	50.0	62.5	69.1	77.8	80.0	79.0	73.5	64.1	51.1	41.0	60.8
Charleston	30	43.6	46.2	55.2	67.9	76.6	83.4	85.6	84.4	79.0	69.1	55.8	45.2	66.0
Elkins	30	40.7	42.4	50.5	62.6	71.1	78.0	80.3	79.3	74.1	64.8	52.2	42.2	61.5
Huntington	30	42.9	45.4	54.7	67.5	76.2	83.4	85.7	84.6	79.0	68.8	55.4	44.6	65.7
Parkersburg	30	41.3	43.7	53.1	65.9	75.5	83.1	85.6	84.6	78.6	68.4	54.2	43.2	64.8
Wisconsin														
Green Bay	30	23.9	27.2	37.1	54.1	65.8	75.8	80.7	79.1	69.8	59.6	41.8	28.6	53.6
La Crosse	30	25.0	29.7	40.0	57.8	69.3	78.4	83.0	81.7	71.8	61.8	43.0	29.6	55.9
Madison	30	25.4	29.5	39.2	56.0	67.3	76.9	81.4	80.0	70.9	60.9	43.0	29.8	55.0
Milwaukee	30	27.3	30.3	39.4	54.6	65.0	75.3	80.4	79.7	71.5	61.4	44.4	31.5	55.1
Wyoming														
Casper	30	33.6	37.7	42.6	55.5	66.1	76.3	87.1	85.6	74.1	61.4	44.8	36.2	58.4
Cheyenne	30	38.2	40.7	43.5	55.4	65.1	74.4	83.7	81.9	72.8	61.8	47.5	40.3	58.8
Lander	30	31.1	37.6	43.5	55.4	65.9	74.9	86.0	84.3	73.0	60.2	43.3	34.3	57.5
Sheridan	30	33.5	38.0	43.1	56.3	66.0	74.3	86.1	85.3	72.9	62.5	46.0	37.6	58.5

Table 5
AVERAGE DAILY MINIMUM TEMPERATURE (°F) IN THE U.S. AND CERTAIN PACIFIC ISLANDS[2] (NORMALS 1941—1970)

Location	Years	Jan.	Feb.	Mar.	April	May	June	July	Aug.	Sept.	Oct.	Nov.	Dec.	Annual
Alabama														
Birmingham	30	34.1	36.1	41.8	51.0	58.4	66.4	69.5	68.7	63.0	50.8	40.1	34.9	51.2
Huntsville	30	31.3	33.2	39.8	50.2	58.3	65.9	68.8	67.7	61.6	49.8	38.9	32.7	49.9
Mobile	30	41.3	43.9	49.2	57.7	64.5	70.7	72.6	72.3	68.4	58.0	47.5	42.8	57.4
Montgomery	30	37.1	39.7	45.2	53.6	61.2	68.6	71.5	70.7	65.5	53.5	42.7	37.7	53.9
Alaska														
Anchorage	30	3.5	8.9	14.6	26.8	37.2	46.2	50.1	48.0	40.4	27.8	13.9	5.3	26.9
Annette	30	29.0	31.8	32.8	36.7	42.6	48.1	51.6	51.9	48.1	42.1	35.5	31.6	40.1
Barrow	30	−21.3	−24.6	−21.8	−8.2	14.0	28.9	33.0	33.1	27.2	10.4	−6.4	−18.1	3.8
Barter Island	30	−21.9	−25.8	−21.9	−8.1	15.7	29.9	34.5	34.3	28.1	11.2	−5.9	−18.3	4.3
Bethel	30	−2.2	0.5	2.5	16.3	31.7	43.0	47.4	46.0	38.2	24.3	10.8	−2.5	21.3
Bettles	30	−21.1	−16.5	−10.0	9.2	31.7	45.1	47.5	42.9	31.7	13.3	−8.6	−20.0	12.1
Big Delta	30	−12.3	−5.6	1.1	19.4	36.3	47.1	50.0	45.4	35.2	18.2	−0.3	−11.5	18.6
Cold Bay	30	23.8	23.6	24.2	28.5	34.8	40.9	45.7	47.3	43.0	35.1	30.0	24.5	33.5
Fairbanks	30	−21.6	−14.3	−4.3	17.3	35.7	47.2	49.6	44.9	34.4	16.9	−6.2	−19.3	15.0

Table 5 (continued)
AVERAGE DAILY MINIMUM TEMPERATURE (°F) IN THE U.S. AND CERTAIN PACIFIC ISLANDS[2] (NORMALS 1941—1970)

Location	Years	Jan.	Feb.	Mar.	April	May	June	July	Aug.	Sept.	Oct.	Nov.	Dec.	Annual
Gulkana	30	−16.4	−6.9	0.8	18.5	32.5	42.4	45.6	42.1	33.5	17.8	−1.6	−13.1	16.3
Homer	30	14.7	17.9	20.2	27.7	34.2	40.7	44.5	44.6	39.2	30.3	21.8	15.2	29.2
Juneau	30	17.8	22.1	25.6	31.3	38.2	44.4	47.7	46.2	42.3	36.4	27.6	22.5	33.5
King Salmon	30	5.7	8.6	12.1	23.8	33.8	41.7	46.4	46.7	39.7	26.3	15.1	3.8	25.3
Kodiak	30	26.3	27.0	27.2	32.2	38.5	44.7	49.1	49.7	45.0	35.8	30.5	25.5	36.0
Kotzebue	30	−10.6	−11.8	−9.1	3.7	23.7	37.3	47.1	45.4	35.7	18.5	2.0	−10.3	14.3
McGrath	30	−18.2	−11.2	−4.3	15.3	33.6	45.2	48.7	44.9	35.2	18.2	−2.9	−17.6	15.6
Nome	30	−1.6	−3.3	−1.9	10.8	28.1	38.8	44.4	43.7	35.9	22.6	9.1	−3.0	18.6
St. Paul Island	30	22.0	18.6	19.0	24.4	31.0	36.7	42.0	44.1	40.6	33.8	29.5	23.8	30.5
Talkeetna	30	−0.4	4.7	7.1	21.2	33.2	44.3	48.2	45.0	36.6	23.6	8.8	−0.1	22.7
Unalakleet	30	−3.5	−2.9	0.7	14.1	30.1	41.6	47.4	45.5	36.4	21.2	7.3	−5.1	19.4
Valdez	30	10.6	15.0	18.2	27.1	35.0	42.8	45.5	44.0	38.7	30.9	19.9	13.1	28.4
Yakutat	30	17.1	20.9	22.6	28.5	35.7	43.0	47.4	46.1	41.4	34.1	26.2	20.7	32.0
Arizona														
Flagstaff	30	14.4	17.0	20.4	27.3	33.5	40.4	50.4	49.3	41.2	31.1	21.8	16.3	30.3
Phoenix	30	37.6	40.8	44.8	51.8	59.6	67.7	77.5	76.0	69.1	56.8	44.8	38.5	55.4
Tucson	30	38.2	39.9	43.6	50.3	57.5	66.2	74.2	72.3	67.1	56.4	44.8	39.1	54.1
Winslow	30	19.6	24.8	29.4	37.3	45.5	53.7	62.9	61.5	53.5	41.3	28.2	20.9	39.9
Yuma	30	43.3	46.1	50.2	56.7	63.9	70.8	81.4	81.2	73.9	62.0	50.5	44.4	60.4
Arkansas														
Fort Smith	30	28.0	32.0	38.5	50.2	58.8	67.0	70.5	69.3	61.7	49.9	38.0	30.8	49.6
Little Rock	30	28.9	31.9	38.7	49.9	58.1	66.8	70.1	68.6	60.8	48.7	38.1	31.1	49.3
California														
Bakersfield	30	37.4	41.4	44.5	49.9	56.0	62.3	68.7	66.6	62.1	53.3	44.2	38.4	52.1
Bishop	30	20.5	25.3	28.8	36.8	43.6	49.7	55.7	52.8	46.2	37.2	27.5	22.3	37.2
Blue Canyon	30	29.8	30.5	30.9	36.1	42.7	50.4	58.7	57.1	53.1	45.3	37.3	12.6	42.0
Eureka	30	41.1	42.3	42.5	44.4	47.8	50.7	52.1	52.7	51.2	48.4	45.5	42.6	45.8
Fresno	30	35.8	39.0	41.2	46.2	51.9	57.5	62.9	60.6	56.5	48.6	40.8	36.9	48.2
Long Beach	30	43.2	44.8	46.7	50.6	54.3	57.8	61.6	62.5	60.2	55.5	48.4	44.1	52.5
Los Angeles International Airport	30	45.4	47.0	48.6	51.7	55.3	58.6	62.1	63.2	61.6	57.5	51.3	47.3	54.1

City	30	46.8	48.5	49.8	52.9	56.1	59.5	63.5	64.4	62.8	58.7	52.1	48.1	55.3
Mount Shasta	30	25.4	28.3	29.8	34.0	40.0	46.0	50.6	48.6	44.5	37.9	31.7	27.3	37.0
Oakland	30	42.7	45.7	47.2	49.4	52.4	55.2	56.4	56.8	56.6	53.4	48.5	44.2	50.7
Red Bluff	30	36.7	40.4	42.5	47.3	54.2	61.7	66.6	64.1	60.0	51.7	43.3	38.1	50.5
Sacramento	30	37.1	40.4	41.9	45.3	49.8	54.6	57.5	56.9	55.3	49.5	42.4	38.3	47.4
Sandberg	30	33.9	35.3	35.7	40.1	45.6	53.0	62.9	61.8	58.5	49.8	41.1	35.8	46.1
San Diego	30	45.8	47.8	50.1	53.8	57.2	59.9	63.9	65.4	63.2	58.4	51.5	47.2	55.4
San Francisco														
International Airport	30	41.2	43.8	44.9	47.0	49.9	53.0	54.0	54.3	54.5	51.6	47.2	42.9	48.7
City	30	45.7	47.9	48.5	49.3	50.9	52.8	53.2	54.0	55.5	54.6	51.5	47.2	50.9
Santa Maria	30	38.3	40.3	41.3	43.7	46.7	49.5	52.4	52.5	51.2	47.5	42.9	39.3	45.5
Stockton	30	36.3	39.2	40.6	44.8	50.0	55.4	58.7	57.8	55.3	48.9	41.5	37.9	47.2
Colorado														
Alamosa	30	−0.7	5.5	14.6	24.5	33.5	41.5	47.9	45.8	36.2	25.0	11.7	1.0	23.9
Colorado Springs	30	16.1	18.9	22.8	33.1	42.6	51.1	57.0	55.8	46.9	36.8	25.1	18.9	35.4
Denver	30	16.2	19.4	23.8	33.9	43.6	51.9	58.6	57.4	47.8	37.2	25.4	18.9	36.2
Grand Junction	30	16.5	23.2	29.6	38.8	48.5	56.6	64.2	61.6	53.0	41.9	28.6	19.6	40.2
Pueblo	30	14.7	19.6	25.0	36.9	46.6	55.6	61.6	60.1	50.8	38.2	25.1	17.7	37.7
Connecticut														
Bridgeport	30	23.4	23.9	30.8	40.3	49.9	59.8	66.1	64.9	58.4	48.5	38.7	27.1	44.3
Hartford	30	16.1	17.9	26.6	36.5	46.2	56.0	61.2	58.9	51.0	40.8	31.9	19.6	38.6
Delaware														
Wilmington	30	23.8	24.9	32.0	41.5	51.6	61.1	66.1	64.3	57.6	46.5	36.2	26.3	44.3
District of Columbia														
Washington														
Dulles Airport	30	23.0	24.1	30.9	41.1	50.6	59.4	64.1	62.1	55.0	43.5	33.7	24.7	42.7
National Airport	30	27.7	28.6	35.2	45.7	55.7	64.6	69.1	67.6	61.0	49.7	38.8	29.5	47.8
Florida														
Apalachicola	30	46.3	48.5	53.8	61.5	68.0	73.6	75.2	75.2	72.4	63.3	53.2	47.6	61.6
Daytona Beach	30	47.6	48.7	53.3	59.2	64.8	70.4	72.3	72.8	72.0	65.2	55.1	49.0	60.9
Fort Myers	30	52.3	53.3	57.3	61.8	66.4	71.7	73.8	74.1	73.4	67.5	58.8	53.6	63.7
Jacksonville	30	44.5	45.7	50.1	57.1	63.9	70.0	72.0	72.3	70.4	61.7	51.0	45.1	58.7
Key West	30	65.8	66.5	69.8	73.6	76.4	79.1	80.0	79.9	78.6	75.2	70.6	66.6	73.5
Lakeland	30	51.0	52.2	56.3	62.1	67.1	71.3	72.7	73.3	72.3	66.1	57.5	52.5	62.9
Miami	30	58.7	59.0	63.0	67.3	70.7	73.9	75.5	75.8	75.0	71.0	64.5	60.0	67.9
Orlando	30	50.0	51.2	55.7	61.1	66.1	71.1	72.9	73.5	72.3	66.0	56.9	51.5	62.4
Pensacola	30	43.0	45.5	50.8	59.4	66.1	72.1	73.9	73.6	70.2	60.0	49.3	44.3	59.0

Table 5 (continued)
AVERAGE DAILY MINIMUM TEMPERATURE (°F) IN THE U.S. AND CERTAIN PACIFIC ISLANDS[2] (NORMALS 1941—1970)

Location	Years	Jan.	Feb.	Mar.	April	May	June	July	Aug.	Sept.	Oct.	Nov.	Dec.	Annual
Tallahassee	30	41.0	43.0	48.4	55.7	62.8	69.6	71.6	71.7	68.7	57.9	46.4	41.3	56.5
Tampa	30	50.1	51.7	55.9	61.6	66.9	72.0	73.7	74.0	72.6	65.5	56.4	51.2	62.6
West Palm Beach	30	55.9	56.2	60.2	64.9	68.9	72.7	74.1	74.4	74.7	70.1	62.5	57.4	66.0
Georgia														
Athens	30	33.4	34.9	40.4	49.7	58.1	65.7	68.7	67.9	62.3	51.1	40.1	33.9	50.5
Atlanta	30	33.4	35.5	41.1	50.7	59.2	66.6	69.4	68.6	63.4	52.3	40.8	34.3	51.3
Augusta	30	34.0	36.1	42.0	50.7	59.1	66.7	69.9	69.0	63.2	51.2	40.2	34.1	51.4
Columbus	30	35.9	37.8	43.3	51.8	60.0	67.5	70.4	69.8	65.1	53.4	41.9	36.3	52.8
Macon	30	36.9	39.0	44.6	53.2	60.9	68.2	70.7	70.0	64.9	53.2	42.4	37.0	53.4
Rome	30	30.0	32.0	37.5	46.9	55.0	63.4	67.1	66.1	59.7	47.1	36.1	30.6	47.6
Savannah	30	38.7	40.5	46.4	54.3	61.8	68.8	71.3	70.9	66.9	55.9	44.9	38.7	54.9
Hawaii														
Hilo	30	62.8	62.6	63.3	64.6	65.6	66.5	67.5	68.2	67.6	66.7	65.6	63.7	65.4
Honolulu	30	65.3	65.3	66.3	68.1	70.2	72.2	73.4	74.0	73.4	72.0	69.8	67.1	69.8
Kahului	30	63.9	63.9	64.1	65.7	66.9	68.6	70.0	70.8	69.8	68.9	67.8	65.5	67.2
Lihue	30	64.4	64.4	65.4	67.3	69.5	71.7	72.7	73.6	72.8	71.3	69.6	66.8	69.1
Idaho														
Boise	30	21.4	27.2	30.5	36.5	44.1	51.2	58.5	56.7	48.5	39.4	30.7	25.0	39.1
Lewiston	30	24.4	30.1	32.9	38.6	45.5	52.1	57.6	55.9	48.9	40.3	32.9	28.6	40.6
Pocatello	30	14.0	20.1	24.9	32.8	40.7	47.1	54.1	52.5	43.1	33.8	25.5	18.3	33.9
Illinois														
Cairo	30	28.8	31.7	39.0	51.0	59.9	68.3	71.6	70.0	62.5	51.8	39.9	32.0	50.6
Chicago														
O'Hare Airport	30	14.7	17.8	26.7	38.2	47.0	57.0	60.7	59.9	52.0	42.0	30.1	19.2	38.8
Midway Airport	30	17.0	20.2	29.0	40.4	49.7	60.3	65.0	64.1	56.0	45.6	32.6	21.6	41.8
Moline	30	13.0	17.0	26.4	39.8	50.2	60.2	63.8	62.0	53.2	42.8	30.2	18.5	39.8
Peoria	30	15.7	19.3	28.1	40.8	50.7	60.9	64.6	62.9	54.6	44.0	31.1	20.3	41.1
Rockford	30	11.5	15.3	25.0	37.3	47.2	57.7	61.4	60.1	51.8	41.1	28.7	16.7	37.8
Springfield	30	18.6	21.8	30.1	42.6	52.6	62.5	65.6	63.7	55.6	45.0	32.7	22.7	42.8
Indiana														

Evansville	30	23.7	26.4	34.0	45.5	54.4	63.4	66.7	64.4	56.7	45.2	34.5	26.5	45.1
Fort Wayne	30	17.9	19.7	27.9	39.0	48.9	58.8	62.4	60.4	53.0	42.5	32.0	21.4	40.3
Indianapolis	30	19.7	22.1	30.3	41.8	51.5	61.1	64.6	62.4	54.9	44.3	32.8	23.1	42.4
South Bend	30	16.5	18.5	26.6	37.8	47.3	57.7	61.7	60.2	52.8	43.0	31.9	21.2	39.6
Iowa														
Burlington	30	14.1	18.4	27.1	40.5	50.9	60.9	64.6	63.1	54.1	43.9	30.3	19.3	40.6
Des Moines	30	11.3	15.8	25.2	39.2	50.9	61.1	65.3	63.4	54.0	43.6	29.2	17.2	39.7
Dubuque	30	9.3	13.4	23.3	37.1	47.4	57.3	61.3	59.9	51.0	41.1	27.4	15.2	37.0
Sioux City	30	7.7	13.4	23.4	37.4	49.3	59.3	63.9	62.1	51.4	40.4	25.6	13.6	37.3
Waterloo	30	6.9	11.3	22.3	36.3	47.5	57.6	61.5	59.5	49.8	39.7	25.6	12.9	35.9
Kansas														
Concordia	30	15.8	20.5	27.4	40.5	51.2	61.2	66.1	65.3	55.2	44.4	29.9	20.1	41.5
Dodge City	30	19.0	23.2	28.4	41.1	51.7	61.4	66.9	65.7	56.3	45.0	30.4	22.2	42.6
Goodland	30	13.7	17.8	22.2	34.1	44.8	54.7	61.0	59.4	49.0	37.2	24.3	16.7	36.2
Topeka	30	17.7	22.7	29.7	42.6	53.2	63.0	67.2	65.9	56.0	44.8	31.5	21.8	43.0
Wichita	30	21.2	25.4	32.1	45.1	55.0	65.0	69.6	68.3	59.2	47.9	33.8	24.6	45.6
Kentucky														
Covington (Cincinnati Airport)	30	22.4	23.8	31.6	42.7	51.9	61.0	64.6	63.0	55.9	45.0	34.3	25.3	43.5
Lexington	30	24.5	26.2	33.7	44.6	53.8	62.5	65.9	64.4	57.6	46.8	35.3	27.2	45.2
Louisville	30	24.5	26.5	34.0	44.8	53.9	62.9	66.4	64.9	57.7	45.9	35.1	27.1	45.3
Louisiana														
Baton Rouge	30	40.5	43.2	48.7	57.7	64.3	70.3	72.7	72.1	67.7	56.6	46.9	42.0	56.9
Lake Charles	30	42.9	45.5	50.5	59.5	66.0	71.8	73.6	73.2	68.8	58.4	49.2	44.3	58.6
New Orleans	30	43.5	46.0	50.9	58.8	65.3	71.2	73.3	73.1	69.7	59.6	49.8	45.3	58.9
Shreveport	30	37.8	40.6	46.2	55.9	63.1	70.2	72.8	72.5	66.8	55.7	45.2	39.4	55.5
Maine														
Caribou	30	1.5	2.7	14.4	27.9	39.0	48.8	54.1	51.5	43.3	34.8	24.6	8.2	29.2
Portland	30	11.7	12.5	22.8	32.5	41.7	51.1	56.9	55.2	47.4	38.0	29.7	16.4	34.7
Maryland														
Baltimore	30	24.9	25.7	32.5	42.4	52.5	61.6	66.5	64.7	57.9	46.4	36.0	26.6	44.8
Massachusetts														
Blue Hill Observatory	30	18.7	19.4	27.2	36.8	46.3	55.6	61.7	60.1	53.4	44.4	34.8	22.3	40.1
Boston	30	22.5	23.3	31.5	40.8	50.1	59.3	65.1	63.3	56.7	47.5	38.7	26.6	43.8
Worcester	30	16.2	17.0	25.3	35.6	45.4	55.1	60.8	59.0	51.9	42.5	32.5	20.1	38.4
Michigan														
Alpena	30	8.7	7.9	15.6	28.5	37.4	47.4	51.9	51.1	44.1	35.7	26.9	15.7	30.9
Detroit														

Table 5 (continued)
AVERAGE DAILY MINIMUM TEMPERATURE (°F) IN THE U.S. AND CERTAIN PACIFIC ISLANDS[2] (NORMALS 1941—1970)

Location	Years	Jan.	Feb.	Mar.	April	May	June	July	Aug.	Sept.	Oct.	Nov.	Dec.	Annual
City Airport	30	19.2	20.1	27.6	38.6	48.3	59.1	63.4	62.1	54.8	45.2	34.4	23.8	41.4
Metro Airport	30	17.3	18.8	26.7	37.3	47.0	57.2	61.1	59.5	52.3	42.1	32.3	21.5	39.4
Flint	30	14.6	15.5	23.7	35.3	44.5	54.6	58.2	56.6	49.7	40.5	30.6	19.7	37.0
Grand Rapids	30	16.0	16.4	24.2	35.6	45.4	55.6	59.6	58.1	50.8	40.8	31.1	20.8	37.9
Houghton Lake	30	9.0	8.3	16.4	30.5	40.3	50.0	53.6	52.6	45.7	37.2	27.1	15.5	32.2
Lansing	30	15.3	16.0	24.1	35.9	45.5	55.8	59.2	57.8	50.4	41.0	30.8	20.2	37.7
Marquette	30	12.0	12.6	20.5	32.4	40.8	50.3	57.3	57.0	49.5	41.0	28.8	17.9	35.0
Muskegon	30	17.8	17.7	24.8	35.7	44.9	55.0	59.6	58.8	51.6	42.4	32.6	22.6	38.7
Sault Ste. Marie	30	6.4	6.7	15.5	29.2	38.5	47.3	52.5	52.9	46.1	37.6	26.5	13.3	31.0
Minnesota														
Duluth	30	−0.6	2.0	14.4	29.3	38.8	48.3	54.7	53.7	44.8	36.2	21.4	6.3	29.1
International Falls	30	−9.1	−5.5	8.9	27.3	37.7	48.3	53.4	50.9	41.7	32.9	17.3	−0.8	25.3
Minneapolis—St. Paul	30	3.2	7.1	19.6	34.7	46.3	56.7	61.4	59.6	49.3	39.2	24.2	10.6	34.3
Rochester	30	3.9	7.4	19.1	34.1	45.3	55.5	59.5	57.7	48.3	38.7	24.2	10.9	33.7
Saint Cloud	30	−1.4	2.2	15.9	32.0	43.3	53.6	58.6	56.6	46.4	36.1	21.4	6.8	31.0
Mississippi														
Jackson	30	35.8	37.8	43.4	53.1	60.4	67.7	70.6	69.8	64.0	51.5	42.0	37.3	52.8
Meridian	30	35.4	37.7	43.4	52.3	59.5	67.0	69.9	69.1	63.6	50.8	40.7	36.0	52.1
Missouri														
Columbia	30	20.6	24.5	32.0	44.6	54.3	63.3	67.1	65.5	57.2	46.7	34.2	24.5	44.5
Kansas City	30	18.4	23.1	30.6	43.7	54.0	63.2	66.9	65.8	57.1	46.9	33.1	23.3	43.8
Saint Joseph	30	16.4	21.1	29.1	42.6	53.7	63.5	67.4	65.4	56.1	45.1	31.5	21.3	42.8
St. Louis	30	22.6	26.0	33.5	46.0	55.5	64.8	68.8	67.1	59.1	48.4	35.9	26.5	48.2
Springfield	30	22.6	26.5	32.8	45.0	54.0	62.9	66.5	65.2	57.3	46.8	34.5	26.3	45.0
Montana														
Billings	30	12.5	17.7	23.1	33.4	43.3	51.5	58.0	56.3	46.5	37.5	26.4	17.7	35.3
Glasgow	30	−0.5	5.0	14.8	30.6	41.7	50.1	56.6	54.9	44.0	33.6	18.9	7.6	29.8
Great Falls	30	11.6	17.2	20.6	32.3	41.5	49.5	54.9	53.0	44.6	37.1	25.7	18.2	33.8
Havre	30	0.5	6.7	16.0	31.5	42.0	49.9	54.8	52.6	42.7	32.5	18.3	7.5	29.6
Helena	30	7.8	14.9	19.4	30.2	39.5	46.7	52.1	50.4	40.9	31.9	20.8	12.3	30.7
Kalispell	30	11.0	15.9	20.3	30.4	38.1	44.0	47.8	45.9	38.6	30.5	22.7	16.6	30.2

Miles City	30	4.7	10.4	18.7	32.7	43.6	52.3	59.8	57.7	46.2	35.5	21.7	11.7	32.9
Missoula	30	12.9	18.6	22.8	31.1	38.5	45.2	48.9	47.5	39.8	31.3	23.7	17.5	31.5
Nebraska														
Grand Island	30	11.2	16.7	24.3	37.6	48.7	58.9	63.9	62.5	51.6	40.1	26.0	16.2	38.1
Lincoln	30	11.7	17.4	26.0	39.2	50.6	60.9	65.7	64.2	53.6	44.2	27.8	16.9	39.7
Norfolk	30	8.1	13.4	22.7	36.6	48.5	58.8	63.9	62.2	51.1	39.4	24.9	13.8	36.9
North Platte	30	10.1	15.2	21.3	34.2	45.2	55.3	61.0	59.3	47.6	35.0	21.8	13.8	35.0
Omaha	30	12.4	17.4	26.4	40.1	51.5	61.3	65.8	64.0	54.0	42.6	29.1	18.1	40.2
Omaha (North)	30	11.2	16.1	25.1	38.9	50.4	60.2	64.8	63.4	53.6	42.8	28.3	17.0	39.3
Scottsbluff	30	11.3	15.9	20.6	32.0	42.8	52.0	58.6	56.3	45.2	34.3	21.9	14.4	33.8
Valentine	30	7.5	11.8	18.9	32.5	43.7	54.1	59.8	58.0	46.5	34.5	21.2	11.9	33.4
Nevada														
Elko	30	10.4	16.8	21.5	28.1	35.2	41.6	48.6	45.8	36.4	28.0	20.6	13.5	28.9
Ely	30	9.2	14.3	18.9	26.2	33.6	39.7	48.0	46.8	37.3	28.4	18.9	12.0	27.8
Las Vegas	30	32.6	36.9	41.7	50.0	59.0	67.4	75.3	73.3	65.4	53.1	40.8	33.7	52.4
Reno	30	18.3	23.0	24.6	29.6	37.0	42.5	47.4	44.8	38.6	30.5	23.9	19.6	31.7
Winnemucca	30	15.7	21.4	23.3	29.1	37.3	44.5	50.9	46.9	38.5	29.3	22.3	17.8	31.4
New Hampshire														
Concord	30	9.9	11.3	22.1	31.7	41.5	51.6	56.7	54.2	46.5	36.3	28.1	14.9	33.7
Mt. Washington	30	−2.8	−3.5	4.5	16.0	28.0	38.5	43.0	41.4	35.4	25.4	14.0	1.2	20.1
New Jersey														
Atlantic City	30	24.0	24.9	31.5	41.0	50.7	59.7	65.4	63.8	56.8	45.9	36.1	26.0	43.8
Newark	30	24.3	24.9	32.4	42.2	52.1	61.6	67.2	65.5	58.6	48.1	38.2	27.4	45.2
Trenton	30	25.3	26.1	33.1	42.5	52.2	61.6	66.8	65.0	58.1	48.2	38.7	28.3	45.5
New Mexico														
Albuquerque	30	23.5	27.4	32.3	41.4	50.7	59.7	65.2	63.4	56.7	44.7	31.8	24.9	43.5
Clayton	30	18.6	21.9	25.4	36.0	45.6	54.9	59.9	59.1	51.1	40.2	27.7	20.9	38.4
Roswell	30	20.8	24.8	30.9	41.2	50.5	59.8	63.7	62.3	54.3	42.2	29.0	21.8	41.8
New York														
Albany	30	12.5	14.3	24.2	35.7	45.7	55.6	60.1	57.8	50.1	40.0	31.1	17.7	37.1
Binghamton	30	15.2	15.4	23.7	35.3	45.2	55.1	59.6	57.8	50.9	41.4	31.8	19.4	37.6
Buffalo	30	17.6	17.7	25.2	36.4	45.9	56.3	60.7	59.1	52.3	42.7	33.5	22.2	39.1
New York														
Central Park	30	25.9	26.5	33.7	43.5	53.1	62.6	68.0	66.4	59.9	50.6	40.8	29.5	46.7
JFK Airport	30	24.8	25.2	32.1	41.7	51.1	60.9	66.9	65.4	58.6	48.7	39.3	28.4	45.3
La Guardia Airport	30	26.4	27.0	34.1	44.0	53.7	63.6	69.3	67.6	60.9	51.0	41.3	30.1	47.4
Rochester	30	16.7	16.9	24.9	36.1	45.7	55.8	60.2	58.5	51.5	42.2	33.1	21.7	38.6

Table 5 (continued)
AVERAGE DAILY MINIMUM TEMPERATURE (°F) IN THE U.S. AND CERTAIN PACIFIC ISLANDS[2]
(NORMALS 1941—1970)

Location	Years	Jan.	Feb.	Mar.	April	May	June	July	Aug.	Sept.	Oct.	Nov.	Dec.	Annual
Syracuse	30	15.8	16.5	24.8	36.4	46.0	56.1	61.0	59.2	52.3	42.5	33.6	21.2	38.8
North Carolina														
Asheville	30	27.3	28.2	33.5	42.4	50.6	58.7	62.6	61.8	55.4	44.5	34.3	28.1	44.0
Cape Hatteras	30	38.2	38.5	43.2	51.5	60.2	68.1	72.1	71.5	67.8	59.1	48.8	40.5	55.0
Charlotte	30	32.1	33.1	39.0	48.9	57.4	65.3	68.7	67.9	61.9	50.3	39.6	32.4	49.7
Greensboro	30	28.5	29.7	36.1	46.2	55.3	63.3	66.9	65.8	58.8	47.2	36.5	29.2	47.0
Raleigh	30	30.0	31.1	37.4	46.7	55.4	63.1	67.2	66.2	59.7	48.0	37.8	30.5	47.8
Wilmington	30	36.2	37.5	43.7	52.2	60.9	68.2	72.0	71.0	65.9	55.1	44.3	36.7	53.7
North Dakota														
Bismarck	30	−2.8	2.4	14.7	31.1	41.7	51.8	57.3	54.9	43.7	33.2	18.3	5.2	29.3
Fargo	30	−3.6	0.8	14.9	31.9	42.3	53.4	58.6	56.8	46.2	35.5	20.0	4.1	30.1
Williston	30	−2.8	2.6	13.8	29.9	41.1	50.4	56.1	54.1	43.1	32.5	17.8	5.0	28.6
Ohio														
Akron	30	18.6	19.4	26.9	37.7	47.5	57.1	60.8	59.3	52.7	42.8	32.7	22.2	39.8
Cincinnati	30	24.3	25.8	33.5	44.6	53.6	62.5	65.8	64.1	57.0	46.7	36.2	27.1	45.1
Cleveland	30	20.3	20.8	28.1	38.5	48.1	57.5	61.2	59.6	53.5	43.9	34.4	24.1	40.8
Columbus	30	20.4	21.4	29.1	39.5	49.3	58.9	62.4	60.1	52.7	42.0	32.4	22.7	40.9
Dayton	30	20.4	22.1	29.9	41.0	51.2	61.0	64.4	62.6	55.5	44.9	33.4	23.4	42.5
Mansfield	30	20.7	21.6	29.3	39.8	49.8	59.5	63.3	61.7	54.8	44.1	34.3	24.0	41.9
Toledo	30	17.2	18.9	26.6	37.2	47.1	57.2	60.8	59.1	51.8	41.2	31.1	20.5	39.1
Youngstown	30	18.3	18.7	26.3	36.9	46.3	55.7	59.6	58.0	51.4	42.2	32.6	22.0	39.0
Oklahoma														
Oklahoma City	30	26.0	30.0	36.5	49.1	57.9	66.6	70.4	69.6	61.3	50.6	37.4	29.2	48.7
Tulsa	30	26.1	30.2	36.9	49.7	58.4	67.3	71.4	70.0	61.7	50.8	38.0	29.5	49.2
Oregon														
Astoria	30	34.6	36.6	36.7	40.0	44.3	49.1	52.2	52.2	49.1	44.5	39.6	36.9	43.0
Burns	30	15.0	20.4	24.2	30.2	37.8	44.1	51.1	48.8	40.9	32.0	24.3	18.3	32.3
Eugene	30	33.1	35.2	36.5	39.4	43.7	48.7	51.1	50.9	47.4	42.3	38.1	35.6	41.8
Medford	30	29.0	30.7	32.8	36.6	42.8	49.1	53.8	52.9	46.7	39.4	34.2	31.1	39.9
Pendleton	30	25.3	31.3	34.3	39.8	46.5	52.8	58.8	57.5	51.1	41.8	33.8	29.6	41.9
Portland	30	32.5	35.5	37.0	40.8	46.3	51.8	55.2	55.0	50.5	44.7	38.5	35.3	43.6

Salem	30	32.2	34.4	35.4	38.5	43.3	48.4	50.7	50.9	47.3	42.3	37.4	34.7	41.3
Sexton Summit	30	29.2	31.4	30.7	33.7	39.3	44.7	51.4	51.4	49.2	42.6	35.2	32.0	39.2
Pacific														
Guam	30	71.5	71.3	71.0	72.5	73.1	72.9	72.6	72.2	72.4	72.3	73.2	72.9	72.3
Johnston Island	30	72.9	72.5	72.7	73.8	75.2	76.4	77.0	77.4	77.0	76.8	75.5	73.8	75.0
Koror Island	30	74.9	75.0	75.2	75.8	76.0	75.6	75.2	75.4	75.7	75.9	75.8	75.4	75.6
Kwajalein Island	30	77.4	77.5	77.6	77.8	77.7	77.7	77.5	77.7	77.7	77.7	77.5	77.8	77.6
Majuro, Marshall Islands	30	77.0	77.3	77.1	76.9	77.0	76.6	76.6	77.0	76.7	76.6	76.8	77.0	76.9
Pago Pago, American Samoa	30	74.0	74.2	73.9	74.2	73.8	73.7	73.1	73.2	73.4	73.8	73.8	74.0	73.7
Ponape Island	30	75.5	75.7	75.4	74.9	74.6	73.9	72.9	72.6	73.5	72.7	73.4	74.7	74.1
Truk, Caroline Islands	30	76.9	76.9	77.1	76.8	76.3	75.8	75.0	75.0	75.1	75.4	76.0	76.8	76.0
Wake Island	30	72.5	72.3	72.9	73.5	75.1	76.8	77.0	77.4	77.7	76.7	76.0	74.1	75.1
Yap Island	30	75.6	75.6	75.8	76.3	76.2	75.8	75.2	75.0	75.2	75.2	75.7	75.9	75.6
Pennsylvania														
Allentown	30	19.8	20.9	28.5	38.5	48.4	57.9	62.7	60.6	53.4	42.5	32.8	22.6	40.7
Erie	30	18.5	17.9	25.4	36.1	45.5	55.6	60.0	58.9	52.6	43.0	33.7	23.3	39.2
Harrisburg	30	22.5	24.0	31.2	41.5	51.6	61.0	65.4	63.2	56.0	44.6	34.7	25.0	43.4
Philadelphia	30	24.4	25.5	32.5	42.3	52.3	61.6	66.7	64.7	57.8	46.9	36.9	27.2	44.9
Pittsburgh														
International Airport	30	20.8	21.3	29.0	39.4	48.7	57.7	61.3	59.4	52.7	42.4	33.3	23.6	40.8
City	30	23.7	24.4	31.8	42.4	52.0	61.5	65.2	63.3	56.2	45.4	36.5	26.8	44.1
Avoca	30	18.4	19.3	27.2	38.0	47.8	56.8	61.3	59.2	52.1	42.2	32.8	22.0	39.8
Williamsport	30	19.4	20.1	27.9	38.4	48.1	57.0	61.3	59.4	52.7	42.0	32.9	22.6	40.1
Puerto Rico														
San Juan	30	68.8	68.4	68.9	70.6	72.8	74.0	74.8	75.1	74.6	73.7	72.3	70.5	72.0
Rhode Island														
Block Island	30	25.4	25.3	31.0	38.8	47.2	56.5	63.3	63.1	57.7	49.2	40.1	29.1	43.9
Providence	30	20.6	21.2	29.0	37.8	46.9	56.5	63.0	61.0	53.6	43.4	34.6	23.4	40.9
South Carolina														
Charleston	30	37.3	39.0	45.1	53.0	61.1	68.1	71.2	70.6	65.9	55.1	44.1	37.7	54.0
Columbia	30	33.9	35.5	41.9	51.3	59.6	67.2	70.3	69.4	63.5	51.3	40.6	34.1	51.5
Greenville	30	33.0	34.7	40.2	49.9	58.3	65.9	69.0	68.1	62.3	50.9	40.1	33.3	50.5
South Dakota														
Aberdeen	30	−1.5	4.1	16.6	31.7	42.8	53.4	58.4	56.4	45.1	33.9	19.4	6.1	30.5
Huron	30	1.6	7.1	18.8	33.6	44.2	55.2	60.6	58.7	47.2	36.1	21.3	9.0	32.8
Rapid City	30	9.6	13.9	19.7	32.0	42.9	52.0	58.8	57.2	46.3	36.4	23.2	14.9	33.9
Sioux Falls	30	3.7	9.0	20.2	34.4	45.7	56.3	61.5	59.8	48.7	37.6	22.7	10.4	34.2

Table 5 (continued)
AVERAGE DAILY MINIMUM TEMPERATURE (°F) IN THE U.S. AND CERTAIN PACIFIC ISLANDS[2]
(NORMALS 1941—1970)

Location	Years	Jan.	Feb.	Mar.	April	May	June	July	Aug.	Sept.	Oct.	Nov.	Dec.	Annual
Tennessee														
Bristol — Johnson City	30	26.7	28.4	34.5	44.2	52.6	60.7	64.4	63.1	56.6	45.3	34.5	28.0	44.9
Chattanooga	30	30.5	32.33	38.4	48.1	56.0	64.5	68.1	67.0	60.4	48.1	37.1	31.4	48.5
Knoxville	30	32.2	33.5	39.4	48.6	56.9	64.8	68.3	67.2	61.2	50.0	39.4	33.1	49.5
Memphis	30	31.6	34.4	41.1	52.3	60.6	68.5	71.5	70.1	62.8	51.1	40.3	33.7	51.5
Nashville	30	29.0	31.0	38.1	48.8	57.3	65.7	69.0	67.7	60.5	48.6	37.7	31.1	48.7
Oak Ridge	30	29.1	30.5	36.4	46.2	54.5	62.9	66.4	65.3	58.6	47.1	36.3	30.1	47.0
Texas														
Abilene	30	31.7	35.9	41.7	52.7	60.8	69.0	72.4	71.9	64.6	54.2	42.0	34.5	52.6
Amarillo	30	22.5	26.4	31.2	42.1	51.9	61.2	65.9	64.7	56.7	46.1	32.5	25.5	43.9
Austin	30	39.3	42.8	48.2	58.2	65.1	71.4	73.7	73.5	68.4	58.9	48.0	41.6	57.4
Brownsville	30	51.0	54.1	58.8	66.7	71.4	75.0	75.9	75.7	73.2	66.6	58.7	53.3	65.0
Corpus Christi	30	46.1	49.3	54.2	63.4	69.1	73.6	75.2	75.4	72.0	63.7	54.6	48.9	62.1
Dallas — Fort Worth	30	33.9	37.6	43.3	54.1	62.1	70.3	74.0	73.7	66.8	56.0	44.1	37.0	54.4
Del Rio	30	38.1	42.8	48.8	58.9	66.1	72.4	74.2	73.6	68.5	59.2	46.9	39.5	57.4
El Paso	30	30.2	34.3	40.3	49.3	57.2	65.7	69.9	68.2	61.0	49.5	37.0	30.9	49.5
Galveston	30	48.3	50.9	55.9	65.0	71.8	77.4	79.0	78.9	75.3	68.1	58.2	51.5	65.0
Houston	30	41.5	44.6	49.8	59.3	65.6	70.9	72.8	72.4	68.2	58.3	49.1	43.4	58.0
Lubbock	30	24.8	28.3	34.0	45.1	54.5	63.6	66.9	65.5	58.2	47.3	34.4	27.4	45.8
Midland — Odessa	30	29.4	33.5	39.2	49.4	58.1	66.9	69.5	69.1	62.8	52.4	39.1	31.6	50.1
Port Arthur	30	42.4	45.1	49.7	59.4	65.7	71.7	74.0	73.6	69.2	58.5	49.4	44.1	58.6
San Angelo	30	33.6	37.5	43.5	54.0	62.4	69.8	72.4	72.0	65.1	54.7	42.5	35.2	53.6
San Antonio	30	39.8	43.4	49.1	58.8	65.7	72.0	73.8	73.4	68.8	59.2	48.2	41.8	57.8
Victoria	30	43.6	47.1	52.1	61.5	67.9	73.5	75.3	75.0	71.0	61.6	51.9	46.0	60.5
Waco	30	36.6	40.3	46.0	56.8	64.5	71.8	75.0	74.7	68.3	57.7	46.2	39.1	56.4
Wichita Falls	30	29.4	33.6	39.2	51.1	59.8	68.6	72.3	71.6	63.6	52.7	39.7	32.2	51.2
Utah														
Milford	30	12.9	18.6	23.7	31.4	39.3	46.8	55.8	54.8	44.1	33.4	22.5	15.8	33.3
Salt Lake City	30	18.5	23.3	28.3	36.6	44.2	51.1	60.5	58.7	49.3	38.4	28.1	21.5	38.2
Vermont														
Burlington	30	7.6	8.9	20.1	32.6	43.5	53.9	58.5	56.4	48.6	38.8	29.7	14.8	34.5

Virginia														
Lynchburg	30	27.3	28.3	34.8	45.0	53.7	61.6	65.5	64.3	57.5	47.0	36.7	28.7	45.9
Norfolk	30	32.2	32.7	38.9	47.9	57.2	65.5	69.9	68.9	63.9	53.3	42.6	34.0	50.6
Richmond	30	27.6	28.8	35.5	45.2	54.5	62.9	67.5	65.9	59.0	47.4	37.3	28.8	46.7
Roanoke	30	27.2	28.3	34.3	43.9	52.7	60.4	64.4	63.3	56.5	45.6	35.8	28.1	45.0
Washington														
Olympia	30	30.4	32.4	32.8	36.5	40.8	45.9	48.7	48.4	45.0	40.0	35.2	33.1	39.1
Quillayute	30	33.3	34.9	34.7	37.9	42.2	47.3	50.3	50.1	47.3	42.7	37.6	35.4	41.1
Seattle	30	34.7	36.9	38.0	41.8	47.1	52.1	55.6	55.3	51.8	46.5	40.4	37.4	44.8
Seattle (International Airport)	30	33.0	36.0	36.6	40.3	45.6	50.6	53.8	53.7	50.4	44.9	38.8	35.5	43.3
Spokane	30	19.6	25.3	28.8	35.2	42.8	49.4	55.1	54.0	46.7	37.5	29.2	24.0	37.3
Stampede Pass	30	18.9	23.5	24.9	29.5	35.9	41.3	46.9	47.0	43.2	35.7	27.2	22.2	33.0
Walla Walla	30	27.5	33.4	37.1	43.0	49.6	55.6	62.3	61.2	53.9	44.9	36.2	31.4	44.7
Yakima	30	18.6	25.2	28.8	34.8	42.6	49.3	53.3	51.2	44.3	35.4	28.3	23.5	36.3
West Virginia														
Beckley	30	22.9	23.4	30.3	40.7	47.9	56.7	59.9	58.5	52.0	41.6	32.4	24.1	40.9
Charleston	30	25.3	26.8	33.8	43.8	52.3	60.6	64.3	62.8	55.9	44.8	35.0	27.2	44.4
Elkins	30	19.2	20.4	27.0	36.7	45.3	53.3	57.1	55.5	48.8	38.3	29.2	21.0	37.7
Huntington	30	25.6	26.8	33.8	43.8	52.7	61.3	64.8	63.1	56.3	45.4	35.6	27.4	44.7
Parkersburg	30	24.4	25.6	32.7	43.2	52.2	61.0	64.7	63.0	56.2	45.5	35.7	26.9	44.3
Wisconsin														
Green Bay	30	6.9	8.8	20.1	33.5	43.1	53.2	57.7	56.3	48.0	38.7	26.4	13.2	33.8
La Crosse	30	7.1	10.3	22.1	37.4	48.7	58.5	62.5	61.0	51.8	41.7	27.8	14.0	36.9
Madison	30	8.2	11.1	21.2	34.6	44.6	54.6	58.8	57.3	48.5	38.9	26.4	14.0	34.8
Milwaukee	30	11.4	14.6	23.4	34.7	43.3	53.6	59.3	58.7	50.7	40.6	28.5	16.8	36.3
Wyoming														
Casper	30	12.7	15.9	19.4	29.9	39.3	47.4	54.9	53.5	43.3	33.9	22.9	16.2	32.4
Cheyenne	30	14.9	17.3	19.6	30.0	39.7	48.1	54.5	53.2	43.5	33.9	23.5	18.1	33.0
Lander	30	8.0	13.3	19.3	30.4	39.7	47.4	55.1	53.7	43.7	33.4	19.7	11.6	31.3
Sheridan	30	8.5	13.8	18.9	30.9	40.2	47.8	54.6	53.0	42.9	33.1	20.8	13.4	31.5

Table 6
HIGHEST TEMPERATURE OF RECORD (°F) IN THE U.S. AND CERTAIN PACIFIC ISLANDS (DATA THROUGH 1977)[2]

Location	Years	Jan.	Feb.	Mar.	April	May	June	July	Aug.	Sept.	Oct.	Nov.	Dec.	Annual
Alabama														
Birmingham	34	81	83	87	90	99	102	106	103	100	94	84	80	106
Huntsville	10	76	77	83	90	92	101	101	101	95	89	83	77	101
Mobile	36	84	82	90	92	100	102	104	102	98	93	87	81	104
Montgomery	33	83	85	89	91	98	105	105	104	101	100	87	83	105
Alaska														
Anchorage	24	50	48	49	65	77	85	81	82	73	61	53	47	85
Annette	30	61	63	62	82	88	89	89	90	80	71	67	62	90
Barrow	57	36	32	33	42	45	70	78	76	62	43	39	34	78
Barter Island	28	39	34	36	43	52	68	78	72	64	46	37	37	78
Bethel	19	48	46	46	53	79	86	83	84	70	56	45	45	86
Bettles	27	42	40	44	56	83	92	92	87	79	53	45	38	92
Big Delta	36	48	51	53	71	90	92	91	88	79	66	51	48	92
Gold Bay	33	50	50	56	60	65	69	77	78	70	69	54	54	78
Fairbanks	26	47	43	51	74	89	96	94	90	84	65	46	42	96
Gulkana	32	46	46	49	67	85	90	91	86	72	65	48	44	91
Homer	34	51	51	53	63	69	80	78	78	68	64	52	50	80
Juneau	34	57	55	55	71	82	86	90	83	72	61	56	54	90
King Salmon	35	53	51	56	65	75	88	86	84	74	67	50	48	88
Kodiak	29	54	56	57	64	80	86	82	83	71	61	54	54	86
Kotzebue	35	39	35	39	46	74	83	85	80	68	51	38	36	85
McGrath	35	54	55	51	67	80	90	89	89	76	59	49	49	90
Nome	31	43	47	42	51	76	81	86	81	65	59	44	43	86
St. Paul Island	60	48	44	44	44	56	62	63	64	59	54	50	52	64
Talkeetna	38	45	50	53	69	82	91	90	85	78	68	46	47	91
Unalakleet	28	47	40	47	62	78	86	87	85	70	57	44	41	87
Valdez	6	40	46	51	61	71	73	80	81	72	56	50	40	81
Yakutat	31	49	54	55	68	79	81	84	86	77	63	55	52	86
Arizona														
Flagstaff	28	66	70	73	78	87	96	97	92	90	83	74	68	97
Phoenix	40	88	89	95	104	113	117	118	116	118	104	93	88	118

Tucson	37	87	92	92	102	107	111	111	109	107	101	90	84	111
Winslow	46	75	78	85	92	101	106	109	103	99	93	78	74	109
Yuma	27	88	89	98	102	109	116	119	117	115	109	98	85	119
Arkansas														
Fort Smith	32	81	86	94	93	98	105	111	110	106	96	86	82	111
Little Rock	36	83	83	91	92	98	104	107	108	106	97	86	80	108
California														
Bakersfield	40	79	85	92	98	107	114	115	112	112	103	91	82	115
Bishop	30	77	78	87	92	101	109	109	107	106	95	84	78	109
Blue Canyon	34	71	73	72	78	86	92	95	94	93	85	78	75	95
Eureka	67	75	85	78	79	84	85	76	82	85	82	77	77	85
Fresno	28	75	79	90	95	104	110	111	108	105	100	89	76	111
Long Beach	25	91	91	89	99	103	108	105	105	110	111	101	92	111
Los Angeles														
International Airport	42	88	92	88	95	97	100	97	98	110	106	101	94	110
City	37	95	91	94	99	102	106	103	103	110	104	100	89	110
Mount Shasta	35	65	71	80	84	91	96	100	103	103	92	80	72	103
Oakland	49	75	82	85	88	98	107	102	95	102	95	86	74	107
Red Bluff	33	79	85	92	98	108	114	119	118	114	104	93	81	119
Sacramento	27	70	76	86	92	104	115	114	108	108	101	87	72	115
Sandberg	43	71	72	80	85	94	99	102	102	98	89	80	74	102
San Diego	37	88	88	88	92	96	97	93	98	111	107	97	88	111
San Francisco														
International Airport	50	72	78	85	88	96	106	104	98	103	95	85	75	106
City	41	79	77	83	86	96	101	92	96	101	94	86	76	101
Santa Maria	35	86	87	88	97	100	102	104	103	102	103	93	90	104
Stockton	18	71	78	87	93	103	111	114	109	105	98	84	71	114
Colorado														
Alamosa	32	62	64	73	77	85	91	93	90	87	79	67	61	93
Colorado Springs	29	72	76	81	81	92	100	100	99	94	86	74	77	100
Denver	43	72	76	84	85	96	104	104	101	97	88	79	74	104
Grand Junction	31	60	66	81	85	95	103	105	103	98	88	75	64	105
Pueblo	37	78	79	86	88	98	105	105	104	99	92	83	77	105
Connecticut														
Bridgeport	29	65	67	70	86	92	96	103	98	99	85	78	65	103
Hartford	23	65	70	87	96	96	100	102	101	96	91	81	65	102
Delaware														

Table 6 (continued)
HIGHEST TEMPERATURE OF RECORD (°F) IN THE U.S. AND CERTAIN PACIFIC ISLANDS (DATA THROUGH 1977)[a]

Location	Years	Jan.	Feb.	Mar.	April	May	June	July	Aug.	Sept.	Oct.	Nov.	Dec.	Annual
Wilmington	30	75	74	86	91	95	99	102	101	100	91	85	72	102
District of Columbia														
Washington														
Dulles Airport	15	75	75	83	90	97	100	103	98	95	90	84	75	103
National Airport	36	79	82	89	95	97	101	103	100	100	94	86	75	103
Florida														
Apalachicola	48	79	80	82	90	96	101	102	99	96	93	87	82	102
Daytona Beach	34	85	88	91	96	100	102	99	99	99	95	89	85	102
Fort Myers	38	88	92	93	95	99	100	101	100	96	94	91	88	101
Jacksonville	36	85	88	91	95	100	103	105	102	100	96	88	84	105
Key West	25	85	85	87	89	91	94	95	95	94	93	88	85	95
Lakeland	37	85	88	91	95	99	100	101	98	97	94	89	86	101
Miami	35	87	89	92	96	94	98	96	98	95	92	89	86	98
Orlando	35	87	90	92	96	102	100	100	99	97	95	89	90	102
Pensacola	14	80	82	85	87	96	98	99	100	97	92	85	81	100
Tallahassee	17	82	85	90	95	98	100	100	100	99	93	88	84	100
Tampa	31	84	88	91	93	98	98	97	98	96	94	90	86	98
West Palm Beach	41	89	90	94	99	96	98	101	98	97	95	91	90	101
Georgia														
Athens	34	80	79	88	90	97	104	104	105	99	98	86	78	105
Atlanta	29	79	79	85	88	95	101	103	102	98	95	84	77	103
Augusta	27	80	86	88	92	99	105	106	104	101	97	90	82	106
Columbus	32	83	83	89	92	97	103	104	101	99	96	86	82	104
Macon	29	84	85	95	94	99	106	104	104	102	100	88	82	106
Rome	30	80	83	86	91	96	102	106	105	101	96	87	78	106
Savannah	27	84	86	91	95	100	103	104	104	97	94	89	83	104
Hawaii														
Hilo	31	90	92	93	89	94	90	89	93	92	91	88	91	94
Honolulu	8	85	87	87	87	88	90	90	92	92	91	89	89	92
Kahului	13	88	87	87	88	91	92	93	94	95	96	92	89	96
Lihue	27	84	86	87	87	88	89	88	89	89	90	86	85	90

Idaho														
Boise	38	63	69	78	92	98	109	111	110	102	91	73	65	111
Lewiston	31	66	66	76	97	96	107	110	115	103	89	74	63	115
Pocatello	28	57	62	71	85	93	99	102	104	98	87	71	59	104
Illinois														
Cairo	35	75	76	82	89	98	104	104	103	103	92	82	73	104
Chicago														
O'Hare Airport	19	61	71	80	87	93	97	98	96	99	91	77	67	99
Midway Airport	35	67	75	82	88	95	104	103	101	101	94	81	71	104
Moline	45	65	71	83	91	104	104	105	106	100	92	80	69	106
Peoria	38	68	72	81	89	92	100	103	101	100	90	81	71	103
Rockford	27	60	68	76	88	95	99	103	101	102	90	76	65	103
Springfield	30	71	74	83	88	95	103	112	103	100	93	83	74	112
Indiana														
Evansville	37	76	79	83	90	95	104	105	102	103	94	83	75	105
Fort Wayne	31	69	69	79	87	92	100	103	101	100	90	79	67	103
Indianapolis	38	71	74	81	89	93	102	104	100	100	90	81	70	104
South Bend	38	68	69	81	91	95	100	101	100	99	92	82	67	101
Iowa														
Burlington	80	68	76	88	93	103	105	111	110	103	95	86	71	111
Des Moines	38	62	73	83	90	98	101	105	102	101	95	76	69	105
Dubuque	27	56	60	74	89	90	96	99	96	97	89	70	64	99
Sioux City	37	68	66	91	94	102	106	107	104	102	93	78	64	107
Waterloo	29	55	62	84	90	94	99	103	97	98	95	79	65	103
Kansas														
Concordia	16	68	86	87	92	99	103	108	108	104	96	79	62	108
Dodge City	35	78	85	93	94	102	107	107	107	106	96	83	86	107
Goodland	57	79	81	89	92	99	109	111	110	105	96	87	83	111
Topeka	31	73	84	88	94	97	107	109	106	109	96	82	70	109
Wichita	25	75	84	89	96	100	106	113	110	103	95	80	83	113
Kentucky														
Covington (Cincinnati Airport)	16	69	73	82	89	93	97	100	102	98	88	80	70	102
Lexington	33	76	76	83	88	92	101	103	99	103	91	81	72	103
Louisville	30	77	77	85	91	95	102	105	101	104	92	84	73	105
Louisiana														
Baton Rouge	27	82	85	91	92	98	103	101	102	97	94	86	83	103
Lake Charles	13	79	83	86	92	94	98	99	97	96	92	86	82	99

Table 6 (continued)
HIGHEST TEMPERATURE OF RECORD (°F) IN THE U.S. AND CERTAIN PACIFIC ISLANDS (DATA THROUGH 1977)[a]

Location	Years	Jan.	Feb.	Mar.	April	May	June	July	Aug.	Sept.	Oct.	Nov.	Dec.	Annual
New Orleans	31	83	85	87	91	96	100	99	100	97	92	86	84	100
Shreveport	25	84	89	92	91	95	101	106	107	103	97	88	84	107
Maine														
Caribou	38	51	49	73	80	96	96	95	95	91	79	68	58	96
Portland	37	64	64	86	85	92	97	99	103	95	88	74	62	103
Maryland														
Baltimore	27	75	76	85	94	98	100	102	102	99	92	83	74	102
Massachusetts														
Blue Hill Observatory	92	68	67	85	94	93	99	100	101	99	88	81	68	101
Boston	26	63	68	81	94	93	100	102	102	100	90	77	70	102
Worcester	22	60	64	80	91	92	93	94	96	89	85	70	66	96
Michigan														
Alpena	19	52	58	73	86	94	98	98	95	93	88	75	62	98
Detroit														
City Airport	43	67	68	82	87	93	104	105	101	100	92	81	66	105
Metro Airport	19	62	65	77	89	92	99	102	97	98	91	77	66	102
Flint	21	60	61	77	85	90	96	99	97	95	89	73	63	99
Grand Rapids	14	62	67	77	88	92	96	97	100	93	87	77	65	100
Houghton Lake	13	53	53	67	84	89	93	96	92	89	85	69	59	96
Lansing	19	66	63	78	86	94	96	99	100	97	89	77	65	100
Marquette	41	57	62	76	86	100	101	104	102	98	86	73	59	104
Muskegon	38	63	62	77	86	93	97	95	99	95	83	76	63	99
Sault Ste. Marie	37	45	45	75	83	89	92	97	98	95	80	66	59	98
Minnesota														
Duluth	36	52	55	78	88	90	93	97	97	95	86	69	55	97
International Falls	38	48	53	76	93	95	98	98	95	95	88	73	57	98
Minneapolis — St. Paul	39	58	59	83	92	95	100	104	102	98	89	75	63	104
Rochester	18	46	57	76	88	92	95	102	96	93	88	72	58	102
Saint Cloud	38	55	55	79	88	97	98	103	103	98	90	73	58	103
Mississippi														
Jackson	14	82	84	88	92	99	103	102	99	98	91	88	81	103

Meridian	32	83	84	90	91	99	103	103	104	102	97	87	82	104
Missouri														
Columbia	8	66	82	83	90	89	99	103	99	101	92	78	71	103
Kansas City	5	61	74	82	85	91	98	107	104	98	92	78	67	107
Saint Joseph	28	69	79	90	96	98	104	107	103	103	97	82	70	107
St. Louis	20	76	85	88	92	92	98	106	105	100	94	81	76	106
Springfield	32	76	81	87	93	93	101	113	103	104	93	80	77	113
Montana														
Billings	43	68	72	77	87	96	102	106	105	100	90	73	69	106
Glasgow	22	55	59	75	88	95	98	104	106	99	88	75	56	106
Great Falls	40	62	67	72	87	90	99	105	106	98	91	76	69	106
Havre	17	63	67	75	87	94	100	106	111	101	88	78	65	111
Helena	37	62	68	74	85	90	98	102	105	99	85	70	62	105
Kalispell	28	53	56	71	84	89	96	104	105	99	81	65	55	105
Miles City	40	62	66	83	91	99	104	109	110	105	93	75	69	110
Missoula	33	59	60	72	86	92	98	105	105	99	85	65	60	105
Nebraska														
Grand Island	32	71	77	88	94	101	107	109	106	104	96	78	76	109
Lincoln	6	61	84	86	91	93	105	106	101	101	93	74	68	106
Norfolk	32	62	73	87	94	103	106	113	107	101	95	82	71	113
North Platte	26	69	79	86	90	97	107	112	105	102	91	77	74	112
Omaha	42	69	78	89	93	99	105	114	110	104	96	80	72	114
Omaha (North)	23	63	76	87	91	100	101	107	103	103	93	77	66	107
Scottsbluff	35	67	77	87	88	95	105	109	103	101	92	77	71	109
Valentine	23	70	75	83	93	98	107	109	108	102	95	82	69	109
Nevada														
Elko	47	62	67	77	84	92	101	104	103	99	86	74	64	104
Ely	39	68	66	73	78	87	99	97	97	93	84	75	67	99
Las Vegas	29	77	82	91	98	109	115	116	115	113	99	85	75	116
Reno	36	70	74	83	88	95	101	104	103	101	91	77	70	104
Winnemucca	28	68	72	81	89	96	105	106	106	103	89	77	66	106
New Hamsphire														
Concord	36	68	66	85	95	97	98	102	101	98	90	80	63	102
Mt. Washington	45	44	42	48	60	66	71	71	72	67	59	51	45	72
New Jersey														
Atlantic City	34	78	73	87	94	99	106	104	102	97	90	84	72	106
Newark	36	74	76	89	93	98	102	105	103	105	92	85	72	105

Table 6 (continued)
HIGHEST TEMPERATURE OF RECORD (°F) IN THE U.S. AND CERTAIN PACIFIC ISLANDS (DATA THROUGH 1977)[2]

Location	Years	Jan.	Feb.	Mar.	April	May	June	July	Aug.	Sept.	Oct.	Nov.	Dec.	Annual
Trenton	45	72	73	86	93	96	100	106	100	100	94	83	72	106
New Mexico														
Albuquerque	38	69	75	85	89	95	105	104	100	98	87	77	72	105
Clayton	33	78	81	85	91	95	104	102	102	99	90	84	83	104
Roswell	5	82	82	89	89	101	107	103	104	100	94	84	80	107
New York														
Albany	31	62	67	85	92	94	99	100	99	100	89	82	65	100
Binghamton	26	63	66	82	85	87	94	95	94	96	82	74	64	96
Buffalo	34	72	64	81	87	90	95	94	99	98	87	80	66	99
New York														
Central Park	109	72	75	86	96	99	101	106	104	102	94	84	70	106
JFK Airport	17	65	67	72	90	99	99	104	98	94	84	77	68	104
La Guardia Airport	16	68	68	78	91	95	97	107	97	94	85	79	66	107
Rochester	37	74	67	84	93	94	100	98	99	99	91	81	69	100
Syracuse	28	70	66	85	89	96	98	97	97	97	87	81	70	98
North Carolina														
Asheville	13	78	77	82	89	91	96	95	94	92	84	81	78	96
Cape Hatteras	20	75	76	79	85	88	94	95	94	90	86	81	76	95
Charlotte	38	78	81	90	93	100	103	103	102	104	98	85	77	104
Greensboro	49	78	81	90	94	98	102	102	101	100	95	85	78	102
Raleigh	33	79	84	92	93	97	104	105	101	104	98	88	78	105
Wilmington	26	82	85	89	95	98	104	102	102	98	95	87	81	104
North Dakota														
Bismarck	38	54	68	81	92	98	100	109	109	105	95	75	62	109
Fargo	25	49	66	78	88	98	99	100	106	102	93	73	57	106
Williston	16	50	55	75	90	98	99	107	103	101	93	71	56	107
Ohio														
Akron	29	70	68	79	85	92	100	100	98	99	86	80	67	100
Cincinnati	62	77	77	88	90	95	102	109	103	101	92	83	71	109
Cleveland	36	73	69	83	88	92	101	103	102	101	90	82	69	103
Columbus	38	74	73	85	89	94	102	100	100	100	90	80	72	102

Dayton	34	71	71	80	89	93	99	102	100	101	89	79	70	102
Mansfield	18	63	67	76	85	92	97	96	95	93	85	78	68	97
Toledo	22	62	68	80	88	95	99	101	98	95	91	78	67	101
Youngstown	34	71	67	80	87	92	99	100	97	99	87	80	69	100
Oklahoma														
Oklahoma City	24	79	84	93	100	96	102	108	107	102	96	84	86	108
Tulsa	39	79	86	96	102	95	103	112	110	109	98	87	80	112
Oregon														
Astoria	24	65	72	73	83	86	93	100	93	95	81	71	63	100
Burns	27	58	64	75	84	93	98	101	103	99	87	71	61	103
Eugene	35	67	71	75	86	91	100	105	106	101	88	76	67	106
Medford	48	70	77	86	92	100	109	115	109	107	97	75	72	115
Pendleton	42	68	68	79	91	99	108	110	113	102	90	77	67	113
Portland	37	62	70	80	87	92	100	107	105	101	90	73	64	107
Salem	40	64	72	80	88	95	102	108	105	103	93	72	64	108
Sexton Summit	33	65	67	74	82	88	95	100	97	97	85	77	64	100
Pacific														
Guam	21	87	88	89	90	91	91	93	91	95	91	89	89	95
Johnston Island	22	88	87	85	85	87	87	89	88	88	89	89	86	89
Koror Island	28	91	92	91	93	93	95	93	93	92	92	92	92	95
Kwajalein Island	25	91	92	91	92	93	92	94	95	93	97	92	90	97
Majuro, Marshall Islands	22	88	88	88	88	89	89	89	91	90	91	89	89	91
Pago Pago, American Samoa	18	92	92	92	92	90	90	89	89	89	90	92	92	92
Ponape Island	27	91	92	95	92	92	91	91	93	96	94	92	91	96
Truk, Caroline Islands	27	91	91	94	90	94	93	92	92	91	92	91	91	94
Wake Island	30	87	88	89	91	91	92	92	92	92	91	90	88	92
Yap Island	29	90	92	90	97	93	94	96	93	93	93	93	94	97
Pennsylvania														
Allentown	34	72	72	86	93	97	100	105	100	99	90	81	70	105
Erie	24	64	67	79	85	89	92	94	92	94	88	80	68	94
Harrisburg	39	73	75	86	92	97	100	107	101	102	97	84	71	107
Philadelphia	36	74	74	87	94	96	100	104	101	100	96	81	72	104
Pittsburgh														
International Airport	25	68	69	80	87	91	96	99	97	97	87	82	72	99
City	43	77	74	83	91	92	99	103	100	100	91	83	73	103
Avoca	22	67	68	83	92	93	97	101	94	95	84	77	65	101
Williamsport	33	69	71	86	92	95	102	100	100	102	91	83	65	102

Table 6 (continued)
HIGHEST TEMPERATURE OF RECORD (°F) IN THE U.S. AND CERTAIN PACIFIC ISLANDS (DATA THROUGH 1977)[2]

Location	Years	Jan.	Feb.	Mar.	April	May	June	July	Aug.	Sept.	Oct.	Nov.	Dec.	Annual
Puerto Rico														
San Juan	23	90	92	93	94	94	96	93	96	94	95	92	90	96
Rhode Island														
Block Island	26	57	62	74	92	82	90	91	91	87	77	70	64	92
Providence	24	66	69	78	98	94	95	100	104	93	85	78	69	104
South Carolina														
Charleston	35	83	86	90	93	98	103	101	102	99	94	88	83	103
Columbia	30	84	84	91	94	101	107	107	106	101	101	90	83	107
Greenville — Spartanburg	15	79	79	88	91	97	98	101	99	96	88	85	75	101
South Dakota														
Aberdeen	18	55	62	82	96	97	103	110	112	103	96	78	62	112
Huron	36	63	71	89	93	99	106	112	110	106	102	77	62	112
Rapid City	35	74	74	82	89	98	106	110	106	104	94	77	75	110
Sioux Falls	32	59	70	87	94	100	101	108	108	104	94	76	61	108
Tennessee														
Bristol—Johnson City	32	79	80	85	88	92	97	102	98	100	90	81	78	102
Chattanooga	38	78	79	87	93	99	104	106	105	102	94	84	78	106
Knoxville	36	77	83	86	92	94	102	103	102	103	91	84	77	103
Memphis	36	78	81	85	91	99	104	106	105	103	95	85	79	106
Nashville	38	78	84	86	90	97	106	107	104	105	94	84	76	107
Oak Ridge	30	75	79	85	91	93	101	105	100	102	90	83	74	105
Texas														
Abilene	38	89	90	97	99	107	106	107	109	106	100	91	89	109
Amarillo	37	81	88	94	98	102	108	104	106	102	95	85	81	108
Austin	36	90	93	98	98	99	103	109	106	103	96	91	90	109
Brownsville	39	93	94	99	100	102	101	102	102	104	96	94	94	104
Corpus Christi	39	91	98	99	102	99	100	104	103	103	98	95	91	104
Dallas — Fort Worth	24	88	88	96	95	96	105	109	108	105	98	89	88	109
Del Rio	15	89	91	101	102	105	108	106	109	105	100	91	90	109
El Paso	38	80	83	88	98	104	109	109	105	103	96	84	80	109
Galveston	107	77	83	85	92	93	99	101	100	96	94	85	80	101

Houston	8	84	84	90	89	94	99	101	101	97	93	88	83	101
Lubbock	31	83	86	94	96	104	107	107	106	103	93	86	81	107
Midland, Odessa	30	84	87	95	99	107	109	106	107	107	98	88	85	109
Port Arthur	24	80	84	87	93	97	100	103	107	99	95	87	82	107
San Angelo	30	90	90	97	103	107	110	111	107	107	100	90	91	111
San Antonio	36	89	92	100	99	101	103	106	106	102	95	91	90	106
Victoria	17	88	90	95	98	101	99	104	107	99	95	92	88	107
Waco	35	88	90	100	101	99	104	108	112	106	100	92	91	112
Wichita Falls	31	87	89	100	102	104	111	111	113	108	102	89	88	113
Utah														
Milford	29	66	73	78	87	94	105	104	102	98	90	76	65	105
Salt Lake City	49	61	69	78	85	93	104	107	103	98	89	75	67	107
Vermont														
Burlington	34	63	60	84	91	93	96	99	101	94	85	75	62	101
Virginia														
Lynchburg	33	76	79	87	91	93	100	103	98	101	93	83	78	103
Norfolk	29	78	81	85	97	97	101	103	102	98	95	86	79	103
Richmond	48	80	83	93	96	100	104	105	102	103	99	86	80	105
Roanoke	30	78	78	86	95	96	100	104	100	101	93	83	75	104
Washington														
Olympia	36	63	72	76	85	92	101	103	100	95	85	74	64	103
Quillayute	11	62	72	69	75	89	92	92	96	92	81	69	64	96
Seattle (Urban Site)	44	66	74	75	87	92	100	100	97	92	78	73	65	100
Seattle (International Airport)	33	61	70	72	85	93	96	97	99	93	81	74	61	99
Spokane	30	59	61	71	90	92	100	103	108	96	85	67	54	108
Stampede Pass	34	59	57	61	76	81	86	88	90	87	73	58	53	90
Walla Walla	63	71	71	79	93	99	106	112	113	103	88	80	73	113
Yakima	31	68	69	80	92	98	103	108	110	100	85	72	67	110
West Virginia														
Beckley	14	67	74	79	86	85	90	90	90	89	81	75	73	90
Charleston	30	79	78	87	91	93	98	102	100	102	92	85	80	102
Elkins	33	76	72	84	88	88	93	95	95	97	86	80	76	97
Huntington	17	74	79	85	90	92	95	99	100	96	86	82	96	100
Parkersburg	89	78	77	89	93	96	99	104	106	102	91	86	77	106
Wisconsin														
Green Bay	28	50	54	73	87	91	97	99	99	95	88	72	62	99
La Crosse	27	52	61	80	92	94	98	101	103	100	93	74	61	103

Table 6 (continued)
HIGHEST TEMPERATURE OF RECORD (°F) IN THE U.S. AND CERTAIN PACIFIC ISLANDS (DATA THROUGH 1977)[2]

Location	Years	Jan.	Feb.	Mar.	April	May	June	July	Aug.	Sept.	Oct.	Nov.	Dec.	Annual
Madison	38	55	60	78	90	93	97	104	101	99	90	76	62	104
Milwaukee	37	62	65	81	85	92	99	101	100	98	89	77	63	101
Wyoming														
Casper	27	60	65	73	81	91	101	104	99	93	85	71	61	104
Cheyenne	42	63	71	73	82	90	100	100	96	93	83	73	69	100
Lander	31	63	68	76	82	91	100	101	98	93	85	68	64	101
Sheridan	37	70	73	76	87	95	100	106	105	100	91	78	71	106

Table 7
LOWEST TEMPERATURE OF RECORD (°F) IN THE U.S. AND CERTAIN PACIFIC ISLANDS[2] (DATA THROUGH 1977)

Location	Years	Jan.	Feb.	Mar.	April	May	June	July	Aug.	Sept.	Oct.	Nov.	Dec.	Annual
Alabama														
Birmingham	34	−4	3	14	26	35	42	51	51	37	27	5	1	−4
Huntsville	10	−1	7	18	30	36	45	54	55	40	30	15	9	−1
Mobile	36	8	11	11	36	43	56	60	59	42	32	22	10	8
Montgomery	33	5	10	20	31	40	49	59	57	39	26	13	5	5
Alaska														
Anchorage	24	−34	−26	−24	3	17	33	38	33	20	−5	−21	−30	−34
Annette	30	2	2	1	21	31	37	40	40	34	25	10	1	1
Barrow	57	−53	−56	−52	−42	−18	4	22	20	1	−32	−40	−55	−56
Barter Island	30	−54	−59	−51	−38	−16	13	24	24	4	−23	−51	−51	−59
Bethel	19	−46	−37	−39	−22	4	28	31	33	18	−5	−22	−37	−46
Bettles	27	−70	−60	−56	−25	−10	27	29	22	5	−32	−57	−59	−70
Big Delta	36	−63	−60	−49	−37	−1	30	32	22	7	−39	−46	−62	−63
Cold Bay	34	−5	−9	−13	4	18	29	36	33	27	10	1	−9	−13
Fairbanks	26	−61	−56	−49	−21	−1	31	35	30	11	−27	−43	−62	−62
Gulkana	32	−60	−65	−48	−42	5	28	29	20	7	−23	−44	−58	−65

Homer	34	−18	−18	−21	−9	6	29	34	31	20	2	−7	−16	−21
Juneau	34	−22	−22	−15	6	25	31	36	27	23	12	−5	−21	−22
King Salmon	35	−46	−41	−42	−12	4	29	33	31	16	−11	−26	−38	−46
Kodiak	29	−8	−12	−6	7	20	30	37	36	26	10	0	−1	−12
Kotzebue	35	−47	−52	−48	−44	−18	20	30	31	15	−19	−36	−47	−52
Mc Grath	35	−64	−64	−51	−31	−2	30	33	27	6	−22	−49	−67	−67
Nome	31	−40	−42	−46	−30	−11	23	31	27	15	−10	−39	−41	−46
St. Paul Island	60	−26	−15	−19	−8	8	24	28	30	25	13	9	−5	−26
Talkeetna	38	−48	−46	−43	−37	−5	28	33	25	15	−21	−41	−45	−48
Unalakleet	28	−48	−50	−50	−30	−6	25	32	28	6	−20	−47	−50	−50
Valdez	6	−20	−3	−6	5	21	31	33	36	27	−8	5	−6	−20
Yakutat	31	−22	−20	−20	3	21	29	35	29	21	6	−6	−24	−24
Arizona														
Flagstaff	28	−22	−16	−16	−2	14	22	32	24	23	−2	−13	−18	−22
Phoenix	40	17	22	25	32	40	50	61	60	47	34	25	22	17
Tucson	37	16	20	20	27	38	47	63	61	44	26	24	16	16
Winslow	46	−18	−7	6	16	23	36	46	41	31	13	−1	−12	−18
Yuma	27	24	28	32	41	46	54	63	63	53	35	30	27	24
Arkansas														
Fort Smith	32	−10	−9	7	22	35	47	50	51	33	22	8	1	−10
Little Rock	36	−4	−5	11	28	40	46	54	52	37	31	17	−1	−5
California														
Bakersfield	40	20	26	31	35	40	45	53	52	45	29	28	21	20
Bishop	30	−7	−2	9	15	25	29	40	37	26	16	5	−4	−7
Blue Canyon	34	5	6	9	17	21	28	40	35	27	17	13	3	3
Eureka	67	25	27	29	32	36	41	45	44	41	32	29	21	21
Fresno	28	19	25	26	32	36	44	50	49	37	27	26	21	19
Long Beach	25	25	33	33	38	40	47	51	52	50	39	34	31	25
Los Angeles														
International Airport	42	23	32	34	39	43	48	49	51	47	41	34	32	23
City	37	28	34	35	39	46	50	54	53	51	41	39	32	28
Mount Shasta	35	−2	1	11	14	21	25	31	34	25	19	9	−5	−5
Oakland	49	24	25	32	31	36	41	46	47	41	34	27	23	23
Red Bluff	33	20	23	26	31	35	42	52	52	42	32	26	20	20
Sacramento	27	23	26	26	32	36	41	49	49	43	36	26	20	20
Sandberg	43	3	13	15	22	26	30	40	40	35	21	21	11	3
San Diego	37	29	36	39	41	48	51	55	57	51	43	38	35	29

Table 7 (continued)
LOWEST TEMPERATURE OF RECORD (°F) IN THE U.S. AND CERTAIN PACIFIC ISLANDS[2] (DATA THROUGH 1977)

Location	Years	Jan.	Feb.	Mar.	April	May	June	July	Aug.	Sept.	Oct.	Nov.	Dec.	Annual
San Francisco														
International Airport	50	24	25	30	31	36	41	43	42	38	34	25	20	20
City	41	30	36	38	40	44	47	47	48	48	45	41	30	30
Santa Maria	35	20	22	24	31	31	36	43	43	36	26	25	21	20
Stockton	18	19	25	27	32	38	45	52	50	45	33	26	21	19
Colorado														
Alamosa	32	−50	−35	−20	−6	11	25	34	29	17	−10	−30	−41	−50
Colorado Springs	29	−26	−27	−11	0	21	32	42	43	25	5	−8	−16	−27
Denver	43	−25	−30	−11	−2	22	30	43	41	20	3	−8	−18	−30
Grand Junction	31	−23	−14	5	11	26	34	48	43	30	18	−2	−12	−23
Pueblo	37	−29	−31	−20	2	25	38	44	40	29	14	−14	−19	−31
Connecticut														
Bridgeport	29	−5	−5	4	20	31	41	51	48	36	26	16	−2	−5
Hartford	23	−26	−21	−6	9	28	37	44	36	30	18	12	−9	−26
Delaware														
Wilmington	30	−4	−4	9	22	32	41	50	46	36	24	14	3	−4
District of Columbia														
Washington														
Dulles Airport	15	−8	−6	10	17	28	36	47	43	30	15	11	0	−8
National Airport	36	2	4	11	24	34	47	55	51	39	29	16	1	1
Florida														
Apalachicola	48	14	21	26	37	50	56	65	66	50	39	24	13	13
Daytona Beach	34	23	24	31	35	44	54	65	65	52	41	27	21	21
Fort Myers	38	28	30	34	39	50	61	65	65	63	45	34	26	26
Jacksonville	36	19	19	25	35	45	56	61	64	50	36	21	12	12
Key West	25	46	47	53	55	66	68	69	68	70	60	49	46	46
Lakeland	37	25	28	30	40	54	63	66	63	61	43	28	20	20
Miami	35	31	32	37	46	53	65	69	68	69	51	39	34	31
Orlando	35	20	28	31	39	49	60	66	64	56	43	29	20	20
Pensacola	14	10	19	27	37	51	57	61	63	43	38	25	24	10

Tallahassee	17	11	14	23	29	34	49	57	61	40	30	13	10	10
Tampa	31	23	24	34	40	49	61	63	67	59	40	23	18	18
West Palm Beach	41	27	34	31	45	53	62	66	65	66	46	36	30	27
Georgia														
Athens	34	−1	5	13	27	37	45	55	54	36	24	7	2	−1
Atlanta	29	−3	5	10	26	37	46	53	56	36	28	3	1	−3
Augusta	27	5	9	20	30	35	47	55	54	36	22	15	6	5
Columbus	32	3	11	19	28	39	44	59	57	38	24	10	4	3
Macon	29	3	9	19	31	40	46	54	55	35	26	10	5	3
Rome	30	−5	−1	8	23	32	43	53	49	32	21	3	0	−5
Savannah	27	9	14	20	32	39	53	61	61	43	28	15	9	9
Hawaii														
Hilo	31	54	53	54	56	58	60	62	63	61	62	58	55	53
Honolulu	8	53	53	55	59	63	65	67	67	66	64	58	54	53
Kahului	13	48	50	55	57	57	60	58	61	60	58	55	53	48
Lihue	27	50	52	51	56	59	64	65	66	65	62	57	62	50
Idaho														
Boise	38	−17	−10	6	19	26	33	41	37	23	11	−3	−23	−23
Lewiston	31	−22	−15	2	20	23	34	41	42	28	16	−3	−22	−22
Pocatello	28	−30	−27	−8	15	20	30	37	32	21	10	−13	−28	−30
Illinois														
Cairo	35	−5	−5	6	28	39	51	54	53	42	29	5	0	−5
Chicago														
O'Hare Airport	19	−20	−17	−8	13	24	36	40	41	28	19	1	−17	−20
Midway Airport	35	−19	−15	−7	16	29	35	46	43	34	20	−2	−14	−19
Moline	45	−26	−23	−19	9	26	39	46	40	24	16	−9	−22	−26
Peoria	38	−25	−16	−10	17	25	39	47	43	26	19	−2	−18	−25
Rockford	27	−22	−22	−11	9	24	38	43	41	28	15	−10	−20	−22
Springfield	30	−21	−22	−12	21	28	40	48	43	33	17	−3	−17	−22
Indiana														
Evansville	37	−21	−23	−9	24	28	41	47	46	31	21	−3	−9	−23
Fort Wayne	31	−19	−18	−10	14	27	38	44	38	29	20	−1	−14	−19
Indianapolis	38	−20	−19	−6	16	28	39	44	41	28	17	−2	−15	−20
South Bend	38	−22	−17	−13	11	24	35	44	40	29	23	−7	−15	−22
Iowa														
Burlington	80	−24	−27	−13	13	26	39	46	41	23	10	−2	−19	−27
Des Moines	38	−24	−20	−22	9	30	38	47	40	26	14	−3	−16	−24

Table 7 (continued)
LOWEST TEMPERATURE OF RECORD (°F) IN THE U.S. AND CERTAIN PACIFIC ISLANDS[2]
(DATA THROUGH 1977)

Location	Years	Jan.	Feb.	Mar.	April	May	June	July	Aug.	Sept.	Oct.	Nov.	Dec.	Annual
Dubuque	27	−28	−23	−20	11	24	36	44	42	28	13	−17	−23	−28
Sioux City	37	−26	−26	−22	−2	25	38	42	37	24	13	−9	−19	−26
Waterloo	29	−31	−29	−34	4	25	38	43	38	22	13	−17	−27	−34
Kansas														
Concordia	16	−17	−14	−2	14	26	41	48	45	34	17	−4	−11	−17
Dodge City	35	−10	−15	−15	15	26	41	47	47	31	20	0	−7	−15
Goodland	57	−26	−22	−20	0	21	31	42	38	21	10	−12	−19	−26
Topeka	31	−20	−20	−6	10	26	44	43	45	30	19	2	−12	−20
Wichita	25	−12	−6	−2	15	31	43	51	48	35	21	1	−5	−12
Kentucky														
Covington (Cincinnati Airport)	16	−25	−10	7	17	27	39	47	43	33	16	1	−11	−25
Lexington	33	−21	−15	−2	22	26	39	47	42	35	20	−3	−6	−21
Louisville	30	−20	−19	−1	24	31	42	50	49	33	23	−1	−9	−20
Louisiana														
Baton Rouge	27	10	20	26	37	44	55	58	59	43	32	21	11	10
Lake Charles	13	18	26	29	34	50	58	61	61	47	36	23	24	18
New Orleans	31	14	19	26	32	41	55	60	60	42	35	24	17	14
Shreveport	25	3	18	22	32	42	52	58	58	42	31	16	9	3
Maine														
Caribou	38	−32	−41	−20	−2	18	30	36	34	23	14	−3	−26	−41
Portland	37	−26	−39	−21	8	23	33	40	33	23	15	5	−21	−39
Maryland														
Baltimore	27	−7	−1	6	20	32	40	52	48	35	25	13	0	−7
Massachusetts														
Blue Hill Observatory	92	−16	−21	−5	6	27	36	46	39	28	21	5	−19	−21
Boston	26	−12	−4	6	17	34	46	54	47	38	28	17	−4	−12
Worcester	22	−19	−12	−4	14	28	38	46	38	30	20	9	−13	−19
Michigan														
Alpena	19	−28	−25	−27	0	20	28	34	31	26	16	−1	−18	−28
Detroit														
City Airport	43	−13	−16	−1	14	30	38	42	43	32	24	5	−5	−16

Metro Airport	19	−18	−12	1	17	25	36	41	40	29	17	9	−9	−18
Flint	21	−25	−22	−11	13	22	33	40	38	27	19	6	−11	−25
Grand Rapids	14	−16	−19	−5	10	22	33	43	39	28	18	5	−14	−19
Houghton Lake	13	−21	−27	−23	5	21	29	33	33	24	16	0	−21	−27
Lansing	19	−25	−24	−12	2	19	30	37	35	26	15	4	−17	−25
Marquette	41	−21	−13	−13	4	22	33	41	41	30	21	−2	−12	−21
Muskegon	38	−13	−14	−10	12	22	31	40	40	28	21	−14	−15	−15
Sault Ste. Marie	37	−30	−28	−24	1	18	28	36	32	25	16	−10	−24	−30
Minnesota														
Duluth	36	−39	−32	−28	−5	17	27	36	33	22	8	−23	−33	−39
International Falls	38	−46	−44	−38	−14	11	23	35	32	20	7	−27	−41	−46
Minneapolis—St. Paul	39	−34	−28	−32	2	18	34	43	39	26	15	−17	−24	−34
Rochester	18	−32	−25	−31	9	21	36	42	37	23	13	−20	−28	−32
Saint Cloud	38	−43	−35	−32	−3	19	33	40	33	18	5	−20	−34	−43
Mississippi														
Jackson	14	6	11	18	30	38	49	51	55	35	30	17	14	6
Meridian	32	0	8	20	29	38	45	55	53	34	24	16	4	0
Missouri														
Columbia	8	−18	−6	0	19	29	45	48	51	39	22	5	−12	−18
Kansas City	5	−13	−2	8	12	30	49	52	50	39	21	1	−11	−13
Saint Joseph	28	−25	−19	−11	20	30	44	41	43	30	18	−2	−15	−25
St. Louis	20	−14	−5	−5	22	31	43	51	47	36	23	1	−10	−14
Springfield	32	−11	−11	−3	18	30	42	44	44	33	21	4	−8	−11
Montana														
Billings	43	−30	−38	−19	−6	14	32	41	40	25	3	−22	−26	−38
Glasgow	22	−47	−37	−27	−3	20	33	41	37	21	5	−21	−38	−47
Great Falls	40	−37	−35	−29	−6	15	31	40	35	23	−3	−23	−43	−43
Havre	17	−52	−30	−24	−14	24	32	39	36	20	0	−24	−46	−52
Helena	37	−42	−31	−30	1	17	30	36	32	18	−3	−39	−38	−42
Kalispell	28	−38	−36	−29	10	19	28	31	31	16	5	−28	−35	−38
Miles City	40	−37	−37	−27	7	15	32	41	35	20	9	−23	−35	−37
Missoula	33	−33	−22	−13	14	21	31	31	32	20	0	−23	−30	−33
Nebraska														
Grand Island	32	−28	−17	−21	7	23	38	42	40	27	16	−11	−20	−28
Lincoln	6	−33	−24	−4	3	25	42	42	45	29	12	−5	−20	−33
Norfolk	32	−27	−26	−20	2	24	38	42	40	27	13	−15	−22	−27
North Platte	26	−23	−22	−22	7	20	29	40	35	22	11	−13	−25	−25

Table 7 (continued)
LOWEST TEMPERATURE OF RECORD (°F) IN THE U.S. AND CERTAIN PACIFIC ISLANDS[2]
(DATA THROUGH 1977)

Location	Years	Jan.	Feb.	Mar.	April	May	June	July	Aug.	Sept.	Oct.	Nov.	Dec.	Annual
Omaha	42	−22	−19	−16	5	27	39	44	43	30	13	−9	−15	−22
Omaha (North)	23	−22	−16	−16	7	25	41	44	45	29	16	−11	−19	−22
Scottsbluff	35	−32	−28	−27	−8	18	30	40	39	23	9	−13	−25	−32
Valentine	23	−29	−28	−27	3	19	30	38	37	22	13	−22	−27	−29
Nevada														
Elko	47	−43	−37	−9	−2	10	23	30	24	9	8	−12	−38	−43
Ely	39	−27	−25	−13	−5	7	18	30	24	15	−3	−15	−28	−28
Las Vegas	29	8	18	23	31	40	49	62	56	46	26	21	15	8
Reno	36	−16	−12	−2	13	18	25	33	24	20	8	1	−16	−16
Winnemucca	28	−24	−9	−3	6	12	23	25	28	12	7	−7	−34	−34
New Hampshire														
Concord	36	−30	−37	−16	8	21	30	35	29	21	10	−1	−22	−37
Mt. Washington	45	−47	−46	−38	−20	−2	8	25	20	11	−5	−20	−46	−47
New Jersey														
Atlantic City	34	−10	−7	7	12	25	37	46	40	32	23	11	−7	−10
Newark	36	−2	−7	6	23	33	43	52	48	35	28	15	−1	−7
Trenton	45	−4	−14	8	24	34	43	53	48	36	27	14	−2	−14
New Mexico														
Albuquerque	38	−17	−5	8	19	28	42	54	52	37	25	−7	3	−17
Clayton	33	−21	−17	−11	9	23	37	45	45	28	17	−10	−6	−21
Roswell	5	3	11	23	23	34	51	59	57	42	30	4	8	3
New York														
Albany	31	−28	−21	−21	10	26	36	41	37	24	16	5	−22	−28
Binghamton	26	−20	−15	−6	11	25	35	39	37	25	17	3	−8	−20
Buffalo	34	−12	−20	−4	12	26	35	43	38	32	20	9	−4	−20
New York														
Central Park	109	−6	−15	3	12	32	44	52	50	39	28	5	−13	−15
JFK Airport	17	−1	−2	7	26	34	45	55	46	40	25	19	5	−2
La Guardia Airport	16	−1	−2	10	25	38	46	56	52	44	30	18	3	−2
Rochester	37	−16	−16	−5	13	26	35	42	36	28	20	5	−16	−16
Syracuse	28	−26	−22	−16	9	25	35	45	40	28	19	5	−20	−26

North Carolina														
Asheville	13	−7	−2	14	24	29	35	46	43	30	21	8	5	−7
Cape Hatteras	20	12	14	19	26	39	44	54	57	45	33	22	19	12
Charlotte	38	−3	5	10	24	32	45	53	53	39	24	11	2	−3
Greensboro	49	−7	−4	5	21	32	42	48	47	35	20	10	0	−7
Raleigh	33	−1	5	11	23	31	38	48	46	39	19	11	4	−1
Wilmington	26	11	11	20	30	41	49	59	57	46	27	20	10	10
North Dakota														
Bismarck	38	−44	−37	−31	−12	15	30	35	33	11	5	−29	−43	−44
Fargo	25	−35	−34	−23	−7	20	30	36	33	19	7	−22	−32	−35
Williston	16	−40	−41	−28	−15	20	30	34	36	17	3	−23	−37	−41
Ohio														
Akron	29	−21	−11	−1	10	24	32	43	42	32	20	−1	−11	−21
Cincinnati	62	−17	−9	3	18	28	40	48	43	32	20	1	−13	−17
Cleveland	36	−19	−15	−5	10	25	31	41	41	32	22	3	−11	−19
Columbus	38	−19	−13	−2	18	25	35	43	39	31	20	5	−10	−19
Dayton	34	−21	−16	−1	15	27	40	44	40	32	21	−2	−15	−21
Mansfield	18	−20	−10	−2	11	25	38	44	40	34	20	2	−10	−20
Toledo	22	−17	−14	−4	11	25	32	43	37	26	15	2	−12	−17
Youngstown	34	−18	−11	−4	11	24	30	42	41	29	20	1	−9	−18
Oklahoma														
Oklahoma City	24	−4	3	3	20	39	47	53	51	37	22	11	1	−4
Tulsa	39	−8	0	−3	22	35	49	51	52	35	26	10	−3	−8
Oregon														
Astoria	24	11	19	22	29	30	38	39	39	33	26	15	6	6
Burns	27	−26	−18	−3	14	19	25	33	31	17	9	−17	−26	−26
Eugene	35	−4	−3	20	27	28	32	39	38	32	19	14	−12	−12
Medford	48	−3	6	16	21	28	31	38	39	29	18	14	−6	−6
Pendleton	42	−22	−18	10	18	25	36	42	41	30	11	−6	−13	−22
Portland	37	−2	−3	19	29	29	39	43	44	34	26	13	6	−3
Salem	40	−10	−4	12	23	25	32	37	38	26	23	9	−12	−12
Sexton Summit	33	−2	9	15	19	22	27	36	36	26	20	11	2	−2
Pacific														
Guam	21	56	59	54	59	62	63	66	67	61	65	62	61	54
Johnston Island	22	65	64	67	65	68	69	70	71	71	66	63	62	62
Koror Island	28	70	71	69	71	72	70	70	71	70	71	70	71	69
Kwajalein Island	25	71	71	71	71	71	71	70	71	72	71	70	69	69

Table 7 (continued)
LOWEST TEMPERATURE OF RECORD (°F) IN THE U.S. AND CERTAIN PACIFIC ISLANDS[2] (DATA THROUGH 1977)

Location	Years	Jan.	Feb.	Mar.	April	May	June	July	Aug.	Sept.	Oct.	Nov.	Dec.	Annual
Majuro, Marshall Islands	22	69	72	70	70	71	70	71	71	72	70	72	70	69
Pago Pago, American Samoa	18	67	67	67	68	66	64	62	65	63	67	67	67	62
Ponape Island	27	66	67	69	68	69	69	68	68	68	68	66	66	66
Truk, Caroline Islands	27	71	70	71	71	71	70	70	70	68	70	71	70	68
Wake Island	30	65	65	65	66	69	71	69	68	69	68	65	64	64
Yap Island	29	69	70	69	70	69	65	70	67	69	69	69	65	65
Pennsylvania														
Allentown	34	−12	−7	−1	18	28	39	48	43	30	22	11	−8	−12
Erie	24	−15	−12	−1	12	26	32	44	41	33	24	7	−2	−15
Harrisburg	39	−5	−2	8	21	31	41	49	45	30	23	13	−8	−8
Philadelphia	36	−5	−4	7	24	28	44	51	45	35	25	15	1	−5
Pittsburgh														
International Airport	25	−18	−9	−1	15	26	34	42	40	31	16	−1	−7	−18
City	43	−14	−5	4	18	29	39	50	45	35	24	5	−7	−14
Avoca	22	−10	−11	−4	15	27	34	45	42	30	19	9	−7	−11
Williamsport	33	−17	−13	−1	18	28	36	43	38	28	20	8	−15	−17
Puerto Rico														
San Juan	23	61	62	60	64	66	69	69	70	69	67	66	63	50
Rhode Island														
Block Island	26	−2	−2	8	21	34	41	51	48	42	30	20	−4	−4
Providence	24	−13	−6	1	14	29	41	49	40	34	20	14	−5	−13
South Carolina														
Charleston	35	11	12	21	29	36	50	58	58	42	27	15	8	8
Columbia	30	5	5	18	29	34	45	54	53	40	23	12	4	4
Greenville-Spartanburg	15	−6	8	17	25	33	40	58	52	36	25	12	11	−6
South Dakota														
Aberdeen	18	−35	−37	−29	−2	19	33	39	35	20	11	−27	−39	−39
Huron	36	−35	−39	−24	−2	17	32	37	36	19	9	−21	−26	−39
Rapid City	35	−27	−22	−17	1	18	31	39	38	22	10	−19	−23	−27
Sioux Falls	32	−36	−31	−23	5	17	33	38	34	22	9	−17	−26	−36

Tennessee														
Bristol—Johnson City	32	−15	−5	3	21	30	38	45	47	34	20	5	−9	−15
Chattanooga	38	−10	1	8	26	34	41	51	50	36	22	4	−2	−10
Knoxville	36	−9	−2	6	26	34	43	51	49	36	25	5	−2	−9
Memphis	36	−4	−11	12	29	38	48	52	48	36	25	9	−13	−13
Nashville	38	−15	−13	5	24	34	42	51	47	36	26	−1	−7	−15
Oak Ridge	30	−9	1	8	24	30	39	50	51	33	21	0	−3	−9
Texas														
Abilene	38	−9	1	7	25	37	47	55	55	35	28	14	9	−9
Amarillo	37	−11	−14	−3	14	28	42	53	49	33	25	0	−3	−14
Austin	36	−2	7	18	35	43	53	64	61	41	32	20	16	−2
Brownsville	39	19	22	32	41	52	60	68	63	55	44	33	27	19
Corpus Christi	39	14	18	28	39	47	58	64	64	50	40	29	18	14
Dallas—Fort Worth	24	4	12	17	30	41	51	59	56	46	29	20	10	4
Del Rio	15	17	21	26	33	45	55	64	64	48	34	22	21	17
El Paso	38	−8	8	14	24	31	50	59	56	41	25	1	5	−8
Galveston	107	11	8	27	38	52	57	66	67	52	41	26	18	8
Houston	8	18	22	25	31	45	52	62	62	48	33	19	21	18
Lubbock	31	−16	−8	2	22	30	44	51	52	38	25	−1	1	−16
Midland—Odessa	30	−8	−1	10	20	34	48	56	56	42	32	13	8	−8
Port Arthur	24	14	22	27	36	46	57	61	62	45	35	22	17	14
San Angelo	30	6	1	13	25	35	48	58	55	44	29	17	8	1
San Antonio	36	0	6	21	33	44	53	62	61	41	33	21	14	0
Victoria	17	15	22	28	36	49	59	62	63	51	38	24	23	16
Waco	35	−5	5	15	27	45	52	61	60	48	32	17	14	−5
Wichita Falls	31	−5	−1	9	24	36	51	54	54	41	25	14	7	−5
Utah														
Milford	29	−28	−27	−14	9	17	24	38	34	24	−2	−13	−32	−32
Salt Lake City	49	−22	−30	2	14	25	35	40	37	27	16	−14	−21	−30
Vermont														
Burlington	34	−30	−26	−20	2	24	33	39	35	25	15	−2	−23	−30
Virginia														
Lynchburg	33	−8	0	7	24	31	40	50	45	36	21	8	3	−8
Norfolk	29	5	8	20	28	36	45	56	52	45	27	20	14	5
Richmond	48	−12	−10	11	25	31	40	51	46	35	21	10	−1	−12
Roanoke	30	−4	1	10	22	31	39	48	43	35	22	9	2	−4
Washington														
Olympia	36	−7	−1	13	23	25	30	35	33	25	20	−1	−3	−7

Table 7 (continued)
LOWEST TEMPERATURE OF RECORD (°F) IN THE U.S. AND CERTAIN PACIFIC ISLANDS[2] (DATA THROUGH 1977)

Location	Years	Jan.	Feb.	Mar.	April	May	June	July	Aug.	Sept.	Oct.	Nov.	Dec.	Annual
Quillayute	11	7	15	19	24	29	33	38	36	28	24	20	7	7
Seattle														
Urban Site	44	11	12	22	31	35	42	48	48	40	30	13	10	10
International Airport	33	0	1	11	29	28	38	43	44	35	28	6	6	0
Spokane	30	−19	−12	−3	17	24	34	38	35	25	13	−11	−25	−25
Stampede Pass	34	−11	−9	1	16	20	28	30	34	26	17	−5	−21	−21
Walla Walla	63	−16	−14	13	19	28	41	46	45	26	15	1	−14	−16
Yakima	31	−21	−25	−1	21	25	30	34	35	26	11	−3	−17	−25
West Virginia														
Beckley	14	−18	−10	4	15	23	32	41	36	32	18	4	3	−18
Charleston	30	−12	−6	7	19	26	33	46	41	34	17	6	−2	−12
Elkins	33	−20	−22	−4	13	20	25	39	34	27	11	0	−11	−22
Huntington	17	−15	−6	10	22	27	40	46	43	35	16	8	−5	−15
Parkersburg	89	−16	−27	−3	15	29	36	47	42	32	20	4	−10	−27
Wisconsin														
Green Bay	28	−31	−26	−29	7	21	32	40	38	24	15	−9	−21	−31
La Crosse	27	−37	−36	−28	10	26	40	44	40	28	15	−9	−22	−37
Madison	38	−37	−28	−29	9	20	31	36	35	25	14	−11	−22	−37
Milwaukee	37	−24	−19	−10	13	21	33	40	44	28	21	−5	−15	−24
Wyoming														
Casper	27	−40	−19	−21	−4	16	28	30	33	19	−3	−14	−31	−40
Cheyenne	42	−27	−34	−21	−8	16	25	38	36	18	2	−13	−24	−34
Lander	31	−37	−28	−16	−2	18	25	39	35	10	0	−15	−26	−37
Sheridan	37	−35	−31	−23	−2	13	27	35	34	15	1	−25	−30	−35

TABLE 8
MEAN DATE OF LAST 32°F TEMPERATURE IN SPRING, FIRST 32°F IN AUTUMN, AND MEAN LENGTH OF FREEZE-FREE PERIOD (DAYS)[3]

State and station	Mean date last 32°F in spring	Mean date first 32°F in fall	Mean freeze-free period (no. days)
Alabama			
Birmingham	Mar. 19	Nov. 14	241
Mobile U[a]	Feb. 17	Dec. 12	298
Montgomery U	Feb. 27	Dec. 3	279
Alaska			
Anchorage	May 18	Sept. 13	118
Barrow	June 27	July 5	8
Cordova	May 10	Oct. 2	145
Fairbanks	May 24	Aug. 29	97
Juneau	Apr. 27	Oct. 19	176
Nome	June 12	Aug. 24	73
Arizona			
Flagstaff	June 8	Oct. 2	116
Phoenix	Jan. 27	Dec. 11	317
Tucson	Mar. 6	Nov. 23	261
Winslow	Apr. 28	Oct. 21	176
Yuma U	Jan. 11	Dec. 27	350
Arkansas			
Fort Smith	Mar. 23	Nov. 9	231
Little Rock	Mar. 16	Nov. 15	244
California			
Bakersfield	Feb. 14	Nov. 28	287
Eureka U	Jan. 24	Dec. 25	335
Fresno	Feb. 3	Dec. 3	303
Los Angeles U	*[b]	*	*
Red Bluff	Feb. 25	Nov. 29	277
Sacramento	Jan. 24	Dec. 11	321
San Diego	*	*	*
San Francisco U	*	*	*
Colorado			
Denver U	May 2	Oct. 14	165
Palisades	Apr. 22	Oct. 17	178
Pueblo	Apr. 28	Oct. 12	167
Connecticut			
Hartford	Apr. 22	Oct. 19	180
New Haven	Apr. 15	Oct. 27	195
District of Columbia			
Washington U	Apr. 10	Oct. 28	200
Florida			
Apalachicola U	Feb. 2	Dec. 21	322
Fort Myers	*	*	*
Jacksonville U	Feb. 6	Dec. 16	313
Key West	*	*	*
Lakeland	Jan. 10	Dec. 25	349
Miami	*	*	*
Orlando	Jan. 31	Dec. 17	319
Pensacola U	Feb. 18	Dec. 15	300
Tallahassee	Feb. 26	Dec. 3	280
Tampa	Jan. 10	Dec. 26	349
Georgia			
Atlanta U	Mar. 20	Nov. 19	244
Augusta	Mar. 7	Nov. 22	260
Macon	Mar. 12	Nov. 19	252
Savannah	Feb. 21	Dec. 9	291
Idaho			
Boise	Apr. 29	Oct. 16	171

TABLE 8 (continued)
MEAN DATE OF LAST 32°F TEMPERATURE IN SPRING, FIRST 32°F IN AUTUMN, AND MEAN LENGTH OF FREEZE-FREE PERIOD (DAYS)[3]

State and station	Mean date last 32°F in spring	Mean date first 32°F in fall	Mean freeze-free period (no. days)
Pocatello	May 8	Sept. 30	145
Salmon	June 4	Sept. 6	94
Illinois			
Cairo U	Mar. 23	Nov. 11	233
Chicago U	Apr. 19	Oct. 28	192
Freeport	May 8	Oct. 4	149
Peoria	Apr. 22	Oct. 16	177
Springfield U	Apr. 8	Oct. 30	205
Indiana			
Evansville	Apr. 2	Nov. 4	216
Fort Wayne	Apr. 24	Oct. 20	179
Indianapolis U	Apr. 17	Oct. 27	193
South Bend	May 3	Oct. 16	165
Des Moines U	Apr. 20	Oct. 19	183
Dubuque U	Apr. 19	Oct. 19	184
Koekuk	Apr. 12	Oct. 26	197
Sioux City	Apr. 28	Oct. 12	167
Kansas			
Concordia U	Apr. 16	Oct. 24	191
Dodge City	Apr. 22	Oct. 24	184
Goodland	May 5	Oct. 9	157
Topeka U	Apr. 9	Oct. 26	200
Wichita	Apr. 5	Nov. 1	210
Kentucky			
Lexington	Apr. 13	Oct. 28	198
Louisville U	Apr. 1	Nov. 7	220
Louisiana			
Lake Charles	Feb. 18	Dec. 6	291
New Orleans	Feb. 13	Dec. 12	302
Shreveport	Mar. 1	Nov. 27	272
Maine			
Greenville	May 27	Sept. 20	116
Portland	Apr. 29	Oct. 15	169
Maryland			
Annapolis	Mar. 4	Nov. 15	225
Baltimore U	Mar. 28	Nov. 17	234
Frederick	Mar. 24	Oct. 17	176
Massachusetts			
Boston	Apr. 16	Oct. 25	192
Nantucket	Apr. 12	Nov. 16	219
Michigan			
Alpena U	May 6	Oct. 9	156
Detroit	Apr. 25	Oct. 23	181
Escanaba U	May 14	Oct. 6	145
Grand Rapids U	Apr. 25	Oct. 27	185
Marquette U	May 14	Oct. 17	156
S. Ste. Marie	May 18	Oct. 3	138
Minnesota			
Albert Lee	May 3	Oct. 6	156
Big Falls R.S	June 4	Sept. 7	95
Brainerd	May 16	Sept. 24	131
Duluth	May 22	Sept. 24	125
Minneapolis	Apr. 30	Oct. 13	166
St. Cloud	May 9	Sept. 29	144
Mississippi			
Jackson	Mar. 10	Nov. 13	248
Meridian	Mar. 13	Nov. 14	246
Vicksburg U	Mar. 8	Nov. 15	252

Missouri			
Columbia	Apr. 9	Oct. 24	198
Kansas City	Apr. 5	Oct. 31	210
St. Louis U	Apr. 2	Nov. 8	220
Springfield	Apr. 10	Oct. 31	203
Montana			
Billings	May 15	Sept. 24	132
Glasgow U	May 19	Sept. 20	124
Great Falls	May 14	Sept. 26	135
Havre U	May 9	Sept. 23	138
Helena	May 12	Sept. 23	134
Kalispell	May 12	Sept. 23	135
Miles City	May 5	Oct. 3	150
Superior	June 5	Aug. 30	85
Nebraska			
Grand Island	Apr. 29	Oct. 6	160
Lincoln	Apr. 20	Oct. 17	180
Norfolk	May 4	Oct. 3	152
North Platte	Apr. 30	Oct. 7	160
Omaha	Apr. 14	Oct. 20	189
Nevada			
Valentine Lakes	May 7	Sept. 30	146
Elko	June 6	Sept. 3	89
Las Vegas	Mar. 13	Nov. 13	245
Reno	May 14	Oct. 2	141
Winnemucca	May 18	Sept. 21	125
New Hampshire			
Concord	May 11	Sept. 30	142
New Jersey			
Cape May	Apr. 4	Nov. 15	225
Trenton U	Apr. 8	Nov. 5	211
New Mexico			
Albuquerque	Apr. 16	Oct. 29	196
Roswell	Apr. 9	Nov. 2	208
New York			
Albany	Apr. 27	Oct. 13	169
Binghamton U	May 4	Oct. 6	154
Buffalo	Apr. 30	Oct. 25	179
New York U	Apr. 7	Nov. 12	219
Rochester	Apr. 28	Oct. 21	176
Syracuse	Apr. 30	Oct. 15	168
N. Carolina			
Asheville U	Apr. 12	Oct. 24	195
Charlotte U	Mar. 21	Nov. 15	239
Greenville	Mar. 28	Nov. 5	222
Hatteras	Feb. 25	Dec. 18	296
Raleigh U	Mar. 24	Nov. 16	237
Wilmington U	Mar. 8	Nov. 24	262
N. Dakota			
Bismarck	May 11	Sept. 24	136
Devils Lake U	May 18	Sept. 22	127
Fargo	May 13	Sept. 27	137
Williston U	May 14	Sept. 23	132
Ohio			
Akron—Canton	Apr. 29	Oct. 20	173
Cincinnati (Abbe)	Apr. 15	Oct. 25	192
Cleveland	Apr. 21	Nov. 2	195
Columbus U	Apr. 17	Oct. 30	196
Dayton	Apr. 20	Oct. 21	184
Toledo	Apr. 24	Oct. 25	184
Oklahoma			
Okla. City U	Mar. 28	Nov. 7	223
Tulsa	Mar. 31	Nov. 2	216
Oregon			
Astoria	Mar. 18	Nov. 24	251
Bend	June 17	Aug. 17	62
Medford	Apr. 25	Oct. 20	178
Pendleton	Apr. 27	Oct. 8	163
Portland U	Feb. 25	Dec. 1	279
Salem	Apr. 14	Oct. 27	197
Pennsylvania			
Allentown	Apr. 20	Oct. 16	180
Harrisburg	Apr. 10	Oct. 28	201
Philadelphia U	Mar. 30	Nov. 17	232

TABLE 8 (continued)
MEAN DATE OF LAST 32°F TEMPERATURE IN SPRING, FIRST 32°F IN AUTUMN, AND MEAN LENGTH OF FREEZE-FREE PERIOD (DAYS)[3]

State and station	Mean date last 32°F in spring	Mean date first 32°F in fall	Mean freeze-free period (no. days)
Pittsburgh	Apr. 20	Oct. 23	187
Scranton U	Apr. 24	Oct. 14	174
Rhode Island			
Providence U	Apr. 13	Oct. 27	197
S. Carolina			
Charleston U	Feb. 19	Dec. 10	294
Columbia U	Mar. 14	Nov. 21	252
Greenville	Mar. 23	Nov. 17	239
S. Dakota			
Huron U	May 4	Sept. 30	149
Rapid City U	May 7	Oct. 4	150
Sioux Falls U	May 5	Oct. 3	152
Tennessee			
Chattanooga U	Mar. 26	Nov. 10	229
Knoxville U	Mar. 31	Nov. 6	220
Memphis U	Mar. 20	Nov. 12	237
Nashville U	Mar. 28	Nov. 7	224
Texas			
Albany	Mar. 30	Nov. 9	224
Balmorhea	Apr. 1	Nov. 12	226
Beeville	Feb. 21	Dec. 6	288
College Station	Mar. 1	Dec. 1	275
Corsicana	Mar. 13	Nov. 27	259
Dalhart Exp. Sta	Apr. 23	Oct. 18	178
Dallas	Mar. 18	Nov. 22	249
Del Rio	Feb. 12	Dec. 9	300
Encinal	Feb. 15	Dec. 12	301
Houston	Feb. 5	Dec. 11	309
Lampasas	Apr. 1	Nov. 10	223
Matagorda	Feb. 12	Dec. 17	308
Midland	Apr. 3	Nov. 6	218
Mission	Jan. 30	Dec. 21	325
Mount Pleasant	Mar. 23	Nov. 12	233
Nacogdoches	Mar. 15	Nov. 13	243
Plainview	Apr. 10	Nov. 6	211
Presidio	Mar. 20	Nov. 13	238
Quanah	Mar. 31	Nov. 7	221
San Angelo	Mar. 25	Nov. 15	235
Ysleta	Apr. 6	Oct. 30	207
Utah			
Blanding	May 18	Oct. 14	148
Salt Lake City	Apr. 12	Nov. 1	202
Vermont			
Burlington	May 8	Oct. 3	148
Virginia			
Lynchburg	Apr. 6	Oct. 27	205
Norfolk U	Mar. 18	Nov. 27	254
Richmond U	Apr. 2	Nov. 8	220
Roanoke	Apr. 20	Oct. 24	187
Washington			
Bumping Lake	June 17	Aug. 16	60
Seattle U	Feb. 23	Dec. 1	281
Spokane	Apr. 20	Oct. 12	175
Tatoosh Island	Jan. 25	Dec. 20	329
Walla Walla U	Mar. 28	Nov. 1	218
Yakima	Apr. 19	Oct. 15	179

W. Virginia			
Charleston	Apr. 18	Oct. 28	193
Parkersburg	Apr. 16	Oct. 21	189
Wisconsin			
Green Bay	May 6	Oct. 13	161
La Crosse U	May 1	Oct. 8	161
Madison U	Apr. 26	Oct. 19	177
Milwaukee U	Apr. 20	Oct. 25	188
Wyoming			
Casper	May 18	Sept. 25	130
Cheyenne	May 20	Sept. 27	130
Lander	May 15	Sept. 20	128
Sheridan	May 21	Sept. 21	123

[a] U indicates urban.

[b] * indicates occurrence in less than 1 year in 10. No freeze of record in Key West, Fla.

Charts and tabulation were derived from the Freeze Data tabulation in Climatography of the United States No. 60 - Climates of the States.

Table 9
MEAN GROWING DEGREE DAYS (ADJUSTED 50°F BASE[a]): ACCUMULATED WEEKLY MARCH 1 TO INDICATED DATES

Station	Mar. 8	Mar. 15	Mar. 22	Mar. 29	Apr. 5	Apr. 12	Apr. 19	Apr. 26	May 3	May 10
West										
Phoenix, Ariz.	76	160	252	352	458	579	710	839	974	1123
Fresno, Calif.	50	102	164	232	309	391	481	570	663	767
Denver, Colo.	13	30	52	83	117	152	199	255	311	378
Pocatello, Idaho	4	7	15	30	51	77	113	148	188	247
Glasgow, Mont.	1	2	7	18	35	54	85	117	159	209
Pendleton, Ore.	8	20	41	66	99	142	187	231	281	345
North Central										
Des Moines, Iowa	5	14	23	41	67	94	138	194	259	340
Minneapolis, Minn.	0	1	4	11	25	40	71	109	162	224
North Platte, Nebr.	10	24	39	67	100	133	180	236	296	366

Table 9 (continued)
MEAN GROWING DEGREE DAYS (ADJUSTED 50°F BASE[a]): ACCUMULATED WEEKLY MARCH 1 TO INDICATED DATES

Station	Mar. 8	Mar. 15	Mar. 22	Mar. 29	Apr. 5	Apr. 12	Apr. 19	Apr. 26	May 3	May 10
Omaha, Nebr.	8	20	33	58	91	127	178	244	316	405
Fargo, N. Dak.	0	0	2	8	17	30	54	85	134	186
Huron, S. Dak.	2	5	12	26	46	70	108	154	210	272
Green Bay, Wis.	0	1	3	7	17	28	55	85	125	176
Madison, Wis.	1	3	8	17	33	53	89	132	186	255
South Central										
Fort Smith, Ark.	41	87	131	191	269	348	437	551	664	794
Dodge City, Kans.	20	46	76	121	172	222	286	365	442	537
Topeka, Kans.	18	42	63	101	148	194	254	337	418	519
Wichita, Kans.	22	50	77	121	177	231	297	388	474	582
Shreveport, La.	61	132	203	282	384	488	593	728	866	1020
Columbia, Mo.	18	39	58	93	137	184	245	330	417	519
St. Louis, Mo.	19	41	60	97	142	190	254	345	436	541
Oklahoma City, Okla.	35	76	119	174	246	318	400	505	605	728
Amarillo, Tex.	33	78	123	180	244	308	385	474	560	667
Corpus Christi, Tex.	95	210	324	445	587	728	878	1049	1229	1411
Fort Worth, Tex.	57	122	191	271	369	470	578	710	838	988
Midland, Tex.	57	128	200	282	376	473	578	700	822	964
Midwest										
Chicago, Ill.	5	10	18	35	36	81	124	177	238	313
Moline, Ill.	5	10	18	36	61	90	135	192	261	344
Peoria, Ill.	8	16	25	46	74	105	152	214	284	367
Indianapolis, Ind.	12	26	39	64	95	128	175	250	324	411
Louisville, Ky.	22	48	71	108	156	207	272	370	466	569
Flint, Mich.	2	5	9	19	35	53	87	127	175	239
Grand Rapids, Mich.	1	5	10	20	37	55	89	131	185	252
Columbus, Ohio	12	26	38	64	95	129	176	246	320	404
Northeast										
Portland, Maine	0	1	1	4	12	24	41	65	94	135
Albany, N.Y.	1	4	6	16	33	51	84	130	179	244
Binghamton, N.Y.	1	3	5	14	28	41	66	105	145	198
Syracuse, N.Y.	2	5	8	19	36	51	83	126	174	236

Harrisburg, Pa.	8	17	28	52	83	116	163	228	295	377
Philadelphia, Pa.	8	20	31	57	91	125	175	238	302	384
Pittsburgh, Pa.	10	21	32	55	82	111	154	217	287	364
Williamsport, Pa.	4	10	15	33	58	84	126	182	243	318
Richmond, Va.	29	61	89	137	194	254	327	425	517	620
Southeast										
Macon, Ga.	58	127	195	278	375	477	580	715	858	999
Raleigh, N.C.	39	82	118	174	240	311	389	497	604	718
Memphis, Tenn.	39	85	127	183	258	337	422	539	659	793
Nashville, Tenn.	35	74	109	158	222	288	363	474	585	705
	May 17	**May 24**	**May 31**	**June 7**	**June 14**	**June 21**	**June 28**	**July 5**	**July 12**	**July 19**
West										
Phoenix, Ariz.	1274	1442	1614	1794	1976	2172	2378	2593	2817	3045
Fresno, Calif.	877	1008	1139	1276	1411	1569	1732	1895	2066	2243
Denver, Colo.	447	524	613	709	819	937	1068	1212	1360	1510
Pocatello, Idaho	305	378	456	540	630	734	846	965	1101	1244
Glasgow, Mont.	274	347	429	517	613	716	821	934	1073	1217
Pendleton, Ore.	418	504	592	698	795	913	1033	1159	1303	1460
North Central										
Des Moines, Iowa	428	525	629	753	897	1046	1201	1369	1535	1705
Minneapolis, Minn.	295	374	458	565	691	822	954	1103	1252	1409
North Platte, Nebr.	440	523	617	722	846	977	1115	1271	1424	1580
Omaha, Nebr.	502	607	722	854	1007	1165	1327	1504	1677	1855
Fargo, N. Dak.	250	325	404	503	612	724	839	968	1108	1253
Huron, S. Dak.	345	425	514	621	743	870	1001	1150	1305	1462
Green Bay, Wis.	234	302	374	466	573	683	797	923	1050	1185
Madison, Wis.	325	406	491	597	719	845	977	1119	1258	1406
South Central										
Fort Smith, Ark.	929	1081	1240	1406	1589	1774	1964	2159	2356	2553
Dodge City, Kans.	634	741	865	997	1154	1318	1487	1667	1847	2026
Topeka, Kans.	623	741	867	1006	1168	1333	1503	1685	1862	2042
Wichita, Kans.	691	816	952	1100	1271	1447	1628	1819	2010	2196
Shreveport, La.	1178	1348	1525	1706	1902	2100	2300	2504	2712	2916
Columbia, Mo.	622	741	864	1003	1168	1330	1501	1681	1857	2039
St. Louis, Mo.	647	770	899	1047	1220	1389	1566	1752	1932	2119
Oklahoma City, Okla.	849	988	1137	1294	1475	1658	1844	2039	2233	2428
Amarillo, Tex.	770	889	1019	1156	1311	1478	1649	1830	2010	2187

Table 9 (continued)
MEAN GROWING DEGREE DAYS (ADJUSTED 50°F BASE):[a] ACCUMULATED WEEKLY MARCH 1 TO INDICATED DATES

Station	May 17	May 24	May 31	June 7	June 14	June 21	June 28	July 5	July 12	July 19
Corpus Christi, Tex.	1600	1794	1998	2200	2412	2626	2840	3055	3270	3487
Fort Worth, Tex.	1140	1308	1486	1670	1871	2072	2278	2488	2702	2916
Midland, Tex.	1110	1263	1431	1603	1788	1978	2171	2364	2558	2754
Midwest										
Chicago, Ill.	391	481	579	701	843	984	1142	1310	1470	1639
Moline, Ill.	429	526	629	755	898	1045	1200	1365	1524	1691
Peoria, Ill.	454	554	660	788	938	1085	1245	1415	1576	1746
Indianapolis, Ind.	499	602	708	834	986	1132	1293	1462	1623	1792
Louisville, Ky.	677	801	932	1073	1239	1399	1572	1752	1926	2110
Flint, Mich.	302	375	455	555	676	795	924	1064	1197	1338
Grand Rapids, Mich.	321	398	482	588	717	841	976	1121	1259	1408
Columbus, Ohio	490	593	697	818	964	1104	1260	1424	1581	1749
Northeast										
Portland, Maine	181	237	304	386	470	563	668	787	903	1032
Albany, N.Y.	313	393	478	586	702	820	963	1110	1248	1402
Binghamton, N.Y.	253	320	387	476	580	683	805	938	1058	1197
Syracuse, N.Y.	300	376	457	559	677	794	930	1075	1212	1360
Harrisburg, Pa.	463	563	667	792	936	1075	1239	1410	1573	1748
Philadelphia, Pa.	475	576	681	811	955	1094	1262	1436	1604	1785
Pittsburgh, Pa.	442	533	625	736	869	994	1140	1295	1438	1595
Williamsport, Pa.	397	485	576	686	814	937	1083	1237	1380	1537
Richmond, Va.	733	858	984	1120	1279	1430	1608	1786	1962	2147
Southeast										
Macon, Ga.	1149	1314	1484	1656	1842	2029	2225	2421	2616	2814
Raleigh, N.C.	839	976	1113	1255	1420	1579	1757	1936	2113	2298
Memphis, Tenn.	935	1094	1257	1427	1618	1804	1999	2198	2398	2599
Nashville, Tenn.	831	974	1120	1276	1454	1626	1810	1998	2134	2376
	July 26	**Aug. 2**	**Aug. 9**	**Aug. 16**	**Aug. 23**	**Aug. 30**	**Sept. 6**	**Sept. 13**	**Sept. 20**	**Sept. 27**
West										
Phoenix, Ariz.	3274	3499	3721	3943	4161	4373	4582	4792	4986	5168
Fresno, Calif.	2422	2603	2773	2946	3115	3274	3436	3592	3734	3878
Denver, Colo.	1663	1818	1971	2118	2256	2394	2512	2630	2730	2815

Pocatello, Idaho	1389	1535	1677	1815	1943	2062	2172	2283	2372	2452
Glasgow, Mont.	1366	1511	1658	1794	1925	2037	2135	2229	2300	2366
Pendleton, Ore.	1611	1768	1919	2071	2214	2337	2465	2584	2685	2785
North Central										
Des Moines, Iowa	1880	2057	2228	2384	2535	2688	2822	2942	3044	3133
Minneapolis, Minn.	1572	1732	1889	2030	2167	2303	2418	2519	2600	2666
North Platte, Nebr.	1740	1906	2068	2216	2357	2503	2625	2740	2840	2923
Omaha, Nebr.	2039	2222	2401	2566	2724	2889	3031	3159	3268	3362
Fargo, N. Dak.	1405	1551	1696	1828	1957	2082	2186	2278	2353	2413
Huron, S. Dak.	1623	1783	1942	2088	2229	2373	2491	2598	2687	2761
Green Bay, Wis.	1324	1463	1597	1714	1828	1948	2057	2151	2228	2291
Madison, Wis.	1557	1708	1853	1986	2117	2249	2368	2471	2560	2633
South Central										
Fort Smith, Ark.	2754	2958	3157	3349	3540	3726	3905	4070	4229	4378
Dodge City, Kans.	2211	2398	2587	2761	2929	3103	3257	3399	3531	3641
Topeka, Kans.	2231	2420	2608	2779	2949	3120	3274	3414	3542	3655
Wichita, Kans.	2392	2590	2788	2971	3148	3328	3493	3645	3784	3906
Shreveport, La.	3124	3333	3542	3747	3951	4150	4345	4529	4709	4881
Columbia, Mo.	2226	2414	2598	2769	2937	3107	3263	3404	3535	3651
St. Louis, Mo.	2313	2507	2696	2872	3046	3220	3381	3527	3661	3784
Oklahoma City, Okla.	2629	2831	3032	3226	3415	3603	3779	3941	4095	4241
Amarillo, Tex.	2368	2552	2736	2911	3080	3250	3407	3553	3690	3813
Corpus Christi, Tex.	3704	3922	4138	4356	4573	4784	4998	5206	5411	5614
Fort Worth, Tex.	3132	3350	3567	3781	3990	4196	4397	4584	4766	4940
Midland, Tex.	2950	3149	3347	3541	3734	3920	4101	4270	4436	4593
Midwest										
Chicago, Ill.	1814	1989	2158	2316	2473	2631	2778	2905	3023	3123
Moline, Ill.	1863	2034	2198	2350	2499	2652	2787	2904	3014	3108
Peoria, Ill.	1922	2098	2266	2421	2574	2730	2871	2996	3110	3209
Indianapolis, Ind.	1967	2142	2308	2464	2618	2770	2915	3041	3161	3269
Louisville, Ky.	2301	2489	2673	2845	3018	2136	3353	3497	3630	3751
Flint, Mich.	1483	1628	1765	1893	2016	2145	2263	2364	2452	2527
Grand Rapids, Mich.	1560	1712	1858	1993	2125	2260	2387	2494	2590	2668
Columbus, Ohio	1917	2085	2247	2399	2553	2703	2849	2973	3087	3193
Northeast										
Portland, Maine	1164	1293	1411	1532	1646	1755	1856	1945	2018	2087
Albany, N.Y.	1557	1710	1851	1990	2123	2252	2378	2481	2568	2648

Table 9 (continued)
MEAN GROWING DEGREE DAYS (ADJUSTED 50°F BASE):[a] ACCUMULATED WEEKLY MARCH 1 TO INDICATED DATES

Station	July 26	Aug. 2	Aug. 9	Aug. 16	Aug. 23	Aug. 30	Sept. 6	Sept. 13	Sept. 20	Sept. 27
Binghamton, N.Y.	1334	1473	1601	1724	1844	1960	2077	2169	2245	2313
Syracuse, N.Y.	1511	1662	1805	1943	2072	2204	2333	2438	2525	2606
Harrisburg, Pa.	1925	2105	2270	2433	2593	2747	2902	3031	3148	3247
Philadelphia, Pa.	1969	2154	2323	2493	2660	2820	2985	3120	3244	3349
Pittsburgh, Pa.	1756	1917	2068	2214	2356	2497	2635	2751	2856	2946
Williamsport, Pa.	1697	1858	2006	2152	2294	2432	2569	2683	2784	2869
Richmond, Va.	2335	2521	2701	2881	3056	3224	3394	3539	3675	3794
Southeast										
Macon, Ga.	3014	3214	3415	3612	3808	3997	4185	4363	4532	4692
Raleigh, N.C.	2489	2678	2863	3046	3227	3399	3572	3725	3868	4995
Memphis, Tenn.	2804	3011	3214	3411	3604	3792	3974	4143	4303	4452
Nashville, Tenn.	2573	2770	2962	3145	3330	3508	3683	3842	3990	4125
	Oct. 4	Oct. 11	Oct. 18	Oct. 25	Nov. 1	Nov. 8	Nov. 15	Nov. 22	Nov. 29	Dec. 6
West										
Phoenix, Ariz.	5347	5511	5664	5802	5930	6040	6141	6218	6297	6372
Fresno, Calif.	4016	4137	4250	4352	4446	4526	4596	4643	4683	4716
Denver, Colo.	2901	2981	3052	3111	3161	3193	3228	3252	3273	3292
Pocatello, Idaho	2525	2608	2663	2707	2739	2757	2771	2776	2780	2781
Glasgow, Mont.	2427	2479	2529	2563	2591	2606	2615	2620	2623	2624
Pendleton, Ore.	2863	2926	2981	3026	3057	3074	3087	3097	3109	3115
North Central										
Des Moines, Iowa	3218	3289	3371	3427	3467	3489	3512	3527	3534	3539
Minneapolis, Minn.	2725	2777	2837	2875	2901	2912	2921	2927	2928	2929
North Platte, Nebr.	3004	3081	3156	3213	3261	3288	3316	3334	3349	3360
Omaha, Nebr.	3450	3527	3614	3674	3721	3748	3775	3792	3802	3810
Fargo, N. Dak.	2468	2515	2565	2595	2616	2627	2632	2635	2635	2635
Huron, S. Dak.	2830	2890	2955	2998	3030	3048	3063	3072	3076	3078
Green Bay, Wis.	2346	2389	2442	2474	2495	2506	2513	2520	2521	2522
Madison, Wis.	2703	2759	2823	2863	2890	2907	2920	2929	2931	2934
South Central										
Fort Smith, Ark.	4508	4628	4727	4843	4925	4981	5045	5097	5138	5169
Dodge City, Kans.	3749	3847	3941	4010	4069	4105	4144	4172	4195	4213

Topeka, Kans.	3756	3849	3947	4020	4077	4111	4150	4178	4197	4211
Wichita, Kans.	4019	4121	4224	4300	4359	4396	4437	4467	4488	4504
Shreveport, La.	5034	5173	5309	5422	5521	5595	5674	5751	5814	5864
Columbia, Mo.	3756	3847	3944	4014	4069	4103	4144	4173	4192	4207
St. Louis, Mo.	3892	3984	4081	4151	4205	4239	4278	4308	4326	4340
Oklahoma City, Okla.	4367	4482	4597	4685	4759	4808	4865	4908	4942	4971
Amarillo, Tex.	3926	4029	4127	4205	4273	4317	4372	4414	4451	4484
Corpus Christi, Tex	5798	5978	6154	6312	6452	6573	6691	6811	6917	7012
Fort Worth, Tex.	5094	5234	5376	5493	5592	5665	5746	5817	5877	5926
Midland, Tex.	4733	4864	4990	5093	5182	5246	5319	5381	5437	5483
Midwest										
Chicago, Ill.	3211	3283	3363	3415	3451	3473	3494	3512	3519	3524
Moline, Ill.	3193	3264	3345	3400	3439	3461	3484	3501	3508	3515
Peoria, Ill.	3295	3368	3448	3503	3542	3564	3589	3607	3615	3621
Indianapolis, Ind.	3360	3437	3519	3578	3618	3644	3672	3695	3706	3714
Louisville, Ky.	3859	3947	4041	4113	4166	4201	4241	4276	4296	4310
Flint, Mich.	2595	2652	2710	2752	2778	2796	2808	2820	2825	2828
Grand Rapids, Mich.	2740	2799	2862	2903	2931	2946	2961	2974	2879	2983
Columbus, Ohio	3282	3357	3436	3495	3533	3561	3585	3608	3620	3627
Northeast										
Portland, Maine	2145	2192	2236	2271	2293	2308	2315	2324	2328	2330
Albany, N.Y.	2717	2776	2833	2877	2905	2923	2934	2945	2950	2953
Binghamton, N.Y.	2372	2417	2468	2503	2524	2539	2548	2558	2562	2564
Syracuse, N.Y.	2677	2733	2790	2835	2863	2882	2896	2910	2917	2921
Harrisburg, Pa.	3336	3412	3485	3540	3578	3609	3631	3650	3659	3667
Philadelphia, Pa.	3444	3530	3606	3669	3712	3749	3775	3800	3815	3825
Pittsburgh, Pa.	3027	3092	3160	3211	3243	3267	3288	3310	3320	3327
Williamsport, Pa.	2946	3011	3076	3126	3158	3182	3197	3209	3215	3219
Richmond, Va.	3901	3997	4087	4160	4219	4272	4318	4363	4393	4415
Southeast										
Macon, Ga.	4839	4972	5095	5203	5292	5370	5441	5517	5574	5617
Raleigh, N.C.	4114	4217	4314	4395	4460	4518	4568	4623	4658	4683
Memphis, Tenn.	4584	4701	4816	4907	4981	5035	5093	5148	5187	5217
Nashville, Tenn.	4248	4354	4458	4540	4604	4650	4700	4746	4776	4800

[a] The mean growing degree day (GDD) values are based on the adjusted 50°F method and were calculated from records of daily maximum and minimum temperature for the period 1949 to 1968.

Table 10
AVERAGE RELATIVE HUMIDITY (%) MORNING (M) AND AFTERNOON (A) IN THE U.S. AND CERTAIN PACIFIC ISLANDS[2] (DATA THROUGH 1977)

Location	Years	Jan		Feb		Mar		Apr		May		Jun		Jul		Aug		Sep		Oct		Nov		Dec		Ann	
		M	A	M	A	M	A	M	A	M	A	M	A	M	A	M	A	M	A	M	A	M	A	M	A	M	A
Alabama																											
Birmingham	14	81	63	78	55	79	54	83	52	87	56	86	57	89	62	90	62	89	61	88	56	84	59	82	63	85	58
Huntsville	10	83	68	81	58	82	58	81	52	87	57	87	56	88	60	89	59	90	63	86	57	84	58	81	64	85	59
Mobile	15	82	63	80	54	83	56	87	54	86	53	87	54	89	60	89	61	88	61	85	52	85	56	84	62	85	57
Montgomery	14	82	62	79	54	82	54	86	52	88	55	88	55	90	61	91	62	90	59	88	54	87	55	83	61	86	57
Alaska																											
Anchorage	16	72	71	74	68	70	58	68	56	64	50	69	57	74	61	78	65	82	65	78	65	77	73	75	75	73	64
Annette	17	78	74	79	73	74	67	73	66	72	63	75	68	77	69	79	71	80	72	81	77	79	76	79	77	77	71
Barrow	33	67	67	65	65	65	67	75	74	88	85	92	89	91	87	93	89	92	89	85	85	77	77	69	69	80	78
Barter Island	30	69	69	69	68	68	68	75	75	87	85	90	88	89	86	92	89	91	89	84	84	75	75	69	70	80	79
Bethel	19	76	75	76	74	78	74	83	75	82	65	82	61	85	66	91	72	92	70	89	78	81	79	75	75	82	72
Bettles	8	66	67	64	61	62	60	65	63	62	50	61	47	67	51	80	60	80	62	77	74	72	72	68	69	69	61
Big Delta	8	71	71	75	69	73	61	66	50	55	43	64	52	73	58	77	58	79	59	80	72	77	75	76	75	72	62
Cold Bay	22	85	82	85	81	85	79	85	78	85	77	86	77	89	80	90	83	88	80	84	77	84	80	84	82	86	80
Fairbanks	14	68	68	64	62	63	53	60	46	53	37	59	41	68	50	77	53	75	51	77	66	71	70	67	67	67	55
Gulkana	7	68	64	76	63	75	56	66	49	59	42	59	42	70	48	74	49	77	53	84	69	83	79	75	74	72	58
Homer	10	77	72	81	73	79	70	76	67	75	67	75	67	78	69	83	71	85	70	83	72	81	75	80	79	79	71
Juneau	34	79	76	81	75	79	70	75	65	74	62	76	65	81	70	84	74	87	78	87	81	85	82	82	81	81	73
King Salmon	14	73	71	74	68	75	66	76	63	73	56	77	59	80	63	85	66	86	66	83	67	80	76	74	73	78	66
Kodiak	25	79	76	80	75	76	69	75	70	77	73	80	74	82	78	82	76	82	75	79	71	80	75	78	76	79	74
Kotzebue	34	71	71	71	71	72	72	78	76	83	79	85	80	84	77	86	78	86	74	82	77	76	76	72	72	79	75
Mc Grath	35	69	69	72	64	71	56	69	53	64	46	67	49	74	55	84	62	85	62	83	70	76	73	72	71	74	61
Nome	11	71	71	70	67	73	70	73	72	73	69	73	69	76	71	78	70	76	64	76	67	73	71	70	70	73	69
St. Paul Island	29	86	85	86	84	86	84	86	84	91	85	93	86	96	90	96	90	92	85	85	81	84	83	83	82	89	85
Talkeetha	10	71	62	75	68	72	61	69	54	64	52	69	55	75	60	81	61	83	60	83	70	74	69	75	71	74	62
Unalakleet	7	68	66	69	67	71	68	77	72	77	70	80	75	79	72	82	72	79	68	78	73	72	70	67	66	75	70
Yakutat	13	80	77	86	80	84	74	82	73	80	74	82	75	85	79	88	80	90	79	89	80	87	82	84	83	85	78
Arizona																											
Flagstaff	23	51	48	48	42	42	38	34	30	28	24	23	21	34	40	39	45	37	38	38	35	42	42	49	53	39	38

Phoenix	17	43	30	37	25	32	22	23	15	18	13	17	12	28	20	33	23	32	23	30	22	37	27	46	34	31	22
Tucson	37	39	32	34	26	28	22	21	16	16	12	17	13	33	29	39	33	32	27	30	25	32	28	39	34	30	25
Winslow	17	62	50	47	34	35	26	25	19	20	16	17	14	30	27	35	30	37	28	37	27	45	34	58	49	37	30
Yuma	13	38	27	34	23	29	20	25	17	23	15	23	14	32	22	35	24	35	24	31	22	34	26	40	32	32	22
Arkansas																											
Fort Smith	13	80	59	78	54	76	51	81	51	86	55	87	56	88	54	88	52	89	57	87	51	81	54	81	60	83	54
Little Rock	17	81	61	79	57	78	56	82	56	86	56	86	54	87	58	88	56	89	60	85	51	82	57	80	62	84	57
California																											
Bakersfield	14	75	60	65	49	53	39	45	32	37	24	34	23	32	20	35	23	41	27	46	33	64	50	75	62	50	37
Bishop	30	48	34	39	27	28	20	22	16	21	16	18	13	19	14	20	13	21	14	25	18	34	26	43	34	28	21
Blue Canyon	29	58	48	58	57	60	53	52	44	47	48	41	35	32	27	32	38	35	46	42	37	51	43	55	50	47	44
Fresno	14	85	67	77	55	63	44	52	34	43	25	40	24	39	23	43	26	46	29	53	35	74	56	66	71	58	41
Long Beach	17	57	50	60	52	59	53	57	50	61	55	64	56	61	53	60	53	61	54	58	52	58	54	60	53	60	53
Los Angeles (International Airport)	18	54	59	58	52	60	65	60	63	66	66	70	68	68	68	69	69	67	68	60	65	57	63	55	61	62	65
Los Angeles (City)	23	51	50	54	52	52	52	53	54	56	55	59	56	54	53	56	55	52	54	55	56	45	49	45	50	53	53
Mount Shasta	25	74	68	69	60	62	52	54	43	51	41	48	37	39	29	40	29	43	33	54	46	69	63	74	69	56	47
Oakland	14	77	71	76	69	71	66	68	63	70	65	73	67	76	67	76	68	74	65	72	64	75	68	76	71	74	67
Red Bluff	33	72	59	66	51	56	44	46	34	40	29	35	23	30	18	33	20	34	22	44	32	61	50	72	61	49	37
Sacramento	17	85	69	79	60	67	51	57	42	51	36	47	31	47	28	49	29	51	31	57	39	75	59	85	71	63	46
Sandberg	36	56	52	56	49	57	51	51	49	46	47	36	37	26	28	27	30	30	36	39	40	46	47	54	55	44	43
San Diego	17	53	54	57	58	58	58	58	58	64	64	69	67	69	66	67	66	65	65	59	62	56	61	54	58	61	61
San Francisco (International Airport)	18	78	66	75	65	69	62	65	60	64	60	64	59	66	60	67	62	65	59	67	59	74	64	78	69	69	62
San Francisco (City)	7	72	63	70	63	61	61	59	61	65	68	70	72	73	74	73	73	64	66	62	60	69	63	71	63	67	66
Santa Maria	14	62	59	62	61	62	64	59	61	62	61	63	61	63	61	66	63	64	63	56	60	58	63	61	61	61	61
Stockton	18	87	69	80	61	67	49	53	39	46	33	42	29	41	26	44	27	47	30	56	38	76	59	87	73	61	44
Colorado																											
Alamosa	32	60	59	54	50	42	37	33	30	29	28	29	26	39	37	42	38	38	34	38	34	48	48	57	58	42	40
Colorado Springs	17	41	46	38	38	39	37	35	33	35	34	35	34	38	42	39	41	40	38	35	36	40	46	40	46	38	39
Denver	17	44	47	43	42	41	40	38	35	38	36	37	36	36	36	36	35	39	36	35	35	44	49	44	50	40	40
Grand Junction	14	66	62	50	43	38	31	34	27	28	22	25	20	28	21	31	22	33	25	38	33	50	46	62	59	40	34
Pueblo	13	48	45	41	34	39	31	34	29	34	29	33	27	38	34	39	34	40	33	37	33	45	43	48	45	40	35
Connecticut																											
Bridgeport	12	72	61	72	60	72	59	68	53	75	59	79	63	77	60	78	60	82	63	79	61	78	63	75	65	76	60
Hartford	18	71	57	73	56	72	53	69	44	73	47	78	52	79	51	84	53	87	56	84	52	78	57	78	62	77	53
Delaware																											

Table 10 (continued)
AVERAGE RELATIVE HUMIDITY (%) MORNING (M) AND AFTERNOON (A) IN THE U.S. AND CERTAIN PACIFIC ISLANDS² (DATA THROUGH 1977)

		Jan		Feb		Mar		Apr		May		Jun		Jul		Aug		Sep		Oct		Nov		Dec		Ann	
Location	Years	M	A	M	A	M	A	M	A	M	A	M	A	M	A	M	A	M	A	M	A	M	A	M	A	M	A
Wilmington	30	75	60	74	58	74	53	73	50	76	53	79	54	80	54	83	56	85	55	84	54	80	56	77	60	78	55
District of Columbia																											
Washington (Dulles Airport)	8	79	59	77	52	78	52	76	46	84	56	87	58	87	55	90	56	91	59	89	57	84	57	81	61	83	58
Washington (National Airport)	17	68	54	68	51	68	49	68	47	72	51	75	52	75	52	79	54	80	55	78	51	73	52	71	57	73	52
Florida																											
Apalachicola	27	86	68	86	66	85	66	85	65	83	65	84	68	85	71	87	79	88	71	86	64	85	64	87	69	86	68
Daytona Beach	33	88	60	87	57	86	56	85	54	85	57	87	63	88	65	90	67	89	67	87	63	87	60	87	61	87	61
Fort Myers	17	89	58	88	55	89	52	88	48	87	50	88	60	88	60	89	61	90	62	88	57	89	55	89	56	89	56
Jacksonville	41	87	57	85	52	85	49	85	48	84	50	86	56	87	58	90	60	91	62	90	58	89	55	88	58	87	55
Key West	29	82	69	80	67	78	66	76	63	76	65	78	68	76	66	78	67	80	70	82	69	82	69	82	69	79	67
Miami	13	84	60	83	57	82	57	79	54	83	61	87	67	86	65	87	66	89	68	87	65	84	61	83	59	84	62
Orlando	14	88	57	85	51	86	50	87	45	87	49	88	56	89	59	92	60	91	60	86	56	87	54	86	56	88	54
Pensacola	12	81	64	79	56	82	60	84	57	85	58	85	59	88	63	89	66	86	63	83	55	82	58	82	65	84	60
Tallahassee	16	87	60	86	54	89	53	91	48	90	50	91	54	93	61	94	62	93	59	91	52	90	54	88	58	90	55
Tampa	14	86	59	83	54	85	54	86	49	84	51	85	59	86	62	89	63	89	61	87	56	86	57	86	57	86	57
West Palm Beach	13	82	59	81	56	80	54	76	52	77	57	83	66	84	64	84	64	86	66	82	62	81	58	80	57	81	60
Georgia																											
Athens	22	81	59	79	55	81	53	83	51	86	55	88	56	91	60	92	60	92	61	89	55	85	53	82	58	86	56
Atlanta	17	79	60	75	54	78	52	80	51	83	55	86	58	90	63	91	62	90	61	84	53	81	54	80	59	83	57
Augusta	13	83	55	80	47	83	47	84	45	86	50	87	53	89	57	91	58	92	56	89	49	87	48	85	54	86	52
Columbus	32	85	60	83	54	84	52	84	48	84	50	85	53	89	59	90	57	90	57	90	52	88	53	85	58	86	55
Macon	13	83	59	82	53	84	57	86	47	87	51	87	53	90	58	92	59	93	59	88	51	86	51	83	56	87	54
Rome	16	81	58	81	54	82	50	83	47	86	50	87	53	88	54	90	53	90	51	90	50	85	50	82	56	85	52
Savannah	13	84	55	81	49	84	48	85	47	87	52	88	56	90	60	91	63	91	61	88	53	86	50	84	54	86	54
Hawaii																											
Hilo	28	79	67	80	67	81	68	82	69	80	68	78	65	81	67	81	68	79	67	80	68	81	70	81	69	80	68
Honolulu	8	81	63	76	59	73	59	70	58	66	54	65	53	65	51	66	53	65	52	67	54	73	58	76	60	70	56
Kahului	13	85	64	81	61	78	60	76	58	72	56	69	53	71	54	71	55	71	55	73	56	77	59	80	60	75	58
Lihue	28	83	68	81	67	79	67	77	68	76	67	74	65	75	65	76	66	77	65	79	68	80	69	81	68	78	67

Idaho																											
Boise	38	74	71	68	61	55	44	47	36	45	34	42	31	33	22	34	23	39	30	49	41	65	61	74	72	52	44
Lewiston	31	76	72	71	63	62	50	55	41	52	39	48	36	37	23	39	25	49	31	66	51	77	71	77	75	59	48
Pocatello	14	71	67	68	59	58	49	48	38	41	32	42	32	33	23	33	23	38	28	47	36	63	56	73	68	51	42
Illinois																											
Cairo	22	80	67	79	64	78	59	77	54	82	57	83	57	84	59	87	59	87	57	85	53	80	58	79	66	82	59
Chicago (O'Hare Airport)	19	76	68	76	65	78	61	76	55	76	54	78	55	81	57	84	57	84	58	81	55	81	64	80	71	79	60
Chicago (Midway Airport)	14	72	65	72	62	74	59	73	55	73	53	74	54	77	54	81	56	81	56	78	55	78	64	79	71	76	59
Moline	17	72	65	74	62	77	60	79	55	80	53	81	53	84	56	88	57	87	58	81	54	79	62	78	70	80	59
Peoria	18	78	68	79	66	81	63	78	56	81	57	81	56	86	58	87	60	87	59	84	58	83	66	83	72	82	61
Rockford	14	77	69	78	65	80	61	80	57	80	54	81	55	85	55	90	57	89	58	84	56	83	65	83	73	82	60
Springfield	18	78	69	79	67	81	64	78	55	79	53	81	53	83	55	88	59	88	57	83	54	83	64	83	73	82	60
Indiana																											
Evansville	16	77	66	77	64	79	61	77	55	81	55	81	55	84	57	88	58	88	59	83	53	78	61	79	69	81	59
Fort Wayne	16	79	71	78	67	79	65	78	57	79	55	80	55	83	54	87	57	88	58	84	57	82	67	82	74	82	61
Indianapolis	18	81	69	80	66	80	63	78	55	82	56	83	58	87	60	90	61	90	60	88	57	85	67	83	73	84	62
South Bend	14	80	73	80	68	80	64	78	57	77	54	81	56	84	57	88	59	89	61	85	61	83	70	84	77	82	63
Iowa																											
Burlington	36	79	69	79	67	80	63	78	56	80	57	82	59	85	58	88	60	87	57	83	54	81	63	81	70	82	61
Des Moines	16	76	69	79	67	79	63	78	57	78	56	80	56	81	56	84	58	84	60	78	55	79	63	80	70	80	61
Dubuque	12	73	66	75	64	77	61	76	57	77	56	82	60	83	59	86	60	86	60	81	58	80	65	79	72	80	61
Sioux City	18	77	69	78	66	80	63	77	52	78	54	83	58	85	58	87	60	86	60	81	53	81	62	81	70	81	60
Waterloo	18	75	68	79	67	82	66	82	56	82	55	84	55	86	58	90	58	90	61	84	57	83	65	80	71	83	62
Kansas																											
Concordia	15	82	67	80	61	80	56	82	54	84	56	85	55	81	51	83	53	87	58	81	52	83	61	81	66	82	57
Dodge City	14	76	57	74	52	73	48	73	46	77	48	77	45	75	44	78	46	81	51	73	45	77	53	76	57	76	49
Goodland	11	56	60	48	44	50	46	48	43	51	46	45	40	45	40	48	41	49	42	47	44	53	58	55	60	50	47
Topeka	13	77	65	77	60	77	55	80	55	84	57	86	61	85	58	86	57	87	59	83	55	82	60	80	65	82	99
Wichita	24	79	63	78	59	76	53	77	52	83	55	82	53	78	49	79	50	82	56	81	53	79	56	79	61	79	55
Kentucky																											
Covington (Cincinnati Airport)	15	78	67	76	62	77	60	76	53	79	53	82	55	85	57	88	57	88	59	83	55	79	63	79	69	81	59
Lexington	14	78	67	77	62	76	58	74	54	79	55	82	57	85	59	86	58	87	58	83	56	79	61	79	68	81	59
Louisville	17	77	64	76	62	76	58	75	53	83	55	84	58	85	58	88	58	89	60	86	55	79	61	77	65	81	59
Louisiana																											
Baton Rouge	18	85	65	84	58	85	58	89	50	90	56	91	57	91	62	92	63	91	61	88	54	88	58	86	63	88	59
Lake Charles	13	88	70	86	61	89	63	90	62	92	62	92	60	93	64	93	65	93	63	90	56	88	59	89	67	90	63
New Orleans	29	86	67	85	63	85	61	88	60	89	60	89	62	91	66	91	66	90	66	88	59	86	61	86	67	88	63

Table 10 (continued)
AVERAGE RELATIVE HUMIDITY (%) MORNING (M) AND AFTERNOON (A) IN THE U.S. AND CERTAIN PACIFIC ISLANDS[2] (DATA THROUGH 1977)

Location	Years	Jan M	Jan A	Feb M	Feb A	Mar M	Mar A	Apr M	Apr A	May M	May A	Jun M	Jun A	Jul M	Jul A	Aug M	Aug A	Sep M	Sep A	Oct M	Oct A	Nov M	Nov A	Dec M	Dec A	Ann M	Ann A
Shreveport	25	83	63	82	58	82	57	87	57	89	58	90	57	90	57	91	56	90	57	88	53	86	57	85	62	87	58
Maine																											
Caribou	33	73	67	74	64	75	61	76	57	73	53	79	57	82	58	85	59	87	61	86	62	85	72	79	71	80	62
Portland	37	77	62	77	60	75	59	73	55	75	58	79	61	80	60	84	59	86	61	85	59	84	63	80	62	80	60
Maryland																											
Baltimore	24	72	58	71	55	71	51	72	49	77	52	80	53	81	53	84	55	85	56	82	54	77	54	74	59	77	54
Masschusetts																											
Blue Hill Observatory	39	76	62	76	60	75	58	72	52	75	55	79	59	81	58	83	56	84	60	82	57	81	62	77	63	78	59
Boston	13	66	57	67	57	68	57	66	53	70	57	74	60	73	56	75	57	79	61	76	57	74	61	72	61	72	58
Worcester	22	74	59	73	58	70	55	66	49	67	49	73	56	74	57	77	57	81	61	78	57	78	61	76	64	74	57
Michigan																											
Alpena	18	78	69	76	64	79	60	78	53	76	50	80	53	83	53	88	57	91	61	85	60	83	69	82	73	82	60
Detroit (City Airport)	44	78	68	78	65	77	60	73	52	71	51	73	53	74	51	80	53	82	55	81	55	79	64	79	69	77	58
Detroit (Metro Airport)	19	79	70	78	64	79	62	79	55	79	54	81	55	82	52	87	55	88	57	84	56	83	66	81	72	82	60
Flint	14	77	69	75	64	76	61	75	56	76	55	80	57	82	55	86	56	87	59	82	59	81	68	80	72	80	61
Grand Rapids	14	81	72	79	67	80	64	79	58	78	53	81	56	82	54	86	56	88	60	85	61	84	71	83	75	82	62
Houghton Lake	13	83	73	81	70	83	65	81	55	78	50	83	56	85	55	90	59	91	62	88	64	88	74	86	78	85	63
Lansing	14	82	74	81	67	81	65	78	57	77	54	80	57	84	55	88	56	90	60	86	62	86	71	84	77	83	63
Marquette	27	75	68	75	67	76	66	73	61	73	62	76	64	76	63	80	64	81	65	77	62	79	70	76	70	76	65
Muskegon	17	80	75	80	70	79	64	77	57	75	54	80	57	83	58	87	60	87	63	83	64	80	70	81	76	81	64
Sault Ste. Marie	36	82	76	82	73	83	68	80	61	79	56	85	62	88	62	92	63	92	67	89	67	87	76	84	78	85	67
Minnesota																											
Duluth	16	75	69	75	64	78	63	77	57	76	54	81	60	84	58	87	62	87	63	81	62	81	70	78	74	80	63
International Falls	35	73	68	72	63	76	61	77	54	78	50	83	56	87	56	90	60	90	63	85	62	84	74	78	73	81	62
Minneapolis—St. Paul	18	73	67	74	65	77	64	76	54	76	53	79	54	81	54	84	55	86	60	82	58	81	66	78	70	79	60
Rochester	17	77	73	79	70	82	70	81	61	79	57	82	58	84	59	87	60	87	62	82	59	84	70	83	76	82	65
Saint Cloud	28	76	68	77	65	80	63	80	53	80	50	84	55	87	54	90	57	90	58	85	57	82	67	79	71	82	60
Mississippi																											
Jackson	14	87	65	86	58	87	57	90	56	91	56	90	55	93	59	94	60	94	61	93	54	90	58	88	65	90	59

Meridian	13	86	61	85	54	86	53	89	51	89	54	89	53	91	58	91	57	90	58	89	51	87	53	86	60	88	55
Missouri																											
Columbia	8	78	67	78	64	77	58	76	54	83	58	83	58	83	52	85	55	86	61	83	58	81	64	81	69	81	60
Kansas City	5	74	64	75	62	77	61	75	56	83	58	83	57	80	52	86	58	87	64	81	59	79	62	77	65	80	60
Saint Joseph	12	74	63	73	58	76	54	76	53	77	51	79	54	82	56	86	59	88	60	83	52	80	59	77	65	79	57
St. Louis	17	82	65	80	61	81	58	79	54	83	56	84	57	86	57	89	57	91	61	85	55	84	62	85	69	84	59
Springfield	17	77	60	78	58	78	55	78	55	84	57	85	59	86	56	87	54	88	61	82	54	80	58	79	62	82	57
Montana																											
Billings	18	60	57	57	51	52	45	50	42	48	42	47	41	39	30	39	29	49	39	49	41	58	53	61	57	51	44
Glasgow	13	68	69	71	71	64	57	53	46	46	40	48	42	44	35	41	32	48	39	52	45	66	62	73	73	56	51
Great Falls	16	62	61	58	54	53	46	49	42	45	40	45	40	37	28	38	30	46	37	46	41	55	54	61	61	50	44
Havre	17	68	68	69	64	59	51	50	42	44	36	45	37	36	27	36	27	45	34	48	42	61	57	71	70	53	46
Helena	14	64	62	62	54	54	46	48	39	44	37	46	39	39	28	41	30	48	36	52	42	62	57	67	63	52	44
Kalispell	18	79	73	77	68	64	54	51	43	50	41	51	45	43	33	45	34	54	42	66	53	78	70	81	77	62	53
Miles City	17	70	70	70	68	61	55	53	47	49	44	47	41	38	31	38	29	48	40	51	47	65	63	72	72	55	50
Missoula	17	81	77	78	67	67	52	54	42	52	42	53	43	42	30	44	30	57	40	69	52	80	72	83	81	63	52
Nebraska																											
Grand Island	16	76	64	76	60	78	56	78	51	80	50	81	50	81	51	84	52	86	56	80	48	80	56	78	63	80	55
Lincoln	5	78	67	81	63	82	62	80	56	83	55	79	49	77	46	83	53	86	57	81	52	81	60	81	66	81	57
Norfolk	32	76	64	79	54	80	61	79	50	80	50	82	52	83	52	85	54	84	52	79	49	79	57	77	64	80	56
North Platte	13	78	63	78	57	78	52	80	48	82	49	85	52	84	52	84	50	83	50	80	46	80	54	80	61	81	53
Omaha	14	76	65	77	61	76	57	76	53	79	54	81	55	83	56	86	59	87	61	82	56	80	62	80	67	80	59
Scottsbluff	13	58	58	51	47	48	43	46	41	44	39	44	40	42	37	43	37	43	37	43	40	51	51	55	57	47	44
Valentine	10	80	68	83	68	84	63	81	54	83	54	84	51	83	50	83	46	82	48	77	47	76	53	78	62	81	55
Nevada																											
Elko	12	69	57	63	48	51	36	42	33	34	28	32	25	25	19	28	21	31	21	40	27	59	45	68	58	45	35
Ely	25	60	54	58	50	48	41	40	34	34	30	28	24	24	22	26	22	29	23	38	30	51	45	58	54	41	36
Las Vegas	17	38	28	34	25	27	19	21	15	18	13	15	11	19	14	23	17	22	16	25	19	33	27	40	32	26	20
Reno	14	67	50	57	39	46	32	38	28	34	25	33	24	28	19	32	21	34	21	40	26	56	40	65	51	44	31
Winnemucca	28	66	57	59	46	48	37	39	29	34	26	31	23	22	15	24	16	28	20	38	28	56	45	66	58	43	33
New Hampshire																											
Concord	12	74	59	75	59	78	57	76	46	79	49	87	56	88	53	90	54	92	58	88	54	85	62	81	66	83	56
Mt. Washington	27	83	83	82	84	84	85	85	84	84	82	89	85	91	87	90	87	85	86	83	82	85	85	84	85	85	85
New Jersey																											
Atlantic City	13	75	57	77	56	77	55	75	50	78	56	83	58	84	58	86	57	87	58	86	56	81	57	76	59	81	56
Newark	12	73	58	72	55	69	51	63	45	69	51	73	54	71	51	74	53	78	55	78	54	77	57	75	61	73	54
New Mexico																											

Table 10 (continued)
AVERAGE RELATIVE HUMIDITY (%) MORNING (M) AND AFTERNOON (A) IN THE U.S. AND CERTAIN PACIFIC ISLANDS[2] (DATA THROUGH 1977)

Location	Years	Jan M	Jan A	Feb M	Feb A	Mar M	Mar A	Apr M	Apr A	May M	May A	Jun M	Jun A	Jul M	Jul A	Aug M	Aug A	Sep M	Sep A	Oct M	Oct A	Nov M	Nov A	Dec M	Dec A	Ann M	Ann A
Albuquerque	17	49	38	43	31	32	23	25	18	23	16	23	16	36	28	39	29	40	31	37	28	42	34	50	42	36	28
Clayton	31	42	47	42	42	39	37	35	32	38	36	38	35	42	42	42	41	42	40	36	39	39	47	41	52	40	41
Roswell	5	48	38	42	32	32	23	32	23	30	22	32	22	45	36	44	34	52	42	44	35	42	35	38	31	40	31
New York																											
Albany	12	79	63	76	58	73	54	69	46	75	52	79	58	80	55	84	56	88	59	85	56	82	63	81	67	79	57
Binghamton	26	80	72	80	69	80	65	77	57	79	56	83	59	85	58	89	60	90	64	85	62	83	70	83	75	83	64
Buffalo	17	78	73	79	70	80	67	76	58	76	56	78	57	79	55	83	58	83	60	82	61	82	71	82	75	80	63
New York (Central Park)	56	68	60	68	58	67	55	67	51	71	53	74	55	75	55	78	57	79	57	76	55	73	59	70	61	72	56
New York (JFK Airport)	16	71	59	70	58	70	57	68	54	72	57	75	61	74	58	77	59	80	59	78	55	75	58	73	62	74	58
New York (La Guardia Airport)	15	65	56	63	54	66	54	64	48	69	51	73	55	71	52	74	54	74	56	72	54	69	57	68	60	69	54
Rochester	14	75	68	76	67	77	63	77	54	77	54	79	56	81	53	87	56	88	61	84	61	82	69	81	75	80	61
Syracuse	14	76	68	77	66	77	61	75	51	75	55	79	57	81	56	86	59	87	62	84	61	82	69	81	72	80	81
North Carolina																											
Asheville	13	86	60	82	55	85	53	86	50	92	57	95	60	96	64	98	64	98	65	95	59	88	56	96	60	91	59
Cape Hatteras	20	80	68	80	64	80	62	78	60	82	56	83	69	85	70	86	70	84	67	82	65	81	63	81	67	82	66
Charlotte	17	78	57	75	51	79	50	79	47	84	53	86	57	88	58	89	59	90	57	88	53	83	52	80	57	83	54
Greensboro	14	81	58	77	51	80	51	79	48	85	56	87	57	90	60	92	61	91	59	89	55	82	51	81	58	84	55
Raleigh	13	78	55	74	48	79	48	80	44	87	55	88	57	90	59	93	61	93	59	90	54	83	49	80	56	84	54
Wilmington	14	81	58	77	51	81	52	79	47	84	57	86	61	87	64	90	66	91	63	88	57	85	52	82	57	84	57
North Dakota																											
Bismarck	18	73	67	76	68	78	62	79	53	79	49	84	53	82	46	81	43	82	49	78	50	79	62	76	69	79	56
Fargo	18	73	71	76	72	82	72	81	62	79	52	83	57	85	54	86	54	86	59	82	60	83	70	78	75	81	63
Williston	16	78	72	82	72	83	65	82	55	79	47	83	50	81	46	79	45	81	51	79	54	81	68	80	73	81	58
Ohio																											
Akron	14	78	69	76	64	76	60	74	54	76	54	80	56	84	56	88	59	88	61	81	57	79	66	78	70	80	60
Cleveland	17	76	69	77	68	77	65	75	57	77	58	80	59	82	58	85	60	84	61	80	60	79	67	77	71	79	63
Columbus	18	76	67	75	64	74	58	75	52	79	54	81	55	84	56	87	58	88	58	82	55	81	64	78	69	80	59
Dayton	14	75	67	76	64	78	62	76	55	76	53	79	54	81	54	85	56	86	58	81	56	80	66	79	71	79	60
Mansfield	12	81	72	80	68	81	63	76	56	78	56	81	58	83	56	86	59	87	61	82	58	83	69	84	75	82	63

Toledo	22	78	68	78	65	80	61	79	54	79	52	82	54	85	54	90	57	90	57	84	55	83	66	83	73	83	60
Youngstown	30	81	73	81	69	81	64	78	56	78	53	82	56	85	55	88	56	88	58	85	58	82	68	82	74	83	62
Oklahoma																											
Oklahoma City	12	79	60	77	54	75	52	78	52	83	57	84	56	81	50	82	51	85	57	80	52	78	54	77	57	80	54
Tulsa	17	78	59	76	54	75	52	78	52	86	57	86	59	83	54	85	54	89	61	84	53	80	57	80	61	82	56
Oregon																											
Astoria	24	84	79	82	74	78	71	74	69	73	70	76	72	75	69	76	70	75	70	81	74	83	78	85	81	79	73
Burns	27	76	68	72	60	60	47	47	35	45	33	43	30	36	21	37	22	41	25	52	37	69	57	77	70	55	42
Eugene	33	87	80	85	72	77	65	69	56	66	53	62	49	56	37	60	40	64	42	80	64	87	79	88	84	73	60
Medford	16	88	73	83	57	71	49	62	43	56	38	46	33	44	26	48	27	52	28	72	45	87	69	88	76	66	47
Pendleton	38	76	74	70	65	57	48	50	41	46	36	42	32	33	22	36	26	42	32	56	48	72	70	77	77	55	48
Portland	37	82	76	79	67	72	60	68	54	66	53	64	49	61	45	65	46	67	49	79	64	82	4	84	79	72	60
Salem	15	83	75	81	67	75	60	69	56	65	52	62	49	57	40	59	41	64	46	78	61	85	77	86	81	72	59
Sexton Summit	28	79	75	77	69	78	69	73	62	69	53	67	50	60	40	60	41	58	44	66	58	76	75	79	78	70	59
Pacific																											
Guam	10	78	77	76	74	76	75	74	73	73	73	75	74	80	78	81	80	80	80	80	80	80	81	78	78	78	77
Johnston Island	16	76	67	76	68	78	70	78	70	77	70	76	69	77	69	78	71	79	71	79	72	78	72	77	71	77	70
Koror Island	26	81	76	81	75	80	74	78	75	80	77	81	78	81	79	81	78	80	77	80	77	80	77	81	77	80	77
Kwajalein Island	17	80	71	80	70	81	71	83	75	85	78	85	78	85	78	84	77	84	77	83	77	83	76	81	74	83	75
Majuro, Marshall Islands	22	81	75	80	74	82	75	84	78	85	78	85	78	84	77	84	77	83	76	83	77	83	77	82	76	83	76
Pago Pago, American Samoa	11	88	75	89	75	90	75	90	76	88	76	86	76	84	75	84	73	84	72	84	75	85	76	86	75	86	75
Ponape Island	26	76	77	76	77	77	79	79	81	80	82	79	80	79	80	79	80	78	80	78	80	79	81	78	79	78	80
Truk, Caroline Islands	26	76	76	76	75	76	76	78	78	80	80	79	79	79	78	79	77	79	78	79	78	79	79	78	79	78	78
Wake Island	30	77	65	78	66	80	67	81	68	82	69	82	69	82	70	83	71	83	72	82	72	80	70	78	68	81	69
Yap Island	29	78	76	77	74	76	74	76	74	77	76	79	77	80	78	81	78	80	79	80	78	79	79	79	78	79	77
Pennsylvania																											
Allentown	27	76	62	76	59	76	55	75	50	78	52	80	54	82	52	86	55	89	57	87	56	83	60	80	64	81	56
Erie	12	76	73	77	71	77	67	75	61	78	64	82	66	83	65	87	67	86	68	79	66	78	72	78	75	80	68
Harrisburg	38	71	58	70	55	71	52	70	49	74	51	76	53	78	52	82	54	84	55	82	55	76	57	73	58	76	54
Philadelphia	18	73	59	71	56	71	53	69	48	75	52	78	55	79	54	81	55	83	56	82	53	77	55	74	60	76	55
Pittsburgh (International Airport)	18	76	66	73	62	74	58	72	50	75	51	79	52	83	53	86	55	86	57	81	54	79	62	76	67	78	57
Avoca	22	76	65	75	63	74	59	72	51	76	52	82	56	84	55	87	58	88	61	84	58	79	64	78	69	80	59
Williamsport	33	77	62	76	59	77	55	75	49	79	51	84	54	87	55	90	57	92	59	88	56	82	62	78	64	82	57
Puerto Rico																											
San Juan	22	80	64	79	62	77	60	74	61	77	64	78	65	78	66	79	65	79	66	80	68	81	66	80	65	78	64

Table 10 (continued)
AVERAGE RELATIVE HUMIDITY (%) MORNING (M) AND AFTERNOON (A) IN THE U.S. AND CERTAIN PACIFIC ISLANDS[2] (DATA THROUGH 1977)

Location	Years	Jan		Feb		Mar		Apr		May		Jun		Jul		Aug		Sep		Oct		Nov		Dec		Ann	
		M	A	M	A	M	A	M	A	M	A	M	A	M	A	M	A	M	A	M	A	M	A	M	A	M	A
Rhode Island																											
Block Island	14	73	65	73	65	75	65	79	65	80	67	84	69	87	72	86	71	84	70	80	66	76	65	72	65	79	67
Providence	14	71	56	70	55	70	54	68	47	71	51	76	57	78	57	79	55	82	57	80	54	77	58	76	60	75	55
South Carolina																											
Charleston	35	84	56	82	51	83	50	84	49	84	54	86	59	88	63	90	63	91	63	89	55	85	51	84	55	86	56
Columbia	11	83	56	81	47	83	48	83	43	88	50	88	53	90	56	93	59	94	57	91	52	88	49	85	55	87	52
Greenville — Spartanburg	15	78	55	74	50	76	49	76	47	82	52	84	55	86	58	88	58	89	59	85	53	80	51	80	57	82	54
South Dakota																											
Aberdeen	13	73	70	76	69	81	65	82	57	80	52	84	56	83	50	84	47	83	52	81	55	82	68	80	73	81	59
Huron	18	73	67	78	68	82	65	83	56	84	54	86	56	85	50	87	50	87	54	81	53	81	62	78	69	82	59
Rapid City	27	59	64	60	62	56	55	49	46	49	46	51	49	45	40	41	35	41	38	40	44	51	58	59	55	50	50
Sioux Falls	14	75	67	78	66	81	62	80	55	80	52	81	55	81	51	83	51	86	57	80	55	82	64	79	70	81	59
Tennessee																											
Bristol—Johnson City	16	80	62	78	57	78	53	81	50	89	57	91	58	92	61	92	61	92	58	88	54	83	57	81	63	85	58
Chattanooga	47	82	63	81	57	81	53	81	49	85	52	86	54	89	57	91	57	90	55	89	53	84	55	84	62	85	56
Knoxville	17	80	64	77	59	78	54	79	50	84	55	88	59	90	61	92	61	91	60	88	56	83	59	81	64	84	59
Memphis	38	79	64	78	59	77	56	78	53	82	55	83	56	85	57	86	56	86	56	83	51	80	55	79	62	81	57
Nashville	12	80	65	78	58	78	53	81	52	87	56	88	56	91	59	92	60	91	61	87	56	80	59	80	64	84	58
Texas																											
Abilene	14	71	53	71	51	70	48	74	49	79	53	77	50	71	46	74	49	80	57	77	53	75	54	71	51	74	51
Amarillo	16	69	48	69	47	66	40	67	37	72	40	75	43	73	43	76	46	80	50	71	42	72	46	68	47	71	44
Austin	16	79	60	78	56	79	55	84	59	88	60	88	55	87	50	87	51	87	57	85	55	82	57	80	59	84	56
Brownsville	11	87	68	86	59	87	61	87	60	89	61	90	61	90	57	90	58	89	61	88	60	85	60	85	63	88	61
Corpus Christi	13	89	71	88	64	87	64	90	65	92	67	93	64	93	58	93	60	91	63	90	61	89	62	87	65	90	64
Dallas-Fort Worth	14	80	60	78	56	79	57	85	59	88	61	86	56	81	50	83	52	88	60	84	56	81	56	79	58	83	57
Del Rio	14	75	55	73	51	72	48	77	52	83	58	82	55	79	53	80	54	84	58	83	58	80	56	76	53	79	54
El Paso	17	42	33	34	25	28	20	22	15	22	15	24	17	39	30	40	31	44	34	36	28	37	31	43	35	34	26
Galveston	90	85	77	84	74	85	74	86	75	84	73	81	70	81	70	81	69	81	68	80	65	84	72	86	76	63	72
Houston	8	88	64	87	55	88	61	91	60	94	61	92	59	93	58	94	61	94	65	93	59	90	60	88	62	91	60

Lubbock	30	72	48	71	48	67	40	69	41	75	43	76	43	76	48	78	48	81	51	78	48	74	45	70	46	74	46
Midland — Odessa	14	69	45	71	42	65	35	68	35	76	38	75	39	71	41	74	43	82	52	80	45	75	44	70	42	73	42
Port Arthur	17	89	70	88	61	89	63	91	64	92	63	93	62	94	64	94	65	93	65	91	58	89	61	90	68	91	64
San Angelo	17	73	51	73	47	69	42	74	44	80	48	79	47	76	43	77	45	84	56	83	53	80	52	76	50	77	48
San Antonio	35	80	60	80	57	79	53	83	56	88	58	88	56	87	51	86	51	86	55	84	54	81	55	80	57	84	55
Victoria	16	86	65	85	58	86	58	88	60	90	60	91	59	92	56	91	56	91	61	89	56	87	57	87	62	89	59
Waco	14	82	63	81	58	79	58	83	62	86	62	83	54	79	48	81	49	84	57	83	56	82	57	81	60	82	57
Wichita Falls	17	81	56	79	52	77	49	81	50	85	52	85	50	78	44	80	46	88	56	84	52	84	54	82	55	82	51
Utah																											
Milford	24	66	60	62	54	45	37	34	28	28	23	23	17	24	20	27	21	28	20	35	28	51	46	65	60	41	35
Salt Lake City	18	69	67	63	57	51	44	44	39	37	31	31	26	26	20	29	22	34	27	43	40	57	58	70	71	46	42
Vermont																											
Burlington	12	69	63	72	63	72	59	74	52	74	51	80	58	81	55	85	59	88	65	83	63	81	70	77	72	78	61
Virginia																											
Lynchburg	14	72	53	70	48	72	48	73	45	81	52	83	55	86	57	89	58	88	57	84	53	77	51	75	55	79	53
Norfolk	29	75	59	74	56	73	54	73	50	78	56	79	57	82	60	85	62	84	62	83	61	78	55	75	59	78	58
Richmond	43	81	57	79	52	78	48	75	45	79	50	82	53	85	56	88	57	89	56	89	53	84	50	81	55	83	53
Roanoke	13	68	52	65	47	68	46	69	45	77	52	82	54	83	55	87	57	88	56	82	53	73	51	72	53	76	52
Washington																											
Olympia	18	88	81	86	70	80	62	71	57	67	54	67	54	64	49	68	51	74	55	86	69	89	80	90	85	77	64
Quillayute	11	89	85	86	76	83	74	75	68	72	66	74	67	73	64	77	68	76	66	84	75	88	83	90	86	81	73
Seattle (International Airport)	18	79	75	76	66	75	63	71	58	68	54	67	53	65	49	70	53	74	58	80	68	80	74	81	77	74	62
Spokane	18	82	77	78	68	67	54	56	43	51	40	48	35	38	25	43	28	50	34	66	49	81	74	84	81	62	51
Stampede Pass	34	92	91	90	87	87	83	81	77	75	70	75	70	66	59	69	63	70	65	80	78	92	91	95	94	81	77
Walla Walla	19	76		68		59		49		49		43		34		39		44		58		72		78		56	
Yakima	31	77	70	69	56	52	40	40	32	37	30	38	30	35	24	39	27	43	31	55	44	71	62	80	74	53	43
West Virginia																											
Beckley	14	80	68	78	64	78	58	75	50	81	53	86	57	90	61	93	62	92	63	87	57	81	60	81	69	83	60
Charleston	30	77	63	76	59	74	53	75	47	82	50	86	54	90	60	92	58	91	55	88	53	80	56	77	62	82	56
Elkins	15	80	64	78	60	81	55	82	51	86	53	91	58	95	61	96	63	96	62	90	54	84	60	83	67	87	59
Huntington	16	79	66	77	61	76	55	77	49	85	53	88	58	91	60	92	60	93	61	87	55	82	60	80	68	84	59
Parkersburg	60	82	66	81	63	79	56	74	49	74	51	78	56	80	52	83	57	84	51	84	52	81	57	81	64	80	56
Wisconsin																											
Green Bay	16	76	70	77	67	80	66	79	58	79	55	82	59	85	56	88	59	88	61	84	61	83	69	80	73	82	63
La Crosse	27	76	66	78	64	81	62	79	53	80	53	85	56	87	57	90	59	90	61	84	58	82	66	81	71	83	60
Madison	18	78	68	79	65	81	63	81	55	80	54	83	56	86	56	90	58	92	61	86	58	86	67	83	73	84	61
Milwaukee	17	75	68	76	67	79	66	79	62	78	61	82	62	82	60	87	62	87	63	82	63	81	67	80	73	81	64

Table 10 (continued)
AVERAGE RELATIVE HUMIDITY (%) MORNING (M) AND AFTERNOON (A) IN THE U.S. AND CERTAIN PACIFIC ISLANDS[2] (DATA THROUGH 1977)

Location	Years	Jan		Feb		Mar		Apr		May		Jun		Jul		Aug		Sep		Oct		Nov		Dec		Ann	
		M	A	M	A	M	A	M	A	M	A	M	A	M	A	M	A	M	A	M	A	M	A	M	A	M	A
Wyoming																											
Casper	13	59	61	57	55	53	48	50	44	43	39	40	34	33	26	31	23	37	30	45	41	53	55	58	61	47	43
Cheyenne	18	44	48	43	45	45	45	41	39	39	40	39	40	34	37	34	35	36	37	36	39	41	47	43	49	40	42
Lander	31	59	59	56	53	51	47	46	41	43	38	41	34	34	28	33	27	39	33	45	42	56	57	59	61	47	43
Sheridan	13	60	63	58	59	52	48	50	45	47	46	48	47	38	34	34	30	43	39	47	46	57	60	61	65	50	48

Table 11
AVERAGE PERCENTAGE OF POSSIBLE SUNSHINE IN THE U.S.AND CERTAIN PACIFIC ISLANDS[2] (DATA THROUGH 1977)

Location	Years	Jan.	Feb.	Mar.	April	May	June	July	Aug.	Sept.	Oct.	Nov.	Dec.	Annual
Alabama														
Birmingham	34	42	50	55	63	66	65	59	63	61	66	55	46	58
Montgomery	27	48	54	57	65	66	66	63	66	63	66	56	50	59
Alaska														
Anchorage	24	42	44	55	52	53	48	45	42	41	40	38	33	46
Juneau	32	32	32	37	38	39	33	31	32	26	20	23	20	31
Nome	29	40	51	52	53	47	41	35	30	32	33	30	34	40
Arizona														
Flagstaff	4	75	72	69	84	89	90	75	81	82	78	77	78	81
Phoenix	82	78	80	83	88	93	94	85	85	89	88	84	77	86
Tucson	30	81	84	86	91	93	93	78	81	87	89	86	80	86
Yuma	27	84	88	91	94	96	97	89	92	94	92	86	82	91
Arkansas														
Fort Smith	32	49	55	56	59	63	69	71	71	65	66	57	51	62
Little Rock	32	46	54	57	61	68	73	71	73	68	69	56	48	63
California														
Eureka	67	42	46	50	55	55	57	52	47	52	49	42	40	50

Fresno	28	51	67	80	85	89	94	96	96	94	88	67	47	82
Los Angeles	32	69	72	73	70	66	65	82	83	79	73	74	71	73
Red Bluff	33	54	63	68	80	84	89	96	94	92	81	61	51	78
Sacramento	29	47	62	71	80	87	92	97	96	93	84	63	46	79
San Diego	37	72	72	71	65	58	56	67	69	68	67	74	71	67
San Francisco (City)	38	56	62	69	73	72	73	66	65	72	70	62	53	67
Colorado														
Denver	28	72	71	70	66	65	71	71	72	74	73	66	68	70
Grand Junction	31	60	65	65	67	71	78	77	75	79	74	64	60	70
Pueblo	37	75	74	74	73	73	79	78	78	79	78	74	73	76
Connecticut														
Hartford	23	58	57	57	57	58	58	62	63	59	58	46	48	57
District of Columbia														
Washington (National Airport)	29	49	52	56	58	59	64	63	63	62	59	52	47	58
Florida														
Apalachicola	43	59	63	65	74	78	72	64	63	64	75	68	57	67
Jacksonville	27	57	61	66	71	69	62	60	58	54	57	60	56	61
Key West	19	72	76	82	85	80	73	77	77	71	70	72	73	76
Lakeland	24	62	65	67	74	72	63	61	61	59	65	66	63	65
Pensacola	5	48	53	61	63	67	67	57	58	60	71	64	49	60
Tampa	31	65	67	71	74	74	65	61	59	60	64	65	62	66
Georgia														
Atlanta	42	48	54	57	66	68	67	62	65	63	67	59	50	61
Macon	29	54	59	61	69	70	69	64	70	64	68	63	56	63
Savannah	27	54	59	63	70	68	65	63	63	58	64	62	55	62
Hawaii														
Hilo	27	45	42	38	33	34	42	42	41	42	39	33	35	39
Honolulu	25	63	65	68	66	69	70	73	75	74	67	60	59	67
Kahului	15	66	68	66	65	72	76	75	74	78	73	67	67	71
Lihue	27	51	53	51	50	56	61	62	64	66	57	49	47	56
Idaho														
Boise	36	41	52	63	68	72	75	88	85	82	67	45	40	67
Pocatello	28	37	52	61	64	67	73	83	81	79	70	48	37	65
Illinois														
Cairo	35	45	52	57	63	67	73	75	77	68	69	53	45	64
Chicago (Midway Airport)	35	44	47	52	54	61	67	70	68	63	61	41	38	57
Moline	34	49	50	50	53	58	64	71	68	63	61	44	41	57

Table 11 (continued)
AVERAGE PERCENTAGE OF POSSIBLE SUNSHINE IN THE U.S. AND CERTAIN PACIFIC ISLANDS[2] (DATA THROUGH 1977)

Location	Years	Jan.	Feb.	Mar.	April	May	June	July	Aug.	Sept.	Oct.	Nov.	Dec.	Annual
Peoria	34	46	50	52	55	59	66	69	67	64	62	45	40	58
Springfield	30	47	51	52	56	64	69	73	72	70	65	50	42	61
Indiana														
Evansville	37	43	49	56	61	67	73	76	77	70	67	49	41	62
Fort Wayne	31	43	48	52	57	65	71	73	72	65	62	41	37	59
Indianapolis	34	42	51	52	56	62	67	70	71	66	64	43	40	58
Iowa														
Burlington	23	53	55	55	60	64	71	77	77	72	71	56	51	65
Des Moines	27	51	54	54	56	60	67	72	69	64	63	50	45	60
Sioux City	37	58	59	59	61	64	68	76	72	67	66	54	51	64
Kansas														
Concordia	15	64	66	65	66	71	77	80	77	69	71	62	59	70
Dodge City	35	68	65	66	68	69	76	79	77	74	74	67	65	71
Topeka	28	55	55	54	56	60	65	69	69	64	64	54	51	61
Wichita	24	60	61	61	62	65	70	74	74	65	66	60	58	65
Kentucky														
Louisville	30	42	47	51	55	62	66	66	68	64	62	47	40	57
Louisiana														
New Orleans	4	47	65	53	69	58	69	58	55	63	68	53	53	60
Shreveport	25	50	56	56	56	65	70	73	71	68	71	61	53	64
Maine														
Portland	37	56	59	56	57	56	59	64	65	61	59	47	53	58
Maryland														
Baltimore	27	52	55	55	56	57	62	65	62	60	59	51	48	57
Massachussetts														
Blue Hill Observatory	91	46	49	48	50	52	55	57	58	56	55	47	46	52
Boston	42	54	56	57	57	59	63	66	67	63	61	51	52	60
Michigan														
Alpena	18	39	46	54	56	62	65	68	63	55	48	30	29	53
Detroit (City Airport)	32	32	43	49	52	59	65	70	65	61	56	35	32	54
Detroit (Metro Airport)	12	39	46	50	55	60	63	69	69	59	53	32	28	54

Grand Rapids	14	33	41	46	54	58	62	66	64	54	47	27	24	50
Lansing	23	36	44	48	54	63	66	71	67	58	54	30	29	53
Marquette	40	35	43	51	55	58	60	67	62	52	46	27	28	51
Sault Ste. Marie	36	35	46	54	55	57	58	63	58	45	41	23	28	48
Minnesota														
Duluth	27	49	54	55	54	56	58	67	62	52	48	34	39	54
Minneapolis—St. Paul	39	50	57	54	56	59	63	71	67	61	57	39	40	58
Mississippi														
Jackson	13	49	57	59	62	62	69	62	62	58	64	53	47	59
Missouri														
Columbia	8	54	56	55	60	65	72	75	67	54	60	49	46	61
Kansas City	5	62	61	65	72	74	76	82	68	57	64	53	59	67
St. Louis	18	54	53	56	57	62	68	72	65	63	63	49	43	59
Springfield	32	51	55	57	61	64	68	72	72	68	66	55	49	63
Montana														
Billings	38	47	54	62	59	61	64	78	77	68	62	46	45	62
Great Falls	35	48	57	67	62	64	65	80	77	68	61	46	45	63
Havre	17	48	61	69	68	71	73	84	81	73	63	49	43	68
Helena	37	46	54	60	58	60	61	79	74	67	60	45	43	61
Missoula	33	29	39	48	52	56	57	79	74	65	50	32	25	54
Nebraska														
Lincoln	21	59	59	57	61	63	71	74	73	65	66	54	52	64
North Platte	25	60	58	60	62	62	70	74	73	69	70	60	60	65
Omaha (North)	41	55	55	55	58	62	68	75	72	67	67	52	49	62
Valentine	22	65	63	61	60	63	69	77	77	70	70	61	61	67
Nevada														
Ely	38	66	66	71	67	71	79	80	81	83	75	66	64	73
Las Vegas	28	78	81	83	86	88	92	87	88	92	86	81	78	86
Reno	35	66	68	75	80	81	85	92	93	92	84	71	64	81
Winnemucca	28	48	54	56	63	69	75	86	84	82	74	55	49	68
New Hampshire														
Concord	36	52	54	52	53	55	57	62	61	55	54	42	47	54
Mt. Washington	39	33	33	33	35	36	31	30	32	35	40	29	30	33
New Jersey														
Atlantic City	17	50	51	53	55	55	58	60	63	58	56	49	43	55
Trenton	45	51	55	55	59	60	62	65	63	61	60	52	48	58
New Mexico														

Table 11 (continued)
AVERAGE PERCENTAGE OF POSSIBLE SUNSHINE IN THE U.S. AND CERTAIN PACIFIC ISLANDS[2] (DATA THROUGH 1977)

Location	Years	Jan.	Feb.	Mar.	April	May	June	July	Aug.	Sept.	Oct.	Nov.	Dec.	Annual
Albuquerque	38	73	73	74	76	80	83	76	76	80	79	78	72	77
Roswell	4	69	71	83	83	86	89	76	80	78	86	80	77	81
New York														
Albany	39	46	51	52	53	55	59	63	61	56	52	36	38	53
Binghamton	26	38	41	44	51	56	62	65	61	56	50	30	27	51
Buffalo	34	33	39	46	53	58	66	69	66	59	53	29	27	52
New York (Central Park)	100	50	55	56	59	61	64	65	64	63	61	52	49	59
Rochester	37	37	42	50	56	59	67	69	67	59	51	31	31	54
Syracuse	28	35	39	46	52	56	60	65	62	54	45	25	25	49
North Carolina														
Asheville	13	56	64	65	68	63	65	60	55	56	62	61	57	81
Cape Hatteras	15	49	55	62	67	63	62	63	63	61	59	58	46	59
Charlotte	27	55	60	63	70	69	70	69	70	68	69	63	58	66
Greensboro	49	52	57	60	64	65	65	62	63	63	66	59	54	61
Raleigh	23	55	59	63	65	59	60	61	60	60	62	62	56	60
Wilmington	26	57	61	63	71	67	65	63	62	62	63	65	59	64
North Dakota														
Bismarck	38	54	55	60	59	63	64	76	73	65	59	45	47	62
Fargo	35	50	58	58	58	59	59	71	68	59	56	39	43	58
Williston	15	48	56	61	56	61	65	75	74	63	58	43	47	61
Ohio														
Cincinnati	62	41	45	50	55	61	67	68	66	66	59	44	38	57
Cleveland	34	32	37	44	53	59	65	68	64	60	55	31	26	52
Columbus	26	37	42	44	52	58	61	63	63	61	57	38	31	52
Dayton	34	42	46	49	54	61	67	68	70	66	62	42	37	57
Toledo	22	43	47	49	53	61	63	68	65	60	58	38	34	55
Oklahoma														
Oklahoma City	23	59	61	63	63	65	73	75	77	69	68	60	59	67
Tulsa	35	53	56	56	57	59	67	72	72	65	65	59	54	62
Oregon														
Portland	28	25	36	43	50	55	52	69	63	59	41	29	20	48

Pacific														
Guam	16	56	61	68	68	65	63	51	46	44	47	51	53	56
Johnston Island	15	68	71	71	65	70	75	78	73	71	64	58	63	69
Koror Island	17	56	59	64	66	57	50	51	50	56	52	54	50	55
Majuro, Marshall Islands	17	59	61	62	54	55	51	53	59	57	51	52	51	56
Pago Pago, American Samoa	10	49	48	52	40	41	40	53	59	68	53	48	43	50
Ponape Island	19	49	49	53	46	45	47	48	49	49	44	47	42	47
Truk, Caroline Islands	17	60	60	63	59	55	53	55	60	54	52	56	53	57
Wake Island	8	70	72	76	74	78	79	75	73	67	70	65	65	72
Yap Island	18	61	64	67	70	63	57	52	52	53	50	58	55	58
Pennsylvania														
Harrisburg	39	48	53	56	58	59	63	68	67	61	57	46	44	58
Philadelphia	35	51	54	56	57	57	62	63	63	59	59	52	49	58
Pittsburgh (International Airport)	25	35	39	46	49	53	58	60	59	59	54	39	30	50
Pittsburgh (City)	42	26	31	40	45	50	53	53	50	52	45	30	23	42
Avoca	22	45	47	49	54	56	59	61	60	54	53	36	35	52
Puerto Rico														
San Juan	22	65	70	74	68	61	59	65	65	59	59	57	58	63
Rhode Island														
Providence	24	57	56	55	56	57	57	60	59	58	59	49	51	56
South Carolina														
Charleston	19	57	61	66	71	71	66	67	64	62	67	65	59	65
Columbia	24	57	61	63	68	65	64	64	66	64	66	64	60	64
Greenville—Spartanburg	15	53	60	63	65	59	59	59	60	61	66	62	54	60
South Dakota														
Huron	38	57	60	58	60	64	68	77	75	67	63	51	49	64
Rapid City	35	55	60	61	59	57	61	72	73	68	66	56	54	62
Tennessee														
Chattanooga	47	42	48	51	60	65	65	61	62	63	62	53	42	57
Knoxville	35	39	46	52	60	62	62	61	61	59	60	49	38	57
Memphis	27	49	55	57	64	70	74	73	76	69	71	58	50	65
Nashville	35	41	48	53	60	63	67	65	66	63	63	50	41	58
Texas														
Abilene	29	63	66	70	69	70	78	78	77	69	71	70	67	71
Amarillo	36	69	69	72	72	72	77	77	77	74	74	73	68	73
Austin	36	49	53	55	53	57	69	76	75	67	65	57	51	61
Brownsville	35	44	50	51	57	66	73	81	76	68	66	52	45	62

Table 11 (continued)
AVERAGE PERCENTAGE OF POSSIBLE SUNSHINE IN THE U.S. AND CERTAIN PACIFIC ISLANDS[2] (DATA THROUGH 1977)

Location	Years	Jan.	Feb.	Mar.	April	May	June	July	Aug.	Sept.	Oct.	Nov.	Dec.	Annual
Corpus Christi	35	47	53	55	57	61	73	82	77	68	69	58	48	63
El Paso	35	78	82	85	87	89	89	79	80	82	84	83	78	83
Galveston	87	48	51	55	61	68	75	72	70	68	72	60	49	63
Houston	8	45	56	45	52	59	65	67	62	60	61	55	65	57
Lubbock	5	70	73	78	73	78	82	74	79	72	76	76	75	76
Port Arthur	24	43	53	51	52	63	68	66	63	63	66	57	47	58
San Antonio	35	49	53	58	55	56	67	75	74	68	65	56	52	61
Utah														
Milford	4	65	65	60	67	77	82	75	84	83	77	68	68	74
Salt Lake City	40	48	56	64	67	73	79	84	83	84	73	54	45	70
Vermont														
Burlington	34	42	48	51	51	56	60	65	62	54	50	30	33	51
Virginia														
Lynchburg	33	51	55	58	62	63	65	62	62	61	61	55	52	59
Norfolk	19	57	58	63	66	67	68	65	65	64	60	60	57	63
Richmond	27	51	56	59	63	63	66	65	64	63	59	56	51	60
Washington														
Quillayute	11	20	31	30	35	37	35	46	44	49	35	21	17	35
Seattle (Urban Site)	31	28	34	42	47	52	49	63	56	53	37	28	23	45
Seattle (International Airport)	11	21	43	49	52	59	55	67	62	61	41	28	17	49
Spokane	29	26	40	53	60	63	65	80	77	70	51	29	20	57
Walla Walla	60	23	35	50	61	67	72	85	83	73	58	31	18	59
West Virginia														
Parkersburg	80	32	37	43	50	56	59	62	60	59	54	37	29	49
Wisconsin														
Green Bay	28	46	50	52	51	58	62	65	62	54	49	36	37	53
Madison	31	49	52	54	52	59	65	69	68	62	57	41	39	57
Milwaukee	37	45	47	51	54	59	64	71	67	60	56	41	38	56
Wyoming														
Cheyenne	42	61	65	65	60	58	64	68	67	69	68	60	59	64
Lander	31	67	70	73	68	66	72	77	77	74	71	62	65	71
Sheridan	37	55	58	61	58	58	63	75	74	67	63	52	53	62

Table 12
MEAN DAILY SOLAR RADIATION (LANGLEYS) AND YEARS OF RECORD USED[3]

States and stations	Jan.	Yrs	Feb	Yrs	Mar	Yrs	Apr	Yrs	May	Yrs	June	Yrs	July	Yrs	Aug	Yrs	Sept	Yrs	Oct	Yrs	Nov	Yrs	Dec	Yrs	Annual
Alaska																									
Annette	63	6	115	6	236	7	364	7	437	6	438	6	438	6	341	6	258	7	122	7	59	7	41	7	24
Barrow	#		38	8	180	8	380	8	513	8	528	8	429	9	255	10	115	10	41	10	#		#		200
Bethel	38	9	108	10	282	9	444	10	457	10	454	10	376	10	252	10	202	10	115	10	44	9	22	0	203
Fairbanks	16	25	71	27	213	25	376	28	461	28	504	29	434	28	317	29	180	29	82	30	26	26	6	26	224
Matanuska	32	6	92	6	242	4	356	7	436	7	462	6	409	6	314	6	198	6	100	6	38	6	15	7	224
Ariz.																									
Page	300	2	382	3	526	3	618	2	695	2	707	2	680	3	596	3	516	3	402	3	310	3	243	3	498
Phoenix	301	11	409	11	526	11	638	11	724	11	739	11	658	11	613	11	566	11	449	11	344	11	281	11	520
Tucson	315	5	391	5	540	4	655	5	729	5	699	5	626	6	588	6	570	6	442	6	356	6	305	6	518
Ark.																									
Little Rock	188	9	260	9	353	10	446	9	523	9	559	9	556	8	518	9	439	7	343	8	244	10	187	10	385
California																									
Davis	174	18	257	17	390	18	528	18	625	18	694	18	682	18	612	18	493	18	347	19	222	19	148	19	431
Fresno	184	31	289	31	427	31	552	31	647	31	702	32	682	32	621	31	510	31	376	32	250	31	161	32	450
Inyokern (China Lake)	306	11	412	11	562	11	683	11	772	11	819	11	772	11	729	10	635	8	467	9	363	11	300	12	568
LaJolla	244	19	302	18	397	19	457	20	506	19	487	21	497	22	464	22	389	22	320	21	277	20	221	20	380
Los Angeles WBAS	248	10	331	10	470	10	515	10	572	9	596	9	641	9	581	10	503	10	373	10	289	10	241	10	463
Los Angeles WBO	243	9	327	9	436	9	483	9	555	9	584	9	651	9	581	10	500	10	362	10	281	10	234	10	436
Riverside	275	8	367	8	478	9	541	9	623	9	680	9	673	9	618	9	535	9	407	9	319	9	270	9	483
Santa Maria	263	11	346	11	482	11	552	10	635	11	694	11	680	11	613	11	524	11	419	11	313	11	252	11	481
Soda Springs	223	4	316	3	374	4	551	4	615	3	691	4	760	3	681	3	510	3	357	4	248	4	182	3	459
Colo.																									
Boulder	201	5	268	4	401	4	460	4	460	4	525	5	520	5	439	5	412	4	310	4	222	4	182	4	367
Grand Junction	227	9	324	9	434	8	546	8	615	9	708	8	676	8	595	8	514	8	373	10	260	10	212	10	456
Grand Lake (Granby)	212	6	313	7	423	7	512	8	552	8	632	8	600	8	505	7	476	6	361	7	234	6	184	7	417
D.C.																									
Washington (C.O.)	174	3	266	3	344	2	411	2	551	2	494	2	536	2	446	3	375	3	299	3	211	3	166	3	356
American University	158	39	231	39	322	39	398	39	467	39	510	39	496	39	440	38	364	38	278	38	192	39	141	39	332
Silver Hill	177	7	247	6	342	7	438	7	513	7	555	7	511	7	457	7	391	8	293	8	202	7	156	6	357
Fla.																									
Apalachicola	298	10	367	10	441	10	535	10	603	9	578	9	529	9	511	9	456	9	413	10	332	10	262	10	444
Belle Isle	297	10	330	10	412	10	463	10	483	10	464	10	488	11	461	10	400	10	366	11	313	11	291	10	397
Gainesville	267	11	343	10	427	12	517	12	579	12	521	10	488	10	483	8	418	9	347	8	300	10	233	10	410
Miami Airport	349	10	415	9	489	9	540	10	553	10	532	10	532	10	505	10	440	10	384	10	353	10	316	10	451
Tallahassee	274	2	311	2	423	3	499	3	547	3	521	3	508	3	542	2	*		*		292	2	230	2	—
Tampa	327	8	391	8	474	8	539	8	596	8	574	9	534	9	494	9	452	9	400	9	356	9	300	9	453
Ga.																									
Atlanta	218	11	290	11	380	11	488	11	533	11	562	11	532	10	508	10	416	10	344	11	268	11	211	11	396

Table 12 (continued)
MEAN DAILY SOLAR RADIATION (LANGLEYS) AND YEARS OF RECORD USED[3]

States and stations	Jan.	Yrs	Feb	Yrs	Mar	Yrs	Apr	Yrs	May	Yrs	June	Yrs	July	Yrs	Aug	Yrs	Sept	Yrs	Oct	Yrs	Nov	Yrs	Dec	Yrs	Annual
Griffin	234	9	295	9	385	10	522	11	570	11	577	11	556	11	522	11	435	11	368	11	283	11	201	11	413
Hawaii																									
Honolulu	363	4	422	4	516	4	559	5	617	5	615	5	615	5	612	5	573	5	507	5	426	5	371	5	516
Mauna Loa Obs.	522	2	576	2	680	2	689	3	727	3	*		703	3	642	2	602	2	560	2	504	2	481	3	—
Pearl Harbor	359	5	400	4	487	4	529	5	573	5	566	5	598	5	567	5	539	5	466	5	386	5	343	5	484
Idaho																									
Boise	138	10	236	9	342	9	485	9	585	10	636	9	670	10	576	10	460	10	301	11	182	11	124	11	395
Twin Falls	163	20	240	20	355	20	462	21	552	20	592	18	602	20	540	20	432	19	286	20	176	20	131	19	378
Ill.																									
Chicago	96	19	147	19	227	19	331	19	424	19	458	18	473	19	403	18	313	19	207	20	120	20	76	20	273
Lemont	170	6	242	6	340	6	402	6	506	6	553	6	540	6	498	6	398	5	275	5	165	5	138	5	352
Ind.																									
Indianapolis	144	10	213	10	316	10	396	10	488	9	543	11	541	10	490	11	405	11	293	11	177	11	132	11	345
Iowa																									
Ames	174	5	253	5	326	5	403	5	480	5	541	5	436	6	460	6	367	6	274	7	187	7	143	7	345
Kans.																									
Dodge City	255	7	316	7	418	7	528	7	568	7	650	7	642	8	592	9	493	9	380	9	285	10	234	10	447
Manhattan	192	3	264	3	345	3	433	3	527	4	551	4	531	4	526	4	410	4	292	4	227	4	156	4	371
Ky.																									
Lexington	172	9	263	9	357	10	480	10	581	10	628	9	617	10	563	10	494	10	357	9	245	9	174	11	411
La.																									
Lake Charles	245	11	306	11	397	11	481	11	555	11	591	11	526	11	511	11	449	11	402	11	300	10	250	10	418
New Orleans	214	14	259	14	335	15	412	16	449	14	443	13	417	15	416	15	383	15	357	13	278	13	198	14	347
Shreveport	232	3	292	3	384	3	446	4	558	4	557	4	578	4	528	4	414	4	354	4	254	4	205	4	400
Maine																									
Caribou	133	8	231	9	364	8	400	10	476	10	470	10	508	11	448	11	336	11	212	11	111	11	107	9	316
Portland	152	7	235	8	352	7	409	8	514	9	539	9	561	9	488	8	383	7	278	9	157	8	137	9	350
Mass.																									
Amherst	116	2	*		300	2	*		431	2	514	2	*		—		—		—		152	2	124	2	—
Blue Hill	153	27	228	27	319	26	389	26	469	27	510	27	502	26	449	27	354	28	266	28	162	28	135	28	328
Boston	129	16	194	17	290	17	350	17	445	16	483	16	486	16	411	16	334	17	235	16	136	16	115	15	301
Cambridge	153	4	235	3	323	3	400	3	420	3	476	3	482	4	464	4	367	4	253	4	164	4	124	4	322
East Wareham	140	13	218	13	305	12	385	14	452	14	508	14	495	14	436	14	365	13	258	14	163	14	140	13	322
Lynn	118	2	209	2	300	2	394	2	454	2	549	4	528	4	432	3	341	2	241	3	135	3	107	3	317
Mich.																									
East Lansing	121	10	210	11	309	11	359	11	483	10	547	11	540	11	466	11	373	11	255	11	136	11	108	11	311
Sault St. Marie	130	10	225	9	356	10	416	10	523	10	557	11	573	11	472	10	322	10	216	9	105	9	96	9	333
Minn.																									
St. Cloud	168	8	260	8	368	8	426	8	496	8	535	8	557	9	486	8	366	8	237	7	146	8	124	8	348
Mo.																									
Columbia (C. O.)	173	10	251	10	340	11	434	11	530	11	574	11	574	10	522	10	453	10	322	10	225	10	158	9	380
University of Missouri	166	5	248	6	324	6	429	6	501	6	560	6	583	6	509	6	417	6	324	5	177	5	146	5	365

Mont.																									
Glasgow	154	6	258	8	385	7	466	8	568	8	605	8	645	9	531	10	410	10	267	8	154	8	116	7	388
Great Falls	140	8	232	9	366	9	434	8	528	8	583	8	639	9	532	9	407	10	264	10	154	10	112	10	366
Summit	122	3	162	2	268	3	414	3	462	3	493	3	560	2	510	2	354	2	216	2	102	2	76	2	312
Nebr.																									
Lincoln	188	39	259	39	350	39	416	39	494	40	544	38	568	38	484	38	396	38	296	36	199	40	159	39	363
North Omaha	193	3	299	3	365	3	463	3	516	3	546	4	568	4	519	4	410	4	298	4	204	4	170	4	379
Nev.																									
Ely	236	7	339	9	468	9	563	9	625	10	712	10	647	11	618	11	518	11	394	10	289	10	218	10	469
Las Vegas	277	11	384	11	519	11	621	11	702	11	748	10	675	11	627	11	551	11	429	11	318	11	258	11	509
N.J.																									
Seabrook	157	8	227	8	318	8	403	8	482	9	527	8	509	8	455	9	385	9	278	7	192	8	140	8	339
N.H.																									
Mt. Washington	117	2	218	2	238	2	*		*		*		—		—		*		*		*		96	2	—
N. Mex.																									
Albuquerque	303	13	386	13	511	13	618	13	686	13	726	13	683	12	626	13	554	14	438	15	334	15	276	14	512
N.Y.																									
Ithaca	116	22	194	21	272	23	334	23	440	24	501	23	515	23	453	23	346	21	231	22	120	23	96	23	302
N.Y. Central Park	130	34	199	34	290	33	369	35	432	35	470	34	459	35	389	35	331	36	242	36	147	36	115	35	298
Sayville	160	11	249	11	335	10	415	10	494	10	565	10	543	10	462	10	385	10	289	10	186	10	142	11	352
Schenectady	130	8	200	9	273	9	338	9	413	9	448	8	441	8	397	8	299	8	218	8	128	8	104	8	282
Upton	155	8	232	8	339	8	428	8	502	8	573	8	543	7	475	7	391	7	293	6	182	7	146	7	355
N.C.																									
Greensboro	200	7	276	9	354	9	469	9	531	10	564	10	544	10	485	10	406	10	322	10	243	10	197	8	383
Hatteras	238	10	317	9	426	8	569	9	635	10	652	10	625	10	562	11	471	11	358	11	282	11	214	11	443
Raleigh	235	3	302	2	*		466	3	494	2	564	2	535	3	476	3	379	3	307	3	235	3	199	3	—-
N.D.																									
Bismarck	157	7	250	8	356	6	447	8	550	8	590	9	617	10	516	11	390	11	272	11	161	10	124	10	369
Ohio																									
Cleveland	125	6	183	6	303	7	286	8	502	8	562	8	562	8	494	8	278	8	289	9	141	9	115	7	335
Columbus	128	7	200	7	297	7	391	7	471	6	562	4	542	5	477	4	422	4	286	4	176	4	129	5	340
Put-in-Bay	126	10	204	9	302	10	386	11	468	11	544	11	561	10	487	10	382	11	275	11	144	11	109	11	332
Okla.																									
Oklahoma City	251	10	319	10	409	9	494	10	536	10	615	7	610	8	593	8	487	9	377	10	291	9	240	9	436
Stillwater	205	8	289	8	390	9	454	9	504	9	600	10	596	10	545	10	455	11	354	10	269	9	209	8	405
Oreg.																									
Astoria	90	7	162	8	270	8	375	8	492	8	469	8	539	8	461	7	354	7	209	8	111	8	79	8	301
Corvallis	89	2	*		287	3	406	3	517	3	570	3	676	4	558	1	397	4	235	4	144	4	80	4	—
Medford	116	11	215	11	336	11	482	11	592	11	652	11	698	10	605	11	447	11	279	11	149	11	93	11	389
Pa.																									
Pittsburgh	94	6	169	5	216	6	317	6	429	6	491	6	497	7	409	6	339	6	207	5	118	6	77	9	280
State College	133	19	201	19	295	20	380	20	456	20	518	20	511	20	444	20	358	20	256	20	149	20	118	20	318
R.I.																									
Newport	155	23	232	22	334	23	405	23	477	23	527	24	513	24	455	24	377	24	271	24	176	24	139	24	338
S.C.																									
Charleston	252	11	314	11	388	11	512	11	551	11	564	11	520	11	501	11	404	11	338	11	286	11	225	11	404

Table 12 (continued)
MEAN DAILY SOLAR RADIATION (LANGLEYS) AND YEARS OF RECORD USED[3]

States and stations	Jan.	Yrs	Feb	Yrs	Mar	Yrs	Apr	Yrs	May	Yrs	June	Yrs	July	Yrs	Aug	Yrs	Sept	Yrs	Oct	Yrs	Nov	Yrs	Dec	Yrs	Annual
S.D.																									
Rapid City	183	11	277	11	400	11	482	11	532	11	585	11	590	11	541	11	435	11	315	10	204	10	158	10	392
Tenn.																									
Nashville	149	18	228	19	322	19	432	19	503	18	551	18	530	17	473	17	403	17	308	19	208	18	150	19	355
Oak Ridge	161	11	239	11	331	11	450	11	518	11	551	11	526	11	478	11	416	11	318	11	213	10	163	11	364
Texas																									
Brownsville	297	10	341	10	402	10	456	11	564	10	610	9	627	8	568	11	475	11	411	11	296	11	263	10	442
El Paso	333	11	430	11	547	10	654	11	714	11	729	11	666	11	640	10	576	11	460	11	372	11	313	11	536
Ft. Worth	250	11	320	11	427	11	488	11	562	11	651	11	613	11	593	11	503	11	403	11	306	11	245	9	445
Midland	283	7	358	8	476	9	550	8	611	8	617	8	608	7	574	8	522	9	396	9	325	8	275	8	466
San Antonio	279	9	347	9	417	9	445	9	541	9	612	9	639	9	585	9	493	10	398	10	295	10	256	8	442
Utah																									
Flaming Gorge	238	2	298	2	443	2	522	2	565	2	650	2	599	3	538	3	425	3	352	3	262	3	215	3	426
Salt Lake City	163	8	256	8	354	8	479	8	570	7	621	7	620	6	551	7	446	8	316	8	204	8	146	9	394
Va.																									
Mt. Weather	172	2	274	2	338	2	414	2	508	2	525	3	510	3	430	3	375	3	281	2	202	2	168	2	350
Wash.																									
North Head	*		167	2	257	3	432	2	509	3	487	3	486	3	436	3	321	3	205	3	122	3	77	3	—
Friday Harbor	87	8	157	7	274	8	418	8	514	9	578	10	586	10	507	11	351	8	194	10	102	10	75	8	320
Prosser	117	4	222	4	351	4	521	5	616	4	680	4	707	4	604	4	458	4	274	4	136	4	100	4	399
Pullman	121	4	205	2	304	2	462	2	558	4	653	5	699	5	562	4	410	4	245	5	146	5	96	5	372
University of Washington	67	9	126	9	245	10	364	9	445	10	461	10	496	11	435	10	299	8	170	9	93	9	59	9	272
Seattle-Tacoma	75	9	139	9	265	9	403	9	503	9	511	9	566	9	452	10	324	10	188	10	104	9	64	10	300
Spokane	119	8	204	8	321	8	474	9	563	9	596	9	665	9	556	9	404	10	225	9	131	9	75	7	361
Wis.																									
Madison	148	46	220	46	313	45	394	47	466	47	514	47	531	47	452	47	348	47	241	47	145	44	115	46	324
Wyo.																									
Lander	226	8	324	9	452	9	548	11	587	11	678	11	651	11	586	10	472	8	354	9	239	9	196	9	443
Laramie	216	3	295	3	424	3	508	3	554	3	643	3	606	3	536	3	438	3	324	3	229	3	186	4	408
Island Stations																									
Canton Island	588	9	626	7	634	7	604	9	561	9	549	8	550	9	597	9	640	9	651	9	600	8	572	8	597
San Juan, P.R.	404	5	481	4	580	4	622	4	519	5	536	6	639	5	549	6	531	6	460	6	411	6	411	6	512
Swan Island	442	6	496	7	615	6	646	6	625	6	544	8	588	8	591	7	535	8	457	7	394	8	382	8	526
Wake Island	438	7	518	7	577	7	627	7	642	8	656	6	629	7	623	7	587	6	525	7	482	7	421	7	560

Notes: * Denotes only one year of data for the month — no means computed. —- No data for the mmonth (or incomplete data for the year). # Barrow is in darkness during the winter months. † Madison data after 1957 not used due to exposure influences. ‡ Riverside data prior to March 1952 not used — instrumental discrepancies. Langley is the unit used to denote one gram calorie per square centimeter.

Table 13
MEAN ANNUAL AND MAY TO OCTOBER CLASS A PAN EVAPORATION[3]

Location	Annual (in.)	May—Oct (in.)
Ala.		
Birmingham	56	40
Ariz.		
Phoenix	105	72
Yuma	127	86
Ark.		
Little Rock	58	42
Calif.		
Fresno	90	72
Los Angeles	59	41
Sacramento	70	53
Colo.		
Denver	60	43
Grand Junction	52	40
Conn.		
Hartford	37	28
D.C.		
Washington	47	33
Fla.		
Jacksonville	59	37
Miami	66	38
Tampa	65	40
Ga.		
Atlanta	55	38
Idaho		
Pocatello	51	41
Ill.		
Chicago	38	30
Springfield	44	35
Ind.		
Indianapolis	43	34
Iowa		
Des Moines	50	40
Kan.		
Dodge City	91	65
Ky.		
Lexington	47	35
La.		
New Orleans	66	43
Shreveport	65	45
Maine		
Caribou	24	20
Portland	31	24
Mass.		
Boston	35	27
Mich.		
Lansing	40	32
Minn.		
Duluth	32	26
Minneapolis	41	33
Mo.		
Kansas City	60	45
St. Louis	48	37

Table 13 (continued)
MEAN ANNUAL AND MAY TO OCTOBER CLASS A PAN EVAPORATION[3]

Location	Annual (in.)	May—Oct (in.)
Mont.		
Billings	53	41
Great Falls	50	40
Neb.		
Lincoln	62	47
North Platte	67	50
Nev.		
Las Vegas	120	85
Winnemucca	69	54
N.J.		
Seabrook	42	31
N.H.		
Concord	33	25
N.M.		
Albuquerque	90	65
N.Y.		
Albany	35	27
N.C.		
Raleigh	54	37
N.D.		
Bismarck	47	39
Fargo	39	33
Ohio		
Columbia	44	34
Okla.		
Oklahoma City	85	60
Ore.		
Burns	56	45
Portland	33	25
Pa.		
Harrisburg	44	32
Pittsburgh	39	29
S.C.		
Columbia	56	38
S.D.		
Huron	49	40
Tenn.		
Nashville	50	37
Texas		
Abilene	97	65
Amarillo	96	65
Brownsville	80	51
El Paso	105	69
Fort Worth	80	55
Houston	73	47
Utah		
Milford	67	53
Salt Lake City	55	44
Va.		
Richmond	52	36
Wash.		
Seattle	30	23
Spokane	55	45

Table 13 (continued) MEAN ANNUAL AND MAY TO OCTOBER CLASS A PAN EVAPORATION[3]

Location	Annual (in.)	May—Oct (in.)
Wis.		
Green Bay	35	28
Madison	40	32
Wyo.		
Cheyenne	56	41

REFERENCES

1. **Trewartha, G. T.,** *An Introduction to Climate,* 4th Ed. McGraw-Hill, New York, 1968.
2. U.S. Department of Commerce, Comparative Climatic Data through 1977, National Oceanic and Atmospheric Administration, Environmental Data Service, Asheville, N.C., 1978.
3. U.S. Department of Commerce, Climatic Atlas of the United States, Environmental Science Services Administration, Environmental Data Service, Washington, D.C., 1968.
4. U.S. Department of Commerce, U.S. Department of Agriculture, Weekly Weather and Crop Bulletin, Vol. 57, No. 26, p. 12—13 and No. 28, p. 12—13, U.S. Department of Agriculture, Washington, D.C., 1970.

SOIL CLASSIFICATION

Roy W. Simonson

INTRODUCTION

Soil classification is largely a 20th Century phenomenon although stirrings began in the preceding century. In the U.S., for example, efforts were made during the 19th Century to classify the soils of a few counties and states, but classification of soils of the entire country was considered out of reach. Nationwide efforts began during the present century.

Prerequisite to the classification of soils is the concept that the soil mantle consists of natural geographic bodies with their own genesis. Such a concept became possible after the science of geology was born, even though the first theory of soil genesis as rock weathering was found inadequate later. Proposed systems of soil classification have all reflected the concepts of soil and theories of genesis held by the architects. Theories are only mentioned in this article despite their great importance to soil classification and to soil science generally. The theory of genesis held by an individual largely governs what he sees in examining soils and the importance he attaches to the observed characteristics. The characteristics selected as class criteria and the relative weights assigned to them directly reflect theories of soil genesis.

Two systems of soil classification used in the past and the one now being generally used in the U.S. are described in this article to illustrate what has been done and trends in the efforts. For each system a few remarks are made about the underlying concept of soil and theories of genesis. Following the section on classification systems are discussions of specimen properties basic to all efforts in soil classification and an example of the application of the current system in a general soil map of the world.

The reasons for soil classification parallel those for biological taxonomy, expressed by Walters[1] as " . . . the practical needs of study of living creatures, especially for medicine and food. . . . ."

SYSTEMS OF CLASSIFICATION

Numerous systems have been devised and many are in use in various parts of the world today. The trend is toward national systems, as illustrated by those developed in recent years in Australia, Canada, France, Germany, Great Britain, the Netherlands, New Zealand, South Africa, the Soviet Union, and the U.S. The relationships of some other systems to the current one in the U.S. are indicated in the monograph describing that system.[2] Existence of other systems is recognized but only those that have been or are used in the U.S. will be described as examples. These illustrate three approaches to the classification of soils.

Earliest System

The program for a systematic soil survey of the U.S. was launched in 1899 before any scheme for classification of soils had been devised in the country.[3] The kinds of soils shown on the earliest maps were called soil types and no effort was made at first to relate the types of one survey area to those of another. Within the first 10 years of the program, however, a de facto system of three categories evolved. This was done on an ad hoc basis with little thought to the formal logic of a multiple-category system.

The three categories in descending order were the physiographic province or region, the soil series, and the soil type.

The country was divided into 13 physiographic provinces or regions, and each soil series was to be restricted to one of those. Each series was to consist of a group of soil types of "similar origin." In other words, each series was to consist of a set of soil types derived from the same source rock but differing in texture. It was believed that all soil types of one series would have the same or very nearly the same chemical composition because they were derived from the same rock.[4]

This early system of classification was a direct reflection of the prevailing concept of soil as the weathered mantle of rock, and the prevailing theory of soil genesis as rock weathering in place or as the deposition of weathered rock. The system was followed in the U.S. with little change for about 30 years. Despite this frame of reference for the classification of soils, however, attention was given to characteristics that were not directly related to weathering or to source rocks. Those characteristics were being introduced as series criteria within the first 10 years of the century, well before a different approach had been proposed.[5]

Intermediate System

The second system of soil classification in the U.S. was based on a proposal by Marbut[3,6] that was modified a few years after completion. The shift from emphasis on source rocks and weathering to soil profile characteristics began with a pair of papers published in the early 1920s.[7,8] The characteristics spelled out by Marbut in 1922 were still in use as class criteria 30 years later.[9] A skeleton outline of the scheme developed by Marbut is given in Table 1. The categories (a category is a set of classes of the same rank), their bases, and their names are listed without the individual classes. Names of some great soil groups were carried over when the scheme devised by Marbut was modified appreciably a few years after it was published.[10] The modified version gradually came into use although it did not fully replace the initial scheme.

The revised version also consisted of six categories called orders, suborders, great soil groups, families, series, and types. The names of categories were the same as those introduced by Marbut.[6] The 1938 system had 3 orders, 9 suborders, and 36 great soil groups, all identified in Table 2. At the time, approximately 2000 series had been recognized in the country, but these were not grouped into families and great soil groups. A specimen series was listed for each great soil group.

From the time the intermediate system was introduced in 1938 until it was replaced in the soil survey program in 1965, a number of changes were made in the list of great soil groups but not in the structure as a whole. The full set of orders, suborders, and great soil groups in use in 1960 is given in Table 3.[11]

The concept of soil and the theories of genesis reflected in the intermediate system of classification are greatly different from those expressed in the initial system. Rather than being considered the weathered rock mantle, soil was looked upon as an organized, natural body closely related to but differing from weathered rock. Prevailing theories of genesis held that there were a number of broad processes active in soil formation. These were named chiefly in terms of prominent great soil groups, e.g., podzolization, laterization, calcification, solonization, and gleyzation.[12] Each of these was believed to be best expressed in one great soil group but largely responsible as well for some other groups. The broad processes were defined in general terms, as were the great soil groups themselves. Podzolization gave rise to Podzols and certain other groups of soils. Calcification gave rise to Chernozems and related groups. Some efforts were made to describe mechanisms in the processes but these did not go far.

Table 1
CATEGORIES IN THE SCHEME OF MARBUT[6] WITH BRIEF DESCRIPTIONS AND THE NAMES

Category	Descriptions	Names
VI	Solum composition groups	Orders
V	Inorganic colloid composition groups	Suborders
IV	Broad environment groups	Great soil groups
III	Local environment groups	Families
II	Soil type groups	Soil series
I	Soil units	Soil types

Table 2
ORDERS, SUBORDERS, AND GREAT SOIL GROUPS OF THE SYSTEM OUTLINED IN 1938[10]

Orders	Suborders	Great soil groups
Zonal soils	Soils of the cold zone	Tundra soils
	Light-colored soils of arid regions	Desert soils
		Red desert soils
		Sierozems
		Brown soils
		Reddish brown soils
	Dark-colored soils of semi-arid, subhumid, and humid grasslands	Chestnut soils
		Reddish chestnut soils
		Chernozems
		Prairie soils
		Reddish prairie soils
	Soils of the forest-grassland transition	Degraded chernozems
		Noncalcic brown soils
	Light-colored podzolized soils of the timbered regions	Podzols
		Brown podzolic soils
		Gray-brown podzolic soils
	Lateritic soils of forested warm-temperate and tropical regions	Yellow podzolic soils
		Red podzolic soils
		Yellowish brown lateritic soils
		Reddish brown lateritic soils
		Laterite soils
Intrazonal soils	Halomorphic (saline and alkali) soils of imperfectly drained arid regions and littoral deposits	Solonchak soils
		Solonetz soils
		Solodj soils
	Hydromorphic soils of marshes, swamps, seep areas, and flats	Wiesenböden (meadow soils)
		Alpine meadow soils
		Bog soils
		Planosols
		Ground-water podzols
		Ground-water laterite soils
	Calcimorphic soils	Brown forest soils
		Rendzinas
Azonal soils	(No suborders)	Lithosols
		Alluvial soils
		Dry sands

Current System

The classification now in general use in the U.S. can be looked upon as an outgrowth of efforts to complete the intermediate system. Seven years after the latter had been outlined, a program was started to place all series of the country into great soil groups

Table 3
ORDERS, SUBORDERS, AND GREAT SOIL GROUPS OF THE SYSTEM IN USE IN 1960[11]

Orders	Suborders	Great soil groups
Zonal soils	Soils of cold zones	Tundra soils
		Subarctic brown forest soils
	Light-colored soils of arid regions	Desert soils
		Red desert soils
		Sierozems
		Brown soils
		Reddish brown soils
	Dark-colored soils of semi-arid, subhumid, and humid grasslands	Chestnut soils
		Reddish chestnut soils
		Chernozems
		Brunizems
		Reddish prairie soils
	Soils of the forest-grassland transition	Noncalcic brown soils
	Light-colored podzolized soils of forested regions	Podzols
		Brown podzolic soils
		Gray wooded soils
		Sols bruns acides
		Gray-brown podzolic soils
	Lateritic soils of forested warm-temperate and tropical regions	Red-yellow podzolic soils
		Reddish brown lateritic soils
		Yellowish brown lateritic soils
		Groups of Latosols
Intrazonal soils	Halomorphic soils	Solonchak soils
		Solonetz soils
		Solodj soils
	Hydromorphic soils	Humic gley soils
		Alpine meadow soils
		Bog soils
		Low-humic gley soils
		Planosols
		Ground-water podzols
		Ground-water laterite soils
	Calcimorphic soils	Brown forest soils
		Rendzinas
		Grumusols
		Calcisols
	Dark soils on volcanic ash	Ando soils
Azonal soils	(No suborders)	Lithosols
		Regosols
		Alluvial soils

and families. Actually, the family and suborder were shadow categories at the time. The program begun in 1945 was intended to give substance to those categories and place the series into appropriate classes in all higher categories.

Difficulties were immediately encountered in placing series into great soil groups because of inadequate descriptions of the former and sketchy definitions of the latter. Efforts were therefore made to improve the concepts and definitions of great soil groups, on the one hand, and the series descriptions, on the other. Efforts were also made to develop family criteria, and trial groupings of series into families were constructed. These efforts demonstrated in a few years that more profound changes were necessary. Consequently, efforts were begun in 1950 to revise the entire system rather than try to improve individual segments.

The work of revising the classification system was carried forward through a sequence of approximations. Departures from the intermediate system were small in the first approximation but became progressively greater as more were prepared. The 7th Approximation was the first issued as a monograph, published at the time of the 7th International Congress of Soil Science in Madison, Wis. in 1960.[13] Copies of the monograph were distributed widely both inside and outside the U.S. On the basis of criticisms of the 7th Approximation, trial modifications were made of the scheme and reports of these were distributed as supplements in 1964, 1967, and 1968. A full description of the system was published in 1975.[2] Earlier, an introduction to the system[14] and explanations of the objectives, basic assumptions, and logic had been published in scientific journals.[15,16]

Like its predecessor, the current system has six categories. These are called orders, suborders, great groups, subgroups, families, and series. The subgroup category has been added and the soil type category dropped as compared to the previous system. The soil type is now considered a kind of phase within a series, one for texture of the surface layer parallel to phases for stoniness and slope.

Unlike its predecessor, the current system has been completed by defining all classes in all categories and by grouping all soil series of the U.S. into classes of higher rank. Thus, in 1975 the system consisted of 10 orders, 47 suborders, 185 great groups, 900 subgroups, 4,500 families, and 10,500 series. The full set of orders and suborders is given in Table 4. A specimen great group is also listed for each suborder except for those in which none has been set apart.

A new nomenclature was developed for classes in the top four categories of the current system, mostly from Latin and Greek morphemes. The nomenclature is designed so that every class name identifies its rank in the system. Moreover, each name of each class in the three categories below the order identifies the class or classes into which the named one is placed at higher levels. Thus, all names of orders end in "sol." All names of suborders consist of two syllables and each ends in a syllable taken from the name of the appropriate order. For example, the name *Aquod* ends in the letters *od*, indicating that it is a suborder of Spodosols. All names of great groups have three or four syllables and were constructed by adding a prefix to the name of the appropriate suborder, thus, Haplaquod. All names of subgroups are binominal and consist of two words, as for example, Typic Haplaquod. The name of orders, suborders, great groups, and subgroups are considered to be proper names and the initial letters are therefore capitalized. The formative elements or morphemes used in constructing the names of the ten orders are given in Table 5, those for suborders in Table 6, and those for great groups in Table 7. These three tables are taken directly from the monograph describing the current classification system.[2]

The formative elements chosen as prefixes for constructing names of suborders and great groups are meant to be connotative. The prefix *aqu-*, for example, indicates that soils are wet. Similarly, the prefix *calci-* indicates the presence of much calcium. The prefix *hapl-* identifies the simplest set of soils in a suborder. Those of all other great groups in that suborder have some feature or features in addition to the minimum required for the class. Thus, the name *Haplaquod* connotes the most simple version (hapl-) among the wet (aqu-) Spodosols (Podzols).

Names of subgroups are constructed by first selecting some set of soils to epitomize the given great group. That set is then considered a norm for the great group and the first word of the binomial name is typic. Other sets of soils in the great group may share some characteristics with another subgroup or subgroups, either in the same order or in another order. These are distinguished at subgroup level as intergrades and named to suggest their offside character. Still other sets of soils may have some offside characteristics not shared with any other soils in the system. A group of adjectives has

Table 4
ORDERS, SUBORDERS, AND SPECIMEN GREAT GROUPS OF THE CURRENT AMERICAN SYSTEM[2]

Orders	Suborders	Specimen great groups
Alfisols	Aqualfs	Albaqualfs
	Boralfs	Cryoboralfs
	Udalfs	Hapludalfs
	Ustalfs	Natrustalfs
	Xeralfs	Durixeralfs
Aridisols	Argids	Haplargids
	Orthids	Camborthids
Entisols	Aquents	Haplaquents
	Arents	(None defined)
	Fluvents	Udifluvents
	Orthents	Troporthents
	Psamments	Quartzipsamments
Histosols	Fibrists	Sphagnofibrists
	Folists	Borofolists
	Hemists	Medihemists
	Saprists	Cryosaprists
Inceptisols	Andepts	Hydrandepts
	Aquepts	Fragiaquepts
	Ochrepts	Eutrochrepts
	Plaggepts	(None defined)
	Tropepts	Humitropepts
	Umbrepts	Haplumbrepts
Mollisols	Albolls	Natralbolls
	Aquolls	Haplaquolls
	Borolls	Calciborolls
	Rendolls	(None defined)
	Udolls	Argiudolls
	Ustolls	Vermustolls
	Xerolls	Palexerolls
Oxisols	Aquox	Gibbsaquox
	Humox	Sombrihumox
	Orthox	Acrorthox
	Torrox	(None defined)
	Ustox	Haplustox
Spodosols	Aquods	Cryaquods
	Ferrods	(None defined)
	Humods	Haplohumods
	Orthods	Troporthods
Ultisols	Aquults	Ochraquults
	Humults	Tropohumults
	Udults	Paleudults
	Ustults	Haplustults
	Xerults	Palexerults
Vertisols	Torrerts	(None defined)
	Uderts	Pelluderts
	Usterts	Chromusterts
	Xererts	Chromoxererts

been selected for naming these sets of soils as subgroups, considered extragrades, as contrasted to those considered intergrades. Extragrades lean toward something outside of the system, whereas intergrades lean toward some other soils in the system. To continue with the example of wet Spodosols, a total of 11 subgroups have been recognized in that one great group of Haplaquods. In addition to the typic are the aeric,

Table 5
FORMATIVE ELEMENTS IN THE NAMES OF THE SOIL ORDERS

Name of order	Formative element in name of order	Derivation of formative element	Pronunciation of formative element
Alfisol	Alf	Meaningless syllable	Pedalfer
Aridisol	Id	L. *aridus,* dry	Arid
Entisol	Ent	Meaningless syllable	Recent
Histosol	Ist	Gr. *histos,* tissue	Histology
Inceptisol	Ept	L. *inceptum,* beginning	Inception
Mollisol	Oll	L. *mollis,* soft	Mollify
Oxisol	Ox	F. *oxide,* oxide	Oxide
Spodosol	Od	Gr. *spodos,* wood ash	Odd
Ultisol	Ult	L. *ultimus,* last	Ultimate
Vertisol	Ert	L. *verto,* turn	Invert

From Soil Survey Staff, *U.S. Dep. Agric. Handb.,* 436, 1975.

alfic, alfic arenic, entic, ferrudalfic, grossarenic, lithic, placic, and ultic subgroups. One of these adjectives or the single pair precedes the word Haplaquod to make the name of a subgroup.

A further departure of the current from the intermediate system lies in the definitions of classes within categories. Class limits are explicit and quantitative. Definitions are also based entirely on soil properties rather than partly on presumed genesis of the soils. In the four upper categories, the definitions are in terms of diagnostic horizons and other features, which are themselves defined with care. For the family classes, definitions are chiefly in terms of particle size distribution, mineralogy, temperature regime, and soil thickness.

For the soil orders collectively, the criteria are gross composition of soils (Histosols vs. others), the presence of certain diagnostic horizons, evidence of cracking and churning of the soil, and a combination of weatherability and stage of weathering.

For subdividing the soil orders into suborders, the chief criteria are moisture regimes, temperature regimes, mineralogy, and total amounts of organic matter. Other characteristics are criteria in some instances. The number of suborders ranges from a minimum of two to a maximum of seven among the ten orders in the system. The number of suborders in each soil order and the primary criteria for differentiating the former are given in Table 8.

The criteria for subdivision of suborders into great groups include the presence or absence of specified diagnostic horizons, the presence of horizons extra to the definitive sequence, and some other features. Extra horizons and other features not diagnostic for suborders distinguish more great groups within suborders than do extra diagnostic horizons. Some indication of the application of different criteria can be gained from the numbers of great groups in the two columns in Table 9. No effort is made here to list the differentiating characteristics used for setting apart all great groups because of their large number.

The approach in subdividing great groups departs from those applied to categories at higher levels. First of all, a norm is selected for a great group. This is meant to typify or epitomize that great group. The soils which fit the norm for a great group are considered a typic subgroup, e.g., Typic Haplaquods and Typic Camborthids. Other soils of a great group are considered intergrades to some other great group either in the same or a different order, or extragrades that lean toward something outside of

Table 6
FORMATIVE ELEMENTS IN THE NAMES OF THE SUBORDERS

Formative element	Derivation	Mnemonicon	Connotation
Alb	L. *albus*, white	Albino	Presence of albic horizon
And	Modified from ando	Ando	Andolike
Aqu	L. *aqua*, water	Aquarium	Aquic moisture regime
Ar	L. *arare*, to plow	Arable	Mixed horizons
Arg	Modified from argillic horizon; L. *argilla*, white clay	Argillite	Presence of argillic horizon
Bor	Gr. *boreas*, northern	Boreal	Cool
Ferr	L. *ferrum*, iron	Ferruginous	Presence of iron
Fibr	L. *fibra*, fiber	Fibrous	Least decomposed stage
Fluv	L. *fluvius*, river	Fluvial	Flood plain
Fol	L. *folia*, leaf	Foliage	Mass of leaves
Hem	Gr. *hemi*, half	Hemisphere	Intermediate stage of decomposition
Hum	L. *humus*, earth	Humus	Presence of organic matter
Ochr	Gr. base of *ochros*, pale	Ocher	Presence of ochric epipedon
Orth	Gr. *orthos*, true	Orthophonic	The common ones
Plagg	Modified from Ger. *plaggen*, sod		Presence of plaggen epipedon
Psamm	Gr. *psammos*, sand	Psammite	Sand texture
Rend	Modified from Rendzina		High carbonate content
Sapr	Gr. *sapros*, rotten	Saprophyte	Most decomposed stage
Torr	L. *torridus*, hot and dry	Torrid	Torric moisture regime
Ud	L. *udus*, humid	Udometer	Udic moisture regime
Umbr	L. *umbra*, shade	Umbrella	Presence of umbric epipedon
Ust	L. *ustus*, burnt	Combustion	Ustic moisture regime
Xer	Gr. *xeros*, dry	Xerophyte	Xeric moisture regime

From Soil Survey Staff, U.S. Dep. Agric. Handb., 436, 1975.

Table 7
FORMATIVE ELEMENTS IN THE NAMES OF GREAT GROUPS

Formative element	Derivation	Mnemonicon	Connotation
Acr	Modified from Gr. *akros*, at the end	Acrolith	Extreme weathering
Agr	L. *ager*, field	Agriculture	An agric horizon
Alb	L. *albus*, white	Albino	An albic horizon
And	Modified from ando	Ando	Andolike
Arg	Modified from argillic horizon; L. *argilla*, white clay	Argillite	An argillic horizon
Bor	Gr. *boreas*, northern	Boreal	Cool
Calc	L. *calcis*, lime	Calcium	A calcic horizon
Camb	L.L. *cambiare*, to exchange	Change	A cambic horizon
Chrom	Gr. *chroma*, color	Chroma	High chroma
Cry	Gr. *kryos*, icy cold	Crystal	Cold
Dur	L. *durus*, hard	Durable	A duripan
Dystr, dys	Modified from Gr. *dys*, ill; dystrophic, infertile	Dystrophic	Low base saturation
Eutr, eu	Modified from Gr. *eu*, good; eutrophic, fertile	Eutrophic	High base saturation
Ferr	L. *ferrum*, iron	Ferric	Presence of iron
Fluv	L. *fluvus*, river	Fluvial	Flood plain
Frag	Modified from L. *fragilis*, brittle	Fragile	Presence of fragipan
Fragloss	Compound of fra(g) and gloss		See the formative elements frag and gloss
Gibbs	Modified from gibbsite	Gibbsite	Presence of gibbsite in sheets or nodules
Gyps	L. *gypsum*, gypsum	Gypsum	Presence of a gypsic horizon
Gloss	Gr. *glossa*, tongue	Glossary	Tongued
Hal	Gr. *hals*, salt	Halophyte	Salty
Hapl	Gr. *haplous*, simple	Haploid	Minimum horizon
Hum	L. *humus*, earth	Humus	Presence of humus
Hydr	Gr. *hydor*, water	Hydrophobia	Presence of water
Luv	Gr. *louo*, to wash	Ablution	Illuvial
Med	L. *media*, middle	Medium	Of temperate climates
Nadur	Compound of na(tr) and dur		See the formative elements natr and dur
Natr	Modified from *natrium*, sodium		Presence of natric horizon
Ochr	Gr. base of *ochros*, pale	Ocher	Presence of ochric epipedon
Pale	Gr. *paleos*, old	Paleosol	Excessive development
Pell	Gr. *pellos*, dusky		Low chroma

Table 7 (continued)
FORMATIVE ELEMENTS IN THE NAMES OF GREAT GROUPS

Formative element	Derivation	Mnemonicon	Connotation
Plac	Gr. base of *plax*, flat stone		Presence of a thin pan
Plagg	Modified from Ger. *plaggen*, sod		Presence of plaggen epipedon
Plinth	Gr. *plinthos*, brick		Presence of plinthite
Psamm	Gr. *psammos*, sand	Psammite	Sand texture
Quartz	Ger. *quarz*, quartz	Quartz	High quartz content
Rhod	Gr. base of *rhodon*. rose	Rhododendron	Dark red color
Sal	L. base of *sal*, salt	Saline	Presence of salic horizon
Sider	Gr. *sideros*, iron	Siderite	Presence of free iron oxides
Sombr	F. *sombre*, dark	Somber	A dark horizon
Sphagn	Gr. *sphagnos*, bog	Sphagnum	Presence of Sphagnum
Sulf	L. *sulfur*, sulfur	Sulfur	Presence of sulfides or their oxidation products
Torr	L. *torridus*, hot and dry	Torrid	Torric moisture regime
Trop	Modified from Gr. *tropikos*, of the solstice	Tropical	Humid and continually warm
Ud	L. *udus*, humid	Udometer	Udic moisture regime
Umbr	L. base of *umbra*, shade	Umbrella	Presence of umbric epipedon
Ust	L. base of *ustus*, burnt	Combustion	Ustic moisture regime
Verm	L. base of *vermes*, worm	Vermiform	Wormy, or mixed by animals
Vitr	L. *vitrum*, glass	Vitreous	Presence of glass
Xer	Gr. *xeros*, dry	Xerophyte	A xeric moisture regime

From Soil Survey Staff, U.S. Dep. Agric. Handb., 436, 1975.

Table 8
CRITERIA FOR SUBDIVISION OF SOIL ORDERS INTO SUBORDERS IN THE CURRENT CLASSIFICATION SYSTEM[2]

Soil orders	Number of suborders	Bases for subdivisions
Alfisols	5	Moisture and temperature regimes
Aridisols	2	Argillic horizon
Entisols	5	Moisture regime, particle sizes, stratification
Histosols	4	Moisture regime, nature of organic materials
Inceptisols	6	Moisture and temperature regimes, mineralogy, epipedon
Mollisols	7	Moisture and temperature regimes, mineralogy, albic horizon
Oxisols	5	Moisture regime, amount of organic matter
Spodosols	4	Moisture regime, nature of spodic horizon
Ultisols	5	Moisture regime, amount of organic matter
Vertisols	4	Moisture regime

Table 9
MAJOR BASES FOR SUBDIVISION OF SUBORDERS INTO GREAT GROUPS IN THE CURRENT AMERICAN SYSTEM[2]

	Great groups set apart on	
Orders	Diagnostic horizons	Other features
Alfisols	18	19
Aridisols	9	2
Entisols		27
Histosols		20
Inceptisols	14	20
Mollisols	16	17
Oxisols	9	5
Spodosols	10	8
Ultisols	10	14
Vertisols		7

the system. Names are devised accordingly, e.g., Entic Haplaquods and Lithic Camborthids. The latter subgroup consists of Camborthids (Aridisols) that are shallow to rock.

The concept of soil and the theory of soil genesis reflected in the current system of classification carry over in part but also differ in important respects from the bases for the intermediate system. Changes have been much smaller in concept of soil than in theories of genesis.

The soil mantle is still considered to consist of natural organized bodies related to but differing from weathered rock. More consideration is given than previously to the three-dimensional character of soils. The mantle over the land surface is considered

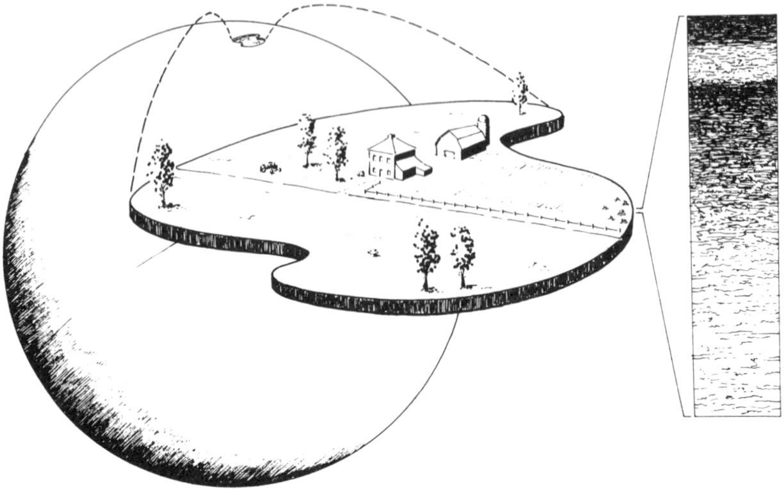

FIGURE 1. Diagram to show relationships between a body of soil that would be represented as a single delineation on a detailed soil map to the rind of the earth, on the one hand, and a profile, on the other. (Sketch by Walter M. Simonson.)

to be a mosaic of small bodies. Relationships between a single mappable body of soil, the rind of the earth, and a pedon are illustrated in Figure 1. The change in concept of soil has therefore been modest, although there is an effort to recognize a small basic entity for which the name *pedon* has been proposed.[13]

The current system of classification does reflect a substantial departure from previous theories of soil genesis, which is now considered to be due to a combination of many relatively narrow rather than a few broad processes. Some of these promote and others retard or offset horizon differentiation. Furthermore, the same processes are believed to operate generally and to be at least potential contributors to the genesis of every kind of soil on earth. The balance among the component processes over time determines the present nature of the soil at any given spot. The number of these narrow or basic processes is large, but they can be grouped under four headings: gains, losses, transfers, and transformations.[17] This multiple process model of soil genesis is an appreciable departure from the early one of broad processes for which the names podzolization, laterization, and the like were used. Those processes were considered to operate uniquely in certain kinds of soils.[18]

One clear-cut expression in the current system of the underlying theory of soil genesis is the recognition of intergrade subgroups. Such subgroups have the dominant characteristics of one great group but have subordinate characteristics normal to some other great group. To state this in another way, the intergrade falls between the norms for a pair of great soil groups, although within the limits of one. The balance among the component processes differs slightly from the balance that holds for the norms of the two great groups. In contrast to the intergrades, marked differences in balance are reflected in contrasting soils such as Aridisols and Ultisols.

WIDELY USED CRITERIA

A large number of properties not mentioned so far are considered in the classification of soils, especially into series and families. As a matter of fact, the full list of characteristics for distinguishing classes in the four top categories of the current American system is not included in this article. Examples have been given instead; readers must go to the monograph describing this system to learn the entire set of properties used as criteria.[2] This section illustrates soil properties used especially to describe and define soil series although not all of those are identified. More complete information is given in the Soil Survey Manual published in the U.S.[19,20]

Soil Profile

Much emphasis in describing soils and in defining soil series is given to the soil profile, a section down through a pedon. The sequence and arrangement of horizons within a profile and their characteristics are keys to the concepts and descriptions of soil series. Horizons are layers parallel to the soil surface and range in thickness from a few millimeters to as much as 1½ m. The nature of the individual horizons is due to the additions, losses, transfers, and transformations that have affected a profile in the past. The sequence and nature of the horizons in a profile are thus a history of its formation, could we but read what is written in them.

A hypothetical soil profile with all horizons is shown in Figure 2. This illustrates the horizons found in mineral soils plus underlying rock which is not part of the soil and not a horizon. All horizons in Figure 2 are seldom present in a single profile but part of the full set always is. As indicated, the notations in Figure 2 apply to mineral soils. A separate notation is applied to organic soils. That notation is not given because the relative extent of Histosols (organic soils) is small as compared to mineral soils. The horizon notations are given in the monograph describing the current classification system.[2]

Major horizons in profiles are identified by the letters A, B, and C. In addition, the letters O and R are used to flag less generally occurring layers. The letter O refers to organic matter on the soil surface, whereas R is used to label rock beneath the soil. Arabic numerals are used with the capital letters for subdivisions of the major horizons, as shown in Figure 2. Further subdivisions of major horizons are indicated by lowercase letters following the suffix numerals. The combination of capital letters, numerals, and lowercase letters provides a form of shorthand to convey some information about the horizons and layers.[20]

The O horizon is a surface layer consisting of organic debris in some stage of decomposition. On the basis of the stage of decay, the horizon is subdivided into O1 and O2 horizons. The O1 horizon consists of leaves, twigs, and other litter which is just beginning to decay. Plant structures are readily evident. The O2 horizon consists of organic residues that have decomposed enough so that plant structures are no longer identifiable. The O horizon is most common in soils under forest cover in humid, cool-temperate climates. A thin O1 horizon may be present in soils under virgin prairie. Many mineral soils, however, lack O horizons.

The A horizon occurs at or near the soil surface. In most soils, it is reached first by the falling rain and by solar radiation. Life is therefore most abundant in that part of the profile, and the A horizon is also highest in organic matter and humus with few exceptions. The immediate surface layer highest in organic matter is also the darkest one in a profile, as a rule, and it is designated the A1 horizon. In well-drained soils, its thickness ranges from 3 cm or less in the eastern U.S. to 30 cm or more in the midwestern prairies. A1 horizons are present in profiles of a great majority of the soils of the world.

Where the A1 horizon is thin, the second layer is usually paler in color and lower in organic matter, bases, and sesquioxides (aluminum and iron oxides). This layer is called an A2 horizon or, in some systems of notation, an E horizon, with that letter standing for *eluviated.* This means that substances such as humus, bases (calcium and the like), and sesquioxides have been washed out of the layer. Substances are lost from both A1 and A2 horizons, but the loss is offset and masked to a greater extent in the

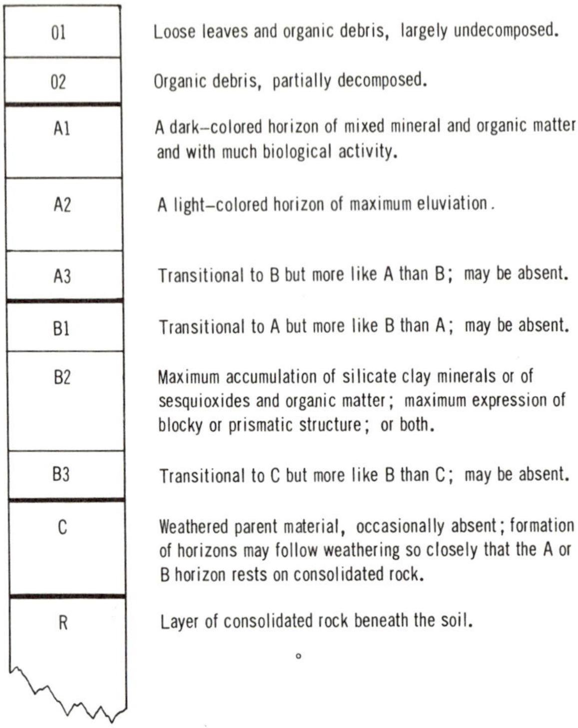

FIGURE 2. A hypothetical profile having all principal horizons and layers. Some but not all are part of every soil profile. Lower case letters are used as suffixes to indicate features subordinate to those identified by the capital letters and numeral suffixes. Meanings of the lower case letters are in the 1962 supplement.[20]

former than in the latter. A2 horizons are commonly present in soils under forest cover in humid climates and are as commonly lacking in soils under prairie vegetation in subhumid and semiarid regions. A2 horizons are best expressed in profiles formed in parent materials with moderate amounts of quartz or its equivalent in the sand and silt fractions.

The principal horizons in the profile of a Spodosol (Podzol) in Quebec are illustrated in Figure 3.

The B horizon is a subsurface layer set apart by one or more characteristics from the A and C horizons. Being between the two, it is transitional from one to the other in many ways. The B horizon may be one of illuviation, a layer into which one or more of organic matter, silicate clays, and sesquioxides have been transferred in suspension, solution, or both. Such B horizons also have blocky or prismatic structure, as a rule (see later description of soil structure). In some soils the B horizon is a layer of residual concentration of either or both sesquioxides and silicate clays. Such B horizons occur in humid and wet-dry tropics and are conspicuous for their bright colors. In many soils, especially in those of grasslands, the B horizon is a subsurface layer with distinct blocky or prismatic structure as the chief evidence of alteration of the original weathered rock material. Little or no illuviation has occurred, although there may have been some losses.

Many of the criteria for defining and differentiating classes in upper categories of the current system in the U.S. are keyed to B horizons. They are less subject to modification than A horizons, which means that placements of soils can be more stable.

FIGURE 3. Profile of a Typic Haplorthod (Podzol) formed in glacial drift in Quebec, Canada, with principal horizons identified. Numbers on scale show inches. This profile represents a subgroup in the great group of Haplorthods, the suborder of Orthods, and the order of Spodosols.

Despite that advantage, the A horizon or that horizon plus the upper part of the B horizon have been used to distinguish and define the order of Mollisols. The relatively thick and dark surface horizon is such a prominent and general feature of Mollisols, the soils of prairie regions, that it was assigned first place in defining the soil order.

Taken together, the A and B horizons are called the solum, meaning that they are the principal genetic horizons. They are sometimes called the "true soil" because those two horizons are the principal reflections of soil genesis.

The C horizon is a subsurface layer of weathered materials, commonly presumed to be like those from which the A and B horizons were developed. This horizon, like the A and B horizons, may be absent with one of the other two resting directly on rock. Conventions have been developed to label C horizons that clearly consist of material different from those in the A and B horizons.[20] As the nature and genesis of soils become progressively better known, more and more profiles are found to have been formed in geologic sections that were not initially uniform. In some stratified deposits the A, B, and C horizons all differ in their lithology. Consequently, many C horizons are simply the weathered materials below the A or B horizons.

The letter R is used to flag a layer of consolidated rock beneath the soil profile. The layer is not considered a soil horizon, but some identification seems desirable wherever rock does occur within fairly shallow depths. The rock may or may not have been the source of the materials in the solum and C horizon above.

Characterization of Horizons

Soil horizons are described in terms of properties such as thickness, color, texture, structure, consistence, porosity, mineralogy, reaction, and chemical composition. These characteristics are in turn the basis for determining whether a given horizon or combination of horizons qualifies as a diagnostic feature for a class in one of the upper categories of the current system. Such characteristics are also the basis for placements of soils into families and series. To describe horizons, classes for describing the properties have been established, even as classes are used for the soils themselves. To illustrate what has been done, the classes for texture and structure will be described. The full ranges in other properties such as color, consistence, and porosity have also been subdivided into classes. All of these classes and other properties that are used in characterizing soil horizons are described and defined in the Soil Survey Manual[19] and supplement.[20]

Texture

Soil texture depends on the relative proportions of different size groups of mineral particles below 2 mm in diameter in a small mass of soil material. The size groups are called soil separates; their names and limits are given in Table 10.

Soil texture classes are defined largely on the basis of relative proportions of separates of different sizes. Permissible proportions of clay (below 0.002 mm), silt (0.002 to 0.05 mm), and sand (0.05 to 2.0 mm) in the basic texture classes are given in Figure 4. Note that the words *clay* and *silt* are used as the names of separates and basic texture classes. "Sand" may also be used as the name for a texture class but usually with some modifier.

Subdivisions are made of some of the basic texture classes shown in Figure 4 according to the relative proportions of the sand separates. Examples are coarse sand and fine sand, loamy fine sand, loamy very fine sand, and coarse sandy loam and fine sandy loam.

The texture of a horizon is a rather permanent property, more so than are structure, porosity, and consistence. All of the latter properties can, for example, be altered by

Table 10
SIZE LIMITS OF SOIL SEPARATES IN SCHEME OF U.S. DEPARTMENT OF AGRICULTURE[19]

Name of separate	Diameter range (mm)
Very coarse sand	2.0—1.0
Coarse sand	1.0—0.5
Medium sand	0.5—0.25
Fine sand	0.25—0.10
Very fine sand	0.10—0.05
Silt	0.05—0.002
Clay	Below 0.002

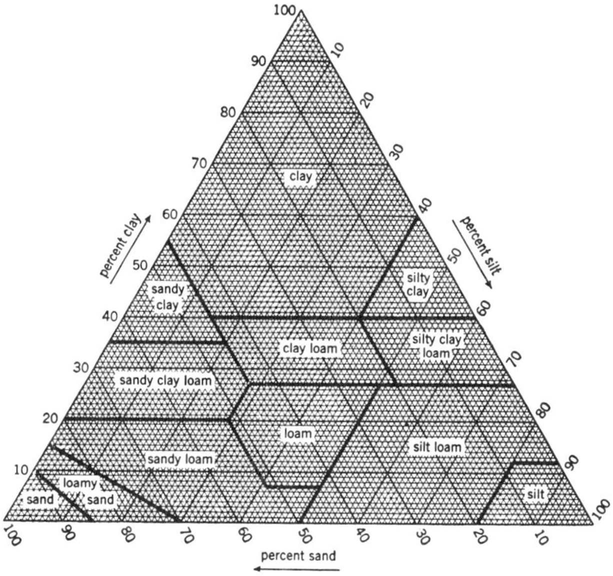

FIGURE 4. Triangular graph showing the proportions of clay (below 0.002 mm), silt (0.05 to 0.002 mm), and sand (2.0 to 0.05 mm) in the basic classes of soil texture.

tillage, whereas much more needs to be done to change texture. Texture was a characteristic used to distinguish soil types in the earliest soil surveys in the U.S. because Whitney[21] believed it to be the best available index to conditions of soil moisture and temperature. Furthermore, he thought those were the primary controls of plant growth.

Textures of horizons are considered in defining soil series and in subdividing them. Texture of the surface layer is used to define and distinguish soil types within a series. The textures of the B and C horizons can be differentiating for soil series, although differences large enough to be considered at series level are commonly associated with other differences.

Particle size distribution in a part of the profile called the "control section" is one basis for differentiating and defining soil families. Particle size distribution or sandiness also sets apart the suborder of Psamments from others in the order of Entisols

Table 11
PRINCIPAL TYPES (SHAPES) AND SIZE CLASSES OF SOIL STRUCTURE[19]

Platelike, with the vertical dimension much smaller than the other two; faces mostly horizontal; size ranges for vertical dimensions	Prismlike, with the vertical dimension much greater than the other two; vertical faces distinct; size ranges for horizontal dimensions	Blocklike, with all 3 dimensions the same; a polyhedron arranged around a point and having plane surfaces	Spheroidal, arranged around a point and without plane surfaces
Very fine platy < 1 mm	Very fine prismatic < 10 mm	Very fine blocky < 5 mm	Very fine granular < 1 mm
Fine platy 1—2 mm	Fine prismatic 10—20 mm	Fine blocky 5—10 mm	Fine granular 1—2 mm
Medium platy 2—5 mm	Medium prismatic 20—50 mm	Medium blocky 10—20 mm	Medium granular 2—5 mm
Coarse platy 5—10 mm	Coarse prismatic 50—100 mm	Coarse blocky 20—50 mm	Coarse granular 5—10 mm
Very coarse platy > 10 mm	Very coarse prismatic > 100 mm	Very coarse blocky > 50 mm	Very coarse granular > 10 mm

and the great group of Psammaquents from others in the suborder of Aquents, the wet Entisols.

Structure

Soil structure is the natural clustering of primary particles into groups that are separated by surfaces of weakness. A single natural cluster is called a ped. Among soils of the world, peds have a variety of shapes and sizes as well as a wide range in distinctness. Classes for shapes and sizes of peds are given in Table 11, which is adapted from the Soil Survey Manual.[19] Distinctness of peds is called structure grade and is normally described as weak, moderate, or strong. Provision has been made for subdivision of each of the weak and strong grades by adding the modifier *very* or *moderately.* If a horizon is massive and lacks discernible structure, as can happen with one high in clay, it is called structureless. At the other end of the scale, an incoherent sand is called single-grained and is also considered structureless.

The four principal types of structure — granular, platy, blocky, and prismatic — are illustrated in Figure 5.

The structures of soil horizons are rarely if ever the sole basis for setting apart soil classes, even those of low rank such as series. Differences in structure of comparable horizons in two sets of otherwise similar soils will be associated with differences in other characteristics. For example, a strong versus weak grade in structures of B horizons of two sets of soils commonly means that peds in the B horizons of one set have coats of some kind. Those may be films of silicate clays, mixtures of clay and humus, or some other substance or substances. In parallel fashion, differences in grade among other types of structure also indicate differences in fabric or composition of soil materials.

Certain types and grades of structure are characteristic of several diagnostic horizons that distinguish classes of high rank in the current system. The mollic epipedon, a dark surface layer that usually coincides with the A1 horizon in a profile, is a requirement for the order of Mollisols. That horizon has granular structure, commonly of moderate or strong grade. Argillic horizons (which are B horizons) are required in the profiles of Alfisols and Ultisols and also in some suborders and great groups in other soil orders. Such B horizons usually have blocky structure of moderate grade, although the type may be prismatic. Natric horizons are a form of argillic horizon high in exchangeable sodium or sodium plus magnesium. That kind of horizon is the basis for

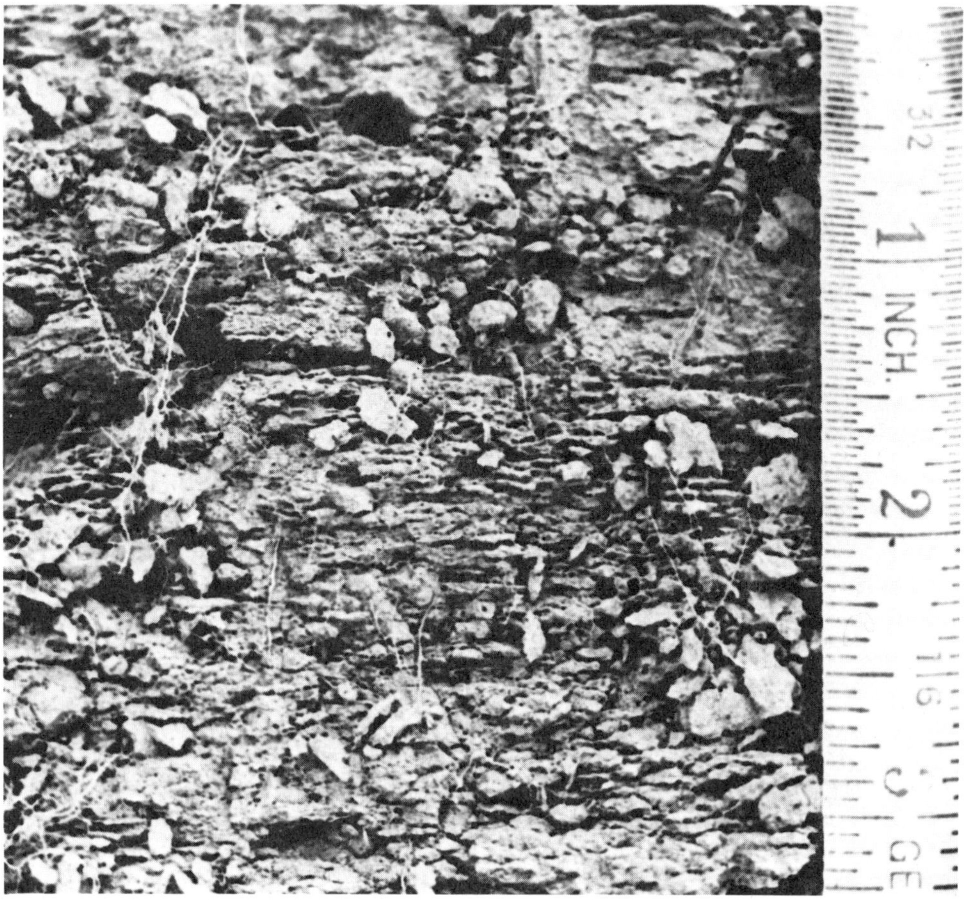

5A

FIGURE 5. Examples of the four types of shapes of soil structure. (A) Platy, (B) granular, (C) blocky, and (D) prismatic. Numbers on scales show inches. Platy structure is common in the A2 horizons of Hapludalfs (Gray-brown podzolic soils), granular structure in the A1 horizons of Mollisols, blocky structure in the B2 horizons of Hapludalfs, and prismatic structure in the B2 horizons of Natrustolls (Solonetz soils).

setting apart great groups within suborders in several soil orders. Natric horizons have prismatic or blocky structure or some combination of the two, commonly of moderate grade or better. These three examples illustrate ways in which the type and grade of structure serve as aids in the identification of horizons that are diagnostic.

APPLICATION OF CURRENT SYSTEM

One purpose in constructing a system with several categories was to provide soil classes appropriate to universes differing in size. For example, a single farm may be the universe for which information about soils is needed. Other possible universes are a township (36 mi^2), a county (few hundred mi^2), the whole of a state, the entire country, a continent, or the world. Classes with narrow ranges in characteristics, such as the soil series, are meant to serve needs within farms. Classes in higher categories have progressively wider ranges in characteristics going up the ladder in the system and those classes are meant for progressively larger universes. How well the system will serve this one purpose remains to be seen as it is applied to universes of different sizes.

FIGURE 5B.

Application of the system to one universe is illustrated by a general soil map of the world in Figure 6. The purpose of the map and the remainder of this section is to provide a general picture of the soils of the world. Soils are identified in terms of the orders, broadest classes in the current system. Nine regions have been delineated on the map, named for the dominant order or orders except for mountainous regions. In such regions, the mountains are so important and the variety of soils so large that names of orders could be misleading. Soil orders in addition to those given in the names are present in every region.

Approximate proportions of the land surface of the earth in the nine broad regions are given in Table 12. For this purpose, the land area is taken to be 51 million mi^2 or 131 million km^2. This total excludes the ice caps of Antarctica and Greenland as well as the large inland lakes and seas.

Some information about each of the broad regions and its soils is given in subsequent pages. Regions are treated in the same sequence as in the legend in Figure 6 and in Table 12.

Inceptisols

This is the region of the tundra, consisting of large bodies in Asia, Europe, and North America. The soil region is the second smallest in the world.

FIGURE 5C.

Prime feature of the region is its cold climate. Because of that, the soils lack distinct horizons. Most are wet and many have permafrost at some depth. Many are also subject to churning by frost action. Moreover, the low temperatures limit plant growth to dwarf vegetation which provides some grazing for caribou or reindeer.

Spodosols and Histosols

The largest bodies are in northern Asia, Europe, and North America. A small additional area lies at the southern tip of South America. The soil region ranks fifth in size among those of the world.

Land surfaces of the region are young, on the whole, because they were recently glaciated. The climate is generally humid and cool. Prevailing temperatures are low but not as low as in the tundra region.

Spodosols, the major soils, have B horizons (spodic horizons) with accumulations of one or more of humus, aluminum, and iron transferred downward from overlying A horizons. As a group, the soils are strongly acid and low in bases such as calcium and magnesium. A Spodosol profile is shown in Figure 3.

Histosols (peats and mucks) form part of the soil pattern throughout most of the region. Those soils are more extensive than Spodosols in a few places in Canada and the Soviet Union.

Most soils of the region remain in forest, although some are cultivated as, for ex-

FIGURE 5D.

ample, in parts of eastern Canada and northeastern Maine. All in all, the prevailing low temperatures place a ceiling on plant growth and on possibilities for utilization of the soils.

Alfisols and Inceptisols

The largest bodies are in Europe and the eastern parts of North America and Asia. Small bodies occur in both islands of New Zealand. The soil region ranks sixth in size among those of the world, being about half as large as that of Spodosols and Histosols.

Land surfaces of the region are relatively young, with many having been affected by the latest glaciations. For the most part, climates are humid and cool-temperate. The region thus falls between the cool climates of Spodosols and Histosols, on the one side, and the warm-temperate climates of Ultisols, on the other.

By and large, the Alfisols and Inceptisols have lost bases, silicate clays, and sesquioxides from their A horizons. Profiles as a whole have been leached but not severely. The Alfisols have argillic horizons, B horizons of silicate clay accumulation; whereas the Inceptisols have cambic horizons, B horizons of alteration and residual concentration without accumulations of transferred materials. Soils of both orders are commonly somewhat acid and low in organic matter.

Most Alfisols and Inceptisols are fairly productive when farmed with simple technology. Such soils have produced food in China and western Europe for many centu-

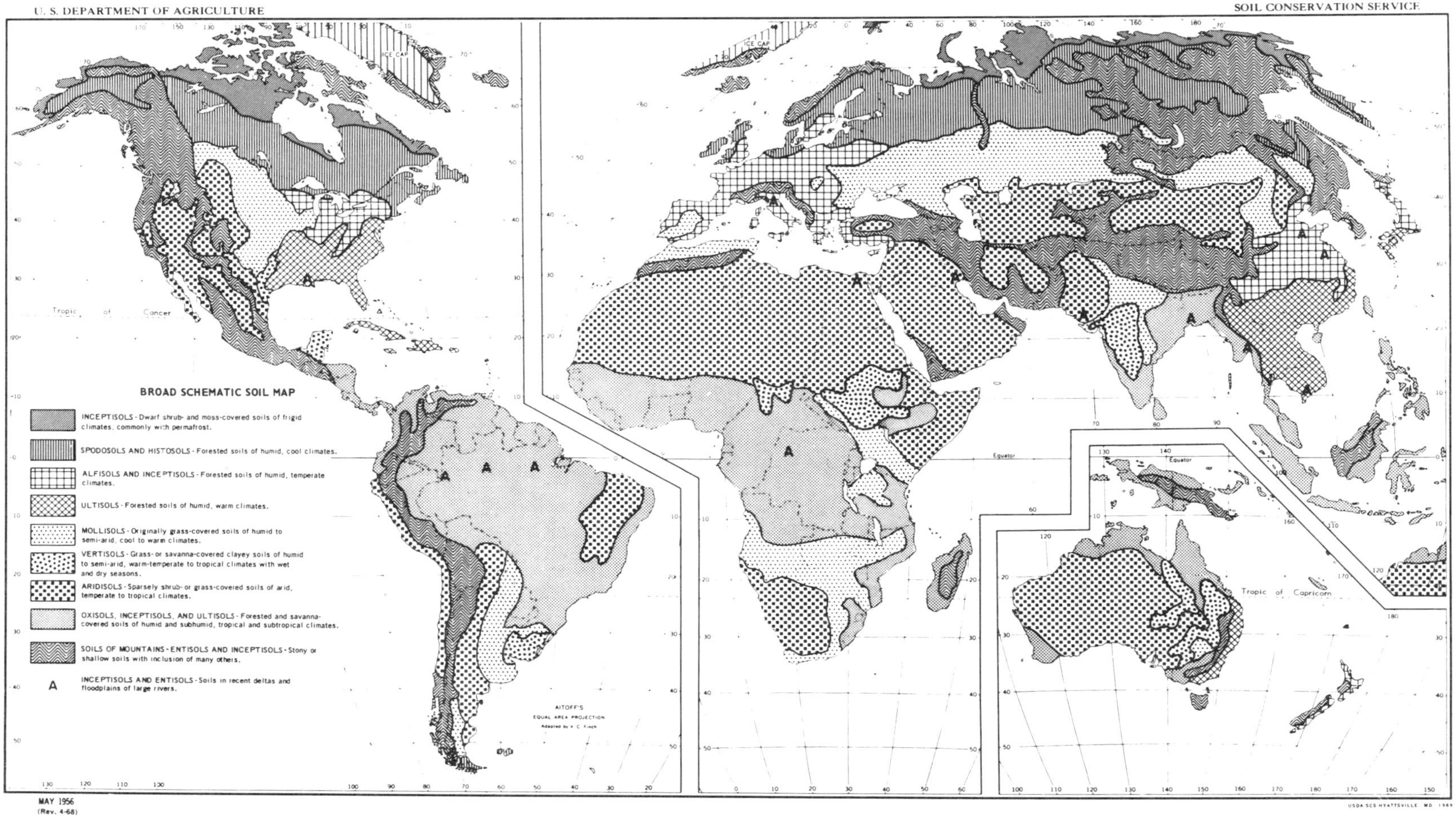

FIGURE 6. Map of the world showing nine major soil regions. Proportions of the land surface of the earth in the broad regions are given in Table 12.

Table 12
APPROXIMATE PERCENTAGES OF THE LAND SURFACE OF THE EARTH IN NINE BROAD SOIL REGIONS

Dominant soil order in region	Percentage of land surface
Inceptisols	3.44
Spodosols and Histosols	9.89
Alfisols and Inceptisols	4.25
Ultisols	3.83
Mollisols	12.38
Vertisols	2.81
Aridisols	25.34
Oxisols, Inceptisols, and Ultisols	20.52
Soils of mountains — Entisols and Inceptisols	17.44

ries. Furthermore, soils of the two orders, especially Alfisols, were being used by Huron Indians in Ontario and New York to produce corn at the time of early explorations of North America by Europeans. The soils can be productive if used with modern technology, some of them highly productive.

Ultisols

The largest bodies are in the southeastern parts of Asia and the U.S. with smaller ones in eastern Australia, southern Japan, and the Caribbean Islands. The soil region ranks seventh in size, being a little smaller than that of Alfisols and Inceptisols.

The Ultisols occur on land surfaces that have been stable for long periods, chiefly under humid, warm-temperate or tropical climates. Consequently, the soils have been strongly leached and tend to be low in nutrient elements. They generally have A2 horizons poor in organic matter, bases, and sesquioxides. They have acid argillic horizons which are brightly colored in well-drained members. Because of the long-time leaching, the soils have low natural fertility although physical properties of many are favorable for plant growth. Generally speaking, the soils are readily subject to erosion when cultivated.

Ultisols have been used to produce food and fiber under shifting cultivation in the past. If nothing more than simple technology is available to cultivators, shifting technology is the most stable system available. In contrast, many Ultisols are highly productive if farmed within systems that can use sophisticated technology. Large unrealized potentials thus hold for the soils.

Mollisols

The largest bodies are in Eurasia, Africa, and North America with smaller ones in South America and Australia. The soil region ranks fourth in size in the world, being one fifth larger than that of Spodosols and Histosols. Mollisols are the principal components of the soil pattern in the steppes of the Soviet Union, Great Plains of North America, and the pampas of Argentina.

The largest areas of these soils are in the temperate zone, but some occur in tropical and subtropical zones. Climate is mostly subhumid to semiarid, falling between the humid and the arid.

The distinguishing mark of Mollisols is the mollic epipedon, a thick dark surface horizon (A1 horizon) high in humus and nitrogen. The B horizon is seldom well expressed, mostly being transitional from the dark overlying A1 to the underlying C horizon. The soils have been subject to limited leaching and the levels of nutrient elements are high. Mollisols are thus among the naturally most fertile soils of the world.

These soils are being used mainly for growing cereals on large farms. Furthermore, they produce a major share of the output of small grains and corn for the world as a whole. Total production goes up and down with the weather from season to season, especially in the areas marginal to arid regions.

Vertisols

The largest bodies are in Africa, Australia, and India with smaller ones in the U.S. and South America. The soil region is smallest among the nine.

These soils occur under subtropical and tropical climates with a wide range in rainfall. Alternate wet and dry seasons are characteristic of all climatic regimes.

These soils are high or very high in clay, are plastic and sticky, and shrink and swell appreciably with changing moisture contents. The shrinking and swelling is accompanied by cracking, which extends to the surface. Mass movement also occurs within the soil body. The movements result in churning and mixing of soil materials throughout the profile. As one consequence, the soils have poorly expressed horizons. The fine texture and the churning tend to preclude or offset leaching, as a result of which the soils are relatively high in nutrient elements. Because of the clay textures, much of the water retained by the soils is not available to plants.

Vertisols are being cultivated in some places, left in native vegetation for livestock grazing in others, and simply left alone in still others. If Vertisols must be cultivated with simple technology dependent on animal power and human labor, production is low. Successful utilization of the soils requires large and powerful machinery as well as other elements of sophisticated technology. Efforts to cultivate the soils are therefore not made in many places.

Aridisols

The largest bodies are the deserts of Africa, Asia, and Australia, with smaller ones in North America and South America. The soil region covers about one fourth of the land surface of the earth and is largest among the nine.

Low rainfall is the hallmark of the region of Aridisols. Shortage of water is general, and the full range in annual precipitation is small. In contrast, the range in temperature extends from hot to cold because deserts occur in the margins of the tundra as well as in the tropics.

Soils of the region lack distinct horizons with few exceptions. The A horizons are generally pale and the B horizons a bit darker. Combined thickness of the two is commonly 1 ft (30 cm) or less. Because of the low rainfall, Aridisols have been leached very little. For the same reason, they are low in organic matter and nitrogen. Levels of plant nutrients except nitrogen tend to be high.

Associated with Aridisols in the deserts are substantial proportions of Entisols, sandy wastes, and barren rock outcrops. The Entisols, on the whole, are comparable to Aridisols in usefulness, whereas the sandy wastes and barren rock have no value for production of food and fiber.

Principal utilization of the region, where the soils are used at all, is for grazing by nomadic herds of livestock. Extensive areas such as parts of the Sahara are not used in any way. Locally, small bodies of Aridisols and Entisols are cultivated under irrigation. The soils can be highly productive where irrigation water and necessary skills are available and temperatures are favorable. Total production from this one fourth of the land surface is, however, small.

Oxisols, Inceptisols, and Ultisols

The largest bodies are in central Africa and northern South America with smaller ones in southeastern Asia, Indonesia, northern Australia, and Central America. The soil region ranks second in size, being smaller than that of Aridisols.

Land surfaces of the region have been stable for long periods, chiefly under humid, tropical climates. Some areas have relatively low rainfall but none is semiarid. The regoliths on the oldest of the surfaces are the most strongly weathered in the world. These oldest surfaces are nearly level, higher parts of local landscapes. Associated with them are slopes, some of which are fairly steep and on which regoliths are younger and have not been subject to as much weathering.

Oxisols and Inceptisols have faint horizons except for darkened surface layers or A horizons. Below the A horizons, profiles are commonly uniform to depths of 2 m or more. Well-drained profiles are red, yellowish red, reddish brown, or yellowish brown in color. Because of the long-continued leaching, the soils are low in nutrient elements, especially the Oxisols. The Inceptisols are not as low and a few have fairly high levels. Phosphorus-fixing capacities are generally high because of the iron and aluminum oxides present. The Ultisols of this region are like those of areas in which the soil order is dominant. Natural fertility of the soils collectively is thus low, with a few exceptions among the Inceptisols.

Soils of the region have produced food under shifting cultivation in the past and still do, for the most part. With simple technology dependent on human labor, continuous cultivation is not feasible. Yet is is also true that some of the soils, especially among the Inceptisols, can be highly productive if used with sophisticated technology that includes modern scientific findings and industrial products.

Soils of Mountains

The largest bodies are in Asia, North America, and South America with smaller ones in Europe, Africa, and Indonesia. The region ranks third in size behind those of Aridisols and of Oxisols, Inceptisols, and Ultisols.

Mountains extend from north to south across continents. Consequently, they occur under a wide range of climates, although temperatures are mostly cool or cold except for some valleys and basins. Rainfall has a wide range.

Mountainous regions have a large variety of soils. Many soils are stony and some are shallow to rocky on mountain slopes. Such soils are largely but not exclusively Entisols and Inceptisols. In contrast to those of the mountain slopes, the soils of intermountain basins and valleys include representatives of all orders in the current system. No effort is therefore made here to identify and describe the different kinds.

Some mountains are kept in forest, some are used for grazing livestock, and some simply provide beautiful scenery. Mountains generally appeal to the esthetic sensibilities of people, especially rugged chains such as the Andes of South America, the Alps of central Europe, and the Rockies of Canada. The esthetic appeal is itself important and has been for a long time, as witness a proverb of ancient Persia — if you give a man two coins, he spends one for bread and one for hyacinths.

REFERENCES

1. **Walters, S. M.**, The shaping of angiosperm taxonomy, *New Phytol.*, 60, 74, 1961.
2. Soil Survey Staff, Soil taxonomy: a basic system of soil classification for making and interpreting soil surveys, *U.S. Dep. Agric. Agric. Handb.*, No. 436, 1975.
3. **Marbut, C. F.**, A scheme of soil classification, in *Proc. 1st Int. Congr. Soil Sci.*, 4, 1, 1927.
4. **Whitney, M.**, Soils of the United States, *U.S. Bur. Soils Bull.*, No. 55, 1909.
5. **Simonson, R. W.**, The soil series as used in the USA, *Trans. 8th Int. Congr. Soil Sci.*, Bucharest, Romania, 5, 17, 1964.
6. **Marbut, C. F.**, Soils of the United States, in Atlas of American Agriculture, Part 3, Baker, O. E., Ed., U.S. Department of Agriculture, Washington, D.C., 1935, 12.
7. **Marbut, C. F.**, The contribution of soil surveys to soil science, in *Proc. Soc. Promotion Agric. Sci.*, 41, 116, 1921.
8. **Marbut, C. F.**, Soil classification, *Am. Soil Surv. Workers Rep.*, 3, 24, 1922.
9. **Whiteside, E. P.**, Changes in the criteria used in soil classification since Marbut, *Soil Sci. Soc. Am. Proc.*, 18, 193, 1954.
10. **Baldwin, M., Kellogg, C. E., and Thorp, J.**, Soil classification, in Soils and Men, U.S. Department of Agriculture, Washington, D.C., 1938, 979.
11. **Simonson, R. W. and Steele, J. G.**, Soil (great soil groups), in *Encyclopedia of Science and Technology*, Vol. 12, 2nd ed., McGraw-Hill, New York, 1966, 433.
12. **Kellogg, C. E.**, Development and Significance of the Great Soil Groups of the United States, *U.S. Dep. Agric. Misc. Publ.*, No. 229, 1936.
13. Soil Survey Staff, *Soil Classification — a Comprehensive System: 7th Approximation*, U.S. Department of Agriculture, Washington, D.C., 1960.
14. **Simonson, R. W.**, Soil classification in the United States, *Science*, 137, 1027, 1962.
15. **Smith, G. D.**, Objectives and basic assumptions of the new soil classification system, *Soil Science*, 96, 6, 1963.
16. **Cline, M. G.**, Logic of the new system of soil classification, *Soil Science*, 96, 17, 1963.
17. **Simonson, R. W.**, Outline of a generalized theory of soil genesis, *Soil Sci. Soc. Am. Proc.*, 23, 152, 1959.
18. **Cline, M. G.**, The changing model of soil, *Soil Sci. Soc. Am. Proc.*, 25, 442, 1961.
19. Soil Survey Staff, Soil Survey Manual, *U.S. Dep. Agric. Agric. Handb.*, No. 18, 1951.
20. Soil Survey Staff, Identification and nomenclature of soil horizons, Supplement to *U.S. Dep. Agric. Agric. Handb.*, No. 18, 1962, replacing pp. 173—188.
21. **Whitney, M.**, Some Physical Properties of Soils in Their Relations to Moisture and Crop Distribution, *U.S. Dep. Agric. Weather Bur. Bull.*, No. 4, 1892.

SOIL PHYSICS

Lewis H. Stolzy and William A. Jury

SOIL AS A PHYSICAL SYSTEM

The soil, from a physical point of view, is composed of three natural materials: solid, liquid, and gas. Soil variables are conveniently expressed in terms of mass and volume ratios of these three components: solid, water, and air. The assumption that these three phases are distinct and homogeneous is an oversimplification. The solid phase consists of crystalline mineral fragments with a wide range of particle sizes and shapes, as well as noncrystalline inorganic material with a size range from colloidal to that of pebbles of amorphous minerals mixed with living and dead organic materials. The liquid phase is a solution containing dissolved gases, soluble salts, and inorganic and organic solutes. The gaseous phase is similar to the atmosphere but differs in its proportions of oxygen, nitrogen, carbon dioxide, and other trace gases. However, the three-phase soil system is a simple way to express commonly used soil properties such as bulk density, water content, pore space, and void ratio.

Various mass and volume ratios can be derived from the three-phase physical model of Figure 1.

SOIL MATERIAL

The solid part of the soil, consisting of both organic and inorganic matter forms the framework of a soil. Soils, like most things in nature, vary in their properties over a wide range with some predictability.[1] Soils are derived from rocks and minerals, and their chemical compositions change along an age scale from those of parent materials to those characteristically imposed by the environment. The processes of mineral alteration and soil development occur at different rates depending upon environmental factors such as temperature, rainfall, nature of parent material, and vegetation. An illustration (Figure 2) of this age scale for different regions of the world shows the type of materials present in the soil.

Soil material is a unit of study in which the characteristics being studied are approximately constant.[2] Preliminary detailed study of the soil is made on soil materials. It is a common practice in soils studies to relate physical measurements to the texture of the soil (sand, silt, and clay). A measurement of particle-size distribution can be used to make qualitative inferences about the behavior of a soil system but often fails because of soil structural development.

The physical properties of large and small particles differ greatly.[3] The coarse separates, composed of rock fragments and such primary minerals as quartz, act as individual particles. Large particles have low specific surface (surface area per unit mass) and so have minimal contribution to physical-chemical reactions in soils. Silt particles are intermediate between those of sand and clays, but are mainly primary minerals acting more like sands.

Specific Surface

The relative reactivity of soils is determined by their specific surface areas as well as by their mineralogy. The clay minerals, with their large surface area, control most of the important properties of soils. Figure 3 shows the relative change in specific surface with change in particle size. Data in Table 1 represents the total specific surface area of certain clay and soil materials. Taylor and Ashcroft[8] have calculated the relation between regular shape particle size and specific surface area. Their calculations show that colloidal clay ($< 0.2\mu$) varies from 1060 m^2g^{-1} for fine material to about 15 m^2g^{-1}

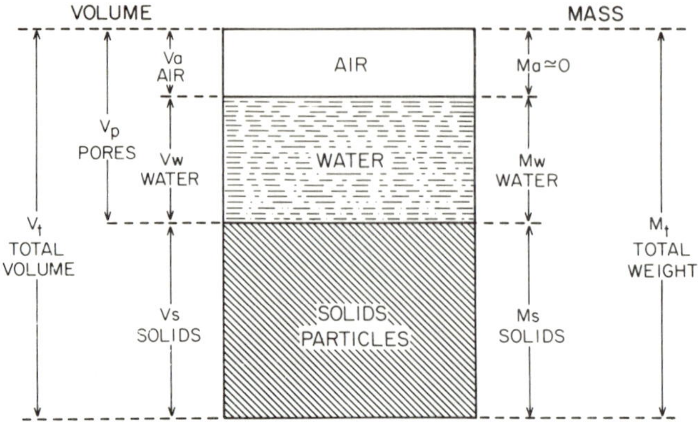

FIGURE 1. Schematic drawing of the soil as a three-phase physical model.

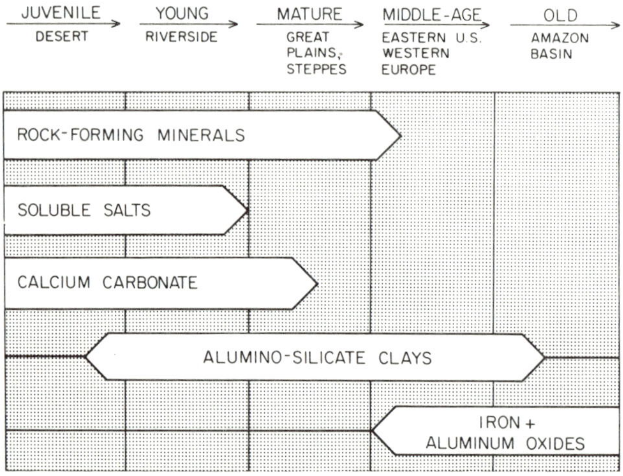

FIGURE 2. Some compositional features of soil. (Redrawn from Coleman[1]).

for the coarse colloidal material. Ormsby et al.[9] reported the surface area of kaolinites to range from 5.0 to 14.5 m^3g^{-1} for different particle size fractions determined by the glycerol adsorption method. Generally zeolites have an open structure providing surface areas similar to 2:1 type expandable layer silicates. They have a surface area in the range of 880 to 750 m^2g^{-1}.[10] Allophane and imogolite clays derived from volcanic ash have marked variation in specific surface areas depending on the nature of adsorbate.[11] Specific surface areas obtained with ethylene glycol monoethylether (EGME) and water give 435 to 534 m^2g^{-1} and with N_2 gas, 145 to 170 m^2g^{-1}. Egashira and Aomine[12] gave EGME surface area values of 700 to 900 m^2g^{-1} and 900 to 1100 m^2g^{-1} for allophane and imogolite, respectively.

Particle Density

In many soils, the mean density of particles is 2.6 to 2.75 g cm^{-3}. The reason most soils have this range of particle density is that they are made up of quartz, feldspar, and the colloidal silicates with densities in this range.

It is difficult to measure the density of certain of the clay minerals in their natural state, because adsorbed water forms an integral part of their structure. The theoretical density of some of these minerals are as follows: kaolinite 2.609 g cm^{-2},[13] halloysite 2.55,[14] illite 2.642 to 2.688,[15] and montmorillonite 2.2 to 2.7. The data in Table 2

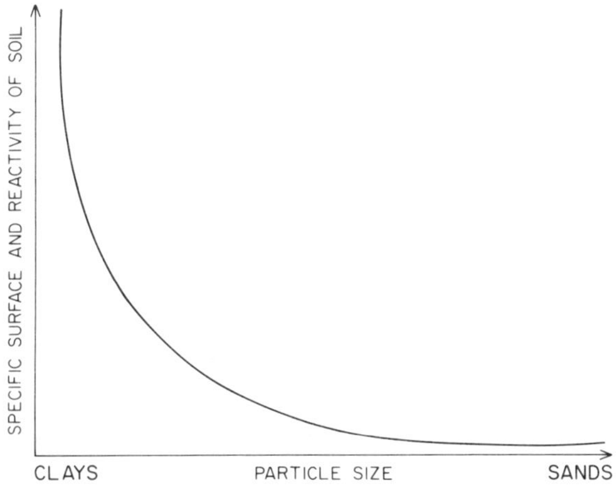

FIGURE 3. Specific surface is important in determining the reactivity of soils. The amount of surface varies inversely with the size of the soil particles. (Adapted from Russell[3]).

Table 1
TOTAL SPECIFIC SURFACE AREA OF CLAY AND SOIL MINERALS

Location	Material	Soil group	m^2g^{-1}[a]
	Kaolinite[b]		55.4
	Illite[b]		208.8
	Montmorillonite[b]		806.7
	Vermiculite[b]		340.0
	Halloysite[b]		170.0
Imperial Valley, Calif.	Silty clay[c]	Holtville	172.3
Imperial Valley, Calif.	Silty clay[c]	Imperial	204.8
Imperial Valley, Calif.	Loam[c]	Mcloland	65.2
Imperial Valley, Calif.	Loam[c]	Indio	90.1
Mississippi	Bentonite[d]		815
Georgia	Kaolinite[d]		25
Georgia	Silt loam[d]	Billings	42
Georgia	Clay laom[d]	Sebree	183
Wyoming	Bentonite[d]		42
Wyoming	Clay loam[d]	Aiken	197

[a] Determined by retention of monomolecular layer of ethylene glycol. Data from Reference 7.
[b] Data from Reference 4.
[c] Data from Reference 5.
[d] Data from Reference 6.

indicate the large change in effective density of hydrous minerals with increases in adsorbed water.

According to Dier et al.,[16] zeolites have a density that ranges between 2.0 and 2.3 g cm^{-3} except for the Ba-rich members that range between 2.5 and 2.8 g cm^{-3}. Allophanes present in soil constitute the most important feature of soils derived from volcanic ash.[11] Comparisons of specific surface and particle density of various soil materials and colloids from these materials are shown in Table 3.

Table 2
CHANGE IN SPECIFIC SURFACE OF SOME HYDROUS MINERALS AND SOILS UPON LOSS OF ESSENTIAL WATER[15]

Material	Treatment (°C)	Loss in weight (%)	Specific surface (m^2g^{-1})	Equivalent sphere diameter (μ)	Plate thickness (μ)
Kaolinite	30		15.5	0.14	0.057
$Al_2Si_2O_5(OH)_4$	200	0.2	15.3	0.14	0.058
< 0.3μ	500	12.7	18.1	0.12	0.049
	700	14.3	16.5	0.13	0.054
	900	14.6	11.5	0.19	0.077
Illite (hydrous mica)	30		97.1	0.023	0.0092
< 0.3μ	200	0.4	92.2	0.024	0.0096
	500	5.8	91.6	0.024	0.0097
	700	7.7	80.4	0.028	0.011
Gibbsite $Al(OH)_3$	30		0.31		
	700	41.1	103.1	0.021	0.0086
	900	42.6	66.2	0.035	0.013
Diaspore AlO(OH)	30		3.9	0.57	0.23
	600		42.8	0.052	0.021
Fraction < 0.3μ from	30		72.7	0.031	0.012
Hagerstown soil,	200	3.2	62.7	0.036	0.014
12—33 in.	500	8.6	66.2	0.033	0.013
	700	11.4	58.8	0.038	0.015
	900	12.2	11.0	0.20	0.081
Fraction < 0.3μ from	30		83.9	0.026	0.011
Dewey soil, 26—43	200	0.0	78.3	0.028	0.011
in.	500	9.9	80.0	0.028	0.011
	700	12.1	61.6	0.036	0.014
	900	12.8	27.1	0.082	0.033
Montmorillonite	30		15.5	0.14	0.057
Brucine montmorillonite	30		12.3	0.18	0.072
Halloysite	30		43.2	0.051	0.021

ARRANGEMENT OF PARTICLES — BULK DENSITY AND POROSITY

The arrangement of the solid particles of soil determines the bulk density and porosity. Bulk density of a soil is the mass (weight) of a known soil volume compared to the mass of an equal volume of water. The soil volume includes both the solid and void space fractions expressed in units of gcm^{-3}. Bulk density data may be used to compute total porosity, if the particle density is known or estimated. Size and shape of pores depend on the way individual soil particles are packed or arranged. Coarse-textured soils such as sands tend to have porosities between 30 and 40%, while fine-textured soils such as clays will have porosities of 40 to 60%.[17] The size frequency distribution of pores in soil materials or the pore-size distribution is usually measured by methods based on the theory of the retention of water in pore spaces of different sizes. Brewer[2] has a lengthy discussion on pores (voids) in soil in his book *Fabric and Mineral Analysis of Soils.* Johnson et al.[18] have discussed conventions for describing soil pores. These conventions, with slight modifications, are used by the Soil Survey Staff of the U.S. Department of Agriculture in making and interpreting soil maps.[19]

Table 3
EXPERIMENTAL DATA ON THE SPECIFIC SURFACE, PARTICLE DENSITY, AND COLLOID PERCENTAGE[14]

Soil material	Soil group	Specific surface[a] ($m^2\ gm^{-1}$)	Density ($gm\ cm^{-3}$)	% > 2μ
Columbiana soil	Lateritic	68.2	2.64	51
Columbiana colloid	Lateritic	82.0	2.65	
Cecil soil	Red and yellow	14.2	2.63	31
Cecil colloid	Red and yellow	44.4	2.53	
Cecil soil	Red and yellow	31.9	2.73	56
Cecil colloid	Red and yellow	52.2	2.82	
Chester soil	Gray brown	24.0	2.74	24
Chester colloid	Podzolic	72.3	2.78	
Carrington soil	Prairie	19.1	2.68	26
Carrington colloid	Prairie	74.1	2.66	
Barnes soil	Chernozem	14.6	2.66	25
Barnes colloid	Chernozem	68.9	2.54	

[a] Based on nitrogen adsorption at −183°C.

SOIL MECHANICS — SOIL DYNAMICS

The term mechanics is defined as "that part of a physical science which studies the action of forces on bodies." The branch of mechanics which deals with forces on soil masses is called soil mechanics.[20] With the publication in 1925 of *Erdbawmechanik* by Dr. Karl Terzaghi,[21] soil engineering became recognized as a distinct branch of engineering science. More recent writings concerning this study are Casagrande,[22] Taylor,[23] and McCarthy.[24]

Soil dynamics, a phase of soil science and mechanics concerned with soils in motion, can be defined as the relation between forces applied to the soil and the resultant soil reaction (Figure 4).[25] The dynamic properties of soil are properties made manifest through movement of the soil. For example, if a block of soil resting on a flat surface is moved, the resultant friction is a dynamic property of the soil; this property cannot be determined until movement occurs. Similarly, soil strength is a dynamic property because its strength increases as it is compacted.

Soil strength can be defined as the capacity of soil to withstand an applied deformation or strain. A strength characterization is usually made by measuring response to a range of applied forces. However, the procedure for characterizing strength as a response to external force has not been satisfactory, both because the properties vary in space and because soil strength will change as the applied force is increased. The formal description of soil strength is usually made through the use of continuum stress-strain equations. Viscoelastic, viscoplastic, and fluid mechanics equations are examples of stress-strain equations.[25]

In soil rheology, that part of soil physics dealing with deformation and flow of matter, soil is usually considered to be a continuous medium in which both stresses and strains are continuous functions of the space coordinates. However, this assumption is not compatible with the particulate nature of soil, or even with the molecular structure of any material.[26-27] According to McMurdie[28] soils demonstrate a number of deformation characteristics, but owing to the complexity of the materials in soil, there is no completely satisfactory theory for predicting response to applied stress. Studies by McMurdie[28] on a Yolo loam, utilizing triaxial compression equipment, showed that deformation is nonlinear, finite, irreversible, that the component ratios such as Young's modulus and Poisson's ratio were not constants during the deforma-

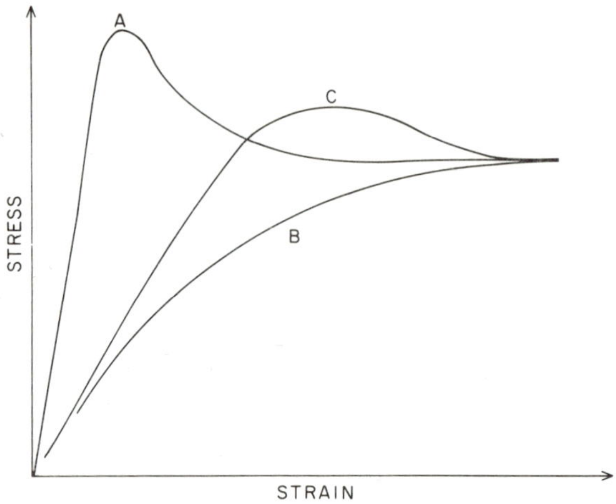

FIGURE 4. Typical shear stress-strain relations for soils in three conditions: (A) Cemented, (B) loose, and (C) dense. Each curve is at a constant normal load. (Redrawn from Gill and Vanden Berg[25]).

tion process, and therefore that elasticity theory is not generally applicable. Plasticity theory was found to be unsatisfactory as a general theory because of its assumption of a region of Hookean behavior or of yield after only small deformations. Neither of these conditions holds for soils. However, tests by McMurdie[28] showed that the basic assumption of viscoelasticity theory, existence of a time-dependent relationship between stress and strain states, is valid for soils.

Soil Consistency

A dynamic property of soil that has been widely researched by soil physicists is soil consistency. Soil consistency, a part of the broader field of general rheology, includes such properties of the soil as resistance to compression, shear, friability, plasticity, and stickiness. These properties are manifested differently as the forces of cohesion and adhesion within the soil mass vary.[29] A review by Bodman[30] lists a number of techniques used by soil scientists to study soil consistency (Table 4).

The methods are grouped according to the kind of deformation to which the soil is subjected and the apparatus used. Methods grouped in 1, 2, and 3 involve disturbance of natural structure, while 6, 7, 8, and 9 involve little or no disturbance of structure before measurements. Groups 4 and 5 make measurements on nonfragmented soil cores of remolded samples. Details of measurements and equations are available in Bodman,[30] Baver et al.,[29] Greacen,[31] and U.S. Department of Agriculture Handbook No. 316.[25]

Resistance to Penetration

The mechanical resistance of the soil to a downward moving probe, if measured in terms of force required to penetrate, or work expended in relation to depth of penetration, is an expression of soil consistency.[30] Penetration can be effected by continuous application of pressure, generally induced by some combination of weights, springs, and gears, which may be mechanically driven or pushed slowly into the soil or by the impact of falling weights.[32-35] Figure 5 shows the relation between the number of impacts and the force required to press the static penetrometer through the soil.[35] The resistance of a soil to the penetration of a probe is a secondary indicator of soil com-

Table 4
GROUPING OF METHODS ACCORDING TO THE KIND OF DEFORMATION TO WHICH THE SOIL IS SUBJECTED[30]

Kind of deformation	Apparatus used	Resistance measured
Mixing by pug-mill	Plastograph	Torque on motor shaft
Plastic flow induced by impact or pressure	Liquid limit or similar device in different forms	None (impact magnitude controlled) or vertical force needed for initiation of flow
Lengthening of slender cylinder by rolling	Pachimeter	Vertical resistance through axis and normal to circumference
Compression of stationary cylinder	Stress-strain apparatus	Resistance to applied normal load
Compression and shearing applied to prism or annulus	Box shear and modified torsion-ring shear apparatus	Resistance to applied normal and shearing stress
Shearing of cylindrical core	Transverse shear testing	Resistance to shear normal to long axis
Shearing core from surrounding mass in position in field	Rotating vanes	Shearing resistance
Compression in position in field	Device for loading soil in place	Resistance to applied normal stress
Deformation by penetration	Penetrometer	Resistance to penetration

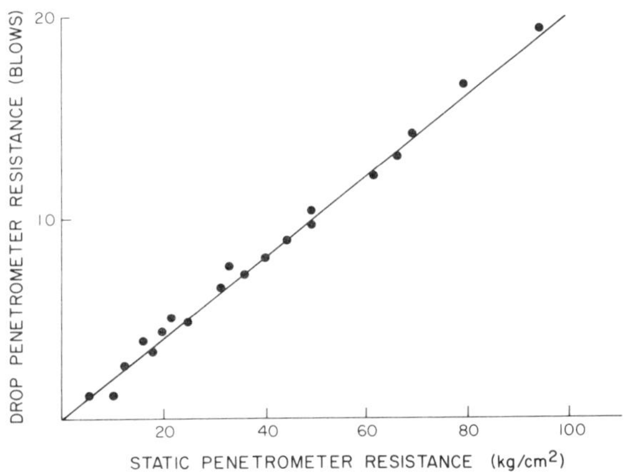

FIGURE 5. Relation between impact and static penetrometer. (Redrawn from Zelenin.[35])

paction. It is affected by many other factors: moisture content, texture, and bulk density (Figure 6, Figure 7, and Table 5).

Farrell and Greacen[38] analyzed forces acting on the point of a fine metal probe, assuming spherical expansion of a cavity in the soil as occurring in two zones, a zone of compression with plastic failure surrounding the probe, and outside this plastic area a zone of elastic compression. They concluded that the magnitude of the resistance to penetration is influenced by both compression and soil-metal friction.

SOIL WATER

The behavior and theory of transport of soil water has been the subject of a number of reviews and books.[17,39-48] Since the liquid phase of the soil shares a large amount of interfacial area with the solid phase and the gas phase, the interactions between

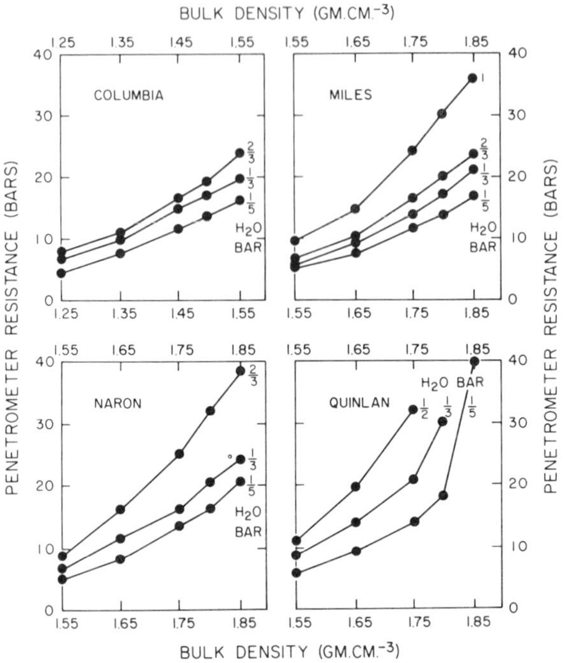

FIGURE 6. Penetrometer resistance vs. dry bulk density relations for four soils at different soil suctions (high suction gives a low moisture content). Penetrometer resistance and density change from soil to soil and at different moisture contents. (Redrawn from Taylor and Bruce[36]).

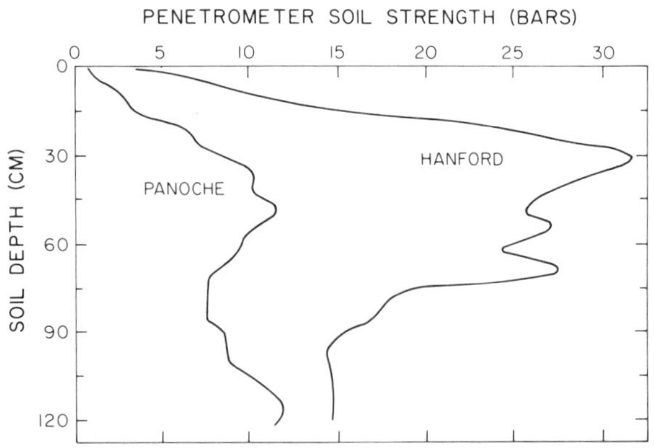

FIGURE 7. Soil strength determined with a recording soil penetrometer in two field soils of the San Joaquin Valley, Calif. (Grimes[37]).

phases have a great influence on the movement and retention of water by the soil. There has been notable success in representing the state and movement of soil water in response to external forces by characterizing the potential energy of soil water due to (1) matric forces (those which result from the presence of the solid phase); (2) osmotic forces (those caused by dissolved solutes in the soil water); and (3) body forces (inertial and gravitational influences).[49] When two states of water in soil at different potential energy levels are connected, flow will occur from the higher state to the lower state until equilibrium is achieved.

Table 5
PHYSICAL CHARACTERISTICS OF THE TWO SOILS IN FIGURE 9: PANOCHE CLAY LOAM (TYPIC TORRIORTHENTS) AND HANFORD SANDY LOAM (TYPIC XERORTHENTS)[37]

Depth range (cm)	Sand (> 50μ) (%)	Silt (2—50μ) (%)	Clay (< 2μ) (%)	Bulk density (g cm^{-3})	Water retention (desorption) 1/3 bar (g g^{-1} × 100)	Water retention (desorption) 15 bars (g g^{-1} × 100)
Panoche						
0—30.5	24	32	44	1.31	30.7	14.1
30.5—61	34	28	38	1.32	27.0	11.1
61—91.5	33	33	34	1.30	26.8	9.3
91.5—122	26	36	38	1.26	28.0	10.7
122—152.5	17	36	47	1.24	34.8	15.5
152.5—183	23	41	36	1.29	31.2	9.6
Hanford						
0—30.5	61	25	14	1.60	10.6	3.0
30.5—61	60	26	14	1.63	10.3	2.9
61—91.5	66	21	13	1.68	9.3	2.6
91.5—122	73	16	11	1.58	7.2	2.5
122—152.5	76	14	10	1.58	5.0	2.4
152.5—183	80	11	9	1.61	4.9	2.3

Water Potential

The potential of soil water, ψ_w (erg cm^{-3} in pressure units; cm in head units), is defined as the work done in moving from a reference state of pure free water at a specified pressure, temperature, and elevation to the state in the soil at the same temperature and elevation. The water potential is made up of the following components.

Matric Potential

The matric potential ψ_m is the component of the water potential arising from the interaction of water molecules with the adsorptive forces of the soil matrix. In unsaturated soil, water is held in crevices and pores and ψ_m is negative with respect to pure, free water at the same elevation and temperature. In saturated soil ψ_m is zero.

Owing to the natural variability of particle size distribution, soil aggregation, and bulk density, the values of ψ_m will vary at equivalent locations under the soil surface. Observation of a vertical grid of matric potential readings from tensiometers, however, will give information on water retention following an input of water at the soil surface. Figure 8 shows matric potential readings in a field soil as a function of depth and time following an irrigation.[50]

Solute Potential

The solute potential ψ_s is the component of the water potential due to dissolved solutes in soil water. It has little or no effect on liquid water flow in soil but can have an important influence on plant water uptake.

Pressure Potential

The pressure potential ψ_p is the component of the water potential due to gas pressure effects on soil water in excess of atmospheric (unsaturated soil) or hydrostatic pressure within the liquid phase (saturated soil).

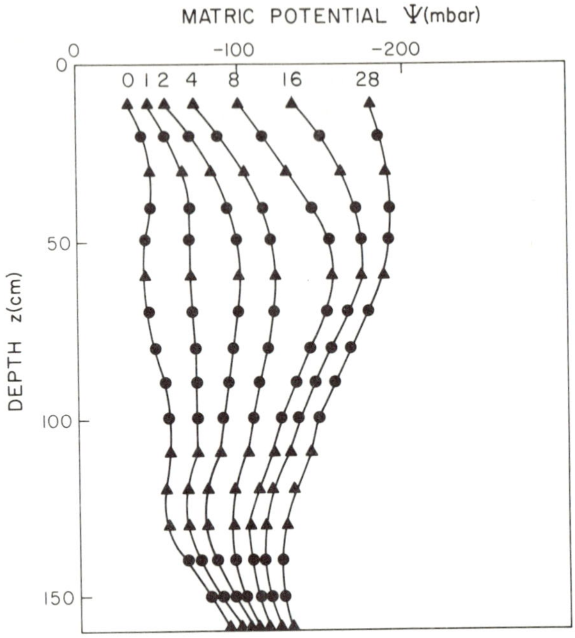

FIGURE 8. Matric potential as a function of depth measured during the transient drainage experiment. The numbers refer to days after irrigation. The values at 10, 30, 60, 110, 120, 130, and 160 cm are measured values (triangles). The others are interpolated over depth (dots). The soil was a Hanford fine sandy loam (Typic xerorthents). (Redrawn from Flühler et al.[50])

Overburden Potential

The overburden potential ψ_b is the component of the water potential due to the weight of the nonrigid solid soil matric on an element of soil water below the surface. It is mainly important in moist clay soils.

Gravitational Potential

The gravitational potential ψ_z of an element of soil water is defined by its elevation with respect to the reference element.

Water Retention in Soil

The quantity of water retained by soil following a wetting event is historically referred to as the field capacity. However, a wetted soil profile will continue to drain at a decreasing rate for an indefinite period of time, so that this concept is not well defined. Figure 9 shows data from a field plot with the surface covered to prevent evaporation. Water storage decreases, due to drainage, continued over the 50 days of observation. Efforts have been made to relate field capacity to water retention at a given matric potential, usually −1/3 bar. Many measurements of the −1/3 bar water content are given in soil survey reports.

Plant roots are able to extract water from soil at moisture contents far below those where drainage becomes insignificant. The range of matric potentials where water is accessible to plants has been approximately taken as −0.1 bar to −15 bars and is called the available water in soil. An approximate relation between regions of available water and matric potential for different soil textures is given in Figure 10.

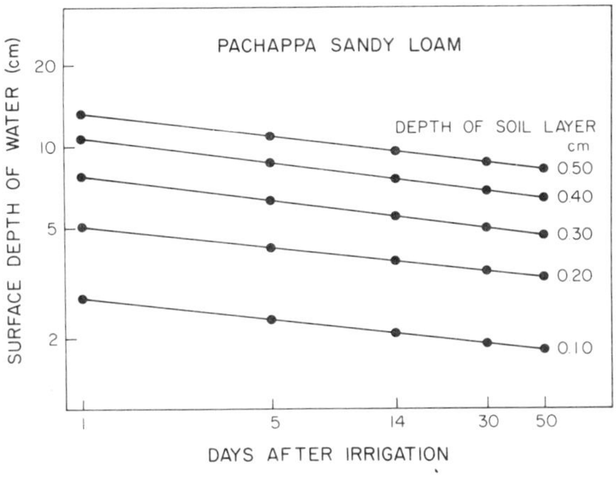

FIGURE 9. Water content changes produced by drainage in a fallow soil that is protected from evaporation. The data points, which are plotted on logarithmic scales, indicate the equivalent surface depths of water contained in the various soil layers. (Richards and Richards[51]).

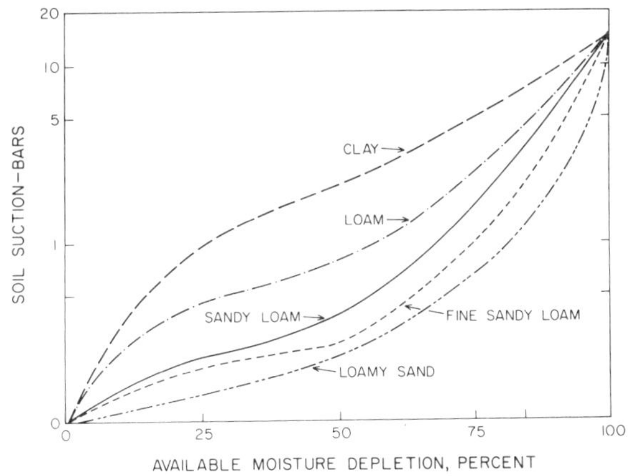

FIGURE 10. Available water depletion is a scale commonly used to express soil water relations. (Redrawn from Richards and Marsh[52]).

Water Flow

The development of a quantitative framework for describing water flow in soil was begun by Henry Darcy[53] in 1856 who discovered that the flow rate of water Q through a saturated sand column of height L was proportional to the hydraulic head difference (hydrostatic head pressure + elevation head) (p + L) times the cross-sectional area A, divided by the height L

$$J = \frac{Q}{A} = -\frac{K_S \Delta (p + L)}{L} \qquad (1)$$

where J is water flux ($cm^3cm^{-2}sec^{-1}$) and the coefficient of proportionality is called the saturated hydraulic conductivity K_s ($cm\ sec^{-1}$). The extension of this equation to unsaturated porous media was later made by Buckingham[54] by changing the driving force to the gradient of matric potential plus gravitational potential and making the coefficient of proportionality a function of the matric potential

$$J = \frac{-K(\psi_m)}{\rho_w g} \frac{d(\psi_m + \psi_z)}{dz} \qquad (2)$$

where $K(\psi_m)$ is called the hydraulic conductivity. The description of soil water flow was extended to time-dependent events by Gardner in 1919[5] by combining the water flux equation (2) with the conservation equation for water (3)

$$\partial\theta/\partial t + \partial J/\partial z = 0 \qquad (3)$$

where θ is volumetric water content. Time-dependent flow is usually written in one dimension but is easily extended to three dimensions.

The combination of Equations 2 and 3 is achieved by writing θ as a function of ψ_m giving

$$C(\psi_m) \frac{\partial\psi_m}{\partial t} = \frac{\partial}{\partial z}\left(\frac{K(\psi_m)}{\rho_w g} \frac{\partial\psi_m}{\partial z}\right) - \frac{\partial K(\psi_m)}{\partial z} \qquad (4)$$

where

$$C(\psi_m) = \frac{\partial\theta}{\partial\psi_m} \qquad (5)$$

is called not giving the differential water capacity ($cm^3 erg^{-1}$). An alternate equation is formed by writing ψ_m as a function of θ, giving

$$\frac{\partial\theta}{\partial t} = \frac{\partial}{\partial z}\left(D(\theta) \frac{\partial\theta}{\partial z}\right) + \frac{\partial K}{\partial z} \qquad (6)$$

where

$$D(\theta) = \frac{1}{\rho_w g} K \frac{\partial\psi_m}{\partial\theta} \qquad (7)$$

is the soil water diffusivity ($cm^2 sec^{-1}$). The functions $\partial\theta/\partial\psi_m$ or $\partial\psi_m/\partial\theta$ are obtained from the slope of the moisture retention curve $\psi_m(\theta)$.

Use of the unsaturated flow Equations 4 or 6 requires the functional relationships $K(\psi_m)$ and $\psi_m(\theta)$. The form of these relations depend strongly on soil texture, aggregation, bulk density, and natural soil features such as cracks, so that measurements should be made on each soil. Soils falling into similar textural groups will bear some similarity to each other, but not enough to justify the use of one functional relationship on a different soil. Figure 11, adapted from Gardner,[56] illustrates $K(\psi_m)$, $D(\psi_m)$, and $\psi_m(\theta)$ relations for three different soil types.

Additional caution must be exercised in the use of Equations 4 or 6 because $\psi_m(\theta)$ is strongly hysteretic, resulting in a different relation for wetting and drying cycles (Figure 12). As a result, the equations and functional relations discussed above are meaningful only for a monotonic wetting or drying process.

The use of transport equations to describe water transport in undisturbed soil profiles is problematical, because the functions $K(\psi_m)$ and $\psi_m(\theta)$ vary from location to location within a field. Extensive soil sampling and averaging is required to obtain meaningful estimates for the field as a whole and *in situ* measurements at a given location yield scattered values rather than smooth functional relationships. An example of a complete set of measurements for K, D, ψ, and C as a function of θ, obtained in a transient column experiment by Weeks and Richards[58] is given in Table 6.

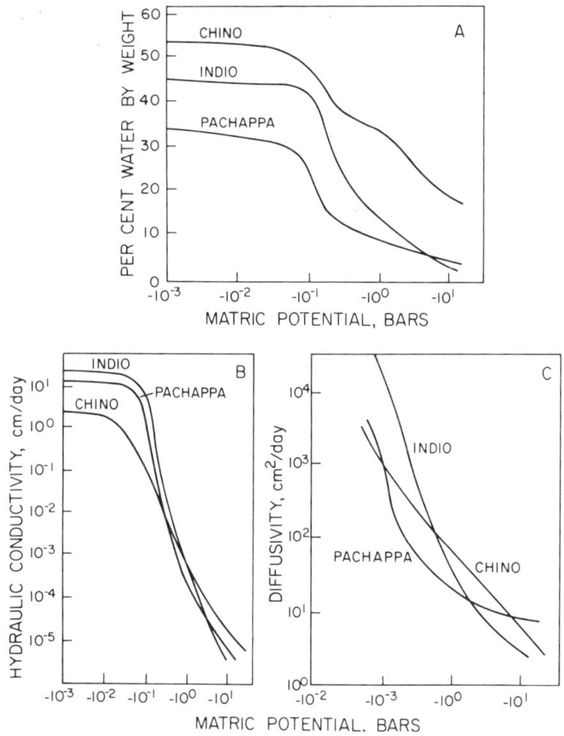

FIGURE 11. (A) Moisture characteristics θ(h), (B) conductivity K(h), and (C) diffusivity D(h), for three soils: Pachappa sandy loam, Indio loam, and Chino clay. (Redrawn from W. R. Gardner[56]).

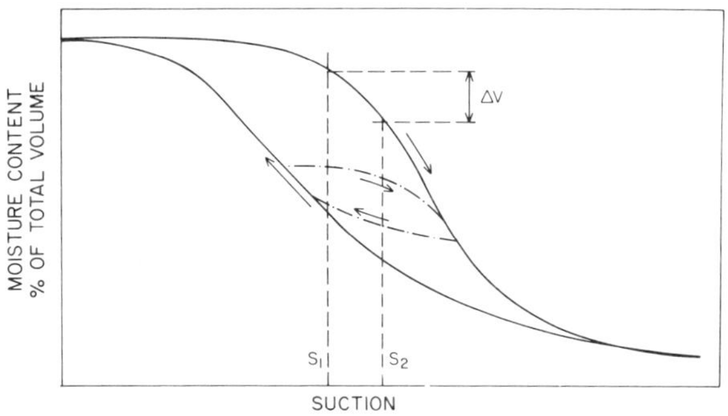

FIGURE 12. The moisture characteristic curve of a non-shrinking soil. The arrows indicate whether the soil is being progressively dried or moistened. The interior or partial hysteresis loop shown by the broken line is a scanning curve. (Redrawn from Childs[57]).

SOIL TEMPERATURE

The oscillations in temperature at the soil surface from high frequency (daily) cycles and low frequency (annual) cycles, are transmitted into the soil with attenuated amplitude and a phase shift. The daily oscillations (Figure 13) are confined to the top 25 cm of soil whereas annual variations may penetrate to 600 cm or more (Figure 14).

Table 6
VALUES OF WATER CONDUCTIVITY AND SOIL-WATER DIFFUSIVITY[a]

Positions x (cm)	Time t (hr)	Suction head h (cm)	Water conductivity K (cm hr¹)	Suction head gradient $\partial h/\partial x$ (cm cm⁻¹)	Soil-water diffusivity D (cm³ hr⁻¹)
			Field Core (Non-Fragmented)		
5	1	52	2.1×10^{-2}	−3.0	40
10	2	53	2.6×10^{-2}	−1.6	51
2.5	4	100	4.2×10^{-3}	−12.9	17
5	7	103	4.2×10^{-3}	−6.6	17
2.5	24	205	5.1×10^{-4}	−29.6	1.6
5	48	222	6.2×10^{-4}	−16.1	2.2
10	96	232	5.1×10^{-4}	−10.3	2.0
2.5	72	330	2.1×10^{-4}	−48.3	1.2
5	96	327	2.6×10^{-4}	−27.8	1.5
10	168	340	2.0×10^{-4}	−18.8	1.3
			Laboratory Core (Fragmented)		
2.5	7	94	3.4×10^{-2}	−1.1	35
5	8	92	4.3×10^{-2}	−0.67	45
5	72	205	1.1×10^{-3}	−13.8	2.8
10	96	198	1.0×10^{-3}	−8.2	2.6
5	120	335	3.8×10^{-4}	−26.5	1.9
10	168	327	3.7×10^{-4}	−14.5	1.8

[a] Values of water conductivity (K) and soil-water diffusivity (D) with related suction head (h) and suction head gradient ($\partial h/\partial x$) values for two soil columns (Placentia loam) calculated for two or three positions (x) along the columns and for different times (t). Data selected to show that similar values of K were found at given h values even though $\partial h/\partial x$ varied as much as two or three to one.

Calculations of temperature changes and quantities of heat flow within soil are made using equations derived to describe the conduction of heat in homogeneous isotropic solids. Heat can be transferred in soils by conduction, mass flow of water, evaporation, condensation of water, and convection.

Conduction is the major method of heat transfer in soils. Two independent thermal properties enter into a quantitative description of heat transfer by conduction, the thermal conductivity λ, and the heat capacity per unit volume C_s. Thermal properties of the major constituents of soils are given in Table 7.

Heat Capacity

Because most soil minerals have nearly the same heat capacity, the heat capacity per unit volume of soil may be written as

$$C_s = 0.46\,\chi_m + 0.60\,\chi_o + \chi_w \quad (\text{cal cm}^{-3}\ {}^\circ\text{C}^{-1}) \tag{8}$$

with χ_m = volume fractions of soil minerals, χ_o = volume fraction of organic matter and χ_w = volume fraction of water. Selected thermal properties of different soils are given in Table 8.[59] Heat capacity of a soil can be calculated from Equation 8 to a high degree of accuracy.[60]

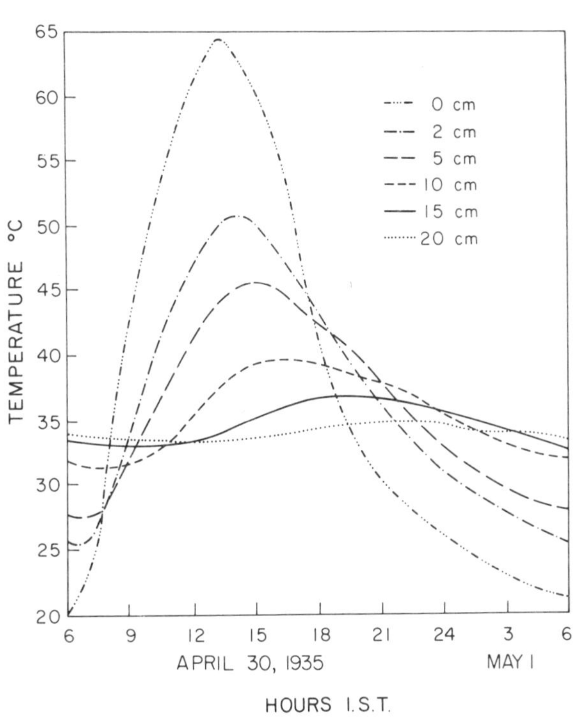

FIGURE 13. Soil temperature variations at depths of 0, 2, 5, 10, 15, and 20 cm taken during a single day (6 a.m. April 30, to 6 a.m. May 1, 1935) at Poona, India. Adapted from Richards et al.[59])

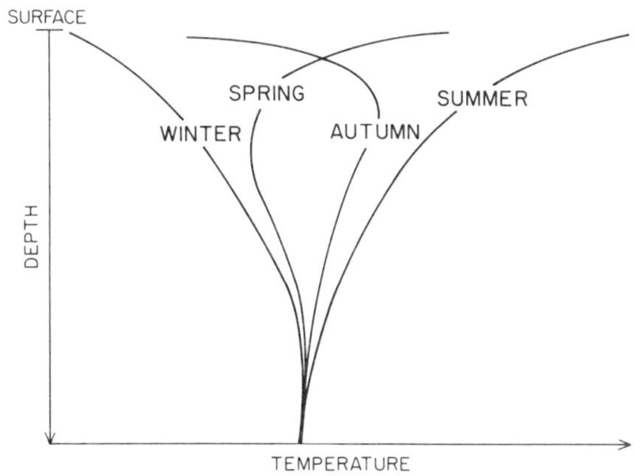

FIGURE 14. Soil temperatures in temperate regions show characteristic seasonal variations, which decrease with depth. (Redrawn from Russell[3]).

Thermal Conductivity

The thermal conductivity, λ,(cal $cm^{-1}K^{-1}sec^{-1}$) of a soil will depend upon the volumetric proportions of the solid, liquid, and air phases; on the size and arrangement of the solid particles; and on the interfacial relationship between the solid and liquid phases.[29] Thermal conductivity of different soils follows the order of sand > loam > clay > peat. An increase in bulk density improves the contact between particles and decreases the air contact which increases the thermal conductivity of the soil. However,

Table 7
THERMAL PROPERTIES AND DENSITIES ϱ OF SOME SOIL MATERIALS, WATER, AND AIR AT 10°C

Substance	λ (mcal cm^{-1} sec^{-1} C^{-1})	C (cal cm^{-3} C^{-1})	ϱ (g cm^{-3})
Quartz	21	0.48	2.66
Clay minerals	7	0.48	2.65
Organic matter	0.6	0.6	1.3
Water	1.37	1.00	1.00
Air	0.060	0.00030	0.00125

Adapted from Reference 60.

an increase in the water content of a soil has a much greater effect on increasing the thermal conductivity of the material.

A study by deVries[60] showed that the thermal conductivity of moist soils with widely different compositions and moisture contents can be computed from theoretical studies, with a fair degree of accuracy. The experimental λ_e and computed values λ_c for a sand in Table 9 agree within 5%, while for a clay soil (55% clay) computed values are systematically higher than that measured below a certain water content. Similar results are shown for a fibrous brown peat. It follows from this study that estimates of thermal conductivity of soil can be made if its mineral composition, porosity, and water content are known. The accuracy is better than 10%.

Thermal Diffusivity

The expression $Y = \lambda/C_s$ is called the thermal diffusivity (cm^2sec^{-1}) and denotes the temperature change that takes place in any portion of the soil as the heat flows into it from an adjacent layer. Figure 15 shows the increase in thermal diffusivity with water content for different soil types. Thermal diffusivity increases rapidly at very low water contents but reaches a maximum and decreases. It may be approximated by a constant at intermediate water contents which allows a linear form of the heat flow equation to be used in making calculations.

WATER VAPOR MOVEMENT IN SOIL

A large number of separate experimental studies of water vapor movement in response to thermal gradients are shown to be independent of water content over a wide range of moisture conditions. Cary[68,69] derived the following steady state equation for the flow of water vapor through soil (g $cm^{-2}sec^{-1}$) in response to a temperature gradient.

$$J_V/\rho_l = \frac{\beta D_V P^* H_V}{R^2 T^3}\, dT/dx \qquad (9)$$

which is valid for all but extremely dry conditions in the soil where H_v is the latent heat of vaporization (erg g^{-1}), R is the water vapor gas constant (erg $cm^{-1}K^{-1}$),P^* is the saturated vapor pressure of the water vapor (erg cm^{-3}) D_v is water vapor diffusion coefficient in air (cm^2sec^{-1}), ϱ_l is liquid water density (g cm^{-3}), and β is a phenomenological coefficient which accounts for all local interactions such as temperature gradient enhancement and liquid island transfer within the porous medium. This formulation does not explain the exact physical transfer mechanisms within the soil. How-

Table 8
SELECTED VALUES FOR THERMAL PROPERTIES OF SOILS[59]

Soil type	Particle size: Sand (%)	Silt (%)	Clay (%)	Bulk density (g cm^{-3})	Water (%)	Specific heat capacity Cal g^{-1} °K^{-1}	Heat capacity/CM^3 Cal cm^{-3} °K^{-1}	Thermal conductivity Cal cm^{-1} °K^{-1}sec^{-1}	Diffusivity (cm^{2}sec^{-1})	Ref.
Hudson River sand	91	8	1	0.90	4.5	0.176	0.20	0.0091	0.0045	61
				0.94	18.1	0.176	0.336	0.0300	0.0089	
Podunk fine sandy loam	90	8	2	0.89	6.6	0.182	0.221	0.0012	0.0058	61
				0.965	20.2	0.182	0.371	0.0026	0.0069	
Leonardtown silt loam	27	58	15	0.99	9.0	0.229	0.316	0.0018	0.0058	61
				0.83	18.4	0.226	0.338	0.0021	0.0061	
Muck soil	—	—	—	0.65	23.0	0.156	0.251	0.00076	0.0030	61
				0.43	59.0	0.156	0.321	0.00108	0.0034	
				0.42	69.4	0.156	0.315	0.00086	0.0024	
Yolo clay	—	—	—	1.18	0	0.194	0.236	0.0014	0.0062	62
				1.47	29.0	0.194	0.720	0.0083	0.0116	
Granitic sandy loam	55	40	5	1.29	0	0.226	0.291	0.0017	0.0059	62
				1.56	22.7	0.226	0.706	0.0071	0.0101	
Fine calcareous loam	72	26	2	1.14	0	0.154	0.175	0.00079	0.0049	62
				1.08	24.4	0.154	0.430	0.0048	0.0110	
Granitic sand	74	23	3	1.32	0	0.204	0.269	0.00137	0.0052	62
				1.90	13.1	0.204	0.636	0.0108	0.0147	
Barns loam				1.16	5.1	0.20	0.29	0.00041	0.0014	63
				1.15	8.0	—	0.33	0.00045	0.0014	
	42	25	33	1.07	13.0	—	0.35	0.00049	0.0014	
				1.18	26.0	—	0.55	0.00086	0.0016	

Table 8 (continued) SELECTED VALUES FOR THERMAL PROPERTIES OF SOILS[59]

Soil type	Particle size: Sand (%)	Silt (%)	Clay (%)	Bulk density (g cm^{-3})	Water (%)	Specific heat capacity Cal g^{-1} $°K^{-1}$	Heat capacity/ CM^3 Cal cm^{-3} $°K^{-1}$	Thermal conductivity Cal cm^{-1} $°K^{-1}sec^{-1}$	Diffusivity (cm^2sec^{-1})	Ref.
Chester loam				1.44	2.0	—	0.32	0.00045	0.0014	63
	27	41	32	1.19	4.7	—	0.29	0.00051	0.0017	
				1.12	13.4	—	0.37	0.00087	0.0023	
Herman sandy loam				1.42	1.3	—	0.30	0.00049	0.0016	63
	80	16	4	1.24	4.4	—	0.30	0.00060	0.0020	
				1.23	8.7	—	0.35	0.00086	0.0024	
Kalkaska loamy sand				1.55	0.8	—	0.32	0.00060	0.0018	63
	—	—	—	1.43	1.6	—	0.32	0.00069	0.0021	
				1.44	5.7	—	0.37	0.00124	0.0033	
Northway silt loam				1.56	6.6	0.18	0.384	0.0013	0.0034	64
				1.54	16.4	—	0.530	0.0021	0.0040	
	14	64	21	1.57	22.5	—	0.636	0.0025	0.0039	
				1.43	16.5	—	0.493	0.0018	0.0036	
				1.82	14.5	—	0.592	0.0030	0.0051	
Fairbanks silty clay loam				1.44	12.3	0.18	0.436	0.0020	0.0046	64
				1.44	18.0	—	0.518	0.0025	0.0048	
	9	64	27	1.44	25.4	—	0.625	0.0028	0.0045	
				1.45	30.0	—	0.696	0.0029	0.0042	
				1.28	12.3	—	0.388	0.0015	0.0039	
				1.62	12.5	—	0.494	0.0026	0.0053	
Dakota sandy loam	58	21	10	1.35	1.9	0.18	0.269	0.00059	0.0022	64
				2.11	4.9	—	0.483	0.0054	0.0112	

Black cotton soil										
Depth in cm:										
0—5				1.45	—	0.22	0.336	0.00037	0.0011	65
5—10				2.00	—	—	0.447	0.00067	0.0015	
10—15				2.2	—	—	0.500	0.00060	0.0012	
15—20				2.3	—	—	0.52	0.00052	0.0010	
Loam[g]	—	—	—	—	—	—	—	0.0024	0.011	66
Sand	—	—	—	—	—	—	—	0.0016	0.0063	
Loamy sand	—	—	—	—	—	—	—	0.0043	0.0079	

Table 9
EXPERIMENTAL (λ_e) AND COMPUTED (λ_c) VALUES OF THE THERMAL CONDUCTIVITY OF TWO SOILS WITH WIDELY DIFFERENT COMPOSITIONS AND MOISTURE CONTENT[60]

X_w	X_s	X_a	λ_e	λ_c	λ_e/λ_c
			($mcal\ cm^{-1}sec^{-1}°C^{-1}$)		
			Fairbanks Sand		
0.212	0.660	0.128	5.49	5.42	1.01
0.203	0.710	0.087	6.07	6.30	0.96
0.184	0.632	0.184	4.97	4.71	1.06
0.117	0.705	0.178	5.24	5.35	0.98
0.112	0.665	0.223	4.85	4.56	1.06
0.102	0.691	0.207	5.31	4.93	1.08
0.101	0.631	0.268	3.75	3.90	0.96
0.050	0.710	0.240	4.28	4.32	0.99
0.047	0.665	0.288	3.45	3.57	0.97
0.043	0.631	0.326	2.92	3.04	0.96
0.026	0.727	0.247	2.87	2.8[a]	1.02
0.025	0.705	0.270	2.20	2.5[a]	0.88
0.024	0.665	0.311	1.93	2.0[a]	0.97
0.021	0.631	0.348	1.38	1.7[a]	0.81
0.004	0.710	0.286	0.125	0.12[a]	1.04
0.004	0.665	0.331	0.90	0.95[a]	0.95
0.003	0.629	0.368	0.79	0.80	0.99
			Healy Clay		
0.359	0.641	0.000	3.68	3.66	1.01
0.450	0.499	0.051	2.93	2.85	1.03
0.402	0.444	0.154	1.97	2.34	0.84
0.334	0.570	0.087	3.22	3.07	1.05
0.304	0.517	0.179	2.18	2.51	0.87
0.270	0.671	0.059	3.89	3.59	1.08
0.256	0.581	0.163	2.84	2.82	1.01
0.231	0.519	0.250	2.27	2.28	1.00
0.163	0.581	0.256	2.11	2.32	0.91
0.144	0.519	0.337	1.53	1.96	0.78
0.095	0.575	0.330	1.42	1.83	0.78
0.091	0.525	0.384	1.00	1.66	0.60
0.041	0.521	0.438	0.71	0.85[a]	0.84
0.038	0.457	0.505	0.49	0.65[a]	0.75
0.033	0.396	0.571	0.38	0.48[a]	0.79

[a] These values are obtained by graphical interpolation.

ever, β is the only parameter that has to be experimentally measured as shown in Table 10.[70] By substituting the average value of 1.83 in the above vapor flow equation, J_v, water vapor flow, can be calculated for different soils.

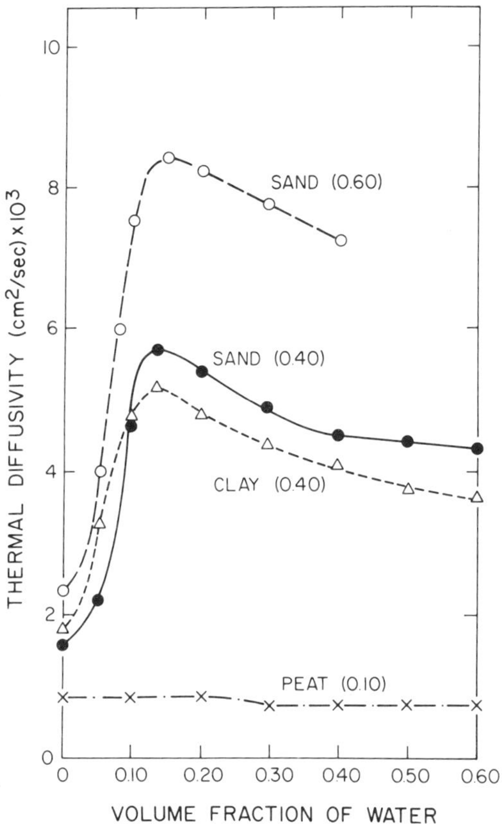

FIGURE 15. The effect of water on diffusivity (Van Duin[67]).

GAS TRANSPORT IN SOIL

Transfer of gases within a soil and between soil and atmosphere occurs principally in an air-filled pore space. The transfer of gas can be by mass flow due to differences in pressure or by concentration gradients. Buckingham[76] concluded that mass flow was of little importance in the exchange of soil gases and that diffusion was the major process for aerating soils. Various other investigators[77-85] arrived at the same conclusion.

Diffusion is a direct consequence of thermal motion. Molecular size particles move randomly in the phase in which they exist from points of high concentration to points of low concentration. Fick's first law can be used to characterize steady-state diffusion of gases through soils.

$$\frac{\Delta Q}{\Delta t} = DA \frac{dc}{dx} \qquad (10)$$

where ΔQ is the quantity diffusing past a certain plane (g), Δt is the time (sec) allowed for the diffusion, A is the area (cm^2) of the diffusion path, C is the concentration in g cm^{-3}, and x is distance in cm. The diffusion coefficient D has the dimensions of cm^2 and sec^{-1}. Factors which cause diffusion coefficient in a porous media to be less than the diffusion coefficient in bulk air or water are discussed by Porter et al.[86]

The diffusivity of gas in soil of unspecified porosity is called apparent or relative diffusivity. The diffusivity is a property of both the medium and the diffusing gas and

Table 10
EXPERIMENTAL MEASUREMENTS OF β[70]

Soil	Temp (°C)	Matric suction (cm)	Water content ($cm^3\ cm^{-3}$)	β	Ref.
Loam	20	239	0.181	1.87	71
	21	268	0.174	1.89	
	22	304	0.167	2.05	
	23	378	0.15	1.99	
	24	596	0.133	1.79	
Loam	15	540		2.87	72, 73
	25	540		2.84	
	35	540		2.52	
	15	370		3.39	
	25	370		3.00	
	35	370		2.34	
Sand	40.6	2×10^4	0.021	1.53	74
	35.4	1000	0.045	1.29	
	30.3	< 100	0.118	1.06	
	26.4	< 100	0.328	1.06	
	35.8	800	0.054	1.21	
	34.6	< 100	0.114	1.15	
	31.6	< 100	0.154	1.07	
	28.6	< 100	0.233	1.11	
Silt loam	41.0	< 10	0.036	0.75	74
	3.0	5×10^4	0.071	1.06	
	33.0	2700	0.178	0.95	
	29.7	950	0.236	1.08	
Loam	7.0	66		1.80	75
	19.0	66		1.72	
	32.0	66		0.72	
	6.0	132		2.22	
	18.0	132		2.28	
	35.0	132		2.21	
	9.0	237		1.95	
	20.0	237		2.21	
	33.0	237		2.37	
Silt loam		34		0.9	17
		270		3.8	
Sandy loam		550		3.1	17
		82		0.9	
Silty clay		310		1.4	17
		420		2.0	
				$\bar{\beta} = 1.83 \pm 0.79$	

is affected by atmospheric pressure and temperature. In order to compare diffusivities of different gases in various media, D is usually expressed as some fraction of D_o, the diffusion coefficient of the gas in free air under the boundary conditions imposed by a particular apparatus used in the study. Various relations between the apparent diffusivities in soil and air D/D_o and the air-filled porosity S have been suggested. Penman[78] reported that $D/D_o = 0.66S$, for the diffusion of vapor through porous solids, whereas Van Bavel[83] found $D/D_o = 0.59S$. Marshall[87] and Millington[88] suggested that $D/D_o = S^{3/2}$ and $D/D_o = S^{4/3}$. Currie[89] pointed out that the reduction in diffusion coefficient is not strictly related to porosity, but requires a shape factor as well. He concluded that the relation would have the general form of $D/D_o = \sigma S^{\mu}$ where $\sigma \leqslant 1$ and $\mu \geqslant 1$. Wesseling and van Wijk[90] suggested that one may assume that 10% by volume of air-filled pores is the lowest value at which air can be exchanged in the soil.

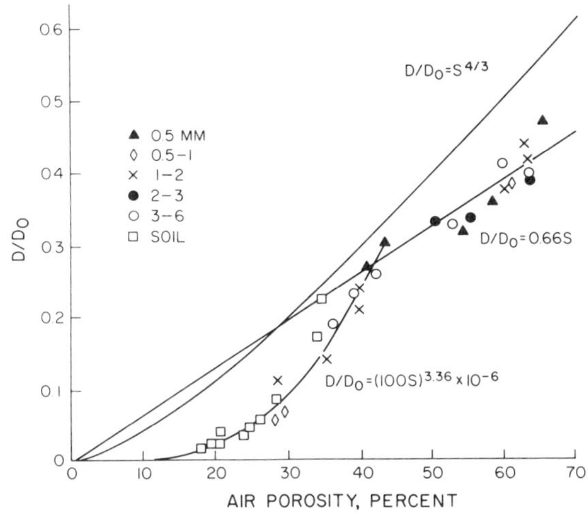

FIGURE 16. Influence of air porosity on relative diffusivity of O_2 in silty clay loam. Data points correspond to the same samples used for Figure 17; points above 45% air porosity represent air-dry aggregates; points below 45%, moist samples of aggregates and compacted soil. S in the lowest equation is percentage of air porosity, but in the upper two equations is on a fractional basis (Adapted from Grable and Siemer[91]).

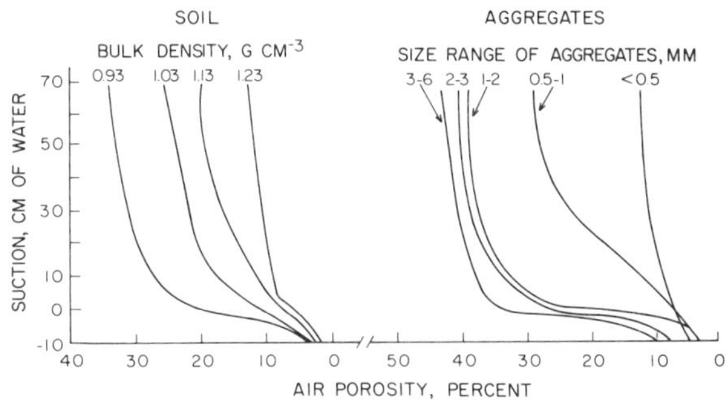

FIGURE 17. Influence of bulk density and size of aggregates on air porosity after desorption of silty clay loam at different suctions. Suctions were developed at the bottom of columns about 10 cm high. Bulk densities of aggregates ranged from 0.84 to 1.07 g/cm^{-3} in largest and smallest fractions, respectively. (Redrawn from Grable and Siemer[91]).

Grable and Siemer[91] consider air porosities of 12 to 15% a safe limit for minimum aeration. They found the empirical equation of $D/D_o = P^{3.36} \times 10^{-6}$ (P = percent air-filled porosity) for aggregates and compacted soil in Figure 16. In Figure 17 they showed the influence of bulk density and size of aggregates on air porosity of silty clay loam at different suctions.

Table 11 shows values of D/D_o for soil with different porosities. Physical properties of some soil gases are given in Table 12. This list does not include volatile organic components such as alcohols, acids, esters, aldehydes, and ketones. Some gases, such as CO_2, NH_3, and H_2S become ionized in water and are more soluble than the two more abundant gases O_2 and H_2.

The composition of soil air in an ideally aerated soil is nearly that of atmospheric air: 78.08% nitrogen, 20.95% oxygen, 0.03% carbon dioxide, and 0.94% other gases.

Table 11
DIFFUSION OF CARBON DISULPHIDE VAPOR IN SOIL MATERIAL[78]

Material	Moisture content (%)	S	Mean temperature (°C)	D/D_o
Rothamsted subsoil	Air-dry	0.518	12.7	0.208
	Air-dry	0.518	13.8	0.304
	Air-dry	0.547	14.3	0.346
	Air-dry	0.550	16.5	0.358
	11.0	0.438	18.4	0.278
	17.0	0.448	16.7	0.297
	25.0	0.626	17.2	0.442
	29.6	0.549	16.8	0.364
Rothamsted subsoil + sand	Air-dry	0.422	15.1	0.273
Natal soil (N 64)	Air-dry	0.496	16.8	0.300
	Air-dry	0.620	16.4	0.417
	Air-dry	0.686	17.1	0.475
	11.0	0.475	17.0	0.312
	19.3	0.355	19.4	0.249
	22.0	0.425	14.9	0.299
	30.6	0.195	17.0	0.118

Table 12
PHYSICAL DATA ON CERTAIN SOIL GASES

Gas	Diameter[92] (cm × 10^{-8})	Mean free path[92] (cm × 10^{-6} 750 mmHg)	Collision[92] frequency (× 10^6/sec, at 20°C)	Average[93] velocity (cm × 10^2/sec, STP)	Density[94] (g/ℓ, STP)	Thermal[95] conductivity (× 10^{-5}cal cm^{-2}sec^{-1} °C^{-1}cm^{-1}, at 0°C)	Solubility[94] (mℓ gas/mℓ sol at STP)
N_2	3.15—3.53	8.50	5070	454	1.251	5.7	0.016
NH_3	2.97—3.08	5.92	9150	583	0.77126	5.2	
N_2O					1.9804	3.6	
CO_2	3.23—3.40	5.56	6120	362	1.977	3.4	0.878
CH_4					0.7168	7.3	0.033
O_2	2.92—2.98	9.05	4430	425	1.429	5.73	0.031
SO_2					2.9262	2.0	39.374
H_2S					1.5392	3.0	

REFERENCES

1. **Coleman, N. T.**, The good and not-so-good earth, *19th Annual Research Lecture*, Riverside, Calif., 1970.
2. **Brewer, R.**, *Fabric and Mineral Analysis of Soils*, Robert E. Krieger, Huntington, New York, 1976, chap. 9.
3. **Russell, M. B.**, Physical properties, in *Soil*, Yearbook of Agriculture, U.S. Department of Agriculture, Washington, D.C., 1957, 31.
4. **Carter, D. L., Heilman, M. D., and Gonzalez, C. L.**, Ethylene glycol monoethyl ether for determining surface area of silicate minerals, *Soil Sci.*, 100, 356, 1965.
5. **Perrier, E. R., MacKenzie, A. J., and Zimmerman, R. P.**, Physical and chemical properties of major Imperial Valley soils, USDA-ARS, W-17, U.S. Department of Agriculture, Washington, D.C., 1974.
6. **Bower, C. A. and Goertzen, J. O.**, Surface area of soils and clays by an equilibrium ethylene glycol method, *Soil Sci.*, 87, 289, 1959.
7. **Dyal, R. S. and Hendricks, S. B.**, Total surface of clays in polar liquids as a characteristic index, *Soil Sci.*, 69, 421, 1950.
8. **Taylor, S. A. and Ashcroft, G. L.**, The soil solids, in *Physical Edaphology*, W. H. Freeman, San Francisco, 1972, chap. 6.
9. **Ormsby, W. C., Shartsis, J. M., and Woodside, K. H.**, Exchange behavior of kaolins and varying degrees of crystallinity, *L. Am. Ceram. Soc.*, 45, 361, 1962.
10. **Zelazny, L. W. and Calhous, F. G.**, Palygorskite, (Attapulgite), Sepiolite, Talc, Pyrophyllite, and Zeolites, in *Minerals in Soil Environments*, Dixon, J. B. and Weed, S. B., Eds., Soil Science Society of America, Madison, Wis., 1977, chap. 13.
11. **Wada, K.**, Allophane and imogolite, in *Minerals in Soil Environments*, Dixon, J. B. and Weed, S. B., Eds., Soil Science Society of America, Madison, Wis., 1977, chap. 16.
12. **Egashira, K. and Aomine, S.**, Effects of drying and heating on the surface area of allophane and imogolite, *Clay Sci.*, 4, 231, 1974.
13. **Gruner, J. W.**, Densities and structural relationships of kaolinites and anauxites, *Am. Miner.*, 22, 85, 1937.
14. **Makower, B., Shaw, T. M., and Alexander, L. T.**, The specific surface and density of some soils and their colloids, *Soil Sci. Soc. Am. Proc.*, 2, 101, 1937.
15. **DeWitt, C. P. and Arens, P. L.**, Moisture content and density of some clay minerals and some remarks on the hydration pattern of clay, *Trans. 4th Int. Congr. Soil Sci.*, 2, 59, 1950.
16. **Deer, W. A., Howie, R. A., and Zussman, J.**, Framework silicates, in *Rock Forming Minerals*, Vol. 4, Longmans, Green, & Co., London, 1963.
17. **Nielsen, D. R., Jackson, R. D., Cary, L. W., and Evans, D. D., Eds.**, Solid and fluid properties of the soil-water system, in *Soil Water*, American Society of Agronomy and Soil Science Society of America, Madison, Wis., 1972.
18. **Johnson, W. M., McClelland, J. E., McCaleb, S. B., Ulrich, R., Harper, W. G., and Hutchings, T. B.**, Classification and description of soil pores, *Soil Sci.*, 89, 321, 1970.
19. Soil Survey Staff, U.S. Soil Conservation Service, Soil Taxonomy, *U.S. Dep. Agric. Agric. Handb.*, No. 436, 1975.
20. **Spangler, M. G. and Handy, R. L.**, *Soil Eng.*, 3rd ed., Harper & Row, New York, 1973.
21. **Terzaghi, K.**, Erdbaumechanik auf bodenphysikalischer Grundlage, *Deuticke*, 1925.
22. **Casagrande, A.**, Classification and identification of soils, *Trans. Am. Soc. of Civ. Eng.*, 113, 901, 1948.
23. **Taylor, D. W.**, *Fundamentals of Soil Mechanics*, John Wiley & Sons, New York, 1948.
24. **McCarthy, D. F.**, *Essentials of Soil Mechanics and Foundations*, Reston Publishing, Reston, Virginia, 1977.
25. **Gill, W. R. and Vanden Berg, G. E.**, Soil Dynamics in Tillage and Traction, *U.S. Dep. Agric. Agric. Handb.*, No. 316, 1967.
26. **Waldron, L. J.**, Soil viscoelasticity: superposition tests, *Soil Sci. Soc. Am. Proc.*, 28, 323, 1964a.
27. **Waldron, L. J.**, Viscoelastic functions for soil, *Soil Sci. Soc. Am. Proc.*, 28, 329, 1964b.
28. **McMurdie, J. L.**, Some characteristics of the soil deformation process, *Soil Sci. Soc. Am. Proc.*, 27, 251, 1963.
29. **Baver, L. D., Gardner, W. H., and Gardner, W. R.**, *Soil Physics*, 4th ed., John Wiley & Sons, New York, 1972.
30. **Bodman, G. B.**, Methods of measuring soil consistency, *Soil Sci.*, 68, 37, 1949.
31. **Greacen, E. L.**, Water content and soil strength, *J. Soil Sci.*, 11, 313, 1960.
32. **Barley, K. P., Farrell, D. A., and Greacen, E. L.**, The influence of soil strength on the penetration of a loam by plant roots, *Aust. J. Soil Res.*, 3, 69, 1965.

33. Carter, L. M., Portable recording penetrometer measures soil strength profiles, *Agr. Eng.*, 48, 343, 1975.
34. **Waldron, L. J. and Constantin, G. K.**, Soil resistance to a slowly moving penetrometer, *Soil Sci.*, 109, 221, 1970.
35. **Zelenin, A. M.**, *Basics of the Theory of Soil Cutting*, Moscow, 1950.
36. **Taylor, H. M. and Bruce, R. R.**, Effects of soil strength on root growth and crop yield in the southern United States, *Trans. 9th Int. Congr. of Soil Sci.*, 1, 803, 1968.
37. **Grimes, D. W., Miller, R. J., and Wiley, P. L.**, Cotton and corn root development in two field soils of different strength characteristics, *Agron. J.*, 67, 519, 1975.
38. **Farrell, D. A. and Greacen, E. L.**, Resistance to penetration of fine probes in compressible soil, *Aust. J. Soil Res.*, 4, 1, 1966.
39. **Bear, J.**, Two-liquid flows in porous media, *Adv. Hydrosci.*, 6, 131, 1970.
40. **Childs, E. C.**, Soil moisture theory, *Adv. Hydrosci.*, 4, 73, 1967.
41. **Childs, E. C.**, *The Physical Basis of Soil Water Phenomena*, John Wiley & Sons, New York, 1969.
42. **Cary, J. W. and Taylor, S. A.**, The dynamics of soil water. II. Temperature and solute effects, in *Irrigation of Agricultural Lands*, Hagan, R. M., Ed., *Agronomy*, 11, 245, 1967.
43. **Klute, A.**, Soil water flow theory and its application in field situations, in *Field Soil Water Regime*, Bruce, R. R., Flach, K. W., and Taylor, H. M., Eds., Soil Science Society of America, Spec. Publ. Ser. No. 5, Madison, Wis., 1973, 9.
44. **Miller, E. E. and Klute, A.**, The dynamics of soil water. I. Mechanical forces, in *Irrigation of Agricultural Lands*, Vol. 11, Hagan, R. M., Haise, H. R., and Edminster, T. W., Eds., American Society of Agronomy, Madison, Wis., 1967, 209.
45. **Philip, J. R.**, Theory of infiltration, *Adv. Hydrosci.*, 5, 216, 1969.
46. **Philip, J. R.**, Flow in porous media, *Annu. Rev. Fluid Mech.*, 2, 177, 1970.
47. **Hillel, D.**, *Soil and Water*, Academic Press, New York, 1970.
48. **Swartzendruber, D.**, Soil-water behavior as described by transport coefficients and functions, *Adv. Agron.*, 18, 327, 1966.
49. **Day, P. R., Bolt, G. H., and Anderson, D. M.**, Nature of soil water, in *Irrigation of Agricultural Lands*, Hagan, R. M., Haise, H. R., and Edminster, T. W., Eds., American Society of Agronomy, Madison, Wis., 1967, 193.
50. **Flühler, H., Ardakani, M. S., Szuszkiewicz, T. E., and Stolzy, L. H.**, Field-measured water uptake of sudangrass roots as affected by fertilization, *Agron. J.*, 69, 269, 1977.
51. **Richards, L. A. and Richards, S. J.**, Soil moisture, in *Soil*, The Yearbook of Agriculture, U.S. Department of Agriculture, Washington, D.C., 1957, 49.
52. **Richards, S. J. and Marsh, A. W.**, Irrigation based on soil suction measurements, *Soil Sci. Soc. Am. Proc.*, 25, 65, 1961.
53. **Darcy, H.**, *Les Fontaines Publique de la Ville de Dijon*, Victor Dalmont, Paris, 1856.
54. **Buckingham, E.**, *Studies on the Movement of Soil Moisture*, Bureau of Soils Bull. 38, U.S. Department of Agriculture, Washington, D.C., 1907.
55. **Gardner, Willard**, The movement of moisture in soil by capillarity, *Soil Sci.*, 7, 313, 1919.
56. **Gardner, W. R.**, Dynamic aspects of water availability to plants, *Soil Sci.*, 89, 63, 1960.
57. **Childs, E. C.**, The physics of land drainage, in *Drainage of Agricultural Lands*, Luthin, J. N., Ed., American Society of Agronomy, Madison, Wis., 21, 1957.
58. **Weeks, L. V. and Richards, S. J.**, Soil-water properties computed from transient flow data, *Soil Sci. Soc. Am. Proc.*, 31, 721, 1967.
59. **Richards, S. J., Hagan, R. M., and McCalla, T. M.**, Soil temperature and plant growth, in *Soil Physical Conditions and Plant Growth*, Vol. 2, Shaw, B. T., Ed., Academic Press, New York, 1952, chap. 5.
60. **DeVries, D. A.**, Thermal properties of soils, in *Physics of Plant Environment*, Van Wijk, W. R., Ed., John Wiley & Co., New York, 1963, chap. 7.
61. **Patten, H. E.**, *Heat Transference in Soils*, Bureau of Soils Bull. 59, U.S. Department of Agriculture, Washington, D.C., 1909, 1.
62. **Johnston, C. N.**, Heat conductivity of soil governs heat losses from heated oil lines, *Pet. Eng.*, 9, 41, 1937.
63. **Smith, A.**, Value of mean and average soil and air temperature, *Soil Sci. Soc. Am. Proc.*, 4, 41, 1939.
64. **Kersten, M. S.**, Thermal Properties of Soils, *Minn. Univ. Eng. Exp. Stn. Bull.*, 28, 1, 1949.
65. **Dravid, R. K.**, Studies on soil temperatures in relation to other factors controlling the disposal of solar radiation, *Indian J. Agr. Sci.*, 10, 352, 1940.
66. **Schubert, J.**, *Handbuch der Bodenlehrer, VI.*, Julius Springer, Berlin, 1930.
67. **Van Duin, R. H. A.**, *The Influence of Soil Management on the Temperature Wave Near the Surface*, Tech. Bull. 19, Institute for Land and Water Management Research, Wageningen, 1963.

68. Cary, J. W., Onsager's relation and the non-isothermal diffusion of water vapor, *J. Phys. Chem.*, 67, 126, 1963.
69. Cary, J. W., An evaporation experiment and its irreversible thermodynamics, *Int. J. Heat Mass Transfer*, 7, 531, 1964.
70. Jury, W. A. and Letey, J., Water vapor movement in soil: reconciliation of theory and experiment, *Soil Sci. Soc. Am. J.*, 40, 823, 1979.
71. Weeks, L. V., Richards, S. J., and Letey, J., Water and salt transfer in soil resulting from thermal gradients, *Soil Sci. Soc. Am. Proc.*, 32, 193, 1968.
72. Cary, J. W. and Taylor, S. A., The interaction of the simultaneous diffusions of heat and water vapor, *Soil Sci. Soc. Am. Proc.*, 26, 413, 1962a.
73. Cary, J. W. and Taylor, S. A., Thermally driven liquid and vapor phase transfer of water and energy in soil, *Soil Sci. Soc. Am. Proc.*, 26, 417, 1962b.
74. Hanks, R. J., Gardner, H. R., and Fairbourn, M. L., Evaporation of water from soils as influenced by drying with wind or radiation, *Soil Sci. Soc. Am. Proc.*, 31, 593, 1967.
75. Cary, J. W., Water flux in moist soil: thermal versus suction gradients, *Soil Sci.*, 100, 168, 1965.
76. Buckingham, E., *Contributions to Our Knowledge of the Aeration of Soils*, Bureau Soils Bull. 25, U.S. Department of Agriculture, Washington, D.C., 1904.
77. Romell, L. G., Luftvaxlungen i marken som ekologisk faktor, *Medd. fran Statens Skoysförsöksanstalt*, 19, 125, 1922.
78. Penman, H. L., Gas and vapor movement in the soil: I. The diffusion of vapors through porous solids, *J. Agr. Sci.*, 30, 437, 1940.
79. Penman, H. L., Gas and vapor movement in the soil: II. The diffusion of carbon dioxide through porous solids, *J. Agr. Sci.*, 30, 570, 1940.
80. Blake, G. R. and Page, J. B., Direct measurement of gaseous diffusion in soils, *Soil Sci. Soc. Am. Proc.*, 13, 37, 1948.
81. Taylor, S. A., Oxygen diffusion in porous media as a measure of soil aeration, *Soil Sci. Soc. Am. Proc.*, 14, 55, 1949.
82. Van Bavel, C. H. M., A soil aeration theory based on diffusion, *Soil Sci.*, 72, 33, 1951.
83. Van Bavel, C. H. M., Gaseous diffusion and porosity in porous media, *Soil Sci.*, 73, 91, 1952.
84. Russell, M. B., Soil aeration and plant growth, in *Soil Physical Conditions and Plant Growth*, Shaw, B. T., Ed., Academic Press, New York, 1952, 253.
85. Wesseling, J., Some solutions of the steady state diffusion of carbon dioxide through soils, *Neth. J. Agr. Sci.*, 10, 109, 1962.
86. Porter, L. K., Kemper, W. D., Jackson, R. D., and Stewart, B. A., Chloride diffusion in soils as influenced by moisture content, *Soil Sci. Soc. Am. Proc.*, 24, 460, 1960.
87. Marshall, T. J., The diffusion of gases through porous media, *J. Soil Sci.*, 10, 79, 1959.
88. Millington, R. J., Gas diffusion in porous media, *Science*, 130, 100, 1959.
89. Currie, J. A., Gaseous diffusion in porous media, *Brit. J. Appl. Phys.*, 11, 318, 1970.
90. Wesseling, J. and van Wijk, W. R., Soil physical conditions in relation to drain depth, in *Drainage of Agricultural Lands*, Luthin, J. N., Ed., American Society of Agronomy, Madison, Wis., 1957, 461.
91. Grable, A. R. and Siemer, E. G., Effects of bulk density, aggregate size, and soil water suction on oxygen diffusion, redox potentials, and elongation of corn roots, *Soil Sci. Soc. Am. Proc.*, 32, 180, 1968.
92. Weast, R. C., Ed., *Handbook of Chemistry and Physics*, CRC Press, Cleveland, 1969.
93. Dittmer, D. S. and Grebe, R. M., Eds., *Handbook of Respiration*, W. B. Saunders, Philadelphia, 1958.
94. Gray, D. E., Ed., *American Institute of Physics Handbook*, McGraw-Hill, New York, 1963.
95. Condon, E. U. and Odishaw, H., Eds., *Handbook of Physics*, McGraw-Hill, New York, 1967.

SOIL CHEMISTRY

G. W. Thomas

The chemical composition of soils is a result of the parent material, the age of the soil, and the climate to which the soil has been subjected. All three of these factors are hard to document because the parent material that formed the existing soil is now gone, age is a wild guess except for very young or very old soils and climate can change. In addition, chronological age of a soil and weathering stage are not the same thing at all.

In Table 1 the chemical composition of whole soils, along with location and climate, are listed. The main constituents, SiO_2, Al_2O_3, and Fe_2O_3, are quite variable and perhaps more closely related to clay content than to soil weathering. Soils with SiO_2 contents above 80% are generally sandy loams; silt loams range between 70 and 80% as do loams. Clays usually contain less than 70% SiO_2. Ratios of Al_2O_3/Fe_2O_3 and $SiO_2/Al_2O_3 + Fe_2O_3$ are not very reliable indicators of weathering or soil age on a whole-soil basis. Percentages of total mono- and divalent bases are also unreliable as weathering indices because of the profound effects of parent material on the amounts remaining.

On a clay basis, some of these trends become more apparent, however. Table 2 shows the total composition of clay-sized fractions of a number of American soils. In a very general way, the SiO_2/R_2O_3 ($R_2O_3 = Al_2O_3 + Fe_2O_3$) ratio increases as the rainfall decreases. There appears to be no trend with mean annual temperature. There also is a trend towards increasing CaO, MgO, and K_2O, and decreasing Fe_2O_3 as rainfall diminishes. There are, however, many exceptions to these trends.

Plants grown annually or grown perennially for a few years do not appear to be affected very much by total compositions of soils or clays. Their nutrition is more closely related to quantities and proportions of exchangeable cations. Table 3 shows pH, exchangeable cations, and cation exchange capacities of soils arranged with increasing temperature and increasing rainfall. There is a remarkable lack of temperature effect and a very small rainfall effect on any of these measurements until rainfall exceeds about 80 cm/year.

Because of parent material and age differences, there appears to be as poor a relationship between climate and the suite of exchangeable cations as there is between climate and total composition. While there are general trends, there is poor predictability of soil pH, exchangeable calcium, and total acidity based on climatic variables.

Soil acidity is one of two common problems which must be overcome before healthy plant growth can occur. Soils become acid when the divalent cations (principally Ca) are removed by excess water saturated with CO_2. Thus, the leachate contains $Ca(HCO_3)_2$ and the soil is left with two hydrogen ions, which then dissolve the Al^{3+} from the clay structure. These reactions are shown in Table 4. In addition to Al^{3+}, other cations such as Mg^{2+} and Fe^{3+} are dissolved from clays by hydrogen ions. The Mg becomes exchangeable while the Fe^{3+} precipitates on the surface as Fe_2O_3, giving a red, yellow, or brownish color to the soil.

The cations found in acid soils are listed in Table 5. The situations under which they are found are also defined. Table 6 gives the pH values which are associated with some of the critical reactions in soils. These pH values are helpful in understanding conditions in soils using a single measurement. The percentage base saturation varies with pH as one goes from one soil to another (Table 7). This is related to clay mineralogy, iron and aluminum oxide, and organic matter content. In general, the higher base saturations at low pH represent montmorillonitic clay, low oxides, and organic matter.

Table 1
PERCENTAGE COMPOSITION OF SOILS FROM THE NORTHEASTERN, SOUTHEASTERN, SOUTHWESTERN AND MIDWESTERN U.S. AND PUERTO RICO

State	N.H.[a]		Pa.[a]		Pa.[a]		N.Y.[a]		W. Va[b]		Va.[b]	N.C.[c]	N.C.[a]	
Soil series name[d]	Gloucester sl		Penn sl		Hagerstown l		Volusia sl		Teas sl		Iredell cl	White Store sl	Cecil sl	
Depth sampled (cm)	0—51	51—228	0—51	51—152	0—51	51—152	0—51	51—228	0—5	18—33	23—61	41—64	0—51	51—228
Constituent (%)														
SiO_2	65.68	73.80	74.33	71.76	70.99	66.49	75.12	74.64	69.2	68.3	50.30	51.10	88.57	55.69
Al_2O_3	14.15	13.24	11.00	14.36	11.39	14.80	10.49	12.26	9.6	15.5	11.50	17.35	5.74	24.42
Fe_2O_3	5.67	4.37	4.65	5.82	4.23	5.99	4.13	5.01	4.3	5.8	17.61	10.47	1.55	8.83
K_2O	2.16	2.22	1.57	1.50	2.71	3.58	1.40	1.99	1.2	2.3	0.21	1.42	0.82	1.06
CaO	1.36	1.19	1.13	1.73	0.93	0.35	0.49	0.37	0.3	< 0.1	1.17	0.35	0.39	0.40
MgO	0.83	0.39	0.69	1.06	1.08	1.93	0.48	0.90	—	—	—	—	0.21	0.29
Na_2O	1.39	1.75	1.53	1.54	0.82	0.66	0.90	0.99	—	—	—	—	0.16	0.14
P_2O_5	0.15	0.11	0.16	0.10	0.19	0.16	0.18	0.15	—	—	—	—	0.08	0.07
SO_3	0.17	0.03	0.15	0.10	0.39	0.14	0.09	0.10	—	—	—	—	0.04	0.09
Mean annual temp (°C)	8	8	12	12	12	12	8	8	11	11	13	18	12	12
Mean annual precipitation (cm)	76	76	101	101	101	101	86	86	129	129	112	132	130	130

State Soil series name[d]	Ala.[a] Decatur cl		Miss.[b] Shubuta sl		Texas Beaumont cl	Texas[c] Houston Black cl	Texas[c] Nacogdoches sc	Ind.[a] Alford sl		Mo.[a] Marshall sl		Minn.[b] Clarion sl		Wis.[a] Carrington l	Puerto Rico[b] Nipe cl	
Depth sampled (cm)	0—25	25—97	0-13	18-36	0-23	0-41	18-74	0-15	33-56	0-97	97-228	0-6	23-38	0-71	0-28	122-157
Constituent (%)																
SiO_2	72.35	74.81	86.2	69.3	65.50	48.00	59.80	80.8	73.4	73.61	71.43	71.3	74.5	73.50	23.7	10.0
Al_2O_3	8.89	12.80	4.8	15.3	11.37	6.68	12.55	8.0	11.7	9.67	13.44	9.5	10.9	9.10	25.3	25.6
Fe_2O_3	4.44	5.28	1.8	7.4	5.99	3.20	21.39	2.5	5.2	3.54	4.28	2.6	3.5	4.30	29.7	44.1
K_2O	0.67	0.75	1.2	1.5	0.37	0.37	0.46	2.0	2.2	2.28	2.03	1.6	1.7	2.03	0.1	< 0.1
CaO	0.63	0.40	0.3	< 0.1	1.10	1.75	0.30	0.5	0.5	1.08	1.40	1.5	1.0	0.94	0.1	< 0.1
MgO	0.39	0.33	—	—	—	—	—	—	—	0.77	1.28	—	—	0.71	—	—
Na_2O	0.24	0.16	—	—	—	—	—	—	—	1.03	0.63	—	—	1.67	—	—
P_2O_5	0.18	0.15	—	—	—	—	—	—	—	0.22	0.16	—	—	0.24	—	—
SO_3	0.13	0.19	—	—	—	—	—	—	—	0.17	0.14	—	—	0.13	—	—
Mean annual temp (°C)	16	16	17	17	17	18	17	13	13	12	12	8	8	8	24	24
Mean annual precipitation (cm)	129	129	140	140	127	97	115	106	106	102	102	85	85	89	192	192

[a] Reference 9
[b] Reference 11
[c] Reference 2
[d] sl - silt loam; l - loam; cl - clay loam; sc - sandy clay.

Table 2
COMPOSITION OF CLAYS IN SOILS FROM VARYING CLIMATES, ARRANGED WITH INCREASING SiO_2/R_2O_3 RATIO

Soil type	Location	Mean ann. rainfall (cm)	Ann. mean temp. (°C)	SiO_2	Al_2O_3	Fe_2O_3 (%)	CaO	MgO	K_2O	$\frac{SiO_2}{R_2O_3}$
Cecil (subsoil)	Ga.	124	17	31.84	38.28	10.04	0.35	0.23	0.46	1.20
Cecil (surface)	Ga.	124	17	33.95	36.06	11.02	0.31	0.40	0.56	1.34
Vega Baja (surface)	P.R.	175	25	36.26	32.85	12.44	0.44	0.18	0.36	1.34
Susquehanna (surface)	Miss.	147	18	36.97	36.91	10.44	0.20	0.42	2.12	1.35
Chester (surface)	Va.	112	13	34.82	27.88	15.93	0.17	2.86	0.85	1.54
Norfolk (surface)	N.C.	127	11	38.25	31.21	11.25	0.54	0.53	0.26	1.66
Orangeburg (subsoil)	Miss.	134	17	40.35	33.27	10.08	0.45	0.67	0.81	1.71
Manor (surface)	Md.	101	12	38.86	31.11	10.33	0.64	1.23	1.37	1.74
Chester (surface)	Md.	109	12	39.00	29.98	11.27	0.93	0.22	1.24	1.77
Chester (subsoil)	Md.	109	12	40.03	30.58	11.34	0.36	0.78	1.14	1.79
Manor (subsoil)	Md.	101	12	40.36	31.20	10.29	0.38	1.35	1.28	1.81
Orangeburg (surface)	Miss.	134	17	40.35	31.04	10.11	0.51	0.73	0.81	1.83
Norfolk (subsoil)	N.C.	127	16	41.60	31.10	11.28	0.34	0.49	0.46	1.84
Sassafras (surface)	Md.	104	13	39.24	28.64	10.19	0.75	1.16	1.17	1.85
Huntington (surface)	Md.	89	12	37.79	27.16	11.28	0.70	1.30	2.67	1.86
Huntington (subsoil)	Md.	89	12	39.54	26.77	13.33	0.54	1.32	2.67	1.89
Sassafras (subsoil)	Md.	104	13	41.14	29.26	12.73	0.53	1.07	1.35	1.89
Hagerstown (subsoil)	Md.	101	12	41.63	29.86	11.31	0.74	1.42	1.84	1.89
Hagerstown (surface)	Md.	101	12	39.93	29.02	9.45	1.25	1.53	1.61	1.91
Iredell (subsoil)	N.C.	132	18	42.58	34.26	4.66	0.61	1.43	0.10	1.93
Dunkirk (surface)	N.Y.	101	8	43.02	27.92	13.16	0.63	2.32	3.71	1.99
Ontario (subsoil)	N.Y.	86	8	42.40	24.71	15.27	1.18	2.59	2.39	2.08
Susquehanna (subsoil)	Md.	104	12	43.71	28.44	10.72	0.24	2.56	0.86	2.10
Clarksville (subsoil)	Ky.	124	14	43.68	27.70	11.51	0.48	1.39	1.65	2.10
Ontario (surface)	N.Y.	86	8	38.01	23.30	10.84	1.37	2.34	2.23	2.13
Clarksville (surface)	Ky.	124	14	42.40	26.49	10.05	0.62	1.22	1.86	2.18
Crowley (surface)	La.	140	20	46.95	27.35	8.77	0.99	1.47	0.85	2.40
Miami (surface)	Ind.	94	10	46.37	23.96	11.47	0.92	2.15	2.47	2.50
Carrington (subsoil)	Iowa	81	8	48.04	25.19	8.80	1.29	1.53	0.89	2.64

Miami (subsoil)	Ind.	94	10	48.48	24.12	10.44	1.39	2.64	2.88	2.72
Stockton (surface)	Calif.	36	16	48.95	21.54	13.79	1.47	2.16	0.28	2.72
Carrington (surface)	Iowa	81	8	44.89	22.59	7.75	1.48	1.44	1.36	2.75
Marshall (surface)	Neb.	76	10	45.93	21.72	8.94	1.19	2.01	2.28	2.82
Stockton (subsoil)	Calif.	71	17	48.61	22.23	10.13	1.70	3.37	0.78	2.85
Stockton (surface)	Calif.	71	17	49.50	22.51	10.57	1.96	2.68	0.26	2.87
Marshall (subsoil)	Neb.	76	10	48.18	22.00	10.03	1.36	1.62	2.07	2.87
Sharkey (surface)	Miss.	137	19	51.34	22.61	8.79	1.41	2.48	—	3.07
Wabash (surface)	Neb.	86	11	51.28	20.28	9.75	2.19	2.09	1.91	3.16
Sharkey (subsoil)	Miss.	137	19	50.63	20.74	9.07	1.92	1.80	1.96	3.23
Lufkin (subsoil)	Miss.	132	18	55.44	22.66	8.98	1.85	2.03	0.65	3.30
Wabash (subsoil)	Neb.	86	11	50.55	20.87	9.57	1.59	2.21	1.97	3.33
Yolo (surface)	Calif.	51	14	46.25	16.42	10.00	1.46	5.64	2.54	3.43
Houston Black (surface)	Tex.	97	18	53.34	20.77	7.15	5.01	1.39	—	3.56
Fallon (surface)	Nev.	13	10	52.57	16.49	10.70	2.20	5.67	1.97	3.82

From Robinson, W. O. and Holmes, R. S., The chemical composition of soil colloids, Bull. 1311, U.S. Department of Agriculture, Washington, D.C. 1924.

Table 3
CATION-EXCHANGE CAPACITIES AND EXCHANGEABLE CATIONS IN SOILS FROM VARIOUS CLIMATES

						Exchangeable cations (meq/100 g)						
Soil Names	Location	Depth of sample (cm)	Mean ann. temp (°C)	Mean ann. precip. (cm)	Soil pH	Ca	Mg	K	Na	H*	CEC meq/100g	% Base saturation
—	Alaska	0 — 8	−10	11	5.4	4.5	4.6	0.2	—	22.0	31.3	30
		28—30			5.4	11.9	10.8	1.3	—	29.7	53.9	45
Tanana	Alaska	0—8	−2	34	5.7	24.3	11.8	0.5	0.3	27.2	64.1	58
		28—53			7.7	13.6	5.6	0.1	0.3	2.8	22.4	88
Kalifonsky	Alaska	0—15	2	43	4.1	1.2	0.6	0.1	0.1	30.3	32.3	6
		15—51			5.2	0.9	< 0.1	0.1	0.1	18.6	19.7	6
Exline	N.D.	0—15	6	45	7.2	13.7	4.8	1.0	0.6	0	19.3	100 +
		43—61			9.4	17.2	12.9	0.5	6.0	0	14.7	100 +

Table 3 (continued)
CATION-EXCHANGE CAPACITIES AND EXCHANGEABLE CATIONS IN SOILS FROM VARIOUS CLIMATES

Soil Names	Location	Depth of sample (cm)	Mean ann. temp (°C)	Mean ann. precip. (cm)	Soil pH	Exchangeable cations (meq/100 g) Ca	Mg	K	Na	H*	CEC meq/100g	% Base saturation
Eakin	S. D.	0—9	8	46	6.8	20.0	5.8	1.7	—	5.9	27.5	—
		20—36			6.9	18.3	6.8	0.7	0.1	3.0	24.6	—
Marlin	N.Y.	1—2	9	89	4.0	1.8	0.6	0.2	0.1	21.7	24.4	11
		5—41			4.4	0.6	0.2	0.1	0.1	15.0	16.0	6
Scituate	N.H.	0—23	8	104	5.4	3.7	0.3	0.2	0.1	12.9	17.2	25
		33—41			5.7	0.4	0.2	0.1	0.1	4.8	5.6	14
Quillayute	Wash.	0—4	9	260	4.9	0.6	0.1	0.4	0.1	15.0	16.2	7
		11—33			5.2	0.4	0.1	0.2	0.1	24.5	25.3	3
Klaus	Wash.	0—20	9	294	5.3	3.0	2.1	0	0.3	52.5	80.8	9
		58—81			5.5	0.4	0.4	0	0.2	41.4	42.1	2
Fillmore	Neb.	0—13	11	58	5.6	12.0	3.9	1.9	0.3	5.1	23.2	78
		25—43			6.6	8.2	2.4	1.5	0.3	1.2	13.2	91
Sharpsburg	Neb.	0—15	11	71	6.1	15.9	6.0	1.0	—	4.1	27.0	85
		25—56			6.5	16.9	7.6	0.5	0.1	1.1	26.2	96
Shelby	Iowa	0—18	10	84	5.6	13.1	3.4	0.4	0.1	9.1	26.1	65
		28—43			5.4	14.1	4.0	0.3	0.1	7.0	25.5	73
Edina	Iowa	0—13	11	85	5.9	13.4	2.9	0.3	0.1	4.0	20.6	81
		51—61			5.6	22.0	12.1	0.3	1.7	7.4	43.5	83
—	Ohio	0—8	11	96	5.4	8.6	4.3	0.6	0	24.3	37.8	36
		28—43			4.6	2.0	8.2	0.4	0.1	21.8	32.6	33
Alford	Ind.	0—15	13	106	6.7	6.7	1.6	0.4	—	4.4	13.1	66
		56—82			5.6	5.4	4.3	0.4	0.1	8.2	18.4	55
Wehadkee	Va.	0—56	14	114	5.3	4.2	2.4	0.1	—	9.8	16.6	41
		56—91			5.9	6.8	6.0	0.3	—	5.7	18.8	69
Teas	W. Va.	0—5	11	129	5.1	6.5	1.6	0.6	0.1	14.7	23.4	37
		33—58			4.7	0.7	1.4	0.2	< 0.1	10.0	12.3	19
Mohave	Ariz.	0—10	21	25	7.2	7.7	8.0	0.9	0.1	—	8.5	100 +
Redding	Calif.	0—20	17	59	5.7	3.2	1.3	0.4	—	2.6	7.5	65
		50—56			5.3	12.6	13.4	0.3	0.3	6.2	32.8	81

Windthorst	Texas	0—8	18	83	5.6	1.9	0.6	0.4	—	2.6	5.6	54
		58—84			5.0	9.1	4.2	0.5	—	7.3	21.1	65
—	Texas	0—15	18	88	7.8	59.4	3.2	1.3	1.1	—	49.2	100 +
					8.2	50.9	4.0	0.9	6.7	—	46.3	100 +
Waimea	Hawaii	0—5	18	104	6.2	34.4	6.6	5.5	0.4	33.9	80.8	58
		13—20			7.6	37.7	5.7	6.3	0.3	24.2	74.1	67
Leon	Ga.	0—8	21	121	4.8	1.5	0	< 0.1	< 0.1	4.3	5.8	26
		43—46			4.7	0.2	0.2	< 0.1	< 0.1	27.1	27.5	1
Shubuta	Miss.	0—13	17	140	5.5	4.6	0.8	0.3	0.1	6.1	11.9	49
		35—48			4.4	0.6	3.4	0.3	< 0.1	19.7	24.0	18
Bladen	N.C.	0—15	18	142	3.9	0.4	0.4	0.3	0.2	24.9	26.2	5
		56—81			4.2	< 0.1	< 0.1	0.1	< 0.1	12.0	12.1	1
Victoria	Texas	0—3	22	71	7.9	43.6	4.9	1.8	1.2	—	37.1	100 +
		30—48			7.9	40.1	6.4	1.2	4.4	—	37.8	100 +
Lakewood	Fla.	0—3	22	136	5.4	1.5	0.1	< 0.1	< 0.1	2.5	4.1	39
		3—30			6.4	0.3	0	< 0.1	0	1.3	1.6	19
Nipe	P.R.	0—28	24	192	5.1	1.4	1.5	0.1	0.1	33.5	36.6	8
		28—46			5.0	0.1	—	—	—	21.4	21.5	< 1
Hilo	Hawaii	0—41	23	356	5.8	2.0	1.8	0.1	< 0.1	63.6	67.6	6
		41—53			6.1	2.2	0.6	0.1	< 0.1	65.5	68.4	4
Los Guineos	P.R.	0—10	21	396	4.8	0.5	0.8	0.1	0.2	28.8	30.4	5
		18—36			5.1	0.3	0.5	< 0.1	0.2	15.5	16.6	7

From Soil Survey Staff, Soil Classification, a Comprehensive System, 7th Approximation, U.S. Department of Agriculture, Washington, D.C., 1960.

Table 4
REACTIONS LEADING TO ACID SOILS

$$\text{Ca-Soil} + 2HCO_3^- + 2H^+ \longrightarrow H_2\text{-Soil} + Ca(HCO_3)_2$$

$$3/2\ H_2\text{-Soil} \longrightarrow Al^a - \text{Soil} + 3H_2O$$

and

$$Al^{3+} - \text{Soil} \rightleftharpoons \underset{H_2\diagup}{Al(OH)_2} - \text{Soil}$$

[a] Al^{3+} ions are dissolved from the clay structure and become exchangeable.

$$(Al(OH)_{3\ \text{lattice}} + 3H^+ \longrightarrow Al^{3+} + 3H_2O)$$

Table 5
CATIONS PRESENT IN ACID SOILS

H +	Present in small quantities in most acid soils unless excess mineral acids are present
Al^{3+}	Present as a major cation in strongly acid soils (pH 4 to 4.5); most is derived from dissolution of the clay lattice
Mg^{2+}	Present in amounts larger than Ca^{2+} in many acid soils due to release from clay structure as soil becomes acid
Mn^{2+}	Present in small but toxic quantities in soils where there is an appreciable quantity present; not important in most sandy, acid soils
$[Al(OH)_{2\ \text{to}\ 2.5}^{+\ \text{to}\ 0.5+}]_n$	A polycation with variable charge and size. Although not exchangeable by neutral salt solutions, it neutralizes the negative charge on soils (blocks CEC); charge on cation will be changed with either OH^- or H^+ titration; this polycation is the source of most "pH dependent" charge on clays
$[Fe(OH)_{2.9-3.0}^{0.1\ \text{to}\ 0}]_n$	A similar polycation with much less residual charge and probably much larger value of n; less important than the Al-analogue per equivalent of metal

Table 6
pH VALUES OF SPECIAL SIGNIFICANCE IN SOILS

pH[a]	Soil conditions inferred from pH
< 4	Free mineral acids present; most often H_2SO_4 from oxidation of pyrite
4	pH of soil largely saturated with exchangeable Al^{3+} and moderately low to very low in exchangeable divalent cations
5.5	pH above which there is essentially no exchangeable Al^{3+}
6.0	Soil pH suitable for a wide variety of plants
7.0	Neutral pH; has no special significance in soils
8.3	Soil in equilibrium with an excess of $CaCO_3$ at the partial pressure of CO_2 in the atmosphere (0.0003 atm)
> 8.3	Na_2CO_3 present in soil

[a] pH in 1:1 soil:water suspension.

These differences are very important in making lime recommendations, which are based on desirable pH rather than desirable percentage base saturation. For example, soils brought to a pH of 6.0 by lime might vary from 92 to 53% base saturation.

The effect of pH and exchangeable Al on plant growth is shown in Table 8. In this

Table 7
THE PERCENTAGE BASE SATURATION OF DIFFERENT SOILS AT SEVERAL VALUES

Soil type	Location	CEC at pH 7 (meq/100 g)	pH 4.8	pH 5.0	pH 5.5	pH 6.0	pH 6.5
			(% Base saturation)				
Delta sil	Miss.	9	77	80	86	92	94
Susquehanna C	Ala.	34	55	70	79	85	94
Oktibbeha C	Ala.	19	52	66	78	82	86
Grundy sil	Ill.	26	57	60	69	80	91
Miami sil	Wisc.	10	43	50	63	72	82
Cory si l	Ill.	8	32	38	56	74	88
Colby si l	Wis.	13	32	37	54	66	78
Cecil s l	S. C.	6	10	20	36	53	69
Greenville fsl	Ala.	7	18	21	34	54	71

From Pierre, W. H. and Scarseth, G. D., *Soil Sci.*, 31, 99, 1931. With permission.

Table 8
ALFALFA YIELD AND % STAND VS. LIME RATE, pH, AND EXCHANGEABLE Al

Lime rate mt/ha	Soil pH	Exch. Al^{3+} meq/100 g	3-Year alfalfa yield mt/ha	% Alfalfa stand after 3 years
0	4.9	0.74	0.58	0
0.56	5.1	0.42	4.30	0
1.12	5.3	0.40	8.94	20
2.24	5.7	0.14	15.55	80
4.48	5.7	0.12	17.18	90
8.96	6.2	0.10	17.49	90
17.92	6.5	0.03	17.47	99
35.84	6.8	0.02	17.92	99

From Moschler, W. W., Jones, G. D., and Thomas, G. W., *Proc. Am. Soc. Soil Sci.*, 24, 507, 1960. With permission.

case maximum alfalfa growth was obtained at the point where exchangeable Al was essentially zero and the pH was near 6.

Alfalfa is one of the more acid sensitive crops grown and lime needs for crops vary greatly as indicated by Table 9. This table shows that legumes in general respond to lime up to about pH 6. This may be partly due to molybdenum availability which affects the rhizobial activity and thus, nitrogen uptake of legumes. Other crops such as banana and coffee appear to grow perfectly well under very acid soil conditions.

In many acid soils which contain oxides of iron and aluminum, anions such as Cl, SO_4, and H_2PO_4 are adsorbed. The effect of SiO_2/R_2O_3 on the amounts of adsorption of these three anions by clays is shown in Table 10. The declining SiO_2/R_2O_3 ratio with increasing adsorption implicates both the oxides and a decreasing cation-exchange capacity in anion adsorption.

Another soil condition preventing healthy plant growth is soil salinity, often combined with sodium saturation of the exchange sites on the soil. Sodium-affected soils are harmful to plants because of the osmotic effects of the salt on plant roots, because sodium-saturated soils tend to disperse and lose structure and because sodium-saturated soils also tend to have high and toxic pH levels. Salt-affected (saline) soils usually

Table 9
pH VALUES ABOVE WHICH NO LIME RESPONSE OCCURS

Crop	Highest reported pH where lime response occurred
Alfalfa	6.5
Banana	>4.6
Barley	6.0
Bermudagrass	5.3
Clover	6.5
Corn	5.7
Cotton	5.8
Fescue	5.2
Lespedeza	6.0
Oats	<5.0
Peanuts	5.7
Pineapple	<4.7
Potato	5.0
Sorghum	5.7
Soybeans	6.0
Sugarcane	5.5
Sweet Potato	5.5
Vetch	5.8
Tropical Grasses	4.7
Coffee	<4.2

Adapted from Adams, F. and Pearson, R. W., *Am. Soc. Agron. Monogr.* 12, 1967.

Table 10
THE RELATION BETWEEN SiO_2/R_2O_3 AND THE QUANTITIES OF ANIONS RETAINED BY SOIL CLAYS

		Anions retained, meq/ 100_s		
Clay	SiO_2/R_2O_3	Cl	SO_4	H_2PO_4
Fallon (Nevada)	3.82	−0.4	−0.3	51.8
Bentonite (S.D.)	3.81	—	—	21.4
Sharkey (Miss.)	3.18	0.4	3.6	70.0
Marshall (Neb.)	2.82	—	4.1	93.2
Sassafras (Md.)	1.89	3.4	15.1	115.2
Norfolk (N.C.)	1.63	2.6	12.9	90.4
Aragon (Costa Rica)	0.55	4.4	26.9	160.1

From Mattson, S., *Int. Congr. Soil Sci. Trans. 1st Congr.*, 2, 199, 1927. With permission.

are located in areas where upward movement of water is important. If that water is salty, then, as the water is evaporated, the salt remains. As the total amount of sodium surrounding the soil builds up, some of the native calcium on the exchange sites is replaced. If the saline condition persists, more and more of this occurs until sodium makes up a significant portion of the exchangeable cations. This condition is aggravated by the presence of bicarbonate and carbonate ions which precipitate the calcium and leave an excess of rather soluble Na_2CO_3, raising the pH and partially dissolving the soil in soda. These gradations in salt-affected soils are shown in Table 11. All of these soils have problems, but the severely affected soils are very hard to correct because soil structure has deteriorated. The most common practice is reclaiming the saline soils is simply to leach them with excess irrigation water. With sodium-affected (natric) soils it is necessary to use water high in calcium, or occasionally to add gypsum or even sulfuric acid to reflocculate the soil and to replace sodium with calcium. The soluble sodium sulfate formed is leached out of the root zone. In many cases, severely affected soils are found only in spots, and they are abandoned because of cost or difficulty of reclamation.

Table 12 shows resistance measurements to be expected with various percentages of salt in soils of varying texture. The textural differences are a reflection of the increasing amounts of water required to obtain a saturated soil paste as texture varies from sand to clay. Table 13 relates the conductivity and osmotic pressure of saturated soil extracts to the concentrations of salt in solution. Measurements of conductivity or resistance are quick and easy to make and reveal the salt concentrations that exist in the soil. Table 14 shows the range in salt tolerance for many important crops. Values in each column vary between the two electrical conductivity values which bracket them. With this table and a measurement of saturated soil paste conductivity, the suitability of crops for saline soils can be ascertained.

Phosphorus is a problem nutrient in soils because no matter how soluble the fertilizer form applied, there is a steady reversion towards less soluble forms. These forms vary depending on the pH and the calcium, iron, and aluminum present. In soils with pH > 7, sparingly soluble calcium phosphates usually predominate, whereas in soils with pH < 7 iron phosphates are more prevalent. In soils with a high degree of weathering and high iron oxide content, an iron phosphate coated with iron oxide (occluded) may predominate. Examples of soils from different soil orders and their phosphorus forms are given in Table 15.

The effect of soil phosphorus level determined by extraction with dilute $HCl\text{-}H_2SO_4$ on yields and phosphorus contents of wheat, corn, and red clover is shown in Table 16. These data show that at an extractable phosphorus level above 45 kg/ha there is no increase in yield and only a small increase in percent of phosphorus. It appears that > 0.20 is adequate for most mature crop plants.

Potassium, to a lesser extent than phosphorus, is also made less available to plants by soil reactions. The most common of these is the tendency for the potassium to be sandwiched in between layers of clay so that a mica-like mineral is formed. This is particularly easy to accomplish with potassium because its size is about right to fit snugly in the voids formed by oxygen rings on the clay faces as shown in Table 17. Thus when potassium fertilizer is added to soils, some of it is "fixed" between layers of clay. On the other hand, when excess available potassium is removed by plants, some of the fixed potassium is "released" to an available form. Both fixation and release also occur more slowly in the absence of fertilization.

It is often difficult to relate potassium in exchangeable form to plant uptake of potassium. This is especially so when soil characteristics such as cation-exchange capacity (CEC) and mineralogy are also changing. The data in Table 18 show that exceptions to increasing % K in sorghum with increasing exchangeable K are traceable mostly to variability in soil solution K. This, in turn, is higher when CEC is low, and lower with high CEC. This relationship is one of importance in making potassium recommendations for crops.

The above discussions outline some of the more prevalent chemical problems in soils. There are many other possible reactions, but those discussed here are most important in their effects on plant growth.

Table 11
CHEMICAL CHARACTERISTICS OF SODIUM-AFFECTED SOILS

				Soluble salts[a] (meq/100 g)							Exchangeable cations (meq/100 g)			%
Soil	State	Depth (cm)	CO_3	HCO_3	Cl	SO_4	Ca	Mg	K	Na	Ca + Mg	K	Na	Na
Fresno sl[a]	Calif.	0—30	1.03	0.75	6.36	1.48	0.09	0.18	0.82	9.60	0	0.61	3.56	85
Jordan sl[a]	Utah	0—30	0.34	0.62	39.84	7.90	0.47	0.99	0.13	47.22	0	3.66	5.87	61
Lahontan c[a]	Nev.	0—15	0.83	0.67	13.84	5.55	0.14	0.20	0.35	21.31	0	0.80	24.48	97
Hanford I[a]	Calif.	0—30	0.85	1.07	22.17	10.40	0.16	0.52	0.66	34.78	0	2.15	20.57	90
Chino I[b]	Calif.	0—30	0.55	1.22	0.10	0.17	0.06	0.14	0.10	1.76	15.88	3.63	6.43	25
Placentia[b]	Calif.	0—30	0	0.60	4.62	1.31	0.57	0.68	0.25	5.07	13.55	1.46	4.09	21
Payette[b]	Idaho	0—30	0	0.35	5.74	10.42	1.10	0.59	0.18	14.80	21.11	1.81	6.52	22
Imperial[c]	Calif.	0—30	—	0.20	77.40	10.40	24.40	12.60	2.00	48.30	15.80	0.70	1.10	6
		30—60	0	0.30	20.50	4.00	4.60	2.60	1.30	15.10	15.40	1.70	2.10	11

[a] From Kelley[5]
[b] Strongly Na-affected soils.
[c] Moderately Na-affected soils.
[d] Saline but not Na-affected.

Table 12
RELATIONSHIP BETWEEN THE ELECTRICAL RESISTANCE OF A SATURATED PASTE AND PERCENT SALT IN SOILS OF DIFFERENT TEXTURE

Resistance soil paste at 15—5°C (Ohms × cm)	Mixed, neutral salts in soil (%)			
	Sand	Loam	Clay loam	Clay
18	3.00	3.00	—	—
19	2.40	2.64	3.00	—
20	2.20	2.42	2.80	3.00
25	1.50	1.70	1.94	2.20
30	1.24	1.34	1.46	1.58
35	1.04	1.14	1.22	1.32
40	0.86	0.94	1.04	1.14
45	0.75	0.78	0.88	0.98
50	0.67	0.71	0.77	0.86
55	0.60	0.64	0.69	0.77
60	0.55	0.58	0.63	0.70
65	0.51	0.54	0.57	0.63
70	0.48	0.50	0.53	0.59
80	0.42	0.44	0.47	0.51
90	0.37	0.39	0.41	0.45
100	0.33	0.35	0.37	0.39
150	0.21	0.21	0.22	0.23

From U.S. Salinity Laboratory Staff, *U.S. Dep. Agric. Handb.*, No. 60, 1954.

Table 13
RELATION BETWEEN CONCENTRATION OF SALTS IN A SATURATED SOIL EXTRACT, CONDUCTIVITY, AND OSMOTIC PRESSURE OF THE SOLUTIONS

Concentration of salt (meq/*l*)	Conductivity of saturated soil extract (EC × 10^3)	Osmotic pressure of saturated soil extract (atm)
10	1	0.32
21	2	0.67
33	3	1.04
44	4	1.41
70	6	2.21
95	8	3.00
122	10	3.80
263	20	7.00
410	30	12.80
560	40	17.20

From U.S. Salinity Laboratory Staff, *U.S. Dep. Agric. Agric. Handb.*, No. 60, 1954.

Table 14
SALT TOLERANCE OF VEGETABLE AND CROP PLANTS

High salt tolerance	Medium salt tolerance	Low salt tolerance
$E \cdot C \times 10^3 = 12$[a]	$E \cdot C \times 10^3 = 10$	$E \cdot C \times 10^3 = 4$
Table beet	Tomato	Radish
Kale	Broccoli	Celery
Asparagus	Cabbage	Green beans
Spinach	Bell pepper	
Barley	Cauliflower	$E \cdot C \times 10^3 = 3$
Sugar beet	Lettuce	
Rape	Sweet corn	$E \cdot C \times 10^3 = 4$
Cotton	Potato	
$E \cdot C \times 10^3 = 10$	Carrot	Field beans
	Onion	White clover
$E \cdot C \times 10^3 = 18$	Pea	Meadow foxtail
	Squash	Alsike clover
Bermudagrass	Cucumber	Red clover
Rhodesgrass		Ladino clover
Western wheatgrass	$E \cdot C \times 10^3 = 4$	
Birdsfoot trefoil		$E \cdot C \times 10^3 = 2$
	$E \cdot C \times 10^3 = 10$	
$E \cdot C \times 10^3 = 12$		
	Rye	
	Wheat	
	Oats	
	Rice	
	Sorghum	
	Corn	
	Flax	
	Sunflower	
	Castor bean	
	$E \cdot C \times 10^3 = 6$	
	$E \cdot C \times 10^3 = 12$	
	Sweet clover	
	Perennial ryegrass	
	Strawberry clover	
	Dallisgrass	
	Sudan grass	
	Alfalfa	
	Tall fescue	
	Orchardgrass	
	Meadow fescue	
	Reed canarygrass	
	Big trefoil	
	Smooth bromegrass	
	$E \cdot C \times 10^3 = 4$	

[a] $E \cdot C \times 10^3$ = mmhos/cm², a common measure of soil salinity. Numbers at top and bottom of each list represent the range within that list. Plants listed in order of decreasing salt tolerance.

From U.S. Salinity Laboratory Staff, U.S. Dep. Agric. Agric. Handb., No. 60, 1954.

Table 15
FORMS OF PHOSPHORUS FOUND IN SOILS FROM DIFFERENT ORDERS

Soil, horizon, and depth (cm)	Location	Soil order	pH	% Free Fe_2O_3([a])	Ca-P	Al-P	Fe-P (ppm)	Occluded Fe-P	Occluded Al-P
Wahiawa A, 0—46	Hawaii	Oxisol	6.7	10.1	10	20	123	504	97
Catalina A, 30—90	Puerto Rico	Oxisol	6.2	18.2	6	0	42	320	40
Miami A_3, 30—40	Wis.	Alfisol	5.8	1.4	45	48	100	73	6
Miami C_1, 107—147	Wis.	Alfisol	7.8	1.2	525	21	10	7	tr
Scott B, 18—71	Neb.	Mollisols	6.7	0.75	162	85	48	16	12
Dewey A_2, 5—30	Tenn.	Ultisol	5.5	3.3	32	2	22	26	19
Rosebud B_2, 25—45	Neb.	Mollisols	7.1	0.82	122	11	8	9	6
Barnes, 8—30	N.D.	Mollisols	7.5	0.77	155	18	8	tr	tr
Barnes, 71—137	N.D.	Mollisols	8.0	0.55	190	11	11	tr	tr
Barnes, 137—162	N.D.	Mollisols	8.0	0.93	185	6	3	tr	tr

From Chang, S. C. and Jackson, M. L.[3]

Table 16
YIELDS AND PHOSPHORUS CONTENTS OF WHEAT, CORN, AND RED CLOVER AS AFFECTED BY SOIL PHOSPHORUS LEVEL (TATUM SIL.)

Extractable P	Wheat—11 year		Corn—13 year		Red Clover—10 year	
kg/ha[a]	Yield (kg/ha)	%P	Yield (kg/ha)	%P	Yield (kg/ha)	% P
6.2	1297	0.16	2465	0.14	3566	0.13
9.9	2056	0.18	3506	0.19	6485	0.15
18.8	2271	0.19	3870	0.23	6987	0.18
44.8	2500	0.23	4045	0.30	7206	0.25
165.3	2419	0.28	4083	0.28	7176	0.28

[a] Method of P Extraction: Mehlich $HC1-H_2SO_4$

Table 17
FIXATION OF CATIONS BY MONTMORILLONITE CLAY AS RELATED TO IONIC RADII

Cation fixed	Cation-exchange capacity after fixation	Differences from original CEC	Ionic radius
Lithium	97.5	+ 3.9	0.60
Sodium	89.6	− 4.0	0.95
Potassium	75.2	−18.4	1.33
Ribidium	83.2	−10.4	1.48
Cesium	96.0	+ 2.4	1.69
Oxygen	—	—	1.40
Original	93.6	—	—

From Wear, J. I. and White, J. L., *Soil Sci.*, 71, 1, 1951. With permission.

Table 18
EFFECTS OF EXCHANGEABLE K, SOIL SOLUTION K, AND CEC ON K CONTENT OF SORGHUM

Soil	Predominant clay minerals	Soil order	Exchangeable K(meq/100g)	Soil solution K (meq/ℓ)	CEC (meq/100 g)	% K in sorghum
Miller c	Mica-Mont	Mollisols	1.17	0.34	39	2.2
San Saba c	Mont	Vertisols	0.94	0.12	44	1.6
Houston Black c	Mont	Vertisols	0.93	0.09	42	1.7
Norwood sil	Mica-Mont	Entisols	0.92	0.30	19	1.9
Clareville cl	Mont	Mollisols	0.78	0.20	29	1.8
Lake Charles c	Mont	Vertisols	0.48	0.10	45	1.4
Nacogdoches fsl	Kaol	Ultisols	0.20	0.34	3.5	1.6
Lufkin fsl	Mont	Alfisols	0.05	0.12	5	1.0

From Hipp, B. W. and Thomas, G. W., *Agron. J.*, 60, 467, 1968. With permission.

REFERENCES

1. **Adams, F. and Pearson, R. W.**, Crop responses to lime in the southern United States and Puerto Rico, in Soil Acidity and Liming, Pearson, R. W. and Adams, F., Eds., *Am. Soc. Agron. Monogr. 12,* 1967.
2. **Barnhisel, R. I., Ed.**, Analyses of clay, silt and sand fractions of selected soils from southeastern United States, *Southern Coop. Ser. Bull.*, 1978, 219.
3. **Chang, S. C. and Jackson, M. L.**, Soil phosphorus fractions in some representative soils, *J. Soil Sci.*, 9, 109, 1958.
4. **Hipp, B. W. and Thomas, G. W.**, Method for predicting potassium uptake by grain sorghum, *Agron. J.*, 60, 467, 1968.
5. **Kelley, W. P.**, *Alkali Soils,* ACS Monogr. No. 3, 1951.
6. **Mattson, S.**, Anionic and cationic adsorption by soil colloidal materials of varying SiO_2/Al_2O_3 + Fe_2O_3 ratio, *Int. Congr. Soil Sci. Trans. 1st Congr.*, 2, 199, 1927.
7. **Moschler, W. W., Jones, G. D., and Thomas, G. W.**, Lime and soil acidity effects on alfalfa growth in a Red-Yellow Podzolic Soil, *Soil Sci. Soc. Am. Proc.*, 24, 507, 1960.
8. **Pierre, W. H. and Scarseth, G. D.**, Determination of the percentage base saturation of soils in different soils at definite pH values, *Soil Sci.*, 31, 99, 1931.
9. **Robinson, W. O. and Holmes, R. S.**, The chemical composition of soil colloids, Bull. 1311, U.S. Department of Agriculture, Washington, D.C., 1924.

10. **Robinson, W. O.**, The inorganic composition of some important American soils, U.S. Department of Agriculture, Washington, D.C., Bull. 122, 1914.
11. Soil Survey Staff, Soil Classification, a Comprehensive System, 7th Approximation, U.S. Department of Agriculture, Washington, D.C., 1960.
12. U.S. Salinity Laboratory Staff, Diagnosis and improvement of saline and alkali soils, *U.S. Dep. Agric. Agric. Handb.*, No. 60, 1954.
13. **Wear, J. I. and White, J. L.**, Potassium fixation in clay minerals as related to crystal structure, *Soil Sci.*, 71, 1, 1951.

SOIL MICROBIOLOGY

James M. Tiedje and Frank B. Dazzo

THE SOIL BIOTA: ITS COMPOSITION, DENSITY, AND BIOMASS

The soil microbiota is comprised of five major groups: bacteria, actinomycetes, fungi, algae, and protozoa. Actinomycetes are a subgroup of bacteria but because of their prevalence in soil, their filamentous morphology, and their historical separation from bacteria by soil microbiologists, they are still commonly treated as a separate group. A sixth but nonliving biological group is the viruses which are also important in soil microbiology since bacterial, plant, and animal viruses are common in soil and can be of practical importance.

Slightly larger organisms, the soil fauna, are also prevalent and important to plant litter decay in soils, although their density is greatly reduced in regularly cultivated soils. More noteworthy fauna include the nematodes, mites, springtails, and earthworms.

Estimates of population densities and biomass of each of the biological groups have been made by several authors. Some are summarized in Table 1. The methods of obtaining these data vary from generalized approximations based on experience to directly measured values obtained using the best methods available. Because of limitations of the methods and spatial variability of soil, data of this type are most useful for obtaining a general concept of the soil community, particularly which groups of organisms are expected to be dominant.

Major advances in measurement of soil biomass have been made in the last decade, although it is still tedious work and often relatively imprecise. Difficulties exist in distinguishing between live and dead biomass and errors of extrapolation from a small sample to a large area basis. The current methods used to measure microbial biomass are summarized in Table 2. They are based either on direct observation of cells or on measurement of a specific chemical entity. Counting live cells by plate counts or by most probable number (MPN) typically gives poor recoveries (1 to 10% or less) of the total biomass of interest, and is inappropriate for estimates of fungal biomass due to the great differences in spore production among genera. Thus the popular counting methods are direct counts of stained cells. The staining helps to distinguish cells from background and perhaps to distinguish live cells from some of the dead cells. The cell count is multiplied by the mean cell volume and by the biomass-to-biovolume conversion factor to obtain the biomass. VanVeen and Paul[23] recently systematically measured the biovolume and conversion factors for representative soil bacteria, yeast, and fungi under different growth conditions. Their results (Table 3) are significantly different from the previously used factors. This probably means that other published microbial biomass values may be underestimated due to use of incorrect conversion values.

Biomass methods based on specific chemical entities require that the only significant source of that chemical in the sample is the population of interest, and that the populations contain a reasonably constant percentage of the chemical. These requirements are often difficult to meet. The constancy of the chemical to biomass ratio usually can be influenced by growth conditions, species, and differential recovery from different types of samples.

The amount of element or cell constituent which is held by the soil biomass can be estimated from biomass information if average cell composition is known. The composition of a well-studied Gram-negative bacterium is given in Table 4. These data are for cells grown in rich media and are probably not accurate for cells under starving

Table 1
DENSITY AND BIOMASS VALUES OF DIFFERENT GROUPS OF SOIL ORGANISMS IN CULTIVATED TEMPERATE SOILS

Group of organisms	Dry biomass[a] (g/m²)[e]	Dry biomass[b] (kg/ha)	Abundance[c] (No./m²)	Live biomass[c] (g/m²)	Abundance[d] (No./g)	Live biomass[d] (kg/ha)
Bacteria }	32.76	36.9	3×10^{14}	300	10^4—10^9	300—3000
Actinomycetes }		0.2			10^5—10^8	
Fungi	84—144	454.0		400	2×10^4—10^6[f]	500—5000
Algae					10^2—5×10^4	7—300
Protozoa		1.0	5×10^8	38	10^4—10^5	50—200
Nematodes		2.0	10^7	12		
Earthworms		12.0	10^5	132		
Enchytraeidae		4.0				
Molluscs		5.0	[g]	[g]		
Acari		1.0	2×10^5	3		
Collembola		2.0	2×10^4	5		
Diptera		3.0	[g]	[g]		
Other arthropods		6.0	2×10^3[g]	36[g]		

[a] From Clark, F. E. and Paul, E. A., *Adv. in Agron.*, Vol. 22, Brady, N. C., Ed., Academic Press, New York, 1970, 405.
[b] Lynch, J. M. and Poole, N. J., *Microbial Ecology, a Conceptual Approach,* Halsted Press, New York, 1979, 78.
[c] From Richards, B. N., *Introduction to the Soil Ecosystem,* Longman Group Limited, New York, 1974, 3; and Macfadyen, A., *Animal Ecology: Aims and Methods,* 2nd ed., Putnam, London, 1963, 234.
[d] From Alexander, M., *Introduction to Soil Microbiology,* 2nd ed., John Wiley & Sons, New York, 1977, chap. 2—6.
[e] Range over non-winter months for this Saskatchewan, Canada site; most precisely measured data of that presented.
[f] Shown as number of propagules/g, which is not a uniform indicator of fungal density; the more accepted approach is hyphal length for which the typical range is 10 to 100 m/g. The biomass estimate is based on the length value.
[g] Organisms of these groups are grouped as other fauna in this case.

Table 2
METHODS FOR DETERMINING MICROBIAL BIOMASS OF SOIL

Microbial class	Principle	Ref.
Fungi		
Agar film-technique	Fungal mycelial length measured from microscopic observation of filaments stained with phenolic aniline blue[7] or by phase contrast; biovolume is then calculated and biomass obtained by use of a conversion factor	6
Chitin content	Chitin is a cell wall component unique to many fungi; after alkaline hydrolysis the glucosamine content is measured colorimetrically and the fungal biomass is obtained by multiplying by a conversion factor; cannot be used if insect exoskelton present	8, 9
Fluorescent staining of hyphae	Hyphal suspension from soil collected on Nucleopore® filters, stained with water-soluble aniline blue which binds β-1,3-glucans in many fungal cell walls and the length of the fluorescing hyphal fragments measured; biomass is obtained from biovolume (see Table 3)	10
Bacteria		
Fluorescent staining		
FITC	Soil smears are stained with FITC which binds well to bacterial cell proteins; the bacterial shapes that fluoresce are counted and their volumes estimated, which is then multiplied by a biomass:biovolume conversion factor (see Table 3)	11
Acridine orange	Other fluorescent dyes that have been tried but do not seem to have proven as useful as FITC for soil; for specific situations or in a particular investigators hands they may show advantages	12
Primulan		13
1-Anilino-8-napthalene sulfuric acid		14
Europium chelate plus fluorescent brightener		15
Electron microscopy	Dilute soil suspension is placed on grid with micropipet; the number of cells observed is multiplied by the appropriate dilution factors from which the total biovolume and biomass can then be calculated;	16 17
Muramic acid (MA)	Muramic acid is a unique component of bacterial cell walls; the soil sample with cell walls is hydrolyzed in 6N HCl to produce the monomer, MA, which is then purified by TLC; the MA is derivatized to form the peracetylated aldonitrile of MA which is quantified by gas chromatography[19]	18 19
Lipopolysaccharide (LPS)	LPS is unique to gram negative bacteria, the dominant bacteria in soil. LPS can be quantified by the extremely sensitive *Limulus* lysate assay. *Limulus* amebocyte lysate (from horseshoe crab blood) forms a turbid solution with LPS; the turbidity can be measured in a spectrophotometer and is proportional to the LPS; the LPS:carbon content of a cell is rather constant; the method has been used in aqueous habitats but the author is not aware of attempts in soils	20
Total microorganisms		
ATP	ATP is present in all living cells; it is extracted by lysing the cells with $CHCl_3$ and saturating the phosphate absorbing sites with $NaHCO_3$; a constant conversion factor of cell carbonP is assumed though it has been found to vary considerably depending on organism and conditions	10
CO_2 mineralized following fumigation	The flush of CO_2 produced following fumigation with $CHCl_3$ is assumed to be derived from decomposition of killed organisms and not from nonbiomass organic matter; the relationship $B = F/k$ is used to determine biomass, B, F = the CO_2−C evolved in the flush and k is a constant describing the fraction of biomass carbon mineralized; for a 10-day incubation k was determined to be 0.5[22]	21

Table 3
CELL CONTENT, VOLUME, AND BIOMASS:BIOVOLUME CONVERSION FACTORS FOR SOIL MICROORGANISMS GROWN UNDER DIFFERENT WATER STRESS AND NUTRIENT CONDITIONS[a]

Microbe	Group	Content (% dry wt) C	N	P	Ash	Volume (μm^3)	Range of conversion factors (g dry wt/cm^3)
Arthrobacter	Bacterium	37	7.3	1.6	4.8	0.64	0.6—1.4
Enterobacter	Bacterium	42	9.6	1.5	12.8	0.58	0.6—1.2
Cryptococcus	Yeast	58	2.0	0.4	3.8	28	0.4—1.4
Trichoderma	Fungus	40	4.0	0.26	0.7	—	0.11—0.41
Penicillium	Fungus	—	—	—	—	—	0.1—0.3
E. coli (From Table 4)		50	8—15	2—6	5—12	1.0	0.22[b]
Mean suggested by authors for bacteria							0.8[b]
Mean suggested by authors for fungi							0.33

[a] Mean values for growth under water stress of 0, 3.25, and 13.75 atm, shake and agar culture, and nutrient or potato dextrose broth vs. soil extract broth. Some of the above parameters vary markedly with growth condition so above mean values shall not be taken as a precise value.

[b] Variance between old and new value is substantial. Authors are not able to explain this and are not convinced that the measured average, 0.8, is reasonable since it would mean a cell of only 51% water and 1.3 specific gravity.

Data from VanVeen, J. A. and Paul, E. A., *Appl. Environ. Microbiol.*, 37, 686, 1979.

Table 4
COMPOSITION OF GRAM NEGATIVE BACTERIA

Element	Composition[a] (% of dry weight)	Component	Composition[a] (% of dry weight)
Carbon	50 ± 5	Protein	50
Nitrogen	8—15	Carbohydrate	10—30
Hydrogen	10	Lipids	10—15
Oxygen	20	RNA	10—20
Ash	5—12	DNA	3—4
Phosphorus	2—6		
Sulfur	1.1		
Sodium	1.4	Cell specific gravity	1.1
Potassium	1.7		
Calcium (as CaO)	1.6	Moisture content of cells	73—86%
Magnesium (as MgO)	0.60	Av cell dry weight	20%
Chloride	0.46		
Iron	0.28		
Manganese	20 ppm		
Copper	80 ppm		
Aluminum	100 ppm		
C:N ratio	5:1		

[a] Data for *E. coli* grown in rich liquid medium. The values have become classical for approximate compositon of bacterial cells but may not be adequate for soil flora (see Table 3 for comparison).

From Luria, S. E., *The Bacteria*, Vol. 1, Gunsalus, I. C. and Stanier, R. Y., Eds., Academic Press, New York, 1960, chap. 1. With permission.

conditions, which is the case for the soil microbiota. The more recent data of VanVeen and Paul[23] (Table 3) shows significantly less N and P contained in these soil forms grown under more typical soil conditions.

GENERA OF SOIL BACTERIA

Genera of soil bacteria most likely to be encountered in soil are listed in Table 5 along with their major distinguishing features and my estimation of their relative prevalence in typical agricultural soils. This list of 60 genera is not exhaustive, as extremely rare or poorly recognized genera are not included.

This table is meant to be a useful guide for identifying isolates from soil. First, the key features which place the isolate within the major group should be determined. Once within the proper group, one can use the distinctive features or tests listed to obtain a tentative identification to genus level.

The table is based on the eighth edition of *Bergey's Manual of Determinative Bacteriology*,[25] but also includes several advances in taxonomic knowledge published since this edition appeared. Once a tentative identification is made, one should consult *Bergey's Manual* to examine the total description and perform other confirmatory tests to be certain of the designation.

MICROBIOLOGICAL MEDIA

Twenty media for enumeration and/or cultivation of most classes of soil microorganisms are given in Table 6. Numerous media have been published for these purposes. I have given the superior medium where this is apparent, but in other cases I have selected one (usually the most common) from among several equally adequate media. For other media and more thorough descriptions, the reader is referred to the following references: Aaronson,[26] Johnson and Curl,[27] Skerman,[28] *The American Type Culture Collection: Catalogue of Strains*.[29]

Four media are given for aerobic heterotrophic bacteria, each varying in complexity and amount of organic substrates. The last two, numbers 3 and 4, are the best media for giving the highest counts of soil bacteria.

Media 18, 19, and 20 are trace mineral, vitamin, and basal salts solutions, respectively. The trace mineral and vitamin solutions can be used to amend any medium where this supplement is desired. The basal salts solution can be amended with carbon source(s) of interest to obtain a defined medium. For example, a pesticide-degrading organism could be enriched, isolated, and grown on this basal salts medium plus pesticide (if, of course, the pesticide could serve as a carbon and energy source).

ENERGETICS OF MICROBIAL TRANSFORMATIONS

Calculation of Energy Yields of Reactions

Microorganisms are successful in soil and carry out numerous chemical transformations because they gain enough energy from many of these processes for growth and maintenance. The amount of energy available from various transformations varies considerably, and thus it is not surprising that the cell yield per amount of substrate transformed also varies. It is often useful to know how much energy is available from a process or particular step in a process so that one can determine whether the reaction is feasible or whether there is sufficient energy for ATP production, which is a requirement for the reaction to support growth.

Chemical reactions are accompanied by changes in energy, of which a portion is

Table 5
MAJOR FEATURES OF THE GENERA OF BACTERIA FOUND IN SOIL

General group and genus	Gram reaction[a]	Cell shape	Motility[b]	Oxygen requirement[c]	Special features and descriptive comments	Prevalence in soil[d]
Gliding bacteria						
Myxococcus	–	Rod	Gliding	Aerobic	Flexible rods which glide and form macroscopic (often colored) fruiting bodies which contain myxospores; found on decaying plant material and dung in soil; this is a common representative of this class of bacteria known as myxobacteria; their importance in soil is not known but their fruiting bodies are often seen; produce a variety of enzymes hydrolyzing macromolecules, including cell walls of procaryotic and eucaryotic microorganisms	4
Cytophaga	–	Rod	Gliding	Most aerobic, some facultative	Decompose cellulose or other polysaccharides; flexible rods or filaments from 5 to 50 μm in length; yellow, orange or red pigment	3
Flexibacter	–	Rod	Gliding	Aerobic	Polymers such as cellulose, chitin, agar not attacked; flexible rods or filaments 5 to 100 μm in length; yellow, orange or red pigment	4
Beggiatoa	–	Rod	Gliding	Aerobic or microaerophilic	Colorless cells in filaments up to 20 μm in length; cells contain sulfur granules when in presence of H_2S; found on wet decaying plant material and in rice rhizosphere	5
Sheathed bacteria						
Leptothrix	–	Rod	+	Aerobic	Rods in sheath which becomes impregnated with hydrated ferric or manganic oxides; free cells are motile; do not appear to obtain energy from Fe^{2+} oxidation; often responsible for rust colored scum at iron springs or tiles	4
Budding or appendaged bacteria						
Hyphomicrobium	0[a]	Rod	+	Aerobic	Oval forms which produce thin (0.3 to 0.4 μm)	3

					hyphae which may branch; reproduction is by budding at tips of hyphae; mature buds become motile, break off and thereafter attach losing motility; some strains denitrify	
Pedomicrobium	–	Rod	+	Aerobic or microaerophilic	Similar to above except one or numerous hyphae grow from sites on cell surface; ferric or manganic deposits often occur on cells or hyphae	3
Caulobacter	–	Rod	+	Aerobic	Rod-shape or fusiform cells with stalk (0.2 μm) extending from one pole with adhesive material at the end; daughter cell becomes swarmer cell; often form rosettes	3
Metallogenium	0	Cocci	–	Aerobic	Coccoid cells which bud and form filaments with tapering ends; colonies dark due to heavy deposits of manganese and/or iron; staining and microscopy are difficult due to the heavy metal deposition	4?
Spiral or curved bacteria						
Bdellovibrio	–	Vibrio	+	Aerobic	Obligately parasitic on other bacteria; broad host specificity; will penetrate host cell and devour it; may play role in soil of limiting bacterial density	2
Gram negative, aerobic rods						
Pseudomonas	–	Rod	+	Aerobic	Strict aerobe though many species can denitrify; motile by polar flagella; acetate can be used by all; can use a wide variety of carbon sources; most commonly isolated motile bacterium from soil	1
Alcaligenes	–	Rod	+ p[b]	Aerobic	Strict aerobe though some species can denitrify; motile by peritrichous flagella; some do not use carbohydrates; closely related to *Pseudomonas*	1
Achromobacter					No longer a recognized genus; most strains are now identified as *Alcaligenes*	
Xanthomonas	–	Rod	+	Aerobic	Strict aerobe; motile by polar flagella; most have complex growth requirements and are plant pathogens; often produce yellow pigment; closely related to *Pseudomonas*	2

Table 5 (continued)
MAJOR FEATURES OF THE GENERA OF BACTERIA FOUND IN SOIL

General group and genus	Gram reaction[a]	Cell shape	Motility[b]	Oxygen requirement[c]	Special features and descriptive comments	Prevalence in soil[d]
Azotobacter	−,v	Rod	+ p,−	Aerobic	Capable of N_2 fixation; forms thick walled cysts; large ovoid cells; colonies often produce copious amounts of slime; grows rapidly	3
Azomonas	−,v	Rod	+ , + p	Aerobic	Capable of N_2 fixation; similar to *Azotobacter* except that no cysts are produced; grows rapidly	3?
Beijerinckia	−	Rod	+ p,−	Aerobic	Capable of N_2 fixation; rods with highly refractile, bipolar, lipoid bodies in cell; grows slowly; tough, tenacious slime produced; acid tolerant; more common in tropical soils	3
Derxia	−	Rod	+	Aerobic	Capable of N_2 fixation; similar to *Beijerinckia* but more numerous lipoid bodies and flagellation is polar; also acid tolerant and more common in tropics	4?
Rhizobium	−	Rod	+ , + p	Aerobic	Is distinguished from other heterotrophic bacteria by the ability to form N_2 fixing nodules in legume plants; commonly produce slime	2
Agrobacterium	−	Rod	+ p	Aerobic	Related to *Rhizobium* except produce plant galls; similar to *Alcaligenes* except for plant infecting property; one species avirulent	2
Methylomonas	−	Rod	+	Aerobic	Methane and methanol are the only known sources of carbon and energy	4?
Gram negative, chemolithotrophic bacteria						
Nitrobacter	−	Rod	−	Aerobic	Oxidize nitrite to nitrate; most strains are obligate chemolithotrophs though some grow slowly heterotrophically; short, often pear shaped rods	2
Nitrosomonas	−	Rod	+ ,−	Aerobic	Oxidize ammonium to nitrite; obligate chem-	2

					olithotroph; ellipsoidal or short rods	
Nitrosospira	–	Spiral	+ p,⁻	Aerobic	Oxidize ammonium to nitrite. Obligate chemolithotroph. Cells appear as tightly wound spirals in the shape of short rods.	2
Notrosolobus	–	Pleomorphic	+ p	Aerobic	Oxidize ammonium to nitrite. Obligate chemolithotroph. Cells pleomorphic and lobate.	3
Nitrosovibrio	–	Vibrio		Aerobic	Oxidize ammonium to nitrite. Obligate chemolithotroph. Cells curved.	4
Thiobacillus	–	Rod	+	Aerobic	Energy derived from oxidation of reduced sulfur compounds. Some autotrophic, others facultatively autotrophic, one species also oxidizes Fe^{+2}, another denitrifies.	3
Sulfolobus	–	Lobed	–	Aerobic	Thermophilic, oxidize sulfur as energy source. Facultative autotroph. Spherical cells with lobes.	5
Gram negative, facultative anaerobic rods						
Escherichia	–	Rod	+ p	Facultative	Normally of intestinal origin but also found in soil. Glucose fermented to acid and H_2; lactose fermented. Nitrate reduced to nitrite. Growth on acetate but not citrate.	3
Klebsiella	–	Rod	–	Facultative	Normally of mammalian origin but certain species, esp. N_2-fixing strains, are readily found in soil. Glucose, lactose and nitrate metabolism as for *Escherichia*. Growth on citrate.	2
Enterobacter	–	Rod	+ p	Facultative	Normally considered of intestinal origin but readily found in soil. N_2-fixing strains found in decaying plant material and in rhizosphere of plants. Glucose, lactose and nitrate metabolism as for *Escherichia*. Growth on both acetate and citrate.	2
Erwinia	–	Rod	+ p	Facultative	Associated with plants as pathogens, saprophytes or ephiphytes; causes fireblight, vascular wilts of plants; gas production from carbohydrates weak or absent	3
Aerobacter					No longer recognized genus; most previous	3

Table 5 (continued)
MAJOR FEATURES OF THE GENERA OF BACTERIA FOUND IN SOIL

General group and genus	Gram reaction[a]	Cell shape	Motility[b]	Oxygen requirement[c]	Special features and descriptive comments	Prevalence in soil[d]
					strains so named are either *Klebsiella* or *Enterobacter*	
Chromobacterium	–	Rod	+p	Aerobic or facultative	Produce nondiffusible violet pigment (violacein); pigment soluble in ethanol but not water or chloroform; more common in tropical soils	3
Flavobacterium	–	Rod	+p,–	Aerobic or facultative	Produce nondiffusible yellow pigments; two major types, one motile and one non-motile; tendency is toward including only strict aerobes in this genus; some strains denitrify	2
Strict anaerobic bacteria						
Desulfovibrio	–	Vibrio	+	Anaerobic	Growth on limited number of simple organic compounds and H_2; reduce SO_4^{2-} to H_2S	4
Desulfotomaculum	–	Rod or vibrio	+p	Anaerobic	Similar to *Desulfovibrio* in nutrition and reduction of SO_4^{2-} to H_2S but forms endospores; will oxidize acetate	4?
Clostridium	+	Rod	+p,–	Anaerobic	Form endospores; some species reduce NO_3^- to NH_4^+; one species fixes N_2; ferment carbohydrates and proteins; most common strict anaerobe in soil	2
Methanobacterium	+,–	Filament, rod	+,–	Anaerobic	Use H_2 and sometimes formate to reduce CO_2 and form methane; most common methanogen	4
Methanosarcina	+	Cocci	–	Anaerobic	Form methane from splitting of acetate; also will grow on methanol; probably present in any methanogenic site to remove acetate produced by fermentation	4
Gram negative, aerobic cocci						
Paracoccus	–	Cocci	–	Aerobic	Similar to *Pseudomonas* and *Alcaligenes* except for shape and motility; the most com-	3

					mon species denitrifies and can grow on H_2; dentrifying species formerly known as *Micrococcus*	
Acinetobacter	−	Coccobacilli	−	Aerobic	Short plump rods nearly coccoid in stationary phase; thought to use wide variety of carbon compounds	3?
Gram positive cocci						
Micrococcus	+	Cocci	−	Aerobic	Cells occurring singly, in pairs or clusters; glucose oxidized with little acid produced; not pathogenic; most common coccus in soil	2
Staphylococcus	+	Cocci	−	Facultative	Mainly associated with skin and mucous membranes of warm-blooded animals; some are pathogens; glucose fermented to acid but not gas; a saprophytic strain easily isolated from soil	4
Streptococcus	+	Cocci	−	Facultative	Strictly fermentative metabolism; cells in pairs or chains; generally has complex nutritional requirements; occurs mostly in mammalian habitat, often pathogenic; a few types thought to exist on plants or in soil	5
Sarcina	+	Cocci	−	Anaerobic	Strictly fermentative metabolism; cells in packets of eight or more; ferment glucose to acid and gas; can be isolated from soil but its prevalence not known	5?
Gram positive rods						
Bacillus	+	Rod	+	Aerobic, facultative	All form endospores; some obligate aerobes, others ferment; some species denitrify; majority are motile with lateral flagella; cells are often larger than cells of the most common rod-shaped organisms in soil	1
Gram positive, irregularly-shaped rods "Coryneform group"						
Corynebacterium	+, v	Rod	−	Aerobic	Straight to curved rods, sometimes club-shaped swellings; uneven staining (sometimes of granules); divided into animal, plant and nonpathogenic groups; the last is common in soil but poorly studied	3?

Table 5 (continued)
MAJOR FEATURES OF THE GENERA OF BACTERIA FOUND IN SOIL

General group and genus	Gram reaction[a]	Cell shape	Motility[b]	Oxygen requirement[c]	Special features and descriptive comments	Prevalence in soil[d]
Arthrobacter	+,v	Rod to coccoid	–	Aerobic	In complex media undergo a marked change in form during the growth cycle; old cultures contain coccoid cells which swell, then elongate in new medium; division produces irregular rods, some at complex angular arrangements showing V-forms and forms suggestive of rudimentary branching, though no true mycelia formed; one of the most common soil organisms; survives starvation and desiccation, can metabolize a wide variety of carbon compounds	1
Brevibacterium					No longer a recognized genus; most strains so named are probably *Arthrobacter* or *Corynebacterium*	
Cellulomonas	+,v	Rod to coccoid	+,–	Aerobic, facultative	Coryneform bacteria with the ability to attack cellulose; aerobic but anaerobic fermentation of glucose to acid is also common; most common strain produces yellow pigment	3
Produce filaments "actinomycete group"						
Produce rudimentary or fragmenting mycelium						
Mycobacterium	+	Rod to filament	–	Aerobic	Rods, sometimes branching and filaments but no true mycelium; when disturbed filaments break into rods; acid-fast; growth slow or very slow; most studied are from animals, many are pathogenic; can be isolated from soil but importance not known	4?
Frankia	+	Hyphae	–	Microaerophilic	N_2-fixing endophyte in nodules of non-leguminous plants such as *Alnus*, *Myrica*, etc.; only recently grown in culture	?

Actinoplanes	+	Spores, hyphae	+	Aerobic	Sporangia produce motile spores; branched or coiled hyphae with few septa; aerial mycelium scanty; most species brightly colored; claimed to be abundant in all soils	?
Nocardia	+	Rod, hyphae	−	Aerobic	Mycelium may fragment early or late, depending on species, into coccoid or bacillary forms; aerial mycelium and spores not formed; can metabolize variety of carbon compounds	2
Produce well-developed mycelium + spores						
Streptomyces	+	Spores, hyphae	−	Aerobic	Mycelium does not fragment; abundant aerial mycelium and long spore chains; no septa in hyphae; produce leathery colonies; desiccation resistant; many are antibiotic producers	1
Streptoverticillium	+	Spores, hyphae	−	Aerobic	Mycelium does not fragment; aerial mycelium is characteristic; 3 to 6 branches produced in a whorl; these branch to produce secondary branches with chains of spores at the end; produce wide variety of pigments	4?
Micromonospora	+	Spores, hyphae	−	Aerobic	Well-developed, branched, septate mycelium; spores born singly; no aerial mycelium; most decompose cellulose	2
Thermoactinomyces	v	Spores, hyphae	−	Aerobic	Similar to *Micromonospora* except also has aerial mycelium with single spores, and all members grow at temperatures between 45 to 60°C; isolated from soils and composts	4
Microbispora	+	Spores, hyphae	−	Aerobic	Longitudinal pairs of spores on aerial mycelium only; aerial mycelium often pink	5

[a] "−", gram negative; "+", gram positive; "v", gram variable; "0", gram reaction not recommended or not recorded.
[b] "+p", motile by means of peritrichous flagella; if "+" motility is generally by polar flagella.
[c] Aerobic means strictly aerobic; i.e., require oxygen as terminal electron acceptor except for the indicated species which can also denitrify.
[d] Range of 1 to 5 with 1 = most prevalent in soil and 5 = infrequently found in soil.

Table 6
MEDIA IN GENERAL USE FOR GROWING, ISOLATING AND/OR ENUMERATING PHYSIOLOGICAL GROUPS OF SOIL MICROORGANISMS

1. Aerobic heterotrophic bacteria
   a. Minimal[a]
      K_2HPO_4 15 g — Agar 15 g
      Soil extract[b] 1000 mℓ
      Adjust to pH 6.8 prior to autoclaving
   b. Moderate[c]
      Glucose 1 g — Agar 15 g
      K_2HPO_4 0.5 g — Tap water 900 mℓ
      Soil extract[b] 100 mℓ
      Adjust to pH 6.8 prior to autoclaving
   c. Complex[d]
      Peptone 1 g — $MgSO_4 \cdot 7H_2O$ 0.05 g
      Yeast extract 1 g — $MgCl_2$ 0.10 g
      Soil extract[b] 250 mℓ — $FeCl_3$ 0.01 g
      K_2HPO_4 0.4 g — $CaCl_2$ 0.10 g
      $(NH_4)_2HPO_4$ 0.5 g — Agar 15 g
      Adjust to pH 7.4 prior to autoclaving
   d. Complex commercial[e]
      Tryptic soy broth 3g — Agar 15 g
      Tap water 1000 mℓ

   Martin found that dilute tryptic soy broth (0.3%) gave no significant difference from the Bunt-Rovira complex medium in recovery of maximum numbers of bacteria from soil.
2. Cellulose degrading bacteria[f]
      Cellulose powder 5.0 g — KCl 0.5 g
      $NaNO_3$ 1.0 g — Yeast extract 0.5 g
      $Na_2HPO_4 \cdot 7H_2O$ 1.2 g — Casein 0.5 g
      KH_2PO_4 0.9 g — Agar 10 g
      $MgSO_4 \cdot 7H_2O$ 0.5 g — Water 1000 mℓ

   Adjust pH to 6.8. Cellulose powder can best be prepared by ball-milling Whatman standard grade cellulose paper. Look for tiny ring of cellulose digestion around colonies after 8 days.

   The inorganic ingredients can be used in MPN tubes with filter paper strips (1 × 9 cm) extending from the liquid to above the surface. Tubes positive for cellulose degradation show disintegration of the paper, especially at the air-water interface.
3. Anaerobic heterotrophic bacteria[g]
      Peptone 10 g — $MgSO_4 \cdot 7H_2O$ 0.11 g
      Yeast extract 10 g — $(NH_4)_2SO_4$ 0.22 g
      Glucose 10 g — NaCl 0.53 g
      K_2HPO_4 0.56 g — $FeCl_3$ 5 mg
      KH_2PO_4 0.33 g — Agar 15 g
      Na_2CO_3 0.22 g — Distilled water 900 ml

   Degas medium of oxygen and autoclave. Dissolve 0.6 g $CaCl_2 \cdot 2H_2O$, 0.5 cysteine hydrochloride (reductant) and 1 mg resazurin (redox indicator) in 50 mℓ distilled water, filter sterilize and add to medium without introduction of oxygen. Medium will be colorless if reduced but will turn pink if oxidized. Plates should be poured in anaerobic glove box. Atmosphere above medium should have 5% CO_2 to establish final pH of near neutrality.
4. Algae (modified Bristol's medium)[h]
      $NaNO_3$ 0.25 g — KH_2PO_4 0.018 g
      $CaCl_2$ 0.025 g — NaCl 0.025 g
      $MgSO_4 \cdot 7H_2O$ 0.075 g — $FeCl_3$ 0.5 mg
      K_2HPO_4 0.075 g — Distilled H_2O 1000 mℓ
      Incubate in diffuse light for 1 month.
5. N_2 fixing cyanobacteria (bluegreen algae) (modified Chu's medium)[h]
      K_2HPO_4 0.01 g — Na_2SiO_3 0.02 g
      $MgSO_4 \cdot 7H_2O$ 0.025 g — Ferric citrate 3 mg
      Na_2CO_3 0.02 g — Citric acid 3 mg
      $CaCl_2$ 0.02 g — Distilled water 1000 mℓ
      Incubate in diffuse light for 1 month.

Table 6 (continued)
MEDIA IN GENERAL USE FOR GROWING, ISOLATING AND/OR ENUMERATING PHYSIOLOGICAL GROUPS OF SOIL MICROORGANISMS

6. Actinomycetes[i]

Starch (or glycerol) 10.0 g	$MgSO_4 \cdot 7H_2O$ 0.5 g
Casein (vitamin free) 0.3 g	$CaCO_3$ 0.02 g
KNO_3 2.0 g	$FeSO_4 \cdot 7H_2O$ 0.01 g
NaCl 2.0 g	Distilled water 1000 mℓ
K_2HPO_4 2.0 g	Agar 18 g

Add 50 μg/mℓ each of the antifungal antibiotics nystatin and actidione to the cooled liquid medium before pouring into dishes. Adjust pH to 7.0 to 7.2 before autoclaving.

7. Sulfate reducers[j]

Sodium lactate (70% w/w) 5.0 g	$MgSO_4 \cdot 7H_2O$ 2.0 g
K_2HPO_4 0.5 g	Ion agar (Oxoid) 10 g
NH_4Cl 1.0 g	Tap water 930 mℓ
$CaSO_4$ 1.0 g	

Degas medium of oxygen and autoclave. Filter sterilize the following solutions and add to the medium in a glove box (or without exposure to oxygen); 50 mℓ of 1% each w/v $FeSO_4$ and $(NH_4)_2SO_4 \cdot 6H_2O$, 10 mℓ of 10% (w/v) yeast extract, and 10 mℓ of 10% (w/v) sodium thioglycollate. Final pH should be 7.2 to 7.6. Atmosphere should be anaerobic. If MPN tubes are used, those that have turned black (FeS) after 4 weeks are considered positive. False positive tubes can also occur from decomposition of the sulfhydral containing reductant.

8. Denitrifiers[k]

Nutrient broth 8 g	Agar 15 g
KNO_3 0.5 g	Distilled water 1000 mℓ

Plates must be incubated under an anaerobic atmosphere.

Medium in tubes need not be degassed or the headspace oxygen removed as the early growth (respiration) will remove the oxygen. In MPN tubes (no agar) presumptive denitrification has occurred when all the nitrate and nitrite have been removed (assayed by colorimetric test, e.g., diphenylamine or diazotization following reduction of nitrate by Zn°). Confirmation of denitrification is by finding 20% of the NO_3^-–N as N_2O in tubes with 10% acetylene.

9. Nitrifiers — ammonia oxidizers[l]

$(NH_4)_2$ 0.5 g	Ferric citrate or EDTA 0.5 mg
KH_2PO_4 0.2 g	Phenol red 0.5 mg
$CaCl_2 \cdot 2H_2O$ 0.04 g	Distilled water 1000 mℓ
$MgSO_4 \cdot 7H_2O$ 0.04 g	

After sterilizing pH is adjusted to 7.5 to 8.0 by adding sterile aqueous solution of 5% sodium carbonate. Plates can be prepared by adding 1.5% purified agar. Incubation of MPN tubes should be 3 to 4 weeks.

10. Nitrifiers — nitrite oxidizers[m]

KNO_2 0.1 g	$CaCl_2 \cdot 2H_2O$ 0.02 g
NaCl 2.0 g	$KHCO_3$ 0.02 g
K_2HPO_4 0.5 g	$NaMoO_4 \cdot 2H_2O$ 2.4 μg
$MgSO_4 \cdot 7H_2O$ 0.05 g	Distilled water 1000 mℓ

In MPN tubes 6 mℓ of sterile medium is added to sterile tubes which contain 0.1 g $CaCO_3$. The pH should be approximately 8.0. Incubation time is 100 to 120 days.

11. Methane oxidizers[n]

$MgSO_4 \cdot 7H_2O$ 1 g	NH_4Cl 0.5 g
$CaCl_2$ 0.2 g	KNO_3 1.0 g
Ferric EDTA 4 mg	Agar 12.5 g
	Distilled Water 1000 mℓ

A trace element solution is added and pH adjusted to 6.8 before autoclaving. After autoclaving 2 mℓ of a solution of sterile phosphate buffer (KH_2PO_4 and Na_2HPO_4, 15 g/300 mℓ, pH 6.8) is added. An atmosphere of methane and air (30:70) is aseptically added.

12. Sulfur oxidizers-autotrophic, nonaciduric[o]

$Na_2S_2O_3 \cdot 5H_2O$ 10 g	$CaCl_2$ 0.1 g
$(NH_4)_2SO_4$ 0.1 g	$FeCl_3 \cdot 6H_2O$ 0.02 g
K_2HPO_4 4.0 g	$MnSO_4 \cdot 4H_2O$ 0.02 g
KH_2PO_4 4.0 g	Distilled water 1000 mℓ
$MgSO_4 \cdot 7H_2O$ 0.1 g	

Adjust to pH 6.6 and sterilize. Thiosulfate is a more convenient energy source than elemental sulfur though it can be used. For the aciduric species the pH should be about 4.0

Table 6 (continued)
MEDIA IN GENERAL USE FOR GROWING, ISOLATING AND/OR ENUMERATING PHYSIOLOGICAL GROUPS OF SOIL MICROORGANISMS

13. Heterotrophic N_2-fixer[p]

Glucose or sucrose 10 g	$FeCl_3 \cdot 6H_2O$ 0.25 g
K_2HPO_4 1 g	$NaMoO_4 \cdot 2H_2O$ 0.5 g
$MgSO_4 \cdot 7H_2O$ 0.2 g	Agar 16 g
$CaCO_3$ 0.5 g	Water 1000 mℓ

Substitution of starch and rhamnose as carbon source will select *Azotobacter vinelandii* and *A. chroococcum* respectively while lowering the pH to 5.0 to 6.0 will favor *Beijerinckia.* Because of the high respiratory rate of these organisms, growth in shaken flasks will be much faster.

14. Rhizobium[q]

Mannitol 10 g	$CaCO_3$ 3.0 g
K_2HPO_4 0.5 g	Yeast extract 0.2 g
$MgSO_4 \cdot 7H_2O$ 0.2 g	Agar 15 g
NaCl 0.1 g	Distilled water 1000 mℓ

15. Trace element solution[r]

EDTA 0.5 g	$MnCl_2 \cdot 4H_2O$ 3 mg
$FeSO_4 \cdot 7H_2O$ 0.2 g	$CuCl_2 \cdot 2H_2O$ 1 mg
$ZnSO_4 \cdot 7H_2O$ 0.01 g	$NiCl_2 \cdot 2H_2O$ 2 mg
H_3BO_3 0.03 g	$Na_2MoO_4 \cdot 2H_2O$ 3 mg
$CoCl_2 \cdot 2H_2O$ 0.02 g	Distilled water 1000 mℓ

Use 10 mℓ of this trace element solution per liter of medium if trace element deficiencies are suspected.

16. Vitamin solution[r]

Thiamine 0.1 g	Vitamin B_{12} 2 mg
Nicotinic acid 0.05 g	p- aminobenzoate 5 mg
Pantothenic acid 0.025	Biotin 1 mg
Pyridoxine 0.05 g	Distilled water 100 mℓ

Use 10 mℓ of this vitamin solution per liter of medium if a vitamin growth factor is needed.

17. Basal salts solution[s]

Solution A	Solution B	Solution C
K_2HPO_4 16 g	$MgSO_4 \cdot 7H_2O$ 2 g	$CaCl_2 \cdot 2H_2O$ 0.25 g
KH_2PO_4 4 g	Distilled water 100 mℓ	$FeCl_3 \cdot 6H_2O$ 0.025 g
NH_4NO_3 5 g		Distilled water 100 mℓ
NaCl 1 g		
Distilled water 100 mℓ		

Use 10 mℓ of each stock solution per liter of water (medium). To obtain clear medium free of precipitate add desired amount of water *before* adding stock salt solution and make sure each solution is well mixed before adding the next solution. In many cases all solutions can be autoclaved together in the medium; a cloudy precipitate will appear after autoclaving but it will dissolve after cooling. If not, solution B should be autoclaved separately and added aseptically to the medium after it cools.

A carbon source of 0.1 to 1.0% (w/v of medium) usually provides suitable carbon and energy for growth.

[a] From Lochhead, A. G.[30]
[b] Soil extract is made by autoclaving for 20 min 500 to 1000 g of fertile soil in 1000 mℓ of tap water. The solution is filtered to clarify, adjusted to pH 6.8 and sterilized. 0.5 g of $CaSO_4$ or $CaCO_3$ as a flocculating agent will aid filtering cloudy extracts.
[c] From Allen, O. N.[31]
[d] From Bunt, J. S. and Rovira, A. D.[32]
[e] From Martin, J. K.[33]
[f] From Riviere, J.[34]
[g] From Molongoski, J. J. and Klug, M. J.[35]
[h] From Pramer, D. and Schmidt, E. L.[36]
[i] From Kuster, E. and Williams, S. T.[37]
[j] From Pankhurst, G. S.[38]
[k] From Focht, D. D. and Joseph, H.[39]
[l] From Belser, L. W. and Schmidt, E. L,[40] and Soriano, S. and Walker, N.[41]
[m] From Matulewich, V. A., Strom, P. F., and Finstein, M. S.[42]

Table 6 (continued)
MEDIA IN GENERAL USE FOR GROWING, ISOLATING AND/OR ENUMERATING PHYSIOLOGICAL GROUPS OF SOIL MICROORGANISMS

[n] From Whittenbury, R., Phillips, K. C., and Wilkinson, J. F.[43]
[o] From Parker, C. D. and Frisk, J.[44]
[p] From Johnson, L. F. and Curl, E. A.[26]
[q] From Allen, O. N.[31] See also last section of this chapter for additional media.
[r] From Reference 29.
[s] From Tiedje, J. M., Mason, B. B., Warren, C. B., and Malec, E. J.[45]

termed free energy, G, which is available for useful work. If the free energy change, $\Delta G^{o\prime}$, is negative, then energy is released and the reaction is termed exergonic. If the $\Delta G^{o\prime}$ is positive, the reaction requires energy and is termed endergonic. However, even if the reaction has a negative $\Delta G^{o\prime}$, it does not mean that the reaction proceeds at a rapid or even measurable rate. This is because an energy barrier, termed energy of activation, prevents the reaction from occurring. Microorganisms contain catalysts known as enzymes which lower the activation energy and allow the reaction to proceed at a much more rapid rate. An example is the fixation of N_2 to ammonia which has a negative $\Delta G^{o\prime}$ ($N_2 + 3H_2 \rightarrow 2NH_3$, $\Delta G^{o\prime} = -19$ kcal/reaction). This reaction does not proceed spontaneously, but with the microbial catalyst nitrogenase, or with chemical catalysts, the high activation energy can be overcome and the reaction proceeds.

The free energy change for microbial reactions can be calculated from differences in oxidation-reduction potentials between the electron-donating and electron-accepting redox pairs using the equation $\Delta G^{o\prime} = -nF\Delta E_o'$, where n = the number of electrons transferred, F = Faraday constant (23 kcal) and E_o' is the difference in the two potentials. The free energy of hydrolysis of ATP is approximately −7.6 kcal/mole, thus it takes a bit more energy than this to synthesize ATP since reactions are not 100% efficient. Thus using the above equation, one can calculate that a potential difference of 250 mV is needed to synthesize one ATP ($-2 \times 23 \times 0.25 = -11.5$ kcal/mole). The oxidation-reduction potentials of redox pairs important in soil microbiology are given in Table 7. As an example, the potential energy available to *Nitrobacter* ($NO_2^- \rightarrow NO_3^-$ step of nitrification) is calculated from the electron-donating NO_2^-/NO_3^- pair and the electron-accepting O_2/H_2O pair, $\Delta G^{o\prime} = -2 \times 23 \times 0.38 = -17.5$ kcal/mole. This is enough energy for synthesis of one ATP but, compared to many other microbial processes, gives a relatively poor energy yield.

The more common way to calculate free energy change of reactions is from the energies of formation, G_f^o, of the elements. The free energy change, $\Delta G^{o\prime}$, for electron-donating and electron-accepting half-reactions of interest are given in Tables 8 and 9, respectively, which have been extracted from Thauer et al.[46] These were calculated using the equation $\Delta G^{o\prime} = \Sigma G_f^o \text{ (products)} - \Sigma G_f^o \text{ (reactants)} - m\Delta G_f' (H^+)$, where m = number of protons in reaction and $\Delta G_f'(H^+)$ is the free energy of formation of a proton at pH 7 (−9.55 kcal at 25°C).

The advantage of using the half-reactions is that one can easily calculate the free energy for any combination of electron-donating and electron-accepting reactions of interest. This is done by first balancing the stoichiometry with respect to H_2, if necessary, and then determining the difference in energy yield by subtraction. An example is given below for the case of dissimilatory nitrate reduction to ammonium with glucose as the substrate:

$$\begin{array}{llll} & \text{Glucose} + 12\,H_2O & \rightarrow 6\,HCO_3^- + 6H^+ + 12\,H_2 & +\ 0.8\text{ kcal} \\ & \times\,3\,[NO_3^- + 4\,H_2 + 2H^+ & \rightarrow NH_4^+ + 3\,H_2O & -\ 143.3\text{ kcal}] \\ \hline \text{Net} & \multicolumn{2}{l}{\text{Glucose} + 3\,H_2O + 3\,NO_3^- \rightarrow 6\,HCO_3^- + 3\,NH_4^+} & -\ 429\text{ kcal/reaction} \\ & & & -\ 143\text{ kcal/mole }NO_3^- \end{array}$$

Table 7
OXIDATION-REDUCTION POTENTIALS OF SOME REDOX PAIRS OF MICROBIOLOGICAL IMPORTANCE

Redox pair	E_o'(mv)[a]
N_2O / N_2	+ 1355[b]
NO / N_2O	+ 1175[b]
NH_2OH / NH_4^+	+ 899
O_2 / H_2O	+ 818
Fe^{3+} / Fe^{2+}	+ 775
NO_3^- / NO_2^-	+ 433
NO_2^- / NO	+ 350[b]
$S_3O_6^{2-}$ / $S_2O_3^{2-}$ + HSO_3^-	+ 225
NO_2^- / NH_2OH	+ 120
$S_4O_6^{2-}$ / $S_2O_3^{2-}$	+ 24
HSO_3^- / HS^-	− 116
HSO_3^- / $S_3O_6^{2-}$	− 173
CO_2 / CH_4	− 244[b]
SO_4^{2-} / HS^-	− 250
S°/ HS^-	− 270
$S_2O_3^{2-}$ / HS^- + HSO_3^-	− 402
H^+ / H_2	− 414[b]
SO_4^{2-} / HSO_3^-	− 516

[a] Standard potential (E_o) of hydrogen electrode at 1.0 atm H_2 and 1.0 M H^+ at 25°C = 0.0 mv; E_o' at pH 7.

[b] Calculated for CO_2, CH_4, H_2, N_2, NO and N_2O in the gaseous state, all other compounds in aqueous solution.

Adapted from Thauer, R. K. Jungerman, K., and Decker, K., *Bacteriol. Rev.*, 41, 100, 1977.

The free energies of formation in the tables are calculated assuming 1 *M* concentrations of reactants and products. However, concentrations change during a reaction. For non-H_2-forming bacteria, changes in the ratio of reactants to products of four orders of magnitude changes the $\Delta G'$ by only 2.8 kcal/mole at most. Hence the concentration factor can usually be disregarded. For H_2-producing bacteria, however, other H_2-consuming bacteria in natural samples can maintain very low H_2 concentrations which significantly improve the free energy change of some H_2-yielding reactions. In fact, it allows certain reactions to proceed which otherwise would not be feasible, (e.g., Table 8, propionate → acetate + H_2). This phenomenon is important only in anaerobic habitats where the H_2 -consuming methane production and/or sulfate reduction reactions occur.

Energies have conventionally been reported in units of calories; however, an international commission has recommended that energy be expressed as joules, J. Hence, both units are given in Tables 8 and 9. One kcalorie is equivalent to 4.184 kJ.

Chemolithotrophic Metabolism

Chemolithotrophs, which use inorganic substrates as energy sources and electron-donors, are the principal agents of nitrogen and sulfur oxidation in soil. Since these organisms obtain their carbon from fixation of CO_2, they are also termed chemoauto-

Table 8
THEORETICAL ENERGY YIELDS OF ELECTRON-DONATING REACTIONS OF IMPORTANCE TO SOIL MICROBIOLOGY

			$\Delta G^{\circ\prime}$	
Process and/or Organism	Substrate	Product	kcal/reaction	kJ/reaction
Dehydrogenations of anaerobic habitats				
Formate lyase[a]	$Formate^- + H_2O$	$\rightarrow HCO_3^- + H_2$	+0.3	+1.3
	$Acetate^- + 4H_2O$	$\rightarrow 2HCO_3^- + 4H_2 + H^+$	+25.0	+104.6
Methanogen/acetate splitting[a]	$Acetate^- + H_2O$	$\rightarrow HCO_3^- + CH_4$	−7.4	−31.0
Acetogen/proton reducer[a]	$Propionate^- + 3H_2O$	$\rightarrow HCO_3^- + acetate^- + H^+ + 3H_2$	+18.2	+76.1
Acetogen/proton reducer[a]	$Butyrate^- + 2H_2O$	$\rightarrow 2\ Acetate^- + H^+ + 2H_2$	+11.5	+48.1
Selected dehydrogenations of heterotrophs				
	$Pyruvate^- + 6H_2O$	$\rightarrow 3HCO_3^- + 2H^+ + 5H_2$	+13.7	
Desulfovibrio[a]	$Lactate + 2H_2O$	$\rightarrow Acetate^- + HCO_3^- + H^+ + 2H_3$	−1.0	−4.2
Acetogen/proton reducer[a]	$Ethanol + H_2O$	$\rightarrow Acetate^- + H^+ + 2H_2$	+2.3	+9.6
	Glycerol	$\rightarrow Pyruvate + H^+ + 2H_2$	−6.2	−25.9
	$Glutamate^- + 3H_2O$	$\rightarrow 2\ Acetate^- + HCO_3^- + H^+ + NH_4 + H_2$	−8.1	−33.9
	$Glucose + 12H_2O$	$\rightarrow 6HCO_3^- + 6H^+ + 12H_2$	+0.8	+3.2
Dehydrogenations of nitrogen				
	$2NH_4^+$	$\rightarrow N_2 + 2H^+ + 3H_2$	+18.8	+78.7
Nitrosomonas[b]	$NH_4^+ + 2H_2O$	$\rightarrow NO_2^- + 2H^+ + 3H_2$	+104.3	+436.4
Nitrification (net)[b]	$NH_4^+ + 3H_2O$	$\rightarrow NO_3^- + 2H^+ + 4H_2$	+143.3	+599.6
Nitrobacter[b]	$NO_2^- + H_2O$	$\rightarrow NO_3^- + H_2$	+39	+163.2
Dehydrogenations of sulfur				
Thiobacilli[b]	$HS^- + H^+$	$\rightarrow S^\circ + H_2$	+6.7	+28.8
Thiobacilli (net)[a]	$HS^- + 4H_2O$	$\rightarrow SO_4^{2-} + H^+ + 4H_2$	+36.4	+152.2
Thiobacilli (net)[b]	$S^\circ + 4H_2O$	$\rightarrow SO_4^{2-} + 2H^+ + 3H_2$	+29.7	+124.3
Thiobacilli[b]	$S_2O_3^{2-} + 5H_2O$	$\rightarrow 2SO_4^{2-} + 5H_2$	+50.2	+210.0
Thiobacilli[b]	$S_2O_3^{2-} + 3H_2O$	$\rightarrow 2SO_3^{2-} + 3H_2$	+60.2	+251.9
Other dehydrogenations				
Thiobacillus ferrooxidans[b]	$2Fe^{2+} + 2H^+$	$\rightarrow 2Fe^{3+} + 3H_2$	+54.6	+228.5

Table 8 (continued)
THEORETICAL ENERGY YIELDS OF ELECTRON-DONATING REACTIONS OF IMPORTANCE TO SOIL MICROBIOLOGY

Note: Most of these electron-donating reactions are typically coupled to O_2 (respiration) as the electron acceptor, though many of the organic reactions also occur in the absence of oxygen.

[a] Examples of organisms for the anaerobic process only.
[b] Examples of organisms which couple this process to O_2 reduction.

Adapted from Thauer, R. K., Jungermann, K., and Decker, K.[46]

Table 9
THEORETICAL ENERGY YIELDS OF ELECTRON-ACCEPTING REACTIONS OF IMPORTANCE TO SOIL MICROBIOLOGY

Biological process and/or type of organism	Substrate	Products	$\Delta G^{\circ\prime}$ kcal/reaction	$\Delta G^{\circ\prime}$ kJ/reaction
Reactions by methanogens				
Methanogen/lithotrophic	$CO_2 + 4H_2$	$\rightarrow CH_4 + 2H_2O$	−31.3	−131
Methanogen/lithotrophic	$HCO_3^- + 4H_2 + H^+$	$\rightarrow CH_4 + 3H_2O$	−32.4	−135.6
Methanogen	Formate $+ H^+ + 3H_2$	$\rightarrow CH_4 + 2H_2O$	−32.1	−134.3
Other reductions of fermentative habitats				
Acetogen/lithotrophic	$2HCO_3^- + 4H_2 + H^+$	$\rightarrow$ Acetate$^- + 4H_2O$	−25	−104.6
	$HCO_3^- + H_2$	$\rightarrow$ Formate$^- + H_2O$	−0.3	−1.3
	Acetate $+ H^+ + 2H_2$	$\rightarrow$ Ethanol $+ H_2O$	−2.3	−9.6
	2 Acetate$^- + H^+ + 2H_2$	$\rightarrow$ Butyrate$^- + 2H_2O$	−11.5	48.1
Clostridia	Glycine $+ H_2$	$\rightarrow$ Acetate$^- + NH_4^+$	−18.6	−77.8
Many fermenters	Fumarate$^{2-} + H_2$	$\rightarrow$ Succinate^{2-}	−20.6	−86.2

Reductions of sulfur				
Desulfuromonas type	$S^\circ + H_2$	$\rightarrow HS^- + H^+$	−6.7	−28.0
Sulfate reduction	$S_2O_3^{2-} + H_2$	$\rightarrow HS^- + SO_3^{2-} + H^+$	−0.3	−1.1
Facultative anaerobes	$S_4O_6^{2-} + H_2$	$\rightarrow 2S_2O_3^{2-} + 2H^+$	−20.2	−84.5
Sulfate reduction	$S_3O_6^{2-} + H_2$	$\rightarrow S_2O_3^{2-} + SO_3^{2-} + 2H^+$	−29.1	−121.8
Sulfate reduction	$SO_3^{2-} + H_2 + 4H^+$	$\rightarrow S_3O_6^{2-} + 3H_2O$	−12.0	−50.2
	$SO_3^{2-} + 2H_2 + 2H^+$	$\rightarrow S^\circ + 3H_2O$	−34.7	−145.1
Sulfate reduction (net)	$SO_3^{2-} + 3H_2 + 2H^+$	$\rightarrow H_2S + 3H_2O$	−41.3	−172.8
Sulfate reduction	$SO_4^{2-} + H_2$	$\rightarrow SO_3^{2-} + H_2O$	+5.0	+20.9
	$SO_4^{2-} + 3H_2 + 2H^+$	$\rightarrow S^\circ + 4H_2O$	−29.7	−124.3
Sulfate reduction (net)	$SO_4^{2-} + 4H_2 + H^+$	$\rightarrow HS^- + 4H_2O$	−36.3	−151.9
Reductions of nitrogen				
Nitrogen fixation	$N_2 + 2H^+ + 3H_2$	$\rightarrow 2NH_4^+$	−18.8	−80.0
Denitrification	$N_2O + H_2$	$\rightarrow N_2 + H_20$	−81.6	−341.4
Denitrification	$2NO + H_2$	$\rightarrow N_2O + H_2O$	−73.2	−306.3
Denitrification	$NO_2^- + 1/2_2H_2 + H^+$	$\rightarrow N + H_2$	−17.5	−73.2
Denitrification (net)	$2NO_2^- + 3H_2 + 2H^+$	$\rightarrow N_2 + 4H_2O$	−189.9	−794.1
Dissimilatory nitrite reduction	$NO_2^- + 3H_2 + 2H^+$	$\rightarrow NH_4^+ + 2H_2O$	−104.3	−436.4
Nitrate respiration and denitrification	$NO_3^- + H_2$	$\rightarrow NO_2^- + H_2O$	−89.0	−163.2
Denitrification (net)	$2NO_3^- + 5H_2 + 2H^+$	$\rightarrow N_2 + 6H_2O$	−267.8	−1120.5
Dissimilatory nitrate reduction (net)	$NO_3^- + 4H_2 + 2H^+$	$\rightarrow NH_4^+ + 3H_2O$	−143.3	−599.6
Other reductions				
Respiration	$O_2 + 2H_2$	$\rightarrow 2H_2O$	−113.4	−474.5
Ferric reduction	$2Fe^{3+} + H_2$	$\rightarrow 2Fe^{2+} + 2H^+$	−54.6	−228.3

Note: The energy-accepting half-reactions can be coupled to any of the electron-donating half-reactions in Table 8 to determine theoretical energy yield of any process of interest.

[a] Examples of organisms for the anaerobic process only.
[b] Examples of organisms which couple this process to O_2 reduction.

Adapted from Thauer, R. K., Jungermann, K. and Decker, K.[46]

trophs. Hydrogen, ferrous iron, and certain C_1 compounds, e.g., HCOOH, CO, also support growth of chemolithotrophic bacteria, but these are of lesser importance in soil.

The general case is that chemolithotrophs require oxygen as their electron acceptor and thus are aerobes. The few exceptions are the denitrifying *Thiobacillus denitrificans* which uses nitrate, and a few hydrogen oxidizing organisms that denitrify or reduce sulfate. Thus for the general case, chemolithotrophic oxidations are coupled to oxygen reduction forming water (respiration). The *net* reaction for these oxidations is often presented, which obscures the role of O_2 in the process. This has led to the misunderstanding that O_2 is incorporated with the element being oxidized. For the *Nitrosomonas* reaction in nitrification, the net reaction is:

$$NH_4^+ + 1\tfrac{1}{2}\,O_2 \rightarrow NO_2^- + 2H^+ + H_2O \qquad -\ 62.1\ \text{kcal/mole} \tag{1}$$

However, if labeled oxygen is used, the true physiological process is apparent as shown by Equation 2.

+6e

$$^{3+}NH_4^+ + 1\tfrac{1}{2}\,{}^{*}O_2 + 2H_2O \rightarrow {}^{-3}NO_2^- + 2H^+ + 6H_2\,{}^{*}O \tag{2}$$

3 (+2e)

This diagram shows that the six electrons released by the oxidation of the N atom are accepted by an O atom of O_2 which yields H_2O. Since each O atom accepts two electrons, it takes three oxygen atoms to accept all six electrons produced from the oxidation of one N atom. The oxygen to form NO_2^- comes from water, not O_2.

The above example shows the importance of oxygen as the electron-accepting reaction for nitrification, although this principle and type of equation can also be written for the other lithotrophic reactions. Though the above is the general principal, it is not entirely correct since it is now known that monooxygenase activity is part of the hydroxylamine (NH_2OH) oxidation step in *Nitrosomonas.* The enzyme incorporates some O directly from O_2 during the formation of nitrite. This, however, is the exception in chemolithotrophic metabolism.

Other Oxidations and Reductions Catalyzed by Bacteria

Microbially catalyzed oxidation and reduction reactions can either be directly linked to the initial or terminal steps of an electron transport chain by enzymes with high specificity for the redox substrate, or indirectly linked such that the redox environment created by the metabolism results in an oxidation or reduction. Many of the minor nutrients and elements foreign to and even toxic to biology undergo redox changes by this indirect mechanism. The direct reactions are usually coupled to energy conservation (ATP synthesis) while the indirect reactions are not. Iron and manganese are the best known examples of elements that are oxidized and reduced by the indirect mechanism. Iron can also be oxidized in very acid environments by a direct chemolithotrophic mechanism (*Thiobacillus ferrooxidans* is such an organism), and may even be a direct electron acceptor as it appears to compete with nitrate for electrons in dissimilatory nitrate reduction.[47] There is also recent evidence that manganous oxidation may be directly coupled to the electron transport chain,[48] although conclusive evidence is hard to obtain since the element is also easily oxidized indirectly. Despite this evidence for direct roles of iron and manganese in microbial metabolism, most oxidation and reduction of these elements in soils is probably indirect.

Table 10
REDUCTIONS OF ELEMENTS OTHER THAN MAJOR NUTRIENTS THAT CAN BE CATALYZED BY BACTERIA[a]

Element	Electron acceptor	Product
Selenium	Se°	HSe^{-}
	$HSeO_3^{-}$	Se°
Tellurium	$HTeO_3^{-}$	Te°
	$HTeO_4^{-}$	Te°
Molybdenum	$Mo_6O_{21}^{6-}$	$6MoO_{2.5}$
Vanadium	H_2VO_4	$VO(OH)_2$
Uranium	$UO_2(OH)_2$	$U(OH)_4$
Bismuth	$Bi(OH)_3$	Bi°
	BiO_3^{-}	Bi°
Arsenic	H_2AsO_4	$HAsO_2$
Osmium	OsO_4	OsO_4^{2-}
	OsO_2	Os°
Copper	$Cu(OH)_2$	$CuOH$
Manganese	MnO_2	Mn^{2+}
Lead	PbO_2	$Pb(OH)_2$
Mercury	Hg^{2+}	Hg°

[a] Evidence largely from Woolfolk and Whitely[52] who used extracts from *Veillonella alcalescens* (formerly *Micrococcus lactilyticus*) and molecular hydrogen as the electron donor.

The *Thiobacillus*-catalyzed reactions, used practically in ore leaching for recovery of metals, can result in some unusual oxidations, e.g., oxidation of copper,[49] antimony,[50] and uranium.[51]

The examples of indirect reductive reactions are more numerous. Examples are given in Table 10. This extensive survey was done using extracts of the obligate anaerobe, *Veillonella alcalescens,* and molecular hydrogen as the reductant.[52] Several of the reductions, e.g., selenium, tellurium, and manganese have been noted in other organisms. Because of the in vitro conditions of the survey, it is not known which of these reductions might occur in soil. However, in poorly drained or flooded soils anaerobic conditions can be intense, and one would expect a variety of indirect reductions of the electron-accepting species present. Whether or not the organisms gain any survival value from these alternate electron acceptors is not known.

STERILIZATION AND INHIBITION OF THE SOIL BIOTA

Sterilization and/or inhibition of the soil community is an essential control in many experiments. It is not easy to achieve a "sterile" control without changing the chemistry of the soil and thereby introducing possible confounding reactions. A sterile soil means one in which *no* viable biological organisms remain. There are not degrees of sterility. In order to state that a soil is sterile, one should confirm at the end of the experiment that no viable organisms remain. This is usually done by placing a soil sample in a sterile, rich medium, e.g., tryptic soy broth or agar, and observing whether growth develops.

Inhibition of biological activity rather than sterilization is adequate for some experiments and has the advantages of causing less change in the soil chemistry and allowing one to manipulate the sample without risking recontamination which can easily happen with a sterilized control. It has the disadvantage that one is less certain that all biological activity is fully inhibited at all microsites.

Though individual treatments will have advantages for particular experiments, the following is a priority ranking of sterilization methods for the general case (best first): gamma irradiation, propylene oxide, autoclaving, chloroform inhibition. Further discussion of soil sterilization and inhibition treatments is given by Powlson.[53]

CALCULATION OF GAS SOLUBILITIES

Many microbial substrates and products are gases. It is usually necessary to know the gas concentration in aqueous solution (substrate concentration) and the total amount of gas formed when it is a product. Generally, the concentration of the gas in the vapor phase is known either from the amount added or from gas chromatographic analysis of the headspace composition. Knowing the vapor phase concentration, one can calculate the amount in solution at any temperature using the Bunsen absorption coefficient, α, and the following equation:

$$C_l = C_g \alpha \tag{3}$$

where C_l = concentration in liquid phase, C_g = concentration in gas phase. Table 12 gives the Bunsen coefficients for gases of microbiological importance.

If one wants to know the total gas produced, i.e., liquid phase plus gas phase, the following equation can be used:

$$M = C_l (V_g + V_l \alpha) \tag{4}$$

Here M = the total amount, V_g = volume of gas phase, and V_l = volume of liquid phase.

Gas solubility will also vary with ionic strength of the solution. For some gases this is insignificant, but for others it can be important if precise values are desired. The reader is referred to Wilhelm et al.,[54] if this correction is deemed important.

OTHER TABLES OF REFERENCE FOR SOIL MICROBIOLOGY

Water Activities That Limit Microbial Growth

Water availability in soil is often reduced by osmotic and matric forces, which can limit microbial metabolism and growth. Table 13 provides an indication of the approximate water activity at which cell division is prevented. In general, fungi can grow at lower water potentials than bacteria.

Ionization Constants of Acids and Bases

Many microbial substrates and products have ionized forms. Whether the compound is in an ionized form can be important to its aqueous solubility, reactivity, volatility, and ease of transport into a cell. Generally ionized species do not pass through the cell membrane (largely lipid) unless a specific membrane carrier exists for that substrate.

Table 14 gives the pKa of inorganic and organic compounds of microbial importance. The pKa is the pH when 50% of the compound is in the form given and 50% in the form with one proton removed. For organics in which more than one proton can be lost, the additional pKs are listed in sequential order.

Table 11
STERILIZATION METHODS AND BIOLOGICAL INHIBITORS USED WITH SOILS

Method	Mechanism of killing	Comments
Physical agents		
Dry heat	Denaturation of proteins and oxidation	Not effective for soil because penetration is slow and killing of occluded spores is difficult; will also oxidize soil organic matter; is useful for sterilizing equipment, though temperature of 160°C for 1 to 2 hr is necessary to kill spores
Moist heat plus pressure (autoclave)	Denaturation of protein and melting of membrane lipids	Steam aids in breaking of hydrogen bonds thus promoting more rapid denaturation of proteins; one of the effective sterilization treatments for soil; the major disadvantage is that the treatment also hydrolyzes substantial amounts of soil organic matter which releases more metal ions, e.g., Mn^{2+}; autoclave treatments on two or three consecutive days will kill occluded spores which may not be killed by a single treatment
Ultraviolet irradiation	Disruption of nucleic acids, especially formation of dimers from adjacent pyrimidines	Short wavelength UV light (e.g., 254 nm from mercury-vapor lamp) is effective in killing microbes on surfaces; UV light is not penetrating so it is ineffective in sterilizing soil; it is often a convenient method for sterilizing working surfaces or smooth, flat equipment; (danger: UV light is damaging to eyes; glasses must be worn; glass is opaque to UV)
Ionizing radiation	Breakage of DNA caused by reactive, short-lived radicals produced by the ionization	Perhaps the best method for soil sterilization because it is effective and does not drastically alter the soil chemistry (at least relative to other methods); it does produce some free radicals which might cause confounding organic reactions (generally not a problem) and does require specialized equipment; enzymes are generally not destroyed by the dosages used; the most common ionizing source is ^{60}Co; 2.5 Mrad is a recommended dosage
Ultrasonic waves	Disruption of cells	Though extensive disruption of cells occurs, it is rarely complete and thus cannot be used for sterilization
Chemical agents		
Metals, especially Hg^{++}, Ag^{+}, Cu^{++}	Disruption of vital functions by binding to sulfhydral groups	Though very effective in aqueous samples, not as reliable in soil due to soil reactions reducing the effective concentration; particularly ineffective in samples with high sulfide (anaerobic) or sulfhydral (high protein) content
Halogens (I_2, Cl_2)	I: iodinates tyrosine, oxidant Cl: $Cl_2 + H_2O \rightleftharpoons HCl + HOCl$ hypochlorite strong oxidant	Effective sterilants but not generally used in soils because the organic matter dilutes the effective concentra-

Table 11 (continued)
STERILIZATION METHODS AND BIOLOGICAL INHIBITORS USED WITH SOILS

Method	Mechanism of killing	Comments
		tion; hypochlorite is a common treatment for surface sterilization of seeds
Alkylating agents Formaldehyde Ethylene oxide Propylene oxide	Replace labile H atoms of $-NH_2$, $-OH$, $-SH$, $-COOH$ groups on proteins and nucleic acids	Probably second to ionizing radiation in being the most generally useful method in sterilizing soils; propylene oxide is the recommended agent since it is easier to work with a liquid; in contrast to other methods, this one is nearly as effective against spores as it is against vegetative cells; very reliable in soils; (danger: the oxides are suspected carcinogens, and because of their volatility should be used only in a hood)
Surfactants (Quarternary compounds)	Disrupt cell membranes	Popular and effective commercial disinfectants but not used in soil probably because the cationic property causes them to bind to soil
Phenols Hexachlorophene Tricresol	Disrupts cells	Traditionally used as disinfectants but not as sterilants; not used in soils; some are good carbon sources for soil bacteria at nontoxic concentrations
Azide/Cyanide	Electron transport inhibitors	Because of mode of action they are more effective against aerobes; probably cannot be relied on as soil sterilants but many examples show sustained inhibition of biological activity in soil; (danger: both volatile under acidic conditions and are extremely poisonous)
Solvents Chloroform Toluene Ether	Dissolves cell membranes	Chloroform especially has been an effective inhibitor of microbial activity in soil though it is probably not a sterilant; chloroform is now recognized as a carcinogen
Fumigants		Commercial fumigants are more effective against fungi and other higher organisms than against bacteria; bacteria will quickly proliferate after fumigation treatment
Antibiotics	Each has specific mode of action	Have been used to inhibit specific groups of microbes in soil but are not sterilants; antibiotics strongly bind to soil reducing the effective concentration

Characteristics of Stable and Radioactive Isotopes

Isotopes are essential experimental tools for soil microbiological research. The natural abundance of the stable isotopes of microbiological importance is given in Table 15. Precise measurement of the quantity of stable isotope in a sample is done with an isotope ratio mass spectrometer. The sensitivity is a function of the total quantity of the measured element present in the sample.

The basic characteristics of useful radioactive isotopes are also given in Table 15.

Table 12
BUNSEN ABSORPTION COEFFICIENTS (α) FOR GASES OF INTEREST IN SOIL MICROBIOLOGICAL STUDIES[a]

	Temperature (°C)						
Gas	5	10	15	20	25	30	35
NH_3	433.3	392.2	354.0	318.9	286.5	257.0	230.2
N_2	0.0211	0.0189	0.0171	0.0157	0.0145	0.0136	0.0128
NO	0.0646	0.0575	0.0517	0.0470	0.0431	0.0399	0.0372
N_2O	1.059	0.882	0.743	0.632	0.544	0.472	0.414
H_2S	4.016	3.436	2.974	2.602	2.299	2.052	1.848
SO_2	64.82	53.84	45.05	37.96	32.19	27.46	23.55
CH_4	0.0495	0.0433	0.0384	0.0342	0.0311	0.0283	0.0262
CO	0.0314	0.0281	0.0343	0.0311	0.0285	0.0264	0.0246
CO_2	1.423	1.194	1.015	0.873	0.758	0.666	0.590
C_2H_4	—	0.152	0.134	0.189	0.106	0.0962	0.0878
C_2H_2	1.51	1.31	1.16	1.03	0.928	0.843	0.771
H_2	0.0206	0.0199	0.0188	0.0181	0.0175	0.0171	0.0167
O_2	0.0430	0.0382	0.0343	0.0311	0.0285	0.0264	0.0246

[a] The Bunsen absorption coefficient is the mℓ of gas at 0°C and 760 mm mercury (S.T.P.) which is absorbed by 1 mℓ of water. This value includes only the dissolved gas and not the ionic species for gases which have an ionic species in equilibrium.

From Wilhelm, E., Battino, R., and Wilcock, R. J., *Chem. Rev.*, 77, 219, 1977. With permission.

Table 13
APPROXIMATE LIMITING WATER ACTIVITIES FOR MICROBIAL GROWTH

Water activity (A_w)	Reference points	Bacteria	Fungi
1.00	Plant wilt Seawater	Water bacteria[a]	
0.95	Bread	Most rods	Basidiomycetes
0.90		Most cocci	Yeasts, *Fusarium*
0.85	Salami		
0.80			*Penicillium*
0.75	Salt Lake	Halophiles	*Aspergillus*
0.70	Cereals		

[a] Certain bacteria prevalent in water with other than rod or coccus morphology, e.g., *Caulobacter* and *Spirillum*.

After Brown, A. D.[55]

The shorter the half-life, the greater the maximum specific activity possible in a substrate and thus the greater the sensitivity. In contrast to stable isotopes, the detection limit for radioactive isotopes is independent of the quantity of that element present in the sample. Isotopes with half-lives shorter than a few days must be produced by the user to avoid the extensive decay during shipment. The "short-lived" isotopes — ^{11}C, ^{13}N, ^{15}O, and ^{18}F — can be produced in a variety of low energy accelerators and have been used for biological research.

Table 14
IONIZATION CONSTANTS OF ACIDS AND BASES OF GENERAL MICROBIOLOGICAL IMPORTANCE[a]

Compound Inorganic	pK_a	Compound Organic	pK_a
H_3BO_3	9.23	Acetic acid	4.76
HCN	9.40	Adipic acid	4.42, 5.41
H_2CO_2	6.37	Benzoic acid	4.20
		Butyric acid	4.82
HCO_3^-	10.33	Caproic acid	4.88
NH_4^+	9.23	Citric acid	3.09, 4.75, 5.41
HN_3	4.72	Formic acid	3.77
HNO_2	3.29	Fumaric acid	3.03, 4.54
		Glyceric acid	3.55
$H_2N_2O_2$	7.05	Lactic acid	3.86
H_3PO_4	2.12	Malic acid	3.40, 5.2
$H_2PO_4^-$	7.21	Oxalic acid	1.25, 4.14
HPO_4^{2-}	12.32	Propionic acid	4.88
		Pyruvic acid	2.50
H_2S	7.00	Succinic acid	4.19, 5.48
HS^-	12.92	Nitrilotriacetic acid (NTA)	1.89, 2.49, 9.73[b]
H_2SO_3	1.9		
HSO_3^-	7.21	EDTA	1.99, 2.67, 6.16, 10.26
H_2SO_4	1.9	Aniline	4.62
HSO_4^-	1.99	Ethylamine	10.67
		Ethylenediamine	7.44, 9.91

[a] Most of the reported values are for 25°C; however, the values do not vary greatly for common room temperatures. Other slight variations can be expected from variations in ionic strength and specific salt effects.

[b] Keys to Chelation, Dow Chemical Co., Midland, Mich., 1969, 54.

Adapted from Jencks, W. P. and Regenstein, J.[56]

RHIZOBIUM: INFECTION PROCESS, IMPORTANT STRAINS, AND SOURCE OF STRAINS*

The genus *Rhizobium* is characterized by a group of Gram negative, non-sporulating motile rods which live in the soil and are capable of infecting, nodulating, and entering a nitrogen-fixing symbiosis with roots of various leguminous plants. Six species are defined according to the legume host which they nodulate.[59] The type species is *R. leguminosarum* (Frank). Legumes nodulated by the same rhizobial species are placed within a "cross-inoculation" group. Organization of rhizobia and legumes into cross-inoculation groups is justified by the high degree of host specificity displayed in this microorganism symbiosis, especially when considering the temperate legumes. The basis for this host specificity is unknown; current theories propose that "recognition" involves unique saccharide sequences on the bacteria which interact specifically with carbohydrate-binding lectins on the host root.[60,61] However, rhizobia which typically

* This section was written by Frank B. Dazzo, Assistant Professor, Michigan State University.

Table 15
CHARACTERISTICS OF STABLE AND RADIOACTIVE ISOTOPES USEFUL IN SOIL MICROBIOLOGICAL RESEARCH

Stable isotopes		Radioactive isotopes			
Isotope	Abundance (%)	Isotope	Type of decay	Energy (MeV)	Half-life (T ½)
^{1}H	99.985	^{3}H	β^-	0.0180	12.262 years
^{2}H	0.015	^{11}C	β^{+a}	0.98	20.4 min
^{12}C	98.892	^{14}C	β^-	0.155	5568 years
^{13}C	1.108	^{13}N	β^{+a}	1.19	9.96 min
^{14}N	99.635	^{15}O	β^{+a}	1.68	2.02 min
^{15}N	0.365	^{18}F	β^{+a}	0.649	110 min
^{16}O	99.759	^{22}Na	β^{+a}	0.542	2.6 years
^{17}O	0.037		γ	1.27	
^{18}O	0.204	^{32}P	β^-	1.707	14.22 days
^{31}P	100.0	^{35}S	β^-	0.167	87 days
^{32}S	95.018	^{36}Cl	β^-	0.714	3.1×10^5 years
^{33}S	0.750	^{45}Ca	β^-	0.256	164 days
^{34}S	4.215	^{54}Mn	γ	0.84	291 days
^{36}S	0.017	^{59}Fe	β^-	0.271	45 days
^{35}Cl	75.53		γ	0.191	
^{37}Cl	24.47	^{65}Zn	β^+	0.324	245 days
			γ	1.119	

[a] Positrons (β^+) annihilate forming two antiparallel 0.511 MeV gamma rays.

From Radiological Health Handbook, Division of Radiological Health, Ed., U.S. Department of Health, Education and Welfare, Washington, D.C., 1960.

Table 16
THE GENUS *RHIZOBIUM*

Species	Strains in common use	Legume cross-inoculation groups
R. leguminosarum	PRE, 300, 128C53	*Pisum, Lathyrus, Vicia, Lens*
R. trifolii	TA1, 0403	*Trifolium*
R. phaseoli	127K17	*Phaseolus* (most species)
R. meliloti	2011, 41, 102F51	*Melilotus, Medicago, Trigonella*
R. japonicum	3I1b110, 3I1b123, 61A76, 3I1b138	*Glycine*
R. lupini	NZP2257, 250	*Lupinus, Ornithopus*
R. species "cowpea miscellany"	32H1	*Vigna, Macroptilium,* etc.

nodulate the tropical legumes display little host specificity and are therefore considered promiscuous. These latter legumes are part of the "cowpea miscellany". Table 16 is a list of rhizobial species, their corresponding legume cross-inoculation groups, and representative strains which are commonly used in research.

Rhizobia are routinely cultivated at neutral pH in either a yeast extract-mannitol medium[62] or in a simple, chemically defined medium[63] containing mannitol as carbon source, glutamate as nitrogen source, a few vitamins, and mineral salts. An improved, defined medium[64] has become the standard choice for many laboratories and is superior in performance because of simplicity in preparation and because of the faster rhizobial growth rates which are achieved in this medium. Rhizobia are classified as "fast-growers" *(R. trifolii, R. leguminosarum, R. phaseoli, R. meliloti)* with typical generation times of 2 to 4 hr or as "slow growers" *(R. japonicum, R. lupini,* and the *Rhizobium* "cowpea miscellany") with generation times of 6 to 8 hr. For primary isolation, root nodules are surface sterilized (by $HgCl_2$ or $NaClO_4$), washed, crushed in sterile media, and the contents streaked for colony isolation. Colonies are purified and reinoculated on sterile seedlings to confirm the identification as a nodulating rhizobial isolate. Identification of rhizobial strains among nodule isolates is accomplished by phage typing, antibiotic resistance markers, or serology. Also, immunofluorescence with specific antisera has been used to identify rhizobial strains in fresh nodule tissue.

The enumeration of rhizobia against a background of the soil microflora can be accomplished by the indirect plate-infection count.[62] This method depends on the ability of a specific *Rhizobium* to nodulate a legume plant. A sample is diluted, inoculated on sterile seedlings, and scored for nodulation following an appropriate growth period. Data from positive and negative cases are used to determine the most probable number after reference to appropriate tables. The assumption that a single rhizobial cell added to the test plant leads to a sufficient population on the root surface to cause nodulation is generally valid.[62] Inoculated seedlings are grown on agar surfaces in test tubes or in cellophane pouches for ease of handling. In either case, the seedlings are provided with a plant nutrient solution lacking fixed nitrogen.

The developmental events of the *Rhizobium*-clover symbiosis can be observed by using a simple slide-culture chamber devised by Fahraeus.[65] Rhizobia accumulate on root hairs which project laterally from the root epidermis. Within 2 to 3 days, the root hairs undergo a marked deformation, particularly at their growing tip. This curled tip is called the "shepherd's crook". The root hair cell wall within the overlap of the shepherd's crook invaginates to form an open refractile tube, called the infection thread, is penetrated by the rhizobia. The tip of the infection thread grows to the base of the root hair cell where, by invagination, it enters the outer cortex of the root. As the infection thread continues to grow through the root tissue, inner cortex cells are stimulated to grow and divide, eventually forming a mass which protrudes from the root surface as a nodule. Rhizobia are released from the infection thread within some nodule cortex cells and eventually become enlarged and pleomorphic to form "bacteroids" which actively fix nitrogen. The developmental events may abort at any stage, resulting in the failure of the plant to derive fixed nitrogen from the symbiosis. Thus rhizobial strains which fail to infect the root hairs, incite nodule formation, or fix nitrogen within nodules are considered noninfective, nonnodulating, or ineffective, respectively.

The competitiveness of a rhizobial strain is determined by its ability to infect and nodulate a legume host in the presence of other nodulating strains within the soil. Many agricultural soils either lack rhizobia or contain rhizobial strains which are competitive but ineffective. Therefore, it is advisable to inoculate the legume at planting with specific rhizobial inoculants to ensure that an efficient N-fixing symbiosis results. The commercially available inoculants consist of competitive and effective rhizobial cultures which have been mixed with a neutralized peat base. The inoculants have a limited shelf life and should be used according to manufacturers' instructions for optimal results.

Table 17 is a list of organizations which maintain culture collections of rhizobia and generously provide subcultures upon request.

Table 17
RHIZOBIUM CULTURE COLLECTIONS

Curator	Location
S. Smith	The Nitragin Co., Milwaukee, Wis., 53209
D. Kaiser	U.S. Department of Agriculture, Beltsville, Md., 20705
J. Halliday	Niftal Project, University of Hawaii, Paia, Maui, Hawaii, 96779
A. Pattison	Soil Microbiology Dept., Rothamsted Experiment Station, Harpenden, Herts, England
F. Bergersen	Microbiology Section, C.S.I.R.O., Division of Plant Industry, Canberra, A. C. J. 2600, Australia

REFERENCES

1. **Clark, F. E. and Paul, E. A.,** The microflora of grasslands, in *Adv. in Agron.,* Vol. 22, Brady, N. C., Ed., Academic Press, New York, 1970, 405.
2. **Lynch, J. M. and Poole, N. J.,** *Microbial Ecology: A Conceptual Approach,* Halsted Press, New York, 1979, 78.
3. **Richards, B. N.,** *Introduction to the Soil Ecosystem,* Longman Group Limited, New York, 1974, 3.
4. **Macfadyen, A.,** *Animal Ecology: Aims and Methods,* 2nd ed., Putnam, London, 1963, 234.
5. **Alexander, M.,** *Introduction to Soil Microbiology,* 2nd ed., John Wiley & Sons, New York, 1977, chap. 2—6.
6. **Jones, P. C. T. and Mollison, J. G.,** A technique for the quantitative estimation of soil micro-organisms, *J. Gen. Microbiol.,* 2, 54, 1948.
7. **Jenkinson, D. S.,** The effects of biocidal treatments on metabolism in soil. III. The relationship between soil biovolume, measured by optical microscopy, and the flush of decomposition caused by fumigation, *Soil Biol. Biochem.,* 8, 189, 1976.
8. **Swift, M. J.,** The estimation of mycelial biomass by determination of the hexosamine content of wood tissue decayed by fungi, *Soil Biol. Biochem.,* 5, 321, 1973.
9. **Swift, M. J.,** Estimation of mycelial growth during decomposition of plant litter, in *Modern Methods in the Study of Microbial Ecology,* Rosswall, T., Ed., NFR Ecological Res. Comm. Bull. 17, Stockholm, 1973, 323.
10. **Paul, E. A. and Johnson, R. L.,** Microscopic counting and adenosine 5′-triphosphate measurement in determining microbial growth in soils, *Appl. Environ. Microbiol.,* 34, 263, 1977.
11. **Babiuk, L. A. and Paul, E. A.,** The use of fluorescein isothiocyanate in the determination of the bacterial biomass of grassland soil, *Can. J. Microbiol.,* 16, 57, 1970.
12. **Trolldenier, G.,** The use of fluorescence microscopy for counting soil microorganisms, in *Modern Methods in the Study of Microbial Ecology,* Rosswall, T., Ed., NFR Ecological Res. Comm. Bull. 17, Stockholm, 1973, 53.
13. **Bernstein, M. E., Howard, H. M., and Carroll, G. C.,** Fluorescence microscopy of Douglas fir foliage epiflora, *Can. J. Microbiol.* , 19, 1129, 1973.
14. **Mayfield, C. I.,** A simple fluorescence staining technique for *in situ* soil microorganisms, *Can. J. Microbiol.,* 21, 727, 1975.
15. **Anderson, J. R. and Slinger, J. M.,** Europium chelate and fluorescent brightener staining of soil propagules and their photomicrographic counting. I. Methods, *Soil Biol. Biochem.,* 7, 1975.
16. **Nikitin, D. I.,** Direct electron microscopic techniques for the observation of microorganisms in soil, in *Modern Methods in the Study of Microbial Ecology,* Rosswall, T., Ed., NFR Ecological Res. Comm. Bull. 17, Stockholm, 1973, 85.
17. **Bae, H. C. and Casida, L. E.,** Response of indigenous microorganisms to soil incubation as viewed by transmission electron microscopy of cell thin sections, *J. Bacteriol.,* 113, 1462, 1973.
18. **Millar, W. N. and Casida, L. E.,** Evidence for muramic acid in soil, *Can. J. Microbiol.,* 16, 299, 1970.
19. **Fazio, S. D., Mayberry, W. R., and White, D. C.,** Muramic acid assay in sediments, *Appl. Environ. Microbiol.,* 38, 349, 1979.
20. **Watson, S. W., Novitsky, T. J., Quinby, H. L., and Valois, F. W.,** Determination of bacterial number and biomass in the marine environment, *Appl. Environ. Microbiol.,* 33, 940, 1977.

21. **Jenkinson, D. S.**, Studies on the decomposition of plant material in soil. II. Partial sterilization of soil and the soil biomass, *J. Soil Sci.*, 17, 280, 1966.
22. **Jenkinson, D. S.**, The effects of biocidal treatments on metabolism in soil. IV. The decomposition of fumigated organisms in soil, *Soil Biol. Biochem.*, 8, 203, 1976.
23. **VanVeen, J. A. and Paul, E. A.**, Conversion of biovolume measurements of soil organisms, grown under various moisture tensions, to biomass and their nutrient content, *Appl. Environ. Microbiol.*, 37, 686, 1979.
24. **Luria, S. E.**, The bacterial protoplasm; composition and organization, in *The Bacteria*, Vol. 1, Gunsalus, I. C. and Stanier, R. Y., Eds., Academic Press, New York, 1960, chap. 1.
25. **Buchanan, R. E. and Gibbons, N. E.**, *Bergey's Manual of Determinative Bacteriology*, 8th ed., Williams and Wilkins Co., Baltimore, 1974.
26. **Aaronson, S.**, *Experimental Microbiol. Ecology*, Academic Press, New York, 1970.
27. **Johnson, L. F. and Curl, E. A.**, *Methods for Research on the Ecology of Soil-Borne Plant Pathogens*, Burgess, Minneapolis, 1972, chap. 16.
28. **Skerman, V. B. D.**, *A Guide to the Identification of the Genera of Bacteria*, 2nd ed., Williams and Wilkins Co., Baltimore, 213, 1967.
29. *The American Type Culture Collection: Catalogue of Strains I*, 12th ed., American Type Culture Collection, Rockville, Md., 1976, 354.
30. **Lochhead, A. G.**, Qualitative studies of soil microorganisms. III. Influence of plant growth on the character of the bacterial flora, *Can. J. Res.*, 18, 42, 1940.
31. **Allen, O. N.**, *Experiments in Soil Bacteriology*, 3rd ed., Burgess, Minneapolis, 1957.
32. **Bunt, J. S. and Rovira, A. D.**, Microbiological studies of some subantarctic soils, *J. Soil Sci.*, 6, 119, 1955.
33. **Martin, J. K.**, Comparison of agar media for counts of viable bacteria, *Soil Biol. Biochem.*, 7, 401, 1975.
34. **Rivière, J.**, Isolement et purification des bacteries cellulolytiques aerobies du sol, *Ann. Inst. Pasteur*, 101, 253, 1961.
35. **Molongoski, J. J. and Klug, M. J.**, Characterization of anaerobic heterotrophic bacteria isolated from freshwater lake sediments, *Appl. Environ. Microbiol.*, 31, 83, 1976.
36. **Pramer, D. and Schmidt, E. L.**, *Experimental Soil Microbiology*, Burgess, Minneapolis, 1964, chap. 9.
37. **Kuster, E. and Williams, S. T.**, Selection of media for isolation of streptomycetes, *Nature*, 202, 928, 1964.
38. **Pankhurst, G. S.**, The isolation and enumeration of sulfate-reducing bacteria, in *Isolation of Anaerobes*, Shapton, D. A. and Parker, C. A., Eds., Academic Press, New York, 1971.
39. **Focht, D. D. and Joseph, H.**, An improved method for the enumeration of denitrifying bacteria, *Soil Sci. Soc. Amer. Proc.*, 37, 698, 1973.
40. **Belser, L. W. and Schmidt, E. L.**, Diversity in the ammonia-oxidizing nitrifier population of a soil, *Appl. Environ. Microbiol.*, 36, 584, 1978.
41. **Soriano, S. and Walker, N.**, Isolation of ammonia oxidizing autotrophic bacteria, *J. Appl. Bacteriol.*, 31, 493, 1968.
42. **Matulewich, V. A., Strom, P. F., and Finstein, M. S.**, Length of incubation for enumerating nitrifying bacteria present in various environments, *Appl. Microbiol.*, 29, 265, 1975.
43. **Whittenbury, R., Phillips, K. C., and Wilkinson, J. F.**, Enrichment isolation and some properties of methane-utilizing bacteria, *J. Gen. Microbiol.*, 61, 205, 1970.
44. **Parker, C. D. and Frisk, J.**, The oxidation of inorganic compounds of sulphur by various bacteria, *J. Gen. Microbiol.*, 8, 344, 1953.
45. **Tiedje, J. M., Mason, B. B., Warren, C. B., and Malec, E. J.**, Metabolism of nitrilotriacetate by cells of *Pseudomonas* species, *Appl. Microbiol.*, 25, 811, 1973.
46. **Thauer, R. K., Jungermann, K., and Decker, K.**, Energy conservation in chemotrophic anaerobic bacteria, *Bacteriol. Rev.*, 41, 100, 1977.
47. **Ottow, J. C. G.**, Selection, characterization, and iron-reducing capacity of nitrate reductaseless (nit^-) mutants of iron-reducing bacteria, *Ztg. Allg. Microbiol.*, 10, 55, 1970.
48. **Arcuri, E. J. and Erlich, H. L.**, Cytochrome involvement in Mn(II) oxidation by two marine bacteria, *Appl. Environ. Microbiol.*, 37, 916, 1979.
49. **Sakaguchi, H., Torma, A. G., and Silver, M.**, Microbiological oxidation of synthetic chalcocite and covelite by *Thiobacillus ferrooxidans*, *Appl. Environ. Microbiol.*, 31, 7, 1976.
50. **Torma, A. E. and Gabra, G. G.**, Oxidation of stibnite by *Thiobacillus ferrooxidans*, Antonie van Leeuwenhoek, 43, 1, 1977.
51. **Quay, R., Silver, M., and Torma, A. E.**, Microbiological leaching of a low-grade uranium ore by *Thiobacillus ferrooxidans*, *Eur. J. Appl. Microbiol.*, 3, 157, 1976.

52. **Woolfolk, C. A. and Whitely, H. R.,** Reduction of inorganic compounds with molecular hydrogen by *Micrococcus lactilyticus, J. Bacteriol.,* 84, 647, 1962.
53. **Powlson, D. S.,** Effects of biocidal treatments on soil organisms, in *Soil Microbiology,* Walker, N., Ed., Halsted Press, New York, 1975, chap. 11.
54. **Wilhelm, E., Battino, R., and Wilcock, R. J.,** Low-pressure solubility of gases in liquid water, *Chem. Rev.,* 77, 219, 1977.
55. **Brown, A. D.,** Microbial water stress, *Bacteriol. Rev.,* 40, 803, 1976.
56. **Jencks, W. P. and Regenstein, J.,** Ionization constants of acids and bases, in *Handbook of Biochemistry,* 2nd ed., Sober, H. A., Ed., CRC Press, Boca Raton, Fla., 1970, J-187.
57. Keys to Chelation, Dow Chemical Co., Midland, Mich., 1969, 54.
58. Radiological Health Handbook, Division of Radiological Health, Ed., U.S. Department Health, Education and Welfare, Washington, D.C., 1960.
59. **Baldwin, I. and Fred, E. B.,** Nomenclature of the root-nodule bacteria of the leguminosae, *J. Bacteriol.,* 17, 141, 1929.
60. **Bohlool, B. and Schmidt, E.,** Lectins: a possible basis for specificity in the Rhizobium-legume root module symbiosis, *Science,* 185, 269, 1974.
61. **Dazzo, F. and Hubbell, D.,** Cross-reactive antigens and lectin as determinants of symbiotic specificity in the *Rhizobium*-clover association, *Appl. Microbiol.,* 30, 1017, 1975.
62. **Vincent, J. M.,** *A Manual for the Practical Study of Root-Nodule Bacteria,* I. B. P. Handbook No. 15, Blackwell Scientific Publications, Oxford, England, 1970, 164.
63. **Bergersen, F.,** The growth of *Rhizobium* in synthetic media, *Aust. J. Biol. Sci.,* 14, 349, 1961.
64. **Bishop, P., Guevara, R. J., Engelke, J., and Evans, H.,** Relation between glutamine synthetase and nitrogenase activities in the symbiotic association between *Rhizobium japonicum* and *Glycine max, Plant Physiol.,* 57, 542, 1976.
65. **Fahraeus, G.,** The infection of clover root hairs by nodule bacteria studied by a simple glass slide technique, *J. Gen. Microbiol.,* 16, 374, 1957.

SOIL ORGANIC MATTER

Bob G. Volk and Richard H. Loeppert

INTRODUCTION

It has long been recognized that organic materials, living and dead, greatly influence the chemical, physical, and biological nature of soil.[1] Some of the most important functions which have been attributed to soil organic matter, especially as related to agriculture, include:

1. Cation and anion exchange which affects the retention, release, and availability of plant nutrients
2. Binding of organic chemicals and pesticides which affects the biological activity and toxicity of these materials
3. Improvement of water percolation into and retention by the soil
4. Formation and maintenance of desirable soil structure
5. Translocation of tightly bound and normally insoluble metals with water-soluble components of soil organic matter
6. Release or immobilization of N, P, and S during decomposition of soil organic matter
7. Provision of C and energy for soil microorganisms
8. Absorption of solar radiation which influences soil temperatures

Addition of fresh organic material to soil provides a source of plant nutrients, increases the exchange capacity, provides energy for microbial activity, helps control erosion or runoff, stabilizes soil structure, and influences the soil, air, water, and temperature relationships.

The organic fraction of soil results from numerous complex formation processes unique to the soil environment. Plant and animal residues are the parent material from which organic matter is formed. Even though there is great variability in the nature of the organic material added to soil, there is a remarkable similarity in the relatively stable portion of organic matter of different soils. The first processes of organic matter formation involve the microbial decomposition of plant residues which serve as a source of nutrients and energy for the microorganisms. Easily decomposed organic compounds are utilized quickly and soon disappear if fresh plant or animal residues are not continually added. Organic compounds that are more resistant to decomposition processes persist for longer time periods. The complexity and interrelationships of some of the components of soil organic matter or humus are illustrated in Figure 1.

FACTORS INFLUENCING AMOUNTS OF ORGANIC MATTER IN SOIL

Climate and Vegetation

Climate determines plant species and the quantity of plant material produced on soils. Generally, total organic matter content of the soil is directly related to quantity of plant residue added annually. Soils of arid regions, as compared to soils of temperate regions, generally contain lower quantities of organic matter due to limited plant growth. Increasing rainfall with subsequent increases in plant production results in soils with a higher organic matter content.

Grassland soils contain about twice as much organic matter in the soil profile as forest soils (Figure 2) and generally have a more gradual decrease of organic matter

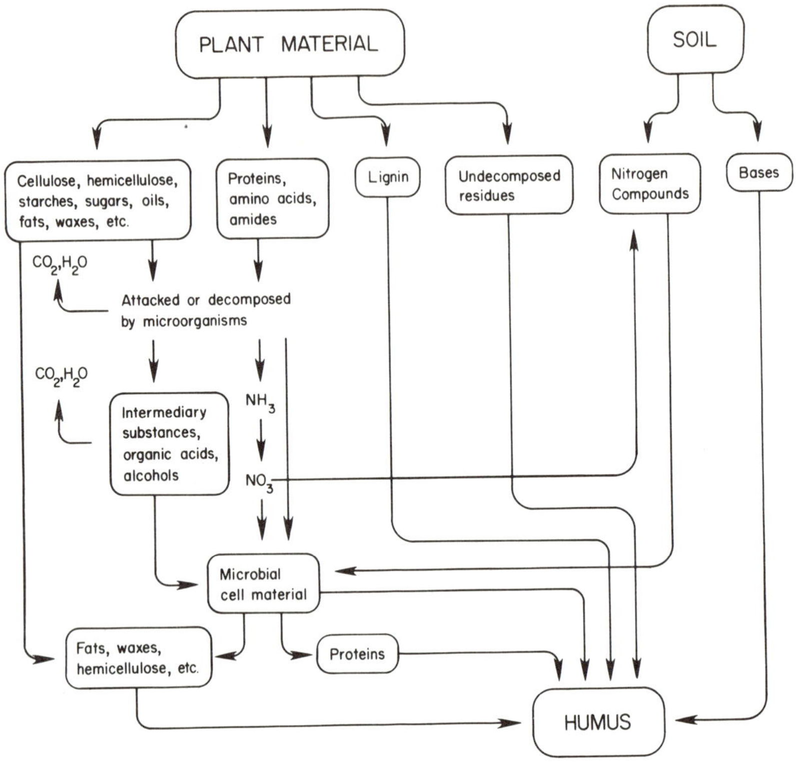

FIGURE 1. Schematic representation of the formation of humus or organic matter from plant material in the soil (adapted from Foth, H. D., 1978).[3]

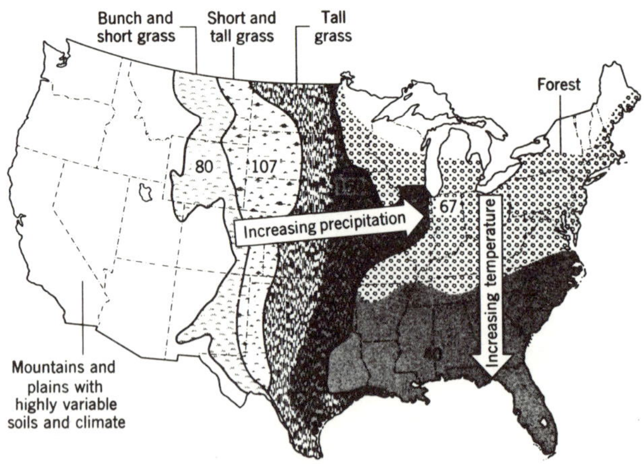

FIGURE 2. Organic matter content of soils (tons per acre to 40 in.) as related to climate and vegetation. (Reproduced from Foth, H. D., *Fundamentals of Soil Science*, 6th ed., Wiley & Sons, New York, 1978. With permission.)

with increasing soil depth (Table 1).[2] Grass roots are short-lived and the decomposition of dead roots each year contributes to the quantity of organic matter. The quantity of grass roots decreases more gradually with depth as compared to tree roots in forest ecosystems. By contrast, the large roots of trees are long-lived. The leaves and dead wood that fall onto the soil surface are primarily responsible for the annual addition

Table 1
DISTRIBUTION OF ORGANIC MATTER IN WELL-DRAINED SILT LOAM SOIL UNDER FOREST AND PRAIRIE ECOSYSTEMS

	Forest (White oak, black oak) (Metric tons)	Prairie (Big bluestem, Indian grass) (Metric tons)
Standing crop	81.7	2.7
Litter		
Feb.	2.7	
Apr.		4.5
Oct.	1.4	1.8
Roots[a]		
0—15 cm	32.6	45.4
15—30	10.0	33.6
30—46	7.3	21.8
46—61	6.4	12.7
61—76	5.4	9.1
76—91	4.5	6.4
91—107	3.6	4.5
Total (Roots)	69.8	133.5
(Total plant biomass)	151.5	136.2

[a] Excludes roots greater than 2.54 cm diameter, estimated to weigh an additional 14.5 t.

Adapted from Nielson, G. A. and Hole, F. D., *Wis. Acad. Sci.*, 52, 213, 1963.

of plant residues. In a forest ecosystem almost half of the organic matter exists in the above-ground parts while in the prairie ecosystem over 90% of the organic matter exists in the roots.[3]

As average annual temperatures increase from north to south, there is an observed decrease in organic matter content of soils. The temperature-induced increase in rate of microbial activity with resultant increase in organic matter decomposition is the major cause of these lower organic matter levels.

After 100 to 200 years of cultivation, the organic matter content of North Central and Great Plains soils has decreased to one half to one third of the original level.[4] Reductions due to cultivation are a result of lower return of plant residues, accelerated oxidation rates, and loss of surface soil by erosion.[5]

Drainage and Texture

Soils formed under restricted drainage do not follow a climatic pattern because the poor aeration retards microbial decomposition. For example, Figures 3 and 4 summarize drainage classes vs. percentage organic matter for selected Entisols and Ultisols. Even though the organic matter content of individual soils varies widely, the trend of increasing amounts of organic matter with more poorly drained soils is obvious. When climatic and drainage conditions are similar for a given group of soils, texture as determined by the sand, silt, and clay content influences the percent of organic matter present. The combined factors of greater supply of water, lower oxygen content, and higher nutrient levels in finer-textured soils favors the production and accumulation of organic matter. Also, organic molecules sorbed to clay surfaces are partially protected from the decomposing microorganisms. As shown in Table 2, finer textured soils generally have higher percentages of N and P, both major components of organic matter.

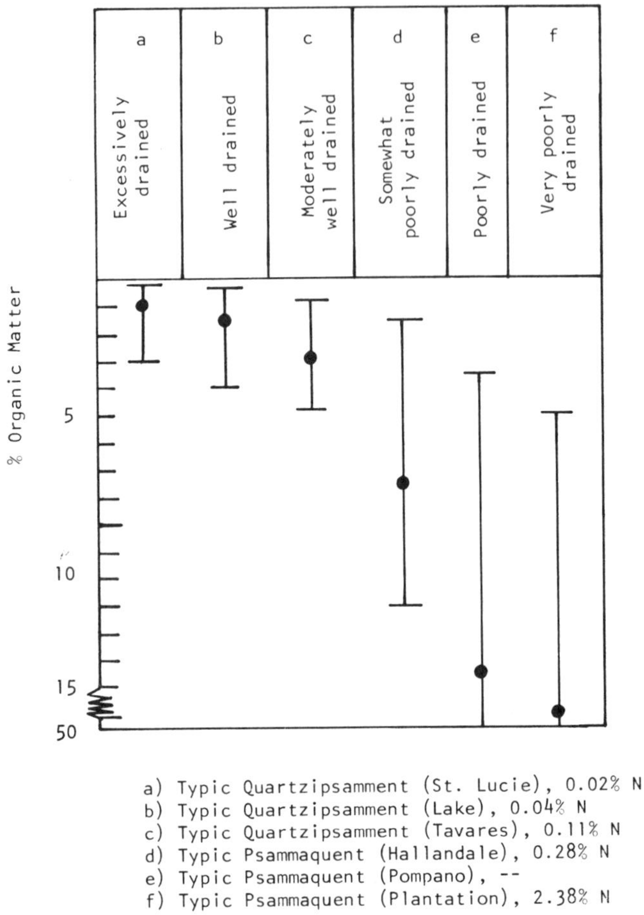

FIGURE 3. Drainage classes vs. percent organic matter for selected Entisols (ranges for similar soils are shown in brackets).

DECOMPOSITION OF PLANT RESIDUES

Organic Matter Level

The soil and climatic environment and the amount of organic matter added to soils determine the long-term equilibrium level of organic matter.[6] Studies[7] using ^{14}C show that smaller amounts of plant residues decomposed more rapidly than large quantities. However, Pinck and Allison[8] found that the decomposition rate was generally independent of residue added if the added C did not exceed 1.5% of the soil dry weight and if decomposition periods were at least 3 to 6 months.

Upon extended incubation, the proportion of plant material retained was very similar when added to a wide variety of soils under different climatic conditions.[9,10] Paul and McLaren[11] have stated that this similarity in quantity of retained organic material is due to mechanisms of production and stabilization of soil organic matter rather than to similarities in decomposition rates.

The loss of ^{14}C-labeled straw in a German sandy loam soil over 11 years is shown in Figure 5.[12] A comparison of the results in Figure 5 with those of investigators from England, Canada, and Austria reveals that a similarity exists in the decomposition behavior of plant materials.[13-16]

The overall straw decomposition can be most closely described by two first-order

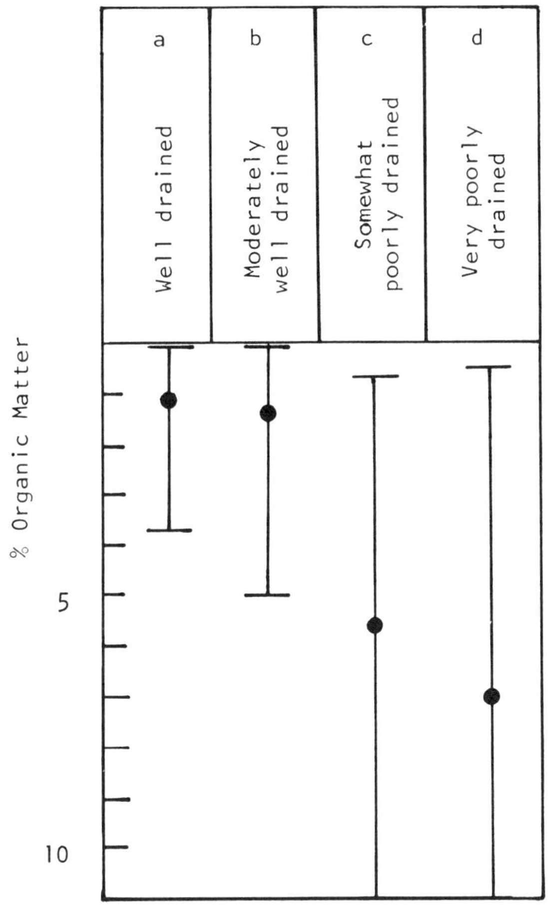

a) Plinthic Paleudault (Dothan)
b) Aquic Paleudault (Angie), 0.05% N
c) Aeric Paleaquuit (Lynchburg)
d) Typic Umbraquult (North Carolina-location), 0.7% N

FIGURE 4. Drainage classes vs. percent organic matter for selected Ultisols (ranges for similar soils are shown in brackets).

reactions — one representing the rapid consumption of original plant material, and the other a much slower breakdown of stabilized residues and decomposition products. The average halflife of humified plant residue carbon is 5 to 6 years.[12]

The curves in Figure 6 show a wide divergence in original decomposition rates. The slower decomposition rates are a result of drought during the growing season and extended frost. Sauerbeck and Gonzalez have found similar rates for many German and Costa Rican soils.[10]

Some equations describing plant-residue decomposition rates are shown in Table 3. Herman et al.[17] developed Equation 1 using C, N, lignin, and carbohydrate contents as variables affecting decomposition rates. Data of Pinck et al.[18] are described by Hunt's[19] Equation 2 in which carbon and nitrogen contents are the variables. Salter and Green's[20] Equation 3 is based on the assumption that the amount of N lost is proportional to the total amount of N in the soil. Jenny[6] proposed Equation 4 which allows for addition of organic materials. He assumed that the losses were proportional

Table 2
NITROGEN AND PHOSPHORUS CONTENTS OF SEVERAL SOIL CLASSES FROM IOWA AND NEW YORK AS RELATED TO SOIL TEXTURE

Soil texture	N content of soil (%)	P content of soil (%)
Sandy and loamy sand	0.02	0.03
Sandy loam	0.16	0.13
Loam	0.17	0.14
Silt loam	0.22	0.15
Silty clay loam	0.21	0.14
Clay	0.24	0.16

Adapted from Foth, H. D., *Fundamentals of Soil Science*, 6th ed., John Wiley & Sons, New York, 1978.

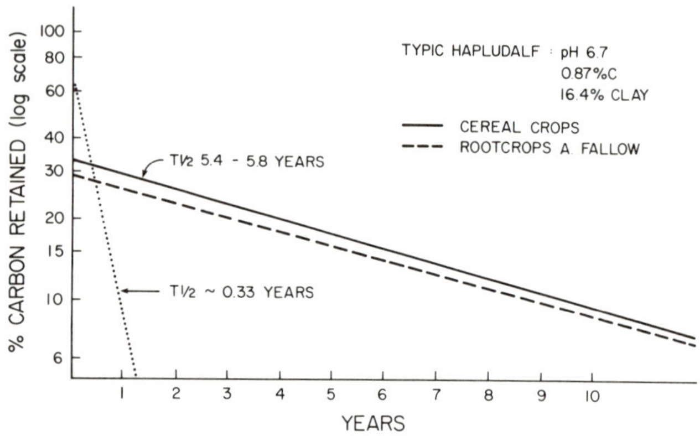

FIGURE 5. Loss of labeled straw carbon during 11 years of decomposition in the field, 1966—1977 (from Sauerbeck and Gonzalez, 1978[12]).

to the amount of organic matter present at any time and that the rate of addition of new N and annual rate of decomposition of N was a constant for all forms of organic matter.

Russell's[21] Equation 5 takes into account that decomposition coefficients may change with time. He also breaks the addition term into two components, including noncrop sources of organic material, e.g., manure and plant residues. The main advantage of Equation 6 is that effects of crop rotation on soil organic matter can be examined.

Woodruff[22] developed a modified form of Jenny's equation, taking into account that soil organic matter components decompose at different rates. Sanborn field plot data at the University of Missouri show average rates of decomposition for nonleguminous meadow crops, small grain, and cultivated crops to be 0.75%, 1%, and 2%, respectively, when calculated with the use of Equation 6.[23] Campbell provides additional examples of the use of Equation 6.

Lucas et al.[24] presented a model (Figure 7) pointing out the major factors affecting soil humus levels. Much of the input data and assumptions used to develop the model were based on research results of radioisotope carbon studies at the Rothamsted Re-

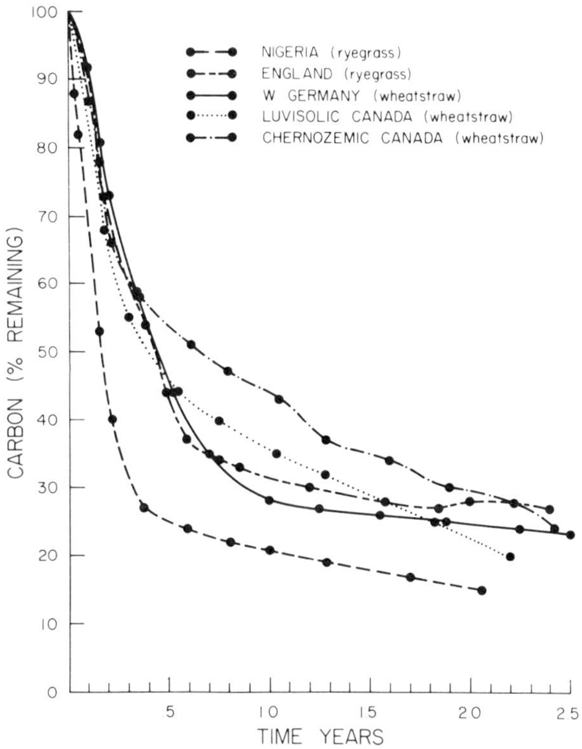

FIGURE 6. Residual carbon remaining after field incubation of wheatstraw and ryegrass residues in different environments (from Paul and Van Veen, 1978[225]).

search Center at Harpenden, England.[25,26] In using the model, Lucas and Vitosh[27] made the following assumptions:

1. All manure and plant residues upon decomposition leave a residue of soil organic C representing 30% of the original carbon content
2. Humus in recently cleared land is estimated to decompose at an annual rate of 2.5% for a loam or clay loam soil, 3.5% for a sandy loam, and 4.5% for a loamy sand
3. Soil weight is 2 million lb/acre
4. Universal Soil Erosion Equation factors are for a Michigan loam, with soil erodibility index of 0.32, "R" value of 100, slope and length factor of 0.3 (2% slope, 200 ft length)
5. Organic matter content in eroded sediments is 50% greater than that in field soil
6. Plant C from root turnover, exudates, and insects is 150% of the amount at harvest.

Use of Lucas and Vitosh's model[27] is illustrated in Figures 8 to 10. The horizontal axis designated as "% soil carbon" represents the carbon in the plow layer. The point for each line at which the humus C change is zero represents a steady-state level for that crop tillage practice or slope. When this point is reached, the soil C level will remain constant. If the soil C content is less than the steady-state value, the change in humus content is positive. The annual change becomes smaller as the steady-state level is approached from above or below. For example, the trend lines in Figure 8 illustrate how corn yields ranging from 50 to 150 bushels of grain per acre influence the steady-

Table 3
EQUATIONS DESCRIBING THE DECOMPOSITION OF PLANT RESIDUES

(1)[a] $$\frac{1}{CO_2 \text{ Evolved}} = \frac{1}{CH_2O} \text{ loss} = \text{lignin} - \text{C loss} = \frac{\left[\left(\frac{C}{N}\text{ straw}\right)\left(\%\text{ lignin}\right)\right]}{\sqrt{\%\text{ carbohydrates}}}$$

(2)[b] $S_o = 0.070 + 1.11\sqrt[3]{(N/C)}$
S_o = initial proportion of easily decomposable constituents

(3)[c] $N_t = N_oK^t$ N_t = amount of organic N per unit mass of soil remaining in soil after t years
N_o = amount of organic N per unit mass of soil initially
K = fraction of original organic N per unit mass of soil remaining after a year

(4)[d] $dN/dt = A - rN$ N = amount of organic N per unit mass of soil present at time (t)
A = rate of addition of new N per unit mass of soil
r = annual rate of decomposition of N per unit mass of soil

(5)[e] $dN/dt = K_1(t) \cdot N + K_2 + K_3(t) \cdot Y(t)$ t = time
$K_1(t)$ = decomposition coefficient
K_2 = additions to soil organic matter from noncrop sources, e.g., manure
Y(t) = crop yield
$K_3(t)$ = coefficient related to a specific crop at time (t)

(6)[f] $N = N_1(\exp - r_1t) + A_1/r_1(1 - \exp - r_1t) + N_2(\exp - r_2t) + A_2/r_2(1 - \exp - r_2t) + \ldots$
Subscripts refer to humus fractions with different decomposition rates

[a] Herman, W. A., McGill, W. B., and Dormaar, J. F.[17]
[b] Hunt, H.W.[19]
[c] Salter, R. M. and Green, T. C.[20]
[d] Jenny, H.[6]
[e] Russell, J. S.[21]
[f] Woodruff, C. M.[22]

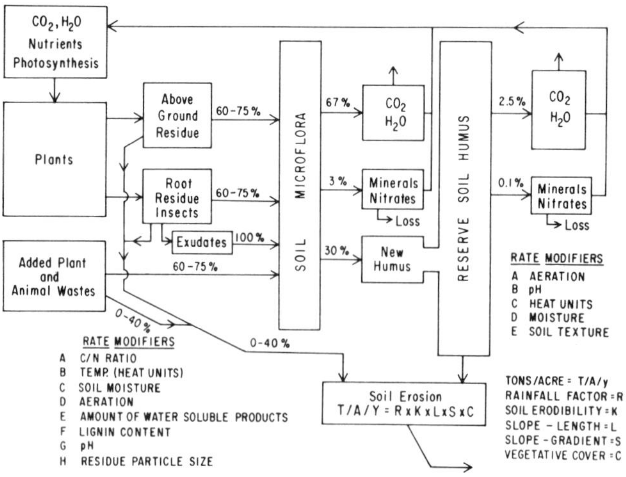

FIGURE 7. Soil humus model, yearly change. (Reproduced from Lucas, R. E., Holtman, J. B., and Connor, L. J., *Agriculture and Energy*, Academic Press, New York, 1977, 333. With permission.)

state point. Figure 10 illustrates how slope and tillage modify soil C levels. The soil C content at the steady-state point could range from 1.4 to 2.4% depending on soil erosion losses.

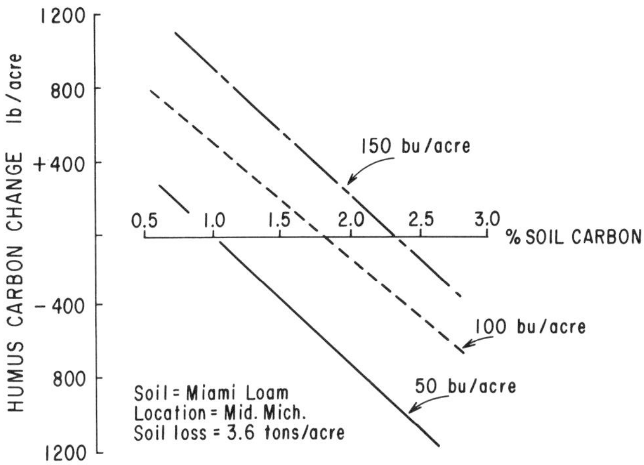

FIGURE 8. Yearly soil humus changes as related to yield of corn and the soil carbon change. (Reproduced from Lucas, R. E. and Vitosh, M. L., Soil Organic Matter Dynamics, Research Report 358, Mich. State Univ. Agric. Exp. Stn., E. Lansing, 1978. With permission.)

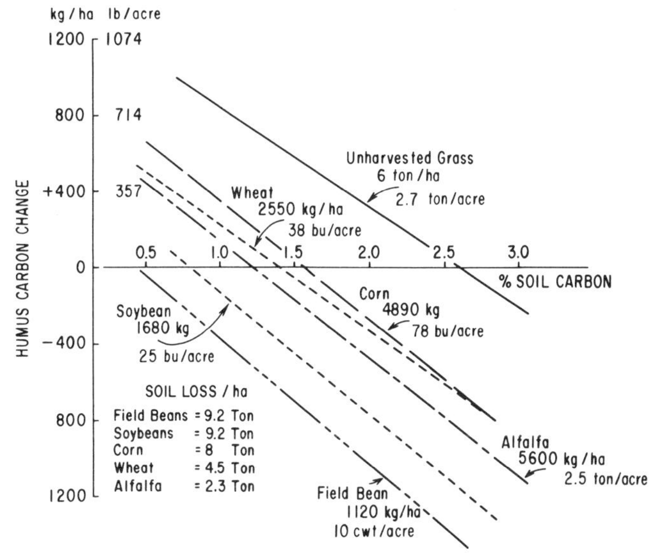

FIGURE 9. Yearly soil humus changes as related to the crop grown and the soil carbon content. (Reproduced from Lucas, R. E. and Vitosh, M. L., Soil Organic Matter Dynamics, Research Report 358, Mich. State Univ. Agric. Exp. Stn., E. Lansing, 1978. With permission.)

On a Typic Hapludoll in Iowa, 0 to 16 metric tons/ha of corn (*Zea mays* L.) and alfalfa (*Medicago sativa* L.) residue were added for 11 consecutive years and plowed under each fall. After 11 years of continuous corn, the organic C content was a linear function of total residue added (Figure 11). Larson et al.[28] estimated that the organic C produced by a 5000 kg/ha corn crop was needed to maintain the 1.8% initial C level in the soil.

The quantity of crop residue produced in the U.S. is shown in Table 4.[5] Larson[5] estimated the quantity of crop residue produced by multiplying total grain production in a given state for 1973[29] by a grain weight to residue weight ratio.

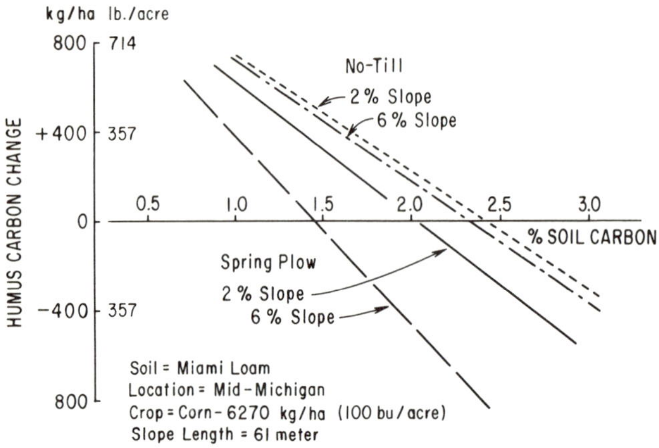

FIGURE 10. Yearly soil humus change as related to slope, tillage, and the soil carbon content. (Reproduced from Lucas, R. E. and Vitosh, M. L., Soil Organic Matter Dynamics, Research Report 358, Mich. State Univ. Agric. Exp. Stn., E. Lansing, 1978. With permission.)

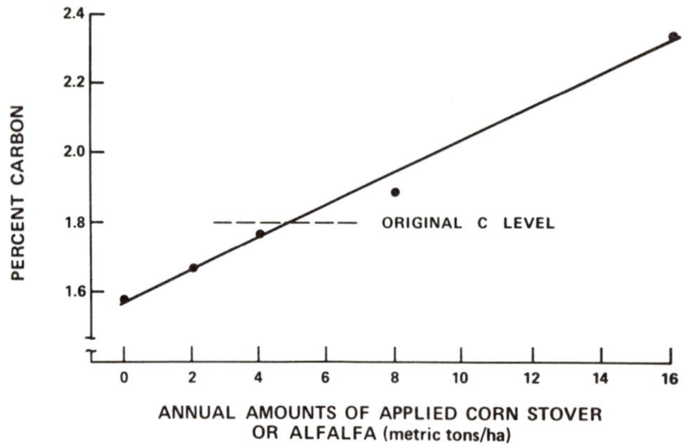

FIGURE 11. Carbon content of a Typic Hapludoll as influenced by amounts of corn stover or alfalfa hay added to the soil for each of 11 consecutive years. (Reproduced from Larson, W. E., Holt, R. F., and Carlson, C. W., *Crop Management Systems,* Oschwald, W. R., Ed., American Society of Agronomy, Special Publication No. 31, Madison, Wis., 1978, chap. 1. With permission.)

Nutrients in Residues and Roots

Quantities of N, P, and K in total crop residues for the 20 leading states in U.S. crop production and the percentages of the respective amounts of commercial fertilizers purchased for all crops are presented in Table 5.[5] Table 6 shows N, P, S, K, Ca, and Mg concentrations in various root materials.[30,31] In Table 7 an estimate is given of the average composition values for N, P, K, Ca, Mg, S, Cu, Mn, and Zn for a variety of plant grains and strains.[32]

Larson et al.,[5] in their discussion of Table 5, point out that any management system that removes residues must take into account the return of the nutrients associated with the residue if crop production levels are to be maintained. Even though P removal is appreciably lower than N removal (due to lower P levels in residues), it is important to keep P in the soil-crop cycle by returning as much residue as possible because P is not recycled after leaving the biosphere. Plant K is very soluble and returns to the soil if the residues are left on the soil surface.

THE EFFECTS OF ORGANIC MATTER ON SOILS

Effects of Carbon on Crop Yields

Some soils have the inherent capacity to produce high yields. In any growing season, yields are influenced by uncontrollable variables, such as weather, and controllable variables, such as nutrients and cultural practices. Yield potential is the maximum yield expected under typical weather conditions.

Soil C plays an important role in yield potential under unirrigated conditions. Figure 8 shows that yield potential increases 21% (relative to yield potential at 1.16% C) for each 1% increase in soil C.[25] The relationship is based upon gross averages and may not accurately describe a particular situation.

Available Moisture Supply

Jamison[33] has stated that except for sandy soils, organic matter does not increase the capacity of a soil to store available water. Much of the water stored in organic matter is held at soil water tensions above the wilting point. In a study of 44 mineral soil profiles including 212 horizons of Florida soils, Calhoun et al.[34] related the contribution of organic matter, clay, and silt + clay to water retention properties (Table 8). The influence of organic matter, as indicated by increasing correlation coefficients (r), became greater with increasing soil water content. Correlation coefficients indicated that as suction decreased, water content was increasingly dependent on soil structure as related to size, shape, and continuity of pores. Multiple regression equations (Table 9) gave closer relationships between water retention and organic matter, clay, and silt + clay than simple linear regression. Analysis of various Florida soils not included in the study showed that these equations provided a reasonable estimate of water contents.[34]

Exchange Capacity

The ability of soil organic matter to adsorb both cations and anions from the soil solution is one of the most important benefits derived from its presence in soil. The release of these ions to the plant root occurs by ion replacement. Cation exchange capacity (CEC) is of great importance, involving most of the cations in the soil solution. However, very few anions are adsorbed by soil organic matter with sufficient strength to make this a major factor in soil productivity.

Charges on soil organic matter arise from the ionization of various functional groups including carboxyl, phenolic hydroxyl, enolic hydroxyl, and possibly other groups.[35,36] The dissociation of these groups is pH-dependent; therefore, pH is an important factor influencing the CEC of soil organic matter (Figure 12).[37] Cation exchange capacities for humus are in the range of 62 to 300 meq/100 g.[38,39] Highly decomposed soil organic matter has a higher exchange capacity than recent additions of organic matter to soils.

The relative contributions of organic matter to CEC of several soil orders are shown in Table 10.[40] Percentage of CEC contributed by soil organic matter usually decreases with depth in the soil profile since the organic matter to clay ratio decreases (Table 11).[41]

Reactions with Metals

Humic materials can react with metals to form organic-metal complexes of widely differing chemical stabilities and characteristics.[42] The stability and abundance of these humic-metal complexes is important when considering metal movement and immobilization in terrestrial or aquatic ecosystems, as well as plant nutrient availability and soil formation.

Table 4
RESIDUE PRODUCTION (METRIC TONS) BY NINE CROPS IN THE 20 LEADING RESIDUE-PRODUCING STATES

State	Corn—1.0[a] (*Zea mays* L.)	Soybeans—1.5 (*Glycine max* [L.] Merr.)	Winter wheat—1.7[b] (*Triticum aestivum* L.)	Total wheat	Cotton—1.0 (*Gossypium hirsutum* L.)	Sorghum—1.0 (*Sorghum bicolor* [L.] Moench.)
Ark.	18,709	4,755,788	281,708	281,708	227,079	170,676
Calif.	628,102	—	1,427,084	1,430,630	381,096	511,391
Colo.	1,137,228	—	2,726,203	2,743,718	—	253,914
Ill.	24,986,373	11,507,778	1,808,196	1,808,196	—	93,165
Ind.	13,605,188	5,528,373	1,140,786	1,140,786	—	45,183
Iowa	30,723,167	10,779,785	50,444	50,444	—	40,881
Kan.	3,920,070	1,080,024	17,840,867	17,840,867	—	5,559,372
Mich.	3,398,497	680,415	921,716	921,716	—	—
Minn.	13,067,579	5,208,252	54,895	2,783,795	—	—
Mo.	5,824,104	5,191,479	1,182,282	1,182,282	39,349	784,014
Mont.	20,440	—	2,555,584	4,030,339	—	—
Neb.	14,117,343	1,485,033	4,348,943	4,348,943	—	3,461,880
N.C.	2,922,234	1,423,668	292,093	292,093	35,737	109,253
N.D.	256,586	209,582	108,306	8,590,915	—	—
Ohio	6,113,273	3,671,673	1,068,227	1,068,227	—	—
Okla.	199,313	188,186	7,316,239	7,316,239	93,134	779,534
S.D.	3,725,085	388,809	988,110	2,363,693	—	286,776
Tex.	1,547,664	347,735	4,571,490	4,571,490	1,023,007	10,614,735
Wash.	143,006	—	3,440,209	3,972,034	—	—
Wis.	4,415,679	246,483	25,964	37,664	—	—
Total for 20 states	130,769,640	52,698,063	52,149,346	62,775,779	1,799,402	22,710,774

[a] Dry weight ratio of residue to grain.
[b] Spring wheat ratio — 1.3.

From Larson, W. E., Holt, R. F., and Carlson, C. W., *Crop Management Systems,* Oschwald, W. R., Ed., American Society of Agronomy, Special Publication No. 31, Madison, Wis., 1978, chap. 1. With permission.

Humic and fulvic acids contain relatively large numbers of functional groups (COOH, C=O, phenolic OH) capable of combining with metals. As pH is increased, a proton is displaced from an acidic OH group of the humic or fulvic acid, thus allowing complexation of the metal to the organic molecule.[43] At a higher pH, a proton dissociates from water bonded covalently to the metal ion, thus forming a hydroxy complex. Metal ions and hydrous oxides have been shown to react with one acidic COOH and one phenolic OH or two acidic COOH groups.[43-46] These structures are responsible for the stability of metal-humic complexes.

Table 12 shows metal-organic stability constants as determined by many workers. Results vary widely because of variation in procedures and uncertainty exists concerning the chemical structure of humic materials. Methods using continuous variation and ion exchange[47] and titration or modified potentiometric titration[48] are commonly employed.

CHEMICAL METHODS OF STUDYING ORGANIC MATTER

Extraction and Fractionation

Procedures for extraction of soil organic matter were reviewed by Mortenson[59] and Stevenson.[60] Dilute aqueous sodium hydroxide, usually 0.5 *N,* is the most commonly

Table 4 (continued)
RESIDUE PRODUCTION (METRIC TONS) BY NINE CROPS IN THE 20 LEADING RESIDUE-PRODUCING STATES

State	Oats—2.0 (*Avena sativa* L.)	Barley—1.5 (*Hordeum vulgare* L.)	Rice—1.5 (*Oryza sativa* L.)	Rye—1.5 (*Secale cereale* L.)	Total
Ark.	107,200	—	1,733,357	—	7,294,517
Calif.	134,400	1,569,076	1,535,437	—	6,190,132
Colo.	56,204	403,495	—	10,881	4,605,440
Ill.	575,420	16,038	—	17,640	39,004,610
Ind.	364,452	15,121	—	11,416	20,710,519
Iowa	1,985,112	—	—	3,971	43,583,360
Kan.	116,364	107,354	—	32,073	28,656,124
Mich.	480,002	29,359	—	31,500	5,541,489
Minn.	4,154,195	1,316,728	—	120,273	26,650,822
Mo.	40,553	14,041	15,408	9,622	13,100,852
Mont.	298,474	1,963,800	—	—	6,313,053
Neb.	612,947	35,348	—	59,564	24,121,058
N.C.	109,091	93,346	—	14,509	4,999,931
N.D.	2,146,916	3,354,498	—	116,035	14,674,532
Ohio	754,039	15,318	—	7,942	11,630,472
Okla.	231,390	253,821	—	50,897	9,112,514
S.D.	2,925,973	722,842	—	303,776	10,716,954
Tex.	775,275	114,882	1,399,694	24,742	20,419,224
Wash.	69,818	465,912	—	4,582	4,655,352
Wis.	1,634,041	25,431	—	7,216	6,366,514
Total for 20 states	17,571,866	10,516,410	4,683,896	826,639	308,347,469

used extractant. The alkali-soluble material is usually separated into the humic acid and fulvic acid fractions. These fractions have considerable historical and chemical significance due to their wide occurrence in the soil science literature and their high concentration of oxygen-containing functional groups. Sodium hydroxide produces high yields of extractable soil organic matter; however, its use has been criticized because of chemical alterations which may occur in alkaline conditions.[61-63] Bremner[64] observed that O_2 was adsorbed from the atmosphere by alkaline soil suspensions. However, Schnitzer[42] has found no evidence that dilute alkali under N_2 modifies the chemical structure or properties of humic substances.

Another commonly used procedure[60,65] is the proximate analyses scheme (Figure 13), which involves sequential extraction with a series of solvents designed to isolate biochemically defined fractions. An application of this technique is summarized in Table 13.[66] Other extraction schemes have been used in which the soil is extracted sequentially with a series of solvents from least polar to most polar.[67] Numerous extracting agents, e.g., neutral salts, organic solvents, and dilute acids, have been used as general extractants of soil organic matter or as extractants for specific fractions. These methods have been extensively reviewed by Mortenson[59] and Stevenson.[60] Of special interest are the dipolar aprotic solvents, dimethylformamide (DMF) and dimethylsulfoxide (DMSO), which provide excellent yields in extraction of H-saturated organic matter. Functional group, elemental, and IR analyses have indicated that DMF-extractable organic matter is chemically similar to material extracted by 0.5 N NaOH,[67,68] and therefore may serve as an excellent complementary solvent to NaOH for chemical studies of extractable soil organic matter. Other organic solvents may be of special interest since they extract organic components with a narrower range of properties. In general, extraction yields obtained with NaOH[61,63] and with organic solvents[69] are increased by pretreatment

Table 5
METRIC TONS OF N, P, AND K IN CROP RESIDUE AND PERCENTAGES OF COMMERCIAL FERTILIZER USED BY STATES THAT THESE AMOUNTS REPRESENT

	N		P		K	
State	Metric tons	% of comm. fert. N	Metric tons	% of comm. fert. P	Metric tons	% of comm. fert. K
Ark.	125,992	91	13,182	29	80,338	112
Calif.	50,780	10	7,090	10	74,489	139
Colo.	37,182	35	4,896	28	51,136	672
Ill.	553,228	84	72,479	38	481,738	106
Ind.	286,003	87	38,132	33	256,939	84
Iowa	596,878	88	82,295	52	555,617	160
Kan.	249,087	49	30,602	43	313,403	671
Mich.	62,609	44	9,097	15	69,792	52
Minn.	317,534	85	45,168	43	341,330	140
Mo.	199,033	62	24,101	32	155,447	95
Mont.	43,839	105	5,495	23	68,838	2,163
Neb.	261,067	49	38,007	65	266,862	618
N.C.	69,692	35	9,134	15	61,676	40
N.D.	104,388	87	14,201	33	167,114	1,258
Ohio	162,532	62	21,072	20	142,909	61
Okla.	69,130	36	7,978	19	94,583	304
S.D.	94,404	100	15,487	63	139,780	1,153
Tex.	202,442	28	27,580	23	254,164	253
Wash.	32,157	17	3,667	13	47,438	235
Wis.	65,333	51	11,167	20	89,009	44
Total for 20 states	3,583,310		480,830		3,712,602	

From Larson, W. E., Holt, R. F., and Carlson, C. W., *Crop Management Systems,* Oschwald, W. R., Ed., American Society of Agronomy, Special Publication No. 31, Madison, Wis., 1978, chap. 1. With permission.

Table 6
CONCENTRATIONS (%) OF SOME ELEMENTS IN SELECTED ROOT MATERIAL[30,31]

Plant	N	P	K	S	Ca	Mg
White clover	3.77	0.22	—	0.39	—	—
Red clover	2.79	0.26	—	0.47	—	—
Perennial grass	1.75	0.13	—	0.21	—	—
Prairie grassland						
Root	0.90	0.09	—	—	—	—
Rhizomes	—	0.15	—	—	—	—
Shoot bases	—	0.12	—	—	—	—
Wheat-maturity	1.01	0.097	—	—	—	—
Corn root						
18 days	3.06	0.79	4.82	—	0.30	0.30
32 days	3.01	0.81	4.31	—	0.30	0.26
46 days	2.61	0.41	2.90	—	0.36	0.41
60 days	2.89	0.61	3.40	—	0.33	0.36
74 days	2.52	0.59	3:69	—	0.35	0.45

with dilute HCl or HCl-HF to remove exchangeable metallic cations and soil clays, respectively.

Table 7
AVERAGE CONCENTRATIONS (%) OF NUTRIENTS CONTAINED IN THE ABOVE GROUND PORTIONS OF CROPS

Crop	Annual yield (MT/ha)	N	P	K	Ca	Mg	S	Cu	Mn	Zn
Barley										
Grain	2.15	1.822	0.341	0.433	0.052	0.104	0.156	0.0016	0.0016	0.0031
Straw	2.24	0.75	0.109	1.245	0.40	0.10	0.20	0.0005	0.016	0.0025
Corn										
Grain	9.41	1.607	0.275	0.395	0.024	0.095	0.119	0.0007	0.0011	0.0018
Stover	10.08	1.111	0.179	1.337	0.289	0.222	0.156	0.0005	0.0166	0.0033
Oats										
Grain	2.87	1.953	0.341	0.486	0.078	0.117	0.195	0.0012	0.0047	0.002
Straw	4.48	0.625	0.164	1.660	0.20	0.20	0.225	0.0008	—	0.0072
Rice										
Grain	4.03	1.388	0.242	0.231	0.083	0.111	0.083	0.0003	0.0022	0.0019
Straw	5.60	0.60	0.087	1.162	0.18	0.10	—	—	0.0316	—
Rye										
Grain	1.88	2.083	0.259	0.494	0.119	0.179	0.417	0.0012	0.0131	0.0018
Straw	3.36	0.50	0.116	0.692	0.267	0.067	0.10	0.003	0.0047	0.0023
Sorghum										
Grain	3.36	1.666	0.363	0.415	0.133	0.167	0.167	0.0003	0.0013	0.0013
Stover	6.72	1.083	0.145	1.314	0.483	0.30	—	—	—	—
Wheat										
Grain	2.68	2.083	0.615	0.519	0.042	0.25	0.125	0.0013	0.0038	0.0058
Straw	3.36	0.667	0.073	0.968	0.20	0.10	0.167	0.0003	0.0053	0.0017
Hay										
Alfalfa	8.96	2.25	0.218	1.868	1.40	0.263	0.238	0.0008	0.0055	0.0053
Bluegrass	4.48	1.5	0.218	1.245	0.40	0.175	0.125	0.0005	0.0075	0.002
Coastal Bermuda	17.92	1.875	0.191	1.401	0.369	0.15	0.219	0.0013	—	—
Cowpea	4.48	3.0	0.273	1.66	1.375	0.375	0.325	—	0.0163	—
Peanut	5.04	2.333	0.242	1.752	1.0	0.378	0.356	—	0.0051	—
Red Clover	5.60	2.0	0.218	1.66	1.38	0.34	0.14	0.0008	0.0108	0.0072
Soybean	4.48	2.25	0.218	1.038	1.0	0.45	0.25	0.001	0.0115	0.0038
Timothy	5.60	1.2	0.218	1.577	0.36	0.12	0.10	0.0006	0.0062	0.004
Fruits & vegetables										
Apples	26.88	0.125	0.018	0.156	0.033	0.021	0.042	0.0001	0.0001	0.0001
Beans, dry	1.01	3.125	0.454	0.864	0.083	0.083	0.208	0.0008	0.0013	0.0025
Cabbage	44.80	0.325	0.038	0.270	0.05	0.02	0.11	0.0001	0.0003	0.0002
Onions	16.80	0.30	0.058	0.222	0.073	0.013	0.12	0.0002	0.005	0.0021
Oranges	62.72	0.002	0.024	0.208	0.059	0.021	0.016	0.0004	0.0001	0.004
Peaches	32.26	0.122	0.030	0.188	0.014	0.028	0.007	—	—	0.001
Potatoes	26.88	0.333	0.055	0.519	0.013	0.025	0.025	0.0002	0.0004	0.0002
Spinach	11.20	0.50	0.065	0.249	0.12	0.05	0.04	0.0002	0.001	0.001
Sweet potatoes	10.80	0.30	0.044	0.415	0.027	0.06	0.04	0.0002	0.0004	0.0002
Tomatoes	44.80	0.30	0.044	0.332	0.018	0.028	0.035	0.0002	0.0003	0.0001
Turnips	22.40	0.225	0.044	0.374	0.06	0.03	—	—	—	—
Other crops										
Cotton seed and lint	1.68	2.666	0.581	0.830	0.133	0.267	0.20	0.004	0.0073	0.0213
Stalks, burs and leaves	2.24	1.75	0.218	1.453	1.4	0.40	0.75	—	—	—

Table 7 (continued)
AVERAGE CONCENTRATIONS (%) OF NUTRIENTS CONTAINED IN THE ABOVE GROUND PORTIONS OF CROPS

Crop	Annual yield (MT/ha)	N	P	K	Ca	Mg	S	Cu	Mn	Zn
Peanuts (nuts)	2.80	3.6	0.174	0.498	0.04	0.12	0.24	0.0008	0.0004	—
Soybeans (grain)	2.69	6.25	0.636	1.902	0.292	0.292	0.167	0.0017	0.0021	0.0017
Sugar beets (roots)	33.60	0.20	0.029	0.139	0.11	0.08	0.033	0.0001	0.0025	—
Sugar cane	67.20	0.16	0.039	0.374	0.047	0.04	0.04	—	—	—
Tobacco (leaves)	2.25	3.75	0.327	4.981	3.75	0.90	0.70	0.0015	0.0275	0.0035

From *Our Land and Its Care*, 5th ed., National Plant Food Institute, Washington, D.C., 1967.

Table 8
SIMPLE CORRELATION COEFFICIENTS RELATING WATER CONTENTS BY WEIGHT AT VARIOUS SUCTIONS TO ORGANIC MATTER (OM), CLAY (cl), AND SILT + CLAY (si + cl)

Soil components	Simple correlation coefficients (r)[a]			
	15-bar	ME[b]	0.33-bar	0.06-bar
OM	0.16	0.34	0.51	0.60
$(OM)^2$	0.16	0.41	0.45	0.51
(OM) (si + cl)	0.25	0.49	0.55	0.60
(OM) (cl)	0.31	0.50	0.61	0.58
si + cl	0.88	0.84	0.67	0.49
$(si + cl)^2$	0.86	0.78	0.58	0.39
cl	0.89	0.83	0.64	0.43
$(cl)^2$	0.86	0.75	0.55	0.35

[a] All values of r are highly significant except those relating OM and $(OM)^2$ with 15-bar water.
[b] ME ≅ 0.5-bar suction.

Calhoun, F. G., Hammond, L. C., and Caldwell, R. E., Influence of particle size and organic matter on water retention in selected Florida soils, *Soil Crop Sci. Soc. Fla.*, 32, 111, 1972. With permission.

Properties of soil organic matter which are useful as a basis for further fractionation include: (1) molecular size, (2) solubility, (3) charge, (4) density, and (5) absorption.

Gel filtration and gel permeation chromatography have been used to fractionate extractable soil organic matter according to molecular size. The Sephadex G gels have been widely utilized.[70-73] With the development of high pressure liquid chromatography there has been interest in the new high-efficiency rigid gels, e.g., porous glass or silica, and semi-rigid gels, such as polystyrene divinylbenzene. The Sephadex G gels or rigid gels are compatible with aqueous solvents; however, with both systems electro-

Table 9

MULTIPLE REGRESSION EQUATIONS AND COEFFICIENTS OF CORRELATION (r) FOR WATER CONTENTS BY WEIGHT AT VARIOUS SUCTIONS AS A FUNCTION OF [1] OM AND CLAY (cl) CONTENT; AND [2] OM AND SILT PLUS CLAY (si + cl) CONTENT

Multiple regression equations	r^a	r^2
[1] 15-bar $H_2O = 0.42 + 0.75(OM) - 0.06(OM)(cl) + 0.0016(cl)^2 + 0.30(cl)$	0.93	0.88
[2] 15-bar $H_2O = 0.14 + 0.86(OM) - 0.04(OM)(si + cl) + 0.17(si + cl) + 0.002(si + cl)^2$	0.91	0.83
[1] Moist. Equiv. $= 0.98 + 0.45(OM) + 0.10(OM)^2 + 0.56(cl) - 0.0017(cl)^2$	0.94	0.90
[2] Moist. Equiv. $= 0.52 + 0.12(OM)^2 + 0.38(si + cl)$	0.93	0.86
[1] 0.33-bar $H_2O = 1.35 + 2.03(OM) + 0.63(cl) - 0.0047(cl)^2$	0.86	0.76
[2] 0.33-bar $H_2O = 0.83 + 2.37(OM) - 0.044(OM)(si + cl) + 0.37(si + cl)$	0.85	0.72
[1] 0.06-bar $H_2O = 7.06 + 2.65(OM) + 0.58(cl) - 0.0056(cl)^2$	0.78	0.61
[2] 0.06-bar $H_2O = 5.82 + 3.03(OM) - 0.05(OM)(si + cl) + 0.48(si + cl) - 0.0034(si + cl)^2$	0.79	0.62

[a] All significant at 0.01 probability level.

From Calhoun, F. G., Hammond, L. C., and Caldwell, R. E., *Soil Crop Sci. Soc. Fla.*, 32, 111, 1972. With permission.

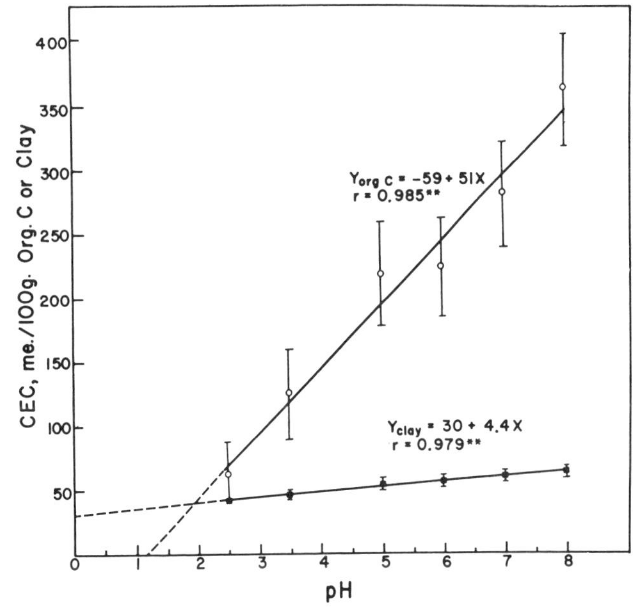

FIGURE 12. Effect of pH on the CEC of organic carbon and clay. (Vertical lines show standard errors of the individual values). (Reproduced from Helling, C. S., Chesters, G., and Corey, R. B., *Soil Sci. Soc. Am. Proc.*, 28, 517, 1964. With permission.)

lyte content must be carefully controlled to minimize absorption and electrostatic exclusion interactions between the organic solute and gel matrix.[73,74] The polystyrene-divinylbenzene gels are not compatible with water; nevertheless, they have demonstrated promise for fractionation of soil organic components that are soluble in solvents such as tetrahydrofuran (THF) and chloroform.

Molecular size fractionations of soil organic materials are affected by solute-solute,

Table 10
CONTRIBUTION OF ORGANIC MATTER AND CLAY TO SOIL CATION EXCHANGE CAPACITY

Soil group	Average CEC (meq/100 g)	Relative contribution(%) O.M.	Relative contribution(%) Clay
Entisols			
Psamments	5.2	74.9	25.1
Aquipsamments	3.8	86.8	15.2
Quartzipsamments	5.6	75.7	24.3
Acid family	3.8	78.7	21.3
Nonacid family	4.2	95.4	4.6
Phosphatic family	10.5	77.4	22.6
Inceptisols			
Aquepts and umbrepts	8.1	69.2	30.8
Mollisols			
Aqualls	12.9	66.4	33.6
Spodosols			
Aquods	5.5	96.5	3.5
All soils	6.7	76.1	23.9

From Yuan, T. L., Gammon, N., Jr., and Leighty, R. G., *Soil Sci.*, 104, 123, 1967. With permission.

Table 11
DISTRIBUTION OF CATION-EXCHANGE CAPACITY WITH DEPTH IN THE ORGANIC FRACTION OF TWO SOIL PROFILES

Horizon	Depth (in.)	Organic matter (%)	Cation-exchange capacity (meq/100 g)	Exchange capacity due to organic matter (meq/100 g)	Total capacity due to organic matter (%)
			Honeoye silt loam		
A_1	0—2.5	7.30	27.1	13.5	50
A_{22}	5—12	3.88	13.8	7.0	51
B_{21}	12—16	1.48	15.7	4.5	29
B_{22}	16—20	1.88	17.3	6.0	35
B_3	20—25	1.61	14.3	3.1	22
C_2	29+	1.39	9.6	2.2	23
			Yates silt loam		
A_1	0—2.5	4.22	12.5	6.2	50
A_{21}	2.5—7	1.46	7.3	2.3	31
A_{22}	7—14	0.47	5.8	0.7	12
B_2	14—20	1.25	8.1	3.4	42
B_3	20—30	1.0	9.4	1.9	21
C_2	30+	0.85	5.1	1.3	26

From Broadbent, F. E., *Soil Sci.*, 79, 107, 1955. With permission.

gel-solute, gel-solvent, and solute-solvent interactions. These interactions are influenced by cations on the exchange complex of the organic solute and properties of the eluting solvent and the gel. Factors affecting fractionation of soil organic matter are discussed by Swift and Posner[73] and Loeppert and Volk[74] for the Sephadex gels and rigid gels, respectively.

Table 12
ASSOCIATION CONSTANTS OF VARIOUS ORGANIC MATERIALS

Complex	Log K (pH)		Ref.
Cu-peat	6.5		49
	μ^a = 0.1	μ = 0.01	
Cu-peat humic acid			
Conventional titration	7.19 (4&5)	—	48
Modified procedure	7.26 (4&5)	8.25 (4&5)	48
Cu-Leonardite humic acid			
Conventional titration	7.52 (4&5)	8.28 (4&5)	48
Modified procedure	7.38 (4&5)	8.01 (4&5)	48
Cu-soil humic acid			
Conventional titration	—	8.54 (4&5)	48
Modified procedure	7.22 (4&5)	7.87 (4&5)	48
Cu-humic acid	7.00(5.0)		50
Cu-fulvic acid	3.23(3.5)		50
Cu-fulvic acid	5.78(3.5), 8.69(5.0)		51
Cu-nondialyzable humus in soil solution	5.5		52
Zn-organic matter	3.4(4.5), 5.6(7.0)		53
Zn-manure (aqueous extract)	7.8(7.0)		54
Zn-humic acid	4.4(3.6), 6.2(5.6), 6.8(7.0)		55
Zn-humic acid	2.87(5.0)		50
Zn-humic acid (4 soils)	1.88—2.64(3.5), 2.60—3.62(5.0)		56
Zn-nondialyzable humus in soil solution	4.3		52
Zn-fulvic acid	2.83(3.5)		50
Zn-fulvic acid (4 soils)	2.02—2.74(3.5), 2.66—3.66(5.0)		56
Zn-fulvic acid	1.73(3.5), 2.34(5.0)		51
Zn-fulvic acid (32 soils)	3.9—9.3, average 6.2(7.0)		57
Zn-humic acid (29 soils)	4.2—10.8, average 7.3(7.0)		57
Zn-humic acid (μ = 0.1)			
Leonardite	3.5(6.5)		58
Peat	3.3(6.5)		58
Soil	3.7(6.5)		58
Zn-humic acid	4.5(5.0)		58
	μ = 0.1	μ = 0.01	
Pb-peat humic acid			
Conventional titration	7.08 (4&5)	8.14 (4&5)	48
Modified procedure	7.00 (4&5)	8.11 (4&5)	48
Pb-Leonardite humic acid			
Conventional titration	7.36 (4&5)	8.18 (4&5)	48
Modified procedure	7.03 (4&5)	7.80 (4&5)	48
Pb-soil humic acid			
Conventional titration	7.14 (4&5)	8.38 (4&5)	48
Modified procedure	7.03 (4&5)	7.65 (4&5)	48
Pb-fulvic acid	2.6—2.7(3.0), 4.0—4.1(5.0)		42
Cd-peat humic acid	μ = 0.1	μ = 0.01	
Conventional titration	5.07 (4&5)	6.39 (4&5)	48
Modified procedure	—	—	48
Cd-Leonardite humic acid			
Conventional titration	5.32 (4&5)	—	48
Modified procedure	6.01 (4&5)	6.69 (4&5)	48
Cd-soil humic acid			
Conventional titration	5.14 (4&5)	6.38 (4&5)	48
Modified procedure	6.06 (4&5)	6.92 (4&5)	48
Ca-humic acid (4 soils)	2.41—2.95(3.5), 3.27—4.14(5.0)		56

Table 12 (continued)
ASSOCIATION CONSTANTS OF VARIOUS ORGANIC MATERIALS

Complex	Log K (pH)	Ref.
Ca-fulvic acid (4 soils)	2.29—2.72(3.5), 3.21—3.77(5.0)	56
Ca-fulvic acid	2.6—2.7(3.0), 3.3—3.4(5.0)	56
Fe-humic acid (4 soils)	4.73—5.13(3.5), 5.65—6.07(5.0)	56
Fe-fulvic acid (4 soils)	4.87—5.01(3.5), 5.62—6.04(5.0)	56
Al-fulvic acid	3.7(2.35)	42
Ni-fulvic acid	3.1—3.2(3.0), 4.2(5.0)	42
Co-fulvic acid	2.8—2.9(3.0), 4.1—4.2(5.0)	42
Mn-fulvic acid	2.1—2.2(3.0), 3.7(5.0)	42
Mg-fulvic acid	1.9(3.0), 2.1—2.2(5.0)	42

[a] μ = ionic strength

Sequential extraction		Nature of extracted fraction
Soil ↓		
cold water or 80% ethanol ↓	⟶	"free" sugars, amino acids, organic acids, phenols, etc. (monomers, oligomers) plus minerals
ether ↓	⟶	fats, waxes } "free" lipids plus associated organics and minerals
absolute ethanol ↓	⟶	resins } "free" lipids plus associated organics and minerals
hot water ↓	⟶	"hemicellulose" fraction = polysaccarides low in uronic acids plus peptides, minerals, etc.
2% HCl reflux ↓	⟶	"polyuronides" plus peptides, aromatics, minerals, etc.
digest in cold, 72% H_2SO_4, dilute reflux ↓	⟶	glucose from cellulose, "hydrolyzable NH_4^{+11} plus complex mixtures of organic and minerals
residue - "ligno protein" Resistant carbon structures and "non-hydrolyzable N." Possibly artifact, but probably "aromatic-based."		

FIGURE 13. Proximate fractionation scheme. (From Wolcott, A. R., *Soil Biochem.*, CCS 881, Mich. State Univ., E. Lansing, 1978; Stevenson, F. J., *Methods of Soil Analyses*, Part 2, Black, C. A., Ed., Am. Soc. of Agron., Madison, Wis., 1965. With permission.)

Table 13
PROXIMATE CHEMICAL COMPOSITION OF SOIL ORGANIC MATTER FROM DIFFERENT SOILS ON THE BASIS OF THE TOTAL SOIL ORGANIC MATTER (C × 1.72)

Ether-soluble material (fats, waxes) (%)	Alcohol-soluble material (resins) (%)	Hydrolysis with 2% HCl (hemicelluloses) (%)	Hydrolysis with 80% H_2SO_4 (Cellulose) (%)	Residue (lignin-humus complex) (%)	N in residue by Kjeldahl (organic nitrogenous complexes) (%)	Sum of the constituents accounted for (%)
3.56	0.58	5.44	3.55	43.37	33.78	90.28
4.71	1.53	8.60	5.22	40.81	34.74	95.61
1.94	3.10	12.59	5.36	35.18	35.77	93.94
0.80	0.82	5.53	4.12	41.87	37.35	90.49
1.02	0.88	6.96	3.50	42.05	33.25	87.66
0.46	0.84	8.54	2.83	42.83	33.36	88.86
0.52	0.63	10.66	3.38	46.50	33.13	94.82
0.62	0.61	8.61	3.64	49.29	30.38	92.75

From Mortenson, J. L. and Himes, F. L., *Chemistry of the Soil*, Bear, F. E., Ed., Reinhold, New York, 1965. With permission.

Fractionation of soil organic matter according to solubility may be accomplished by sequential extraction and differential precipitation. Examples of applications of these techniques are summarized in Figure 14.

Two major methods of fractionation involve separation according to charge: (1) ion exchange, and (2) electrophoresis or electrofocusing. The ion-exchange method involves elution of a solution of soil humic material through a column containing a cation- or anion-exchange resin. The material which is retained by the resin can then be eluted with a suitable solvent. The electrophoretic technique involves placing the organic solute dissolved in a suitable buffer in an electric field. Rate and direction of movement are dependent on properties of the buffer, equilibrium pH, and strength of the electric field.

Density fractionation techniques are commonly used to separate undecomposed or partially decomposed organic matter from the mineral and more highly decomposed organic components of the soil. The method involves flotation of undecomposed plant material in water or heavy liquid. Another density-based method involves fractionation of organic components using the ultracentrifuge.

The final major property of soil organic matter which can be utilized as a basis for fractionation is its adsorption potential. The most commonly used methods involving adsorption behavior are liquid-solid and liquid-liquid chromatography. However, the latter method is more realistically related to solubility, since the separation involves partitioning of the organic solute between the solvent and the liquid coating of the packing material. Adsorbents used include silica-gel, alumina,[75] diatomaceous earth, charcoal,[76] and sucrose. Currently, there is renewed interest in separation method based on adsorption potential due to the advent of high pressure liquid chromatography and the new high-efficiency, reverse-phase packing materials.

Elemental and Functional Group Analyses

The generally preferred methods for elemental and functional group analyses are summarized in Tables 14 and 15. Reviews of these and other methods are given by Schnitzer and Khan[75] and Hayes and Swift.[68]

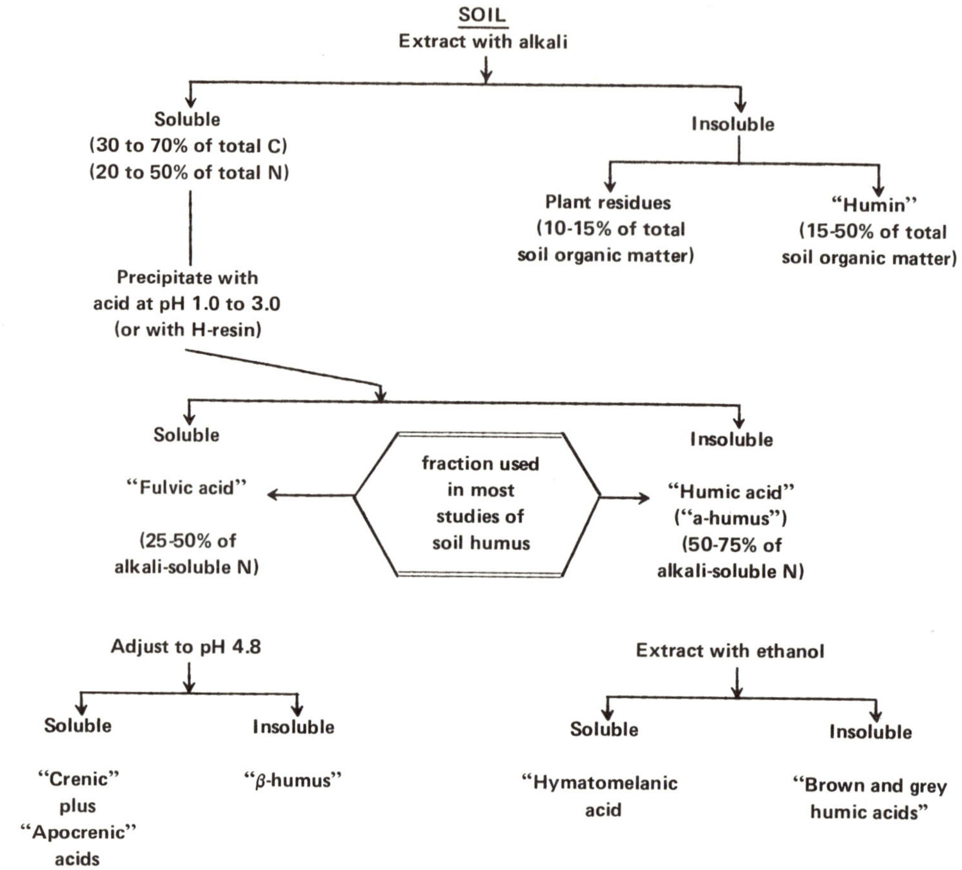

Numerous modifications:

1. Pretreatments: Removal of unhumified plant residues by flotation or dissolution; grinding or ball-milling; solvent extraction to remove lipids, fats, waxes, resins; HCl leaching to remove complexing polyvalent cations; treatment with HCl-HF to dissolve complexing silicate minerals (clays); mild reducing agents (hydrosulfite, dithionite) to remove complexing Fe.

2. Minimizing artifactual alteration of humic materials: Extraction in absence of O_2; extraction with neutral salt solutions (0.1 *M* sodium pyrophosphate); repeated alternate extraction with dilute acids and bases or salt solutions.

3. Reducing ash contents: Dialysis (osmotic and electrolytic), electrophoresis, ion exchange.

FIGURE 14. Scheme for fractionating soil organic matter (From Wolcott, A. R., *Soil Biochem.*, CCS 881, Mich. State Univ., E. Lansing, 1978. With permission.)

Spectroscopy

Spectroscopic methods provide valuable information concerning the chemical and physical properties of soil organic matter. Specific methods are discussed in greater detail below.

Ultraviolet and Visible

Ultraviolet (200 to 400 mμ wavelength) and visible (400 to 800 mμ) spectra of soil humic materials are generally broad band, with the optical density generally decreasing as the wavelength increases. Absorption in the UV region of the spectrum is due to

Table 14
ELEMENTAL ANALYSIS METHODOLOGY

Element	Method	Ref.
C and H	Combustion in an oxygen stream; water is collected in anhydrous magnesium perchlorate and CO_2 in soda lime	79
N	Micro-Kjeldahl with selenium mercury catalysts or by Dumas digestion	78
S	Oxygen flask	77
O	$\%O = 100 - [\%(C + H + N + S)]$	

Table 15
FUNCTIONAL GROUP METHODOLOGY

Functional group	Method	Ref.
Total acidity	Diborane (B_2H_6)	80
	$Ba(OH)_2$: $Ba(OH)_2 + 2H^+$ humate $\rightarrow$	81
	$Ba^{2+}(humate)_2 + H_2O$	82
	Diazomethane:	
	$RCO_2H + CH_2NH_2 \rightarrow RCO_2CH_3 + N_2$	75
	$\phi OH + CH_2N_2 \rightarrow \phi OCH_3 + N_2$	140
Carboxyl	Calcium Acetate:	83
	$2RCO_2H + (CH_3CO_2)_2Ca \rightarrow (RCO_2)_2Ca + 2CH_3CO_2H$	
	$CuSO_4$ − quinoline	84
Phenolic hydroxyl	Total acidity minus carboxyl	85
Alcoholic hydroxyl	Total hydroxyl minus phenolic hydroxyl	88
Total hydroxyl	Acetylation:	
	$ROH + (CH_3CO)_2O \rightarrow CH_3COOR +$	81
	CH_3COOH	85
Carbonyl	Hydroxylamine:	86
	$R_2C{=}O + NH_2OH\ HCl \rightarrow$	
	$R_2C{=}NOH + H_2O + HCl$	
	2,4-dinitrophenylhydrazine	87
	Polarograph	88
Quinone	$SnCl_2{-}K_2Cr_2O_7$	89
	Infrared, 1630—1700 cm^{-1}	90
	Electron spin resonance	91
	Titanometric reduction	92
Methoxyl	Zeisel hydriodic acid	93

unshared electrons, multiple bonds, and aromatic or heterocyclic structures. Spectra are obtained with samples in solution, with the most frequently used solvents being 0.1 N NaOH and 0.2N or 0.25N Na_2CO_3. Organic solvents may also be used, but the solvent must not absorb UV radiation in the spectral regions of interest. Absorption of radiation in the UV and visible regions of the spectra has frequently been used to quantitate soluble organic matter; however, light absorption of humic compounds appears to increase with increases in (1) degree of condensation of aromatic rings, (2)

Table 16
MAJOR IR ABSORPTION BONDS[75,94-96]

Frequency (cm^{-1})	Assignment
3400	H-bonded OH groups
3030—3077	Aromatic C−H stretching
2850—2950	Aliphatic C−H stretching
1725	C=O stretching of COOH or ketone or aldehyde
1600—1650	Aromatic C=C, conjugated or H-bonded C=O
1575—	COO^- ions
1450	Aliphatic CH deformation
1400	COO^- symmetric stretching
1200	C−O stretching or OH deformation of COOH
840	Aromatic C−H

ratio of C in aromatic nuclei to C in aliphatic and/or alicyclic side chains, (3) total C content, and (4) molecular weight. Therefore, it is probably not reasonable to use UV extinction coefficients to compare quantities of soluble organic matter obtained from different sources. The term E_4/E_6 ratio is frequently used in the literature and refers to the ratio of optical densities at 465 and 665 mμ.

Infrared

Infrared spectroscopy of the soil organic material has been extensively reviewed by Schnitzer and Khan,[75] Stevenson and Goh,[94] and Flaig et al.[95] Major resonant bands are summarized in Table 16.

Samples for IR spectroscopy are usually prepared by the KBr-pellet technique.[97] Samples may also be prepared by dissolving the organic material in a suitable solvent (dimethylformamide, dimethylsulfoxide), pipetting onto a window which is transparent to infrared radiation (AgBr, Irtran II), and allowing the solvent to evaporate.[98] Last traces of residual solvent may be removed from the sample by placing the sample over water in a desiccator.[98] The latter procedure has several advantages over the KBr technique. The KBr is hygroscopic, and therefore it is difficult to remove traces of water which would interfere with the IR spectra. Also, the latter method allows preparation of homoionic samples, which would not be possible by the KBr-disk method.

The broad band spectra of soil humic materials provides information about the general chemical character rather than the exact detailed structural character. Infrared spectroscopy may be used to (1) compare general structural characteristics of samples from different sources or from different extraction or fractionation procedures, or (2) monitor changes in samples following chemical reactions or ion exchange. Specific applications have been reviewed by Schnitzer.[99]

Nuclear Magnetic Resonance

The application of ^{1}H-NMR spectroscopy[100-104] and ^{13}C-NMR spectroscopy[100,105-108] to studies of soil organic matter has received attention in the past few years. The sample for NMR analysis is dissolved in a suitable solvent such as D_6-dimethylsulfoxide,[102-104,106] D_4-pyridine,[104,106] D_1-chloroform,[107,108] or sodium deuteroxide in D_2O.[101,105,108] Generally, concentrations of at least 5% by weight are desirable. Paramagnetic ions reduce the quality of NMR spectra. Therefore, samples should be purified to obtain negligible concentrations of paramagnetic ions on the exchange complex.

Table 17
MAJOR RESONANCE BANDS OBSERVED IN ^1H NMR SPECTRA OF SOIL HUMIC MATERIAL

δ (ppm)	Resonance assignment
0.8—0.9	Terminal CH_3 groups
1.2—1.3	CH_2 groups of aliphatic chains
3.2—4.8	Exchangeable proton or methyl or methylene groups deshielded by electron withdrawing substituents in the alpha position
6.8—8.8	Aromatic or heterocyclic protons
10.0—15.0	Strongly deshielded exchangeable protons (COOH)

Table 18
MAJOR RESONANCE BANDS OBSERVED IN ^{13}C NMR SPECTRA OF SOIL HUMIC MATERIAL

δ (ppm)	Resonance assignment
0—50	Aliphatic C (methyl, methylene)
60—100	C−O carbon
105—155	Aromatic carbon
170—200	C=O carbon (COOH and esters)

Major resonant bands observed in ^1H- and ^{13}C-NMR spectra of soil humic materials are summarized in Tables 17 and 18, respectively. Bands are generally broad due to the complexity and heterogeneity of soil organic matter; nevertheless, samples obtained from different sources[100,105] or by different extraction techniques[101,102,104,107] give distinctly different patterns. Therefore NMR techniques have a potential application in comparisons of soil organic matter.

Loeppert and Volk[104] compared ^1H-NMR spectra for a series of solvent extracts of Terra Ceia muck. Relative intensities of the δ0- to 1.5-ppm band decreased according to the following sample order: acetone-extractable material > 2-propanol-extractable material > methanol-extractable material > DMF-extractable material > humic acid > water-extractable material > fulvic acid. NMR patterns, corroborated by IR spectra and functional group analyses, provided evidence that the predominance of aliphatic constituents decreased according to the sample extraction order as presented above. Ratios of peak integrals for the δ6.8- to 8.8-ppm band to the δ0.0- to 1.5-ppm band increased according to the same sample order, thus providing evidence for increasing aromatic character. As illustrated in this study, ^1H-NMR spectroscopy can be used to obtain relative estimates of aliphatic vs. aromatic character of soil organic matter samples. ^{13}C-NMR cannot be used as a quantitative tool in the same way as the ^1H-NMR, due to the variable relaxation time for the ^{13}C nuclei during the NMR experiment. However, ^{13}C-NMR remains an important tool for possible identification of specific types of carbon atoms occurring in organic matter or qualitative comparisons of organic matter from different sources. NMR techniques are valuable tools when used in conjunction with IR spectroscopy and traditional chemical methods.

Degradation

Degradation is a helpful approach to obtaining information on chemical structures of complex materials such as humic and fulvic acids. Degradation may produce simpler identifiable compounds or structures which can then be related to the structures of the starting materials. Caution in interpretation is important. In many instances, the naturally occurring units are altered during degradation of the polymaic material. Hayes and Swift[68] present an extensive review on degradative processes.

Oxidative Degradation

Rates and extent of oxidation of different humic substances depend upon the elemental and functional group composition, stereochemistry, and solvation. Generally, the ease of oxidation of molecules decreases from anions to neutral molecules to cations. Oxidation involves the loss of electrons from the molecule, with subsequent rupture of carbon-hydrogen and carbon-carbon bonds.

Permanganate

Potassium permanganate ($KMnO_4$) is extensively used for oxidation of organic substances. Coal chemists have used permanganate oxidation for a long time[92,109] and Schnitzer and his colleagues applied it to humic substances (Table 19). Their data are particularly useful for comparisons of humic substances from different soil areas. High yields of degradation products are thought to be representative of a lower degree of soil organic matter humification.[68]

Alkaline Nitrobenzene

Gottlieb and Hendricks[111] and Morrison[112,113] have used the alkaline nitrobenzene technique to degrade soil organic matter. Wildung et al.[114] indicate that yields of aldehydes decreased in the order, plant material > peat and eutro-amorphous peat (muck) > organic matter from mineral soils.

Alkaline and Alkaline Cupric Oxide

Although lignin has been degraded using hot sodium hydroxide,[115] alkaline cupric oxide has received much more attention for the degradation of both lignin[116] and humic substances.[117] Yield comparisons of solvent-extractable materials show that hydrolysis in NaOH produces as much soluble material as oxidation by CuO-NaOH; however, the oxidation produces simpler compounds than does hydrolysis.[68] Table 20 shows some degradation products produced by CuO-NaOH from soils of widely different climatic regions. It appears that CuO-NaOH oxidative procedures preserve more of the aliphatic and phenolic structures than does $KMnO_4$, thereby allowing better comparisons to be made of humic substances from different environments.[68]

Nitric Acid

Nitric acid has been used extensively in degradation studies of lignin[118] and coal,[92] but only occasionally for degradation of humic substances.[119,120] Nitric acid digestions produce a variety of aromatic substitution and replacement reactions in addition to the oxidation reactions. The evolution of CO_2 from digests[121] indicates that nitric acid causes the degradation products to be extensively decomposed, particularly if unmethylated.

Reductive Degradation

Reduction reactions add H to a compound to form new covalent linkages. Metals such as Zn and Na, which transfer electrons to the substrate, and metal hydrides, which transfer hydride (H^-), are generally used.

Sodium Amalgam

Sodium amalgam (Na/Hg), introduced by Burges et al.,[122] has been extensively employed to degrade humic acids.[123-127] Degradation of model compounds by Piper and Posner[128] established the importance of the ratio of reactant to reagent and time of reaction.

Despite the reducing conditions of the sodium amalgam degradative reaction, many of the reaction products contain a carboxyl group, thus providing evidence that this functional group as well as hydroxyl, methoxyl, and carbonyl, exists on the aromatic nuclei of humic substances.

Table 19
MAJOR TYPES OF PRODUCTS (mg) RESULTING FROM THE $KMnO_4$ OXIDATION OF 1.0 g OF METHYLATED HUMIC AND FULVIC ACIDS FROM SOILS IN DIFFERENT CLIMATES[110]

	Climatic zone								
	Arctic	Cool, temperate				Subtropical		Tropical	
		Acid soils		Neutral soils					
Type of product	HA	HA	FA	HA	FA	HA	FA	HA	FA
Aliphatic	129.2	9:8—51.6	9.2	7.0—15.8	6.8—8.9	0.6—1.0	0—1.6	8.3—116.7	5.7—27.1
Phenolic	36.8	70.4—79.0	30.5	68.1—96.5	43.1—46.5	16.2—31.2	7.4—60.0	32.3—111.7	15.9—26.3
Benzene-carboxylic	50.8	80.2—122.4	46.9	147.5—173.5	77.6—95.4	9.2—164.1	20.6—86.9	49.6—183.3	37.9—68.4
Total identified	224.2	194.4—282.4	86.6	248.6—281.3	129.6—147.8	118.1—198.5	28.0—148.1	100.2—350.4	64.6—108.0
Weight ratio Benezene-carboxylic: phenolic	1.4	1.1—1.6	1.5	1.8—2.2	1.8—2.1	5.3—6.1	1.2—2.8	1.2—2.4	1.4—3.5

From Schnitzer, M., *Proc. IAEA-FAO-Agrochemica Symp.*, 2, 117, 1977.

Table 20
MAJOR TYPES OF PRODUCTS (mg) RESULTING FROM THE CuO-NaOH OXIDATION OF 1.0 g OF HUMIC AND FULVIC ACID EXTRACTED FROM SOILS IN DIFFERENT CLIMATES[110]

	Climatic zone					
	Arctic	Cool, temperate			Tropical	
		Acid soils		Neutral soils		
Type of product	HA	HA	FA	HA	HA	FA
Aliphatic	104.2	26.2	51.6	46.2	19.6—55.3	15.6—44.2
Phenolic	63.6	51.0	45.2	68.1	62.7—98.0	22.7—55.4
Benzene-carboxylic	24.8	28.4	24.0	20.2	20.9—34.9	19.5—48.8
Total identified	192.6	105.6	120.8	135.0	103.2—188.2	57.8—143.4
Weight ratio:						
Benezene-carboxylic: phenolic	0.4	0.6	0.5	0.3	0.3—0.5	0.8—0.9

From Schnitzer, M., *Proc. IAEA-FAO-Agrochemica Symp.*, 2, 117, 1977.

Zinc-Dust Distillations and Fusions

Cheshire et al.[129,130] and Hansen and Schnitzer[131,132] have used zinc-dust distillation techniques and have isolated products composed of fused aromatic structures such as naphthalene, anthracene, or pyrene. Very low yields and the severity of the degradations make interpretation of results very difficult.

Others

Cheshire et al.[130] have used HI and red phosphorus to produce a volatile and nonvolatile oil which, when dehydrogenated, yielded a number of fused aromatic structures. Felbeck[133,134] has used hydrogenolysis to degrade an organic soil, and Maximov and Krasaskaya[135] reduced hydrolyzed humic acids from a forest soil by metallic Na in liquid NH_3. The phenol-paratoluenesulphonic acid method for the degradation of soil humic acids was introduced by Jackson et al.[136] Separates identified included various xanthene and xanthone structures. The phenolic degradation technique helps to identify interaromatic linkages.

Swift,[137] Craggs et al.,[138] and Hayes et al.[139] have examined the potential value of sodium solutions for degradations of humic materials. Aliphatic mono- and dicarboxylic acids and keto acids have been identified in the more volatile components of a humic acid-sodium sulfide digest. This technique strongly indicates that aliphatic substituents are attached to the aromatic structures in humic materials.

Peracetic acid oxidations,[140,141] electrochemical oxidation,[142] and pyrolysis[143-149] procedures have provided additional useful information on the nature of humic substances. The use of Curie-point pyrolysis with a fast-scanning quadrupole mass spectrometer[149,150] has shown definite similarities and differences between humic acids from soils, peats, model phenolic polymers, fungal humic acids, and lignins.

COMPONENTS

Humic and Fulvic Acids

Organic matter can be partitioned into three major fractions: (1) humic acid, which is soluble in dilute alkali but insoluble upon acidification of the alkali extract; (2) fulvic

Table 21
SOME PROPERTIES OF FULVIC AND HUMIC ACIDS

Fulvic acid		**Humic acid**		
Crenic acid Light yellow	Apocrenic acid Yellow-brown	Hymatomelanic acid Brown	Brown humic acid Dark brown	Grey humic acid Grey-black
Soluble in acetyl bromide (includes exometabolites)			Insoluble in acetyl bromide	
Soluble in concentrated alkali (5%)				Salts out in conc. alkali or 2 *N*NaCl
Acid soluble (pH 1 to 3)		Partially precipitated by acids	Precipitated by acid at pH 1 to 3	
Soluble in water	Soluble in water plus alcohol	Soluble in alcohol	Soluble in alkline solvents	
<50% C		58—62% C	50—60% C	58—62% C

Note: ——— Increasing: →
color, degree of polymerization and cross-linking, molecular weight, carbon and nitrogen content.

——— Decreasing: →
solubility, acidity, CEC, oxygen content.

From Wolcott, A. R., *Soil Biochem.*, CCS 881, Michigan State Univ., E. Lansing, 1978. With permission.

acid, which is soluble in dilute alkali and remains in solution after acidification of the alkali extract; and (3) humin, which is not soluble in dilute acid or alkali. Some properties of humic and fulvic acids based on their solubilities are presented in Table 21. Table 22 presents data from Hayes et al.[151] illustrating the effects of various extractants of soil organic matter on the yields and elemental composition of fulvic and humic acid fractions. Extraction of an organic soil by solvents of different polarities yielded compounds of various elemental and functional group composition (Table 23).[152]

Elemental analyses of humic and fulvic acids extracted from soils of widely differing climatic areas (Table 24) showed that fulvic acids, as compared to humic acids, contain more O, S, and H but less C and N. On a weight basis, C and O are the major elements in both humic and fulvic acids.

Table 25 presents data on the major oxygen-containing functional groups. Total acidity, carboxyl, and alcoholic OH content of the fulvic acids appear to be higher than that of the humic acids. Nitrogen distribution in humic and fulvic acids as hydrolyzable- and nonhydrolyzable-N and amino acids is shown in Tables 30 and 31.

Haider and Martin,[153] Martin and Haider,[154,155] and Haider et al.,[156] in studies involving ^{14}C-ring labeled, or carboxyl or side-chain labeled hydroxybenzenecarboxylic acids, demonstrated that rapid decarboxylation occurs during microbial degradation. Phenolic compounds were found to be more resistant to microbial decomposition than was glucose. Martin and Haider showed that aromatic compounds are relatively resistant to decomposition and can be protected from microbial decomposition by incorporation into humic acid-like polymers.

Oxidative and reductive degradations of soil organic matter have yielded aliphatic mono- and dicarboxylic acids, benzenecarboxylic acids, phenolic acids, and aromatic

Table 22
YIELDS AND COMPOSITIONS OF HUMIC ACID (HA) AND FULVIC ACID (FA)[a] FRACTIONS EXTRACTED WITH DIFFERENT SOLVENT SYSTEMS FROM H^+-SOIL

Extractant	Method	Fraction	Yield (%)	Elemental composition (%) C	H	N	S	Ash
DMF	1 time followed by water	HA	16.0	53.4	4.5	2.6	1.7	1.8
		FA	2.0	49.9	4.0	3.1	1.5	4.7
Sulpholane	Same	HA	10.0	54.0	4.8	3.2	2.4	0.8
		FA	12.0	51.8	4.3	3.2	1.7	3.3
DMSO	Same	HA	17.0	53.5	4.1	3.2	1.9	2.7
		FA	6.0	51.5	4.1	2.1	1.2	6.5
Pyridine	Same	HA	34.0	54.7	5.0	4.3	n.d.[b]	2.2
		FA	2.0	45.3	5.1	5.8	n.d.	3.8
EDTA	Same	HA	49.0	54.7	5.7	6.2	n.d.	3.7
		FA	14.0	48.2	5.4	10.5	n.d.	5.8
EDTA-HCl (2.5M)	Exhaustive	HA	5.5	52.6	5.0	8.3	n.d.	2.6
		FA	7.8	44.8	5.9	9.4	n.d.	2.0
EDTA (anhydrous)	3 times	HA	2.0	57.0	7.0	6.7	n.d.	0.0
		FA	3.0	56.7	6.0	9.1	n.d.	0.2
0.1 *M* $Na_4P_2O_7$	Exhaustive	HA	13.7	50.1	3.3	3.0	n.d.	1.6
		FA	0.8	28.9	4.0	3.9	n.d.	22.5
0.5 *N* NaOH	1 time followed by water	HA	58.0	52.0	5.9	2.8	n.d.	2.0
		FA	2.0	43.9	5.9	4.2	n.d.	2.5
Dowex A−1 (Na^+−)	Exhaustive	HA	17.8	48.9	3.7	1.7	n.d.	3.3
		FA	1.6	46.8	4.7	1.9	n.d.	7.0
1 *M* EDTA	1 time followed by water	HA	12.5	50.8	4.0	n.d.	n.d.	2.6
		FA	3.8	45.7	4.0	n.d.	n.d.	5.6

[a] Not including material lost during dialysis.
[b] n.d. = not determined.

From Hayes, M. H. B., Swift, R. S., Wardel, R. E., and Brown, J. K., *Geoderma*, 13, 231, 1975. With permission.

Table 23
YIELDS, ELEMENTAL, AND FUNCTIONAL GROUP ANALYSIS OF FRACTIONS EXTRACTED WITH SELECTED SOLVENTS FROM TERRA CEIA MUCK CONTAINING 4.1% ASH

Extractant	Yield (%)	Elemental composition (%) C	H	N	S	O	Functional groups (meq/g) Total acidity	COOH	C=O	Phenolic OH	Alcoholic OH
0.5 *N* NaOH	49.2	55.1	5.5	3.7	0.7	35.0	7.3	3.5	2.8	3.8	3.1
0.1 *M* $Na_4P_2O_7$	29.9	52.6	4.3	3.4	0.7	39.0	9.1	4.3	3.5	4.8	3.5
H_2O	1.7	45.6	5.1	2.2	0.7	46.4	10.5	5.5	2.7	5.0	3.3
DMF	25.3	55.6	6.2	6.5	0.5	31.2	6.7	3.1	4.0	3.6	2.6
Acetonitrile	0.06	72.8	9.4	1.7	0.2	15.9	1.9	1.1	—	0.9	1.5
Methanol	1.5	67.5	8.5	1.3	0.4	22.3	4.3	2.1	2.1	2.3	2.8
Pyridine	3.4	64.9	7.9	2.5	0.4	24.3	4.6	2.3	2.0	2.3	1.7
Acetone	0.2	70.5	8.1	0.6	0.1	20.7	n.d.[a]	n.d.	n.d.	n.d.	n.d.
Chloroform	0.6	75.3	7.7	0.6	0.1	16.3	n.d.	n.d.	n.d.	n.d.	n.d.
Benzene	0.05	76.2	8.6	0.3	0.1	14.8	n.d.	n.d.	n.d.	n.d.	n.d.

[a] Not determined due to low yields.

From Loeppert, R. H. and Volk, B. G., *Soil Crop Sci. Soc. Fla. Proc.*, 33, 160, 1973. With permission.

Table 24
ELEMENTARY ANALYSIS OF HUMIC AND FULVIC ACIDS EXTRACTED FROM SOILS OF WIDELY DIFFERING CLIMATIC ZONES (%)[110]

	Climatic zone										Overall range and mean	
	Arctic		Cool, temperate				Subtropical		Tropical			
			Acid soils		Neutral soils							
Element	HA	FA	HA	FA	HA	FA	HA	FA	HA	FA	HA	FA
C	56.2	47.7	53.8—58.7	47.6—49.9	55.7—56.7	40.7—42.5	53.6—55.0	42.2—44.3	54.4—54.9	42.8—50.6	53.6—58.7 (56.2)	40.7—50.6 (45.7)
H	6.2	5.4	3.2—5.8	4.1—4.7	4.4—5.5	5.9—6.3	4.4—5.0	5.9—7.0	4.8—5.6	3.8—5.3	3.2—6.2 (4.7)	3.8—7.0 (5.4)
N	4.3	1.1	0.8—2.4	0.9—1.3	4.5—5.0	2.3—2.8	3.3—4.6	3.1—3.2	4.1—5.5	2.0—3.3	0.8—5.5 (3.2)	0.9—3.3 (2.1)
S	0.5	1.6	0.1—0.5	0.1—0.5	0.6—0.9	0.8—1.7	0.8—1.5	2.5	0.6—0.8	1.3—3.6	0.1—1.5 (0.8)	0.1—3.6 (1.9)
O	32.8	44.2	35.4—38.3	43.6—47.0	32.7—34.7	47.1—49.8	34.8—36.3	43.1—46.2	34.1—35.2	39.7—47.8	32.7—38.3 (35.5)	39.7—49.8 (44.8)

From Schnitzer, M., *Proc. IAEA-FAO-Agrochemica Symp.*, 2, 117, 1977.

Table 25
FUNCTIONAL GROUPS ANALYSIS AND E_4/E_6 RATIOS OF HUMIC AND FULVIC ACIDS EXTRACTED FROM SOILS OF WIDELY DIFFERING CLIMATIC ZONES (meq/g)[110]

	Climatic zone										Overall range and mean	
	Arctic		Cool, temperate				Subtropical		Tropical			
			Acid soils		Neutral soils							
Functional group	HA	FA	HA	FA	HA	FA	HA	FA	HA	FA	HA	FA
Total acidity	5.6	11.0	5.7—8.9	8.9—14.2	6.2—6.6	ND	6.3—7.7	6.4—12.3	6.2—7.5	8.2—10.3	5.6—7.7(6.7)	6.4—14.2(10.3)
Carboxyl	3.2	8.8	1.5—5.7	6.1—8.5	3.9—4.5	ND	4.2—5.2	5.2—9.6	3.8—4.5	7.2—11.2	1.5—5.7(3.6)	5.2—11.2(8.2)
Phenolic OH	2.4	2.2	3.2—5.7	2.8—5.7	2.1—2.5	ND	2.1—2.5	1.2—2.7	2.2—3.0	0.3—2.5	2.1—5.7(3.9)	0.3—5.7(3.0)
Alcoholic OH	4.9	3.8	2.7—3.5	3.4—4.6	2.4—3.2	ND	2.9	6.9—9.5	0.2—1.6	2.6—5.2	0.2—4.9(2.6)	2.6—9.5(6.1)
Quinonoid C=O	2.3	2.0				ND			1.4—2.6	0.3—1.5		
			{ 0.1—1.8	{ 1.7—3.1	{ 4.5—5.6		{ 0.8—1.5	{ 1.2—2.6			{ 0.1—5.6(2.9)	{ 1.2—4.2(2.7)
Ketonic C=O	1.7	2.0							0.3—1.4	1.6—2.7		
Methoxyl	0.4	0.6	0.4	0.3—0.4	0.3	ND	0.3—0.5	0.8—0.9	0.6—0.8	0.9—1.2	0.3—0.8(0.6)	0.3—1.2(0.8)
E_4/E_6	5.3	11.5	3.8—5.0	9.0	4.0—4.3	ND	3.9—5.1	8.4—9.5	5.0—5.8	7.6—11.2	3.8—5.8(4.8)	7.6—11.5(9.6)

From Schnitzer, M., *Proc. IAEA-FAO-Agrochemica Symp.*, 2, 117, 1977.

Table 26
CARBOHYDRATES FOUND IN SOILS

Monosaccharides
  Hexoses — glucose, galactose, mannose, fructose
  Pentoses — arabinase, xylose, ribose, fucose, rhamnose
Disaccharides — sucrose, cellubiose, gentiobiose
Oligosaccharides — cellotriose
Polysaccharides — cellulose, hemicellulose
Amino sugars — glucosamine, galactosamine, n-acetyl-D-glucosamine
Sugar alcohols — mannitol, inositol
Sugar acids — galacturonic, glucuronic acid
Methylated sugars — 2-0-methyl-D-xylose, 2-0-methyl-D-arabinose, 2-0-methyl-rhamnose, 4-0-methyl-galactose

compounds including N- and S-containing heterocyclic compounds. Occurrence of these degradation products plus evidence from electron microscopy and viscosity studies[42] provide evidence that humic substances are not single molecules but rather associations of molecules of plant and microbial origins. Even though compounds in the higher molecular weight fractions may be more complex than compounds in lower molecular weight fractions, the molecular forces holding the compounds together are similar and consist primarily of hydrogen bonds, Van der Waal's forces, and π-bonding.[42]

Carbohydrates

Source

Dead plant material contains 50 to 75% carbohydrates in the form of different polymer saccharides, especially cellulose and hemicellulose.[157] Cellulose may account for 30 to 70% and hemicellulose for 5% of plant residue. Root exudates may amount to as much as 10% of the annual production of the plant carbohydrates. These polymer saccharides are attacked by the soil fauna and flora, which in turn synthesize other saccharides. Carbohydrates account for 5 to 20% of soil organic matter.[158] The major groups of saccharides added to the soils from higher plants, microorganisms, and animals are summarized in Table 26.

A number of reviews deal with various aspects of isolation, composition, and importance of carbohydrates in the soil.[68,159-163]

Carbohydrate Levels in Soils

Many methods have been employed for measuring carbohydrate levels in soils. Most methods are dependent on an initial hydrolysis to release monosaccharides, followed by measurement of the free saccharides. Table 27 outlines yields of crude polysaccharides obtained from soils by various procedures.[163] A commonly used procedure outlined by Gupta and Sowden[164] and Cheshire and Mundie[165] involves pretreatment of the sample with 72% H_2SO_4 followed by dilution to 1 *N* acid and refluxing.

Success in extraction of polysaccharides[159,161,163] has been obtained by refluxing soils with 98% formic acid,[166] or with 0.5 *N* NaOH. Pretreatment with 1 *N* HCl or HF increases yields.[167] Acetylation of the residue further increases yields. Other extraction techniques are shown in Table 28. Polysaccharides isolated from soil can be fractionated by paper and column electrophoresis techniques,[168] gel chromatography,[169-171] and anion exchange chromatography.[172] Table 28 lists sugar contents of polysaccharide materials isolated from selected soils. Techniques for the isolation of saccharides and polysaccharides from soils are not 100% efficient due to the strong interaction between inorganic soil materials and the saccharide fraction of soil organic matter.

Table 27
YIELDS OF CRUDE POLYSACCHARIDES OBTAINED FROM SOILS BY VARIOUS PROCEDURES

Soil	Organic matter (%)	Extractant	Method	Yield[a] Soil (%)	Yield[a] Organic matter (%)
Scottish	n.d.[b]	Water	2 × 4 hr at 85°C	0.3	—
Sphagnum peats	98.3—99.3	Water	3 × 3 hr at 60—65°C after benzene and methanol	1.1—2.0	1.1—2.0
Scottish peats	96.3—99.1	Water	1 × 4 hr at 100°C	n.d.	—
American	3.38	Water	1 × 4 hr + 1 × 6 hr at 85°C	0.05	1.45 (after purification with charcoal, 0.2)
American muck soil	n.d.	Water	1 × 24 hr in Soxhlet	—	1.8
Fen peat	80	Water	n.d.	0.008	0.009
Red-brown earth	4.3	Water	1 × 16 hr at 70°C	0.06	1.5
Five forest soils	n.d.	pH 7 buffers with 5% KCl	10 hr at 20°C	0.01—0.039	—
Fen peat	80	7% disodium EDTA	n.d.	0.45	0.56
Scottish peats	96.3—99.1	0.09 *N* HCl	1 hr at 60—70°C	0.15	0.15
Five British soils	2.3—58.9	Anhydrous HCOOH containing 0.2 N LiBr	"2 hot extractions"	0.4—5.6	4.0—11.5
Fen peat	80	0.6 *N* H_2SO_4	24 hr at 3°C	0.03	0.04
Fen peat	80	Amberlite IRC H+-form	n.d.	0.13	0.17
Red-brown earth	4.3	1 *N* HCl	16 hr at 20°C	0.04	1.0
Four Scottish soils	n.d.	0.5 *N* NaOH	48 hr at 15°C	—	1.32—1.94
Wisconsin soils	0.8—2.6	0.5 *N* NaOH	3 hr at 20°C	0.02—0.25	—
Swiss soils	3.0—32.5	0.5 *N* NaOH	"Extracted at room temperature"	0.02—0.13	0.1—1.0
Saskatchewan soils	4.4—9.8	0.5 *N* NaOH	3 hr at 20°C	0.14—0.78	3.2—14.0
Red-brown earth	4.3	0.5 *N* NaOH	16 hr at 20°C	1.13	3.2

Canadian soils	n.d.	Schweitzer's reagent	2 × 6 hr at 20°C	—	0.3—1.9
Fen peat	80	N-methyl-2-pyrrolidone	n.d.	0.03	0.04
Fen peat	80	8 *M* urea	n.d.	0.46	0.57
Fen peat	80	Dimethylformamide	n.d.	3.0	3.8

[a] These data are for impure polysaccharides, some containing up to 80% of ash and other nonsaccharide materials.
[b] n.d. = no data given.

From Greenland, D. J. and Oades, J. M., *Soil Components,* Vol. 1, Gieseking, J. E., Ed., Springer Verlag, New York, 1975, 213. With permission.

Table 28
NEUTRAL SUGAR DISTRIBUTION IN SOIL HYDROLYSATES AND OF POLYSACCHARIDE FRACTIONS

			Sugar composition (% of total neutral sugars recovered)										
Soil (source of description)	OM (%)	Extractant	Glucose	Galactose	Mannose	Arabinose	Xylose	Rhamnose	Fucose	Ribose	Uronic acid	Amino sugars	Ref.
Scottish agricultural soil	8.5	0.5 *N* NaOH	20.8	20.0	21.9	11.7	23.6	0	0	1.5	15.8	0	96,192
Indiana soil	3.4	Water	21.2	16.6	18.5	10.4	12.6	14.2	0	+	9.1	+	193
Sandy loam	3.5		33.6	19.1	18.0	7.9	7.9	13.5	—	—	15.8	5.2	
Meadow	17.6	Anhydrous	37.7	20.1	18.2	7.6	8.2	8.2	—	—	17.9	5.0	
Peat	31.9	Formic acid	26.8	20.8	15.8	7.9	8.9	18.8	—	—	16.0	3.3	166
Podzol Ao	39.7	containing	33.6	23.1	17.9	6.7	7.5	11.2	—	—	16.3	3.1	
Podzol B	2.3	LiBr	31.8	18.2	20.5	6.8	9.1	13.6	—	—	13.5	4.3	
Compost													
11 months	58.9		42.3	15.4	15.4	11.5	15.4	—	—	—	8.4	5.0	
48 months	—		47.2	16.7	13.9	8.3	13.9	—	—	—	6.7	4.3	
Podzol with permafrost													
O_1	80.2		42	15	15	10	8	7	3		—	—	
O_2	59.7		50	13	14	10	7	4	3		—	—	
B_2	1.8	72%	30	9	17	13	12	10	9		—	—	
C_2	2.1	H_2SO_4	22	16	20	8	11	13	10		—	—	194

Table 28 (continued)
NEUTRAL SUGAR DISTRIBUTION IN SOIL HYDROLYSATES AND OF POLYSACCHARIDE FRACTIONS

Soil (source of description)	OM (%)	Extractant	Sugar composition (% of total neutral sugars recovered) Glucose	Galactose	Mannose	Arabinose	Xylose	Rhamnose	Fucose	Ribose	Uronic acid	Amino sugars	Ref.
Ap no treatment	2.4		31	15	15	15	11	8		5	—	—	
Ap manured	4.0		28	14	20	15	10	8		5	—	—	
Ap fertilized	2.4		31	20	15	13	10	7		4	—	—	
Urrbrae fine sandy loam, pasture (red brown earth) 0—6 cm	4.1	Total soil carbohydrate	39.3	12.2	19.4	11.7	7.5	4.4	3.6	1.9	—	—	
Urrbrae fine sandy loam, fallow wheat (red brown earth)	2.1	1 *N*HCl + 0.5 *N*NaOH	43.1	13.0	18.5	9.8	8.7	3.5	2.6	0.8	—	—	167
Urrbrae fine sandy loam pasture (red brown earth)	4.1	1 *N*HCl	29.2	19.2	19.9	7.5	5.5	11.3	7.0	0.5	—	—	
		0.5 *N*NaOH	27.9	12.3	24.5	10.9	11.3	5.2	4.1	3.7	—	—	
		Acetylation	57.5	8.6	8.5	12.7	7.8	2.4	2.1	0.4	—	—	
		Residue	43.4	11.7	21.4	14.6	3.1	1.8	2.3	1.7	—	—	
		1 *N*HCl:											
		4000-100,000MW	28.1	21.0	19.4	7.7	5.1	10.5	7.8	0.4	—	—	
		>100,000MW	31.5	15.5	20.8	7.1	6.5	12.5	5.5	0.5	—	—	
		0.5 *N*NaOH:											
		<4,000MW	27.4	11.1	26.8	12.2	11.0	9.1	7.5	4.7	—	—	
		4000-100,000MW	26.8	13.4	19.3	10.7	15.9	5.4	3.8	2.8	—	—	
		>100,000MW	32.9	15.3	17.2	6.5	11.0	9.1	7.5	0.4	—	—	
Urrbrae fine sandy loam, fallow wheat (red brown earth)	2.1	1 *N*HCl	41.2	19.5	21.8	2.3	4.6	5.3	4.5	0.2	—	—	
		0.5 *N*NaOH	36.7	13.6	22.1	8.0	10.7	4.6	2.8	1.5	—	—	
		Acetylation	53.1	7.9	10.3	13.7	10.8	2.0	1.8	0.4	—	—	
		Residue	45.5	11.6	19.9	14.2	3.4	2.0	2.4	1.0	—	—	
		1 *N*HCl:											
		<4000MW	63.1	9.7	7.3	12.6	3.1	1.4	1.4	1.4	—	—	
		4000-100,000MW	44.4	22.1	21.7	1.7	3.0	3.7	3.3	0.1	—	—	
		>100,000MW	30.2	10.3	22.9	5.2	10.6	11.7	8.7	0.5	—	—	
		0.5 *N*NaOH:											

		<4000MW	36.6	10.1	24.7	9.0	8.3	4.1	3.4	3.4	—	—	
		4000-100,000MW	36.5	15.2	20.9	9.2	12.6	2.8	1.7	1.1	—	—	
		>100,000MW	36.0	13.6	23.0	4.5	8.8	9.2	4.4	0.5	—	—	
											Total org. C and N (%)		
Continuous cultivation (England)	1.6	H_2SO_4	41.3	14.4	10.8	12.7	14.7	5.1	1.1	—	1.1	4.6	
Continuous pasture (England)	8.2	H_2SO_4	39.2	14.3	10.4	11.5	16.9	5.8	2.0	—	1.4	4.3	195
Organic soil	80	H_2SO_4											
		Fraction 1	68.9	9.0	9.0	2.6	3.6	5.3	1.9	—	—	—	
		2	32.3	21.6	8.7	10.3	18.8	4.4	3.3	—	—	—	
		3	70.8	5.9	5.5	5.0	8.1	2.3	1.9	—	—	—	
		4	19.5	31.8	17.3	4.0	7.9	12.7	3.5	1.5	—	—	151
		5	18.1	33.2	16.7	4.7	8.9	11.5	5.2	0.7	—	—	
		6	21.3	24.3	16.9	8.3	11.2	9.1	7.5	0.6	—	—	
Amiata soil cambisol (Italy)	6.4	Fulvic acid FA	42.2	9.6	20.6	5.0	11.9	5.1	4.8	0.8	—	—	
		Polyamide I	37.6	9.9	23.8	4.8	12.3	5.2	5.5	0.9	—	—	196
		Chromatograph II	35.0	11.5	19.4	6.9	14.9	6.1	5.2	1.0	—	—	

Significance of Carbohydrates

Polysaccharides serve as a ready source of energy and C for soil microbes and therefore undergo continual transformations in soils. This dynamic nature of soil carbohydrates presents a major obstacle to their study. Carbohydrates extracted from soil may be either products of microbial activity or products derived directly from higher plants.

The behavior of polysaccharides is primarily a function of their reactive groups, especially hydroxyl and carboxyl, or in the case of hexosamines, the amino groups. Interactions of polysaccharides with metals results in changes in solubilities and polysaccharide decomposition rates. The adsorption of monosaccharides onto clay mineral surfaces by H-bonding is well documented.[123] If polysaccharides bond to more than one mineral particle, aggregation of soil particles can occur, thus influencing many physical properties of the soil.

While much evidence has been accumulated to support the concept that aggregate stability is highly correlated with higher amounts of polysaccharides,[175-178,180,181] no evidence establishes conclusively that polysaccharides are in fact responsible for aggregation.[182,183] Polysaccharides probably contribute to aggregate stability in certain soils, particularly cultivated soils with low organic matter contents, but other binding agents, organic and inorganic, are important as well.

With respect to microbial activity, soil carbohydrates are extremely important substrates for many soil organisms. Martin,[184] Cheshire et al.,[185] Lynch and Cotnoir,[186] and Tan and McCreery[187] present factors influencing rates of decomposition of soil polysaccharides and discuss mechanisms of polysaccharide protection in soils.

In addition to their effects on microbial populations and physical properties of the soil, carbohydrates can affect nutrient availability to plants.[158] The polysaccharide contribution to cation-exchange capacity (CEC) and formation of soluble complexes may enhance nutrient availability. Sugars can prevent the precipitation of phosphate by Fe or Al.[188] High levels of easily metabolized carbohydrates can increase microbial immobilization of N and enhance either denitrification or N mineralization, depending upon soil conditions.

The translocation of Fe, Al, or clays may be influenced by carbohydrates.[189,190] However, Dormaar[191] found no correlation between polysaccharide or polyphenol contents of soil extracts and Fe solubility. The importance of polysaccharides may be in the production of organic acids which are known chelators or complexors of metals.

Nitrogen

Source

Organic N-containing compounds, the products of microbial decomposition of plant and animal remains, account for over 90% of the total N in most soils. The concentration of N in mineral surface soils varies considerably, but generally ranges from 0.1 to 0.6%. However, organic soil can contain as much as 3% N. Concentrations of N in subsurface horizons are generally lower than in surface horizons (Figure 15). The largest contribution to N in the soil is plant material, which contains from 1 to 4% N, depending upon plant age and species. Nitrogen content of microbial cells ranges from 5 to 10% of the total dry matter. Much of this N is in the form of structural protein. Tables 6 and 7 show amounts of N in above ground plant parts and roots.

The decomposition of N-containing compounds in plant materials which have been freshly added to the soil depends on the activities of the microorganisms. Microflora activities are controlled by the available food supply, as well as by environmental factors such as temperature, water, and availability of O_2. Figure 16 illustrates the complex interrelationships of N in the environment. A similar summary of S transformations is shown in Figure 17. Some of the N in the ecosystems may be present in living higher plant tissue, including roots; in living soil fauna; in soil microorganisms; in

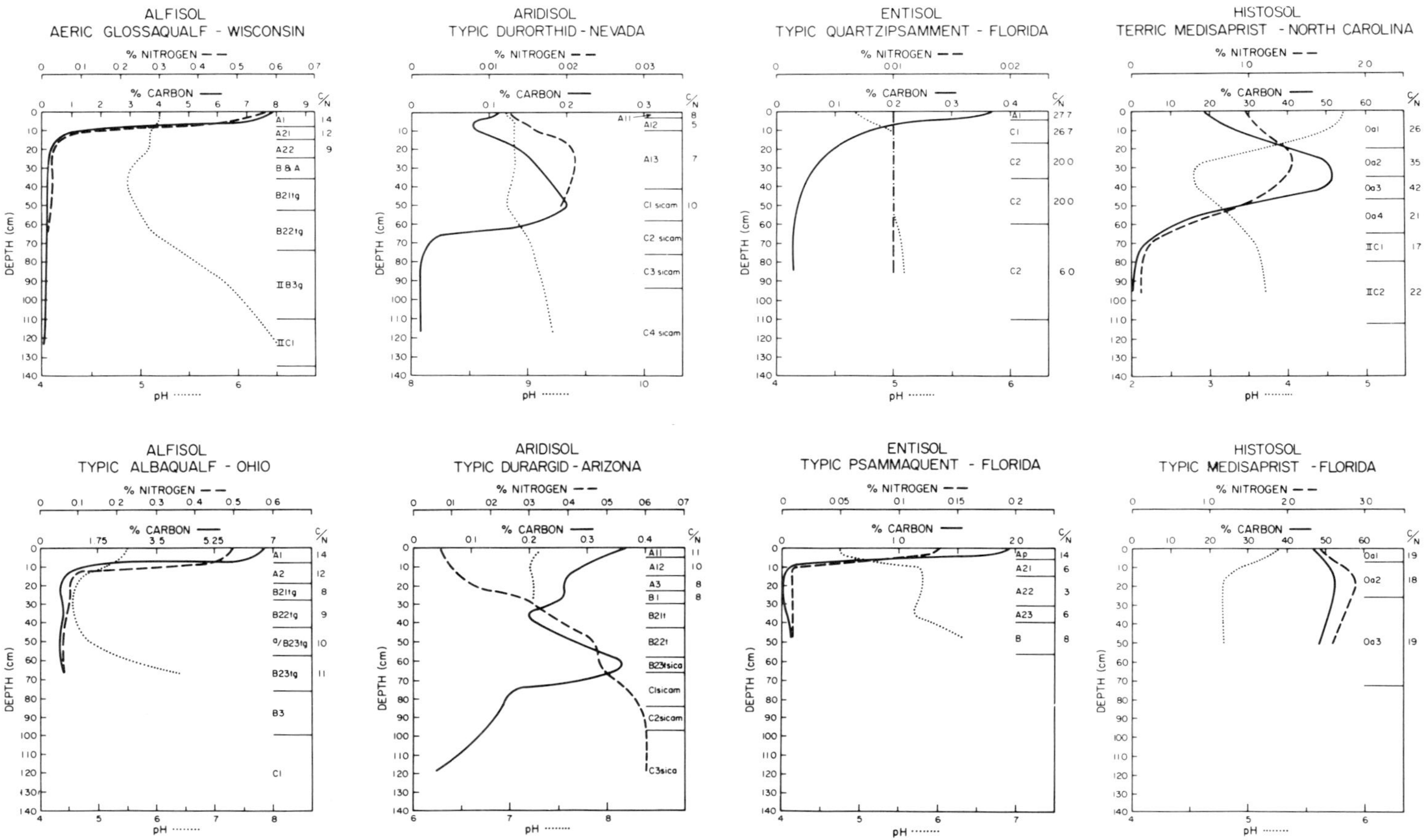

FIGURE 15. Percent carbon and nitrogen, C:N ratio, and pH plotted vs. depth for selected soils in the ten soil orders.

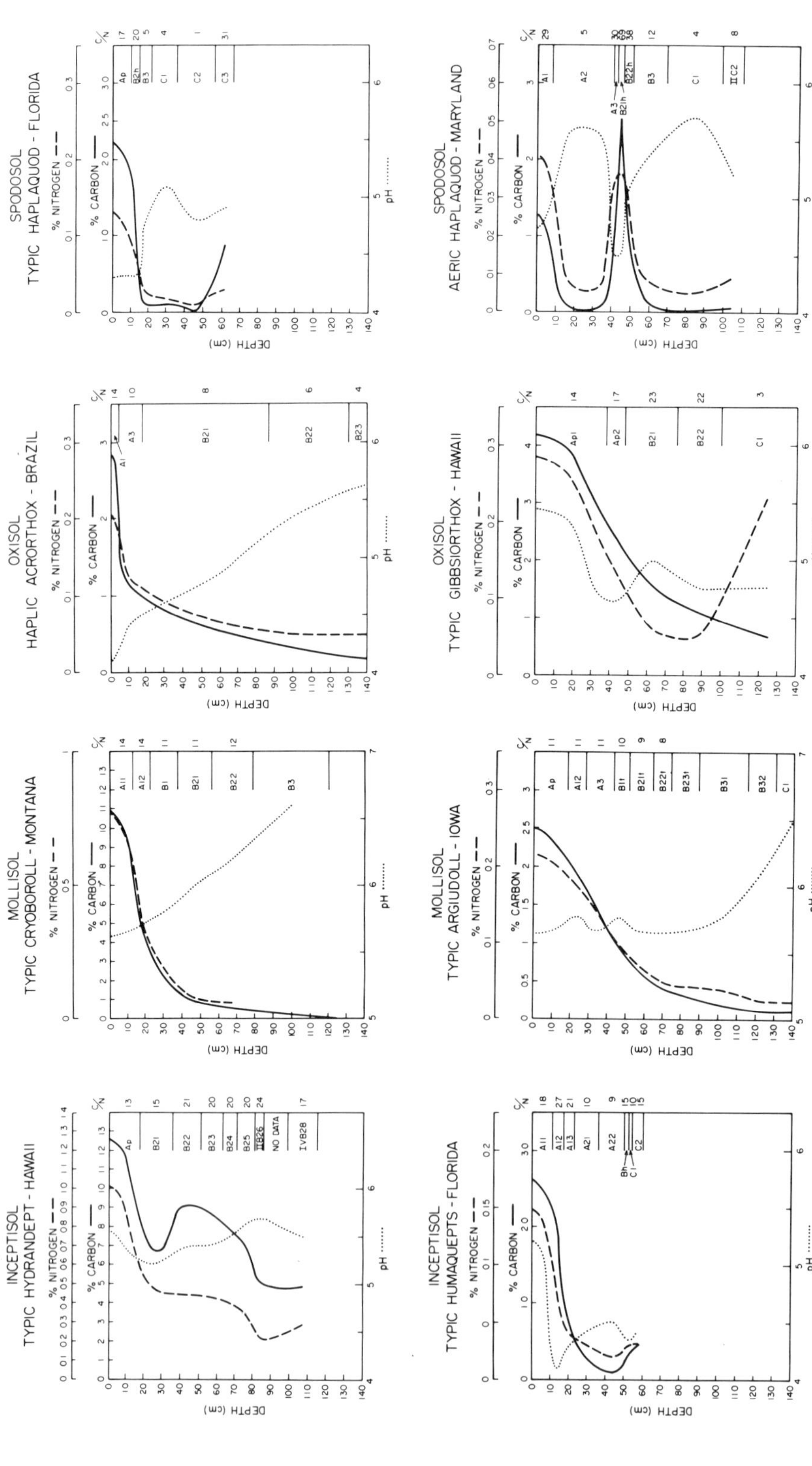

FIGURE 15 (continued)

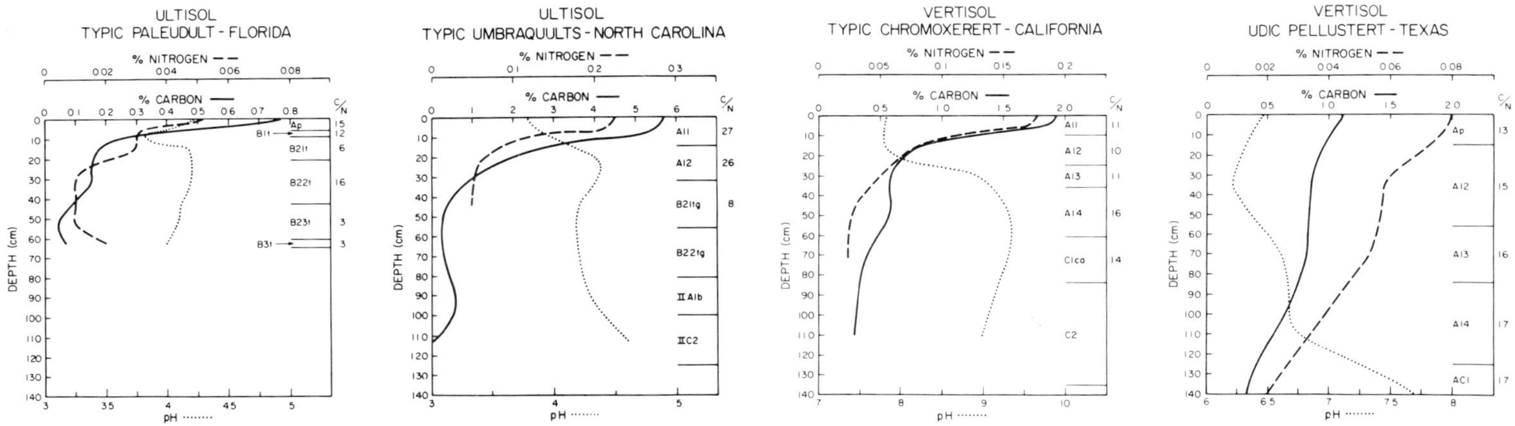

FIGURE 15 (continued)

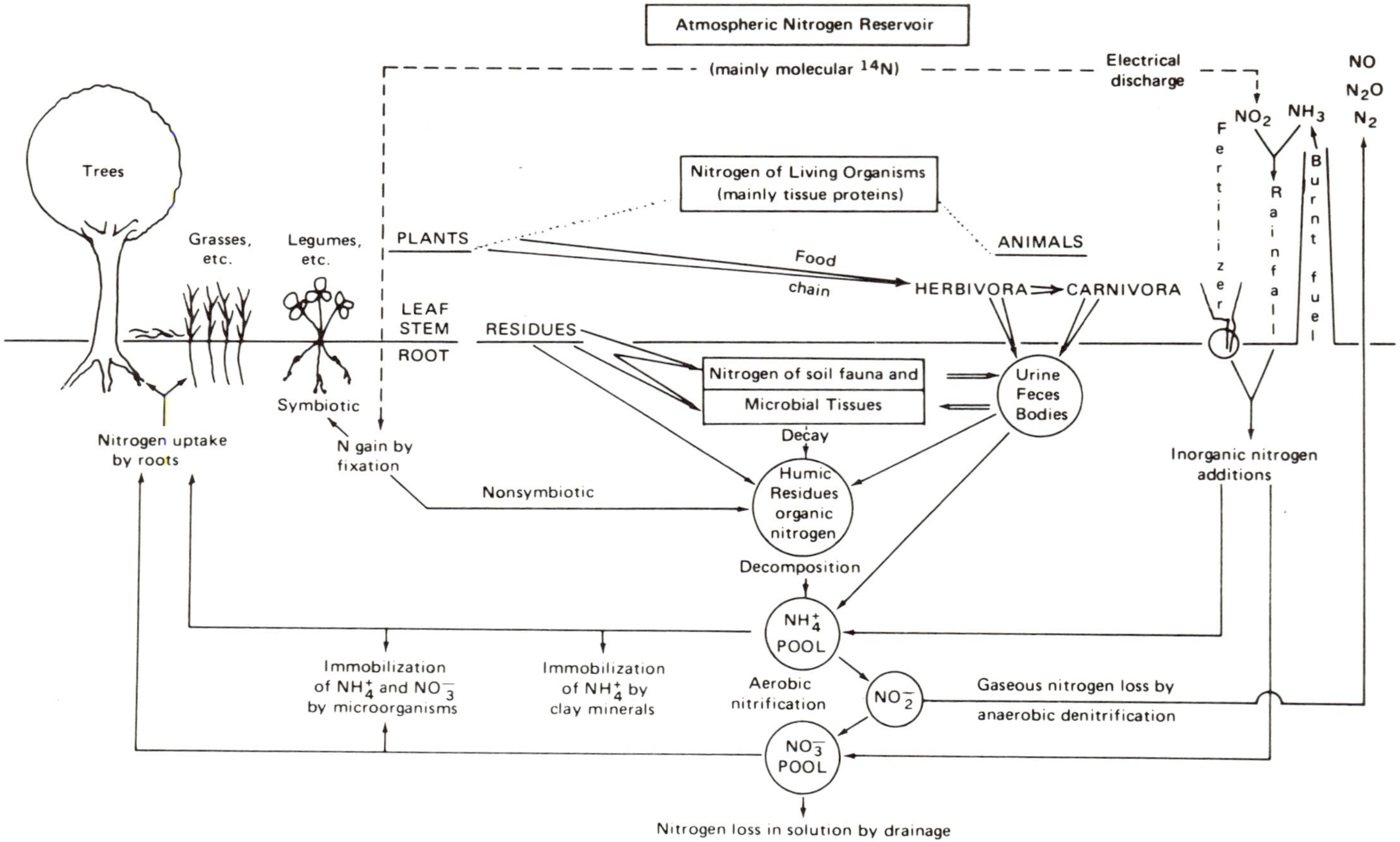

FIGURE 16. Nitrogen cycle (Reproduced from Parsons, J. W. and Tinsley, J., *Soil Components*, Vol. 1, Gieseking, J. E., Ed., Springer-Verlag, New York, 1975, chap. 3. With permission.)

simple organic or inorganic compounds representing the end products of N metabolism from the entire soil population; or in accumulated soil organic matter. One of the obvious problems in studying soil N compounds is to distinguish sources in undecomposed tissue from sources in the more highly decomposed organic material.

Methods

Methods commonly used for measuring total soil N and N in various organic fractions are summarized in Table 29.[60,65,78] Strong acid hydrolysis, i.e., 6*N* HCl, is commonly used to hydrolyze N-containing polymers to monomeric components for subsequent extraction, quantitation and/or identification. Modifications of this procedure, e.g., mixtures of acids,[166,201] alkaline hydrolysis,[200] or partial hydrolysis,[201] are sometimes used. Autoclaving[197] and pretreatments with HF[202] may also offer advantages.

Discussion

Bremner[203] states that between 20 and 50% of total N in surface soils is in the form of amino acids, its proportion decreasing with depth. The distribution of individual amino acids is affected by fertilization, climate, and cultivation. For example, Stevenson[204] found that as organic reserves are depleted (Table 30, IV, continuous corn), the amino acid content falls but that the basic amino acids are more persistent than the neutral and acidic amino acids. Because of their basicity, they may become attached to other soil constituents and therefore protected from microbial attack. Amino sugars (hexosamine) usually account for 5 to 15% of the total N in surface soils,[203] but may be as high as 23 to 40% in soils of arid regions.

Amino acids not found in proteins may be found in soil. These include such compounds as ornithine, β-alanine, α-amino-n-butyric acid, γ-amino-n-butyric acid, and α, ε-diaminopimelic acid (Table 30). Ornithine is found in large quantities in bacteria. Sowden et al.[205] have shown an accumulation of acidic amino acids in amorphous allophanic soils in tropical climates.

The data of Table 30 show that the two acidic amino acids, aspartic and glutamic, and glycine and alanine occur in the soils in the highest concentration, except for the soil from the Morrow plots, which are also high in lysine. Sulfur-containing amino acids do not account for a large portion of the total N.

The nature and source of unidentified acid-soluble N + insoluble N + ammonium N are largely unknown (Table 31). Piper and Posner,[128,207] Griffith et al.,[208] and Freney[209] have examined the acid-insoluble material. Biederbeck and Paul[210] have reviewed the work on "ligno" and "humo" protein and conclude humate fractions contain proteinaceous material.

Nitrogen compounds contained in humic and fulvic acids are summarized in Tables 30 and 31. Most of the N in soils is in the humic acid and humin fractions.[205] Sequi et al.[211] and Guidi et al.[196] fractionated fulvic acid by polyamide and Sephadex chromatography into several fractions. Molecular weight fractionations of fulvic acid show in general that the N content decreases as the molecular weight decreases.

Whether or not N-containing organic compounds are a part of the humic compound "core" is controversial.[212,213] However, most workers agree that organic N is an important component associated in some way with humic materials.

Bonding between clay particles and organic matter is well recognized. Amino- and amide-N groups appear to be important in this bonding.[214] The N content of clay-organic complexes extracted from soils increases and the C:N ratio decreases as particle size decreases.[195,215-218] An accumulation of amino acids and amino sugars may account for the increased N concentration.

Table 29
ANALYTICAL METHODS FOR NITROGEN[60,65,78]

Fraction		Method	
Total N	A.	Kjeldhal (macro, semi-micro, micro)	1. Pretreatment with H_2O (30 min) increase N release *or* extract soil with 2*N* KCl to recover NH_4^+, NO_2^-, and NO_3^-, in which case Kjeldhal determination gives direct measure of insoluble and nonexchangeable organic N plus clay-fixed NH_4^+ 2. Digestion (1 to 4 hr after digest clears) in concentrated H_2SO_4 + K_2SO_4 (raises boiling temperature to 400°C) plus catalysts: a) Cu^{++}, Se or HgO (for oxidation and clearing), if HgO is used, add NaS or $Na_2S_2O_3$ thus preventing $Hg(NH_3)_2^{++}$ complex b) salicylic acid (to include NO_3^-) c) $KMnO_4$, then reduced Fe (to include NO_2^- and NO_3^-) 3. Distillation after adding excess NaOH, collecting NH_3 in boric for titration with standard acid (0.01 to 0.05 N) using mixed indicator (bromcresol green + methyl red)
	B.	Dumas	1. Oxidation at 600°C with CuO as catalyst → N_2O, other N oxides, CO_2, SO_2, CO 2. Passage over hot Cu: $N_2O \rightarrow N_2$, $CO \rightarrow CH_4$ 3. Passage over hot CuO: $SO_2 \rightarrow SO_3$, $CH_4 \rightarrow CO_2$ 4. Passage through concentrated alkali to remove CO_2 and SO_3 5. N_2 measured volumetrically
Amino compounds	A.	Most accurate estimates are based on summation of individual amino acids or amino sugars quantitated separately in hydrolysates cleaned up by various methods (dialysis, electrolytic desalting, phosphotungstate precipitation, fractional separation on exchange resin)	
	B.	General procedures for separation, identification, and quantitation:	1. Chromatographic including paper, thin layer, ion-exchange resins, gas-liquid, high pressure liquid can be used; ion exchange-chromatography coupled with autoanalyzer systems[a] and different detector systems[b] are available; in gas-liquid, volatile derivatives must be made by acetylation or silylation
	C.	Ninhydrin reactions for α-amino N	1. In alkaline medium, determinine colored complex colorimetrically 2. In acid medium, quantitate by collecting CO_2 or make alkaline and

			distill NH_3 into boric acid and titrate
	D.	Amino sugar N	1. Elson-Morgan method Acetylate with acetylacetone in alkali; react acetylated derivatives with N, N-dimethyl-p-aminobenzaldehyde in acid → red complex; quantitate colorimetrically 2. Tracey's method Hydrolyze in alkali (pH 11.2) and collect NH_3
Hydrolyzable organic N fractions	A.	Total N in soil or soil fractions by Kjeldahl	
	B.	Hydrolyze sample by:	1. Finely grinding, <100 mesh 2. Reflux 12 hr in 20 ml 6 *N* HCl, 2 drops octanol prevents foaming
	C.	Total N in hydrolyate by Kjeldhal	
	D.	Hydrolyzable NH_4^+:	1. Steam distill with MgO into boric acid
	E.	Hydrolyzable NH_4^+ + amino sugar N:	1. Steam distill with phosphate-borate buffer (pH 11.2) into boric acid
	F.	Hydroxyamino acids (serine, threonine, etc.):	1. Treat with periodate (HIO_4) to deaminate 2. Treat with arsenite ($NaAsO_2$) to decompose excess $NaIO_4$ 3. Make alkaline and steam distill into boric acid
	G.	α-amino N:	1. Make alkaline, boil to remove hydrolyzable NH_4^+ and amino sugar N 2. Acidify with citric acid and buffer to pH 2.5 3. Add ninhydrin, react 10 min 4. Make alkaline with NaOH and buffer to pH 11.2 with phosphate-borate buffer 5. Steam distill into boric acid
	H.	Hydrolyzable NH_4^+ + amino sugar N + α-amino N:	1. Buffer to pH 2.5 and react with ninhydrin (see G 3 & 4) 2. Steam distill into boric acid
	I.	Calculation:	1. Hydrolyzable N-determined directly by Kjeldhal (see C) 2. Non-hydrolyzable N = A − C 3. Hydrolyzable NH_4^+ = D 4. Amino sugar N = E − D 5. Hydroxyamino N = F 6. Total α-amino N = G 7. α-amino N other than hydroxyamino N = G − F

[a] Reference 197
[b] References 198, 199

Table 30
RELATIVE COMPOSITION, AS MOLAR RATIO OR CONCENTRATION, OF AMINO ACID IN SEVERAL SOILS OR ORGANIC MATTER FRACTIONS[30,75,157,205,209]

Amino acids	Canadian soils 32 surface soils (molar ratio)	Soil extracted with 0.1—0.5 *N*NaOH (molar ratio)						Morrow plots (mg/100 g soil)			
		Humic acid	Fulvic acid	Humin	I[a] (mg/100 g soil)	II[b] (mg/100 g soil)	III[c] (mg/100 g soil)	IV[d]	V[e]	VI[f]	VII[g]
Acidic											
Aspartic	11.9 ± 1.6	12.4, 13.0	24.6, 12.9	17.5, 12.0	32[h]	43[h]	182[h]	26.0	22.4	65.8	66.7
Glutamic	8.4 ± 0.6	8.6, 7.3	18.0, 9.1	9.2, 7.1	26	33	137	20.0	32.8	59.4	94.4
γ-aminobutyric acid	—	—	—	—	—	—	—	3.5	4.1	8.0	12.0
α-amino-n-butyric acid	—	—	—	—	—	—	—	1.0	1.1	2.2	3.5
α, ε-diaminopimelic acid	—	—	—	—	—	—	—	2.5	3.0	6.6	9.9
Basic											
Arginine	2.6 ± 0.3	2.2, 2.6	0.4, 1.2	1.9, 2.5	1	1	12	6.4	8.5	11.8	17.0
Histidine	1.2 ± 0.4	1.9, 1.7	0.6, 0.9	1.8, 1.5	0.8	3.5	13	5.8	7.2	9.4	6.9
Lysine	3.6 ± 0.3	3.4, 3.3	1.6, 2.1	2.6, 3.2	13	17	69	44.4	51.0	65.9	70.0
Ornithine	0.8 ± 0.3	0.7, 0.7	0.8, 0.8	0.8, 0.8	8	5	35	8.5	7.1	8.0	8.6
Neutral											
Phenglalanine	2.4 ± 0.2	3.0, 2.5	1.1, 1.9	2.3, 2.7	5	7	34	7.4	9.1	16.8	23.6
Tyrosine	1.1 ± 0.3	1.4, 1.0	0.7, 0.8	1.4, 1.0	Traces	1.5	17	8.0	9.5	14.9	23.3
Glycine	11.8 ± 0.9	11.2, 12.2	13.0, 13.8	12.2, 12.1	10	12	80	21.8	26.2	49.8	73.7
Alanine	8.5 ± 0.5	8.0, 7.9	8.2, 10.8	8.4, 8.3	18	22	109	31.7	38.8	69.5	76.4
β-alanine	—	—	—	—	—	—	—	5.0	6.1	11.0	17.3
Valine	5.2 ± 0.5	5.6, 5.4	3.8, 6.7	4.9, 5.8	11	16	81	21.3	23.5	43.6	62.0
Leucine	5.1 ± 0.6	5.1, 4.6	2.4, 2.7	4.8, 4.9	7	9	58	19.5	24.0	50.5	63.4
Isoleucine	3.0 ± 0.3	3.4, 2.9	1.8, 3.8	2.6, 3.1	8	9	33	16.1	19.0	36.8	47.7
Serine	5.6 ± 0.6	5.0, 5.0	4.8, 6.0	5.0, 5.1	[h]	[h]	[h]	14.0	20.6	43.0	73.4
Threonine	5.6 ± 0.4	5.0, 5.3	4.2, 6.8	4.6, 5.7	13	16	80	18.2	27.8	42.7	81.0
Proline	4.1 ± 0.5	4.6, 4.6	3.9, 4.2	3.4, 4.7	10	12	48	10.3	14.5	30.3	51.7
Hydroxyproline	0.4 ± 0.2	0.7, —	Trace, —	0.6, —	5	4	33	—	—	—	—

Sulfur containing											
Methionine	0.4 ± 0.2	0.6, 0.1	0.4, 0.1	0.5, 0.2	2.0	1.1	10	Trace	Trace	Trace	Trace
Cystine	0.4 ± 0.2	0.2, 0.3	0.2, 0.2	0.1, 0.4	—	—	—	Trace	Trace	Trace	Trace
Cysteic acid	0.8 ± 0.3	0.5, —	2.6, —	0.6, —	3.5	5	11	5.3	6.4	10.3	15.9
Methionine sulfoxide	0.3 ± 0.1	0.4, —	0.6, —	0.2, —	0.6	1.0	6.4	11.3	10.4	11.2	22.7
Miscellaneous[i]		1.6, 2.7	1.9, 3.9	1.8, 2.3	2.3	2.7	11.7				
% N of total soil N		41.6	6.5	51.9[j]	—	—	—				

[a] Wooster silt loam, cultivated soil.
[b] Wooster silt loam, virgin soil.
[c] Hoytville clay loam, deciduous forest.
[d] Flanagan silt loam, Plot 3 NE-Untreated continuous corn.
[e] Flanagan silt loam, Plot 4 NE-Untreated corn oats rotation.
[f] Flanagan silt loam, Plot 5 SE-Treated corn oats clover rotation.
[g] Flanagan silt loam, grass border surrounding plots.
[h] Includes aspartic acid and serine.
[i] Includes allo-isoleucine, γ-NH_2–butyric acid, α, 4-disaminobutyric acid, diaminopimelic acid, β-alanine, ethanolamine, and unidentified compounds.
[j] Calculated by difference, organics lost during purification would be included.

Table 31
DISTRIBUTION OF N IN SOILS AND SOIL FRACTIONS[65,206]

				Soil extracted with 0.1 to 0.5 *N* NaOH		
6 *N* HCl reflux	**Virgin soil (% of soil N)**	**Cultivated soil (% of soil N)**	**Canadian soils (32 surface horizons — % of soil N)**	**Fulvic fraction (% of fulvic N)**	**Humic fraction (% of humic N)**	**Humin (% of humin N)**
Nonhydrolyzable	18—37(25)[a]	19—34(24)	15 ± 6	4—8	15—60	16
Hydrolyzable						
Total	63—82(75)	66—81(76)	85	?—100	40—85	84
Ammonium	19—26(22)	19—29(25)	21 ± 5	15—50	10—30	16—22
Hexosamine	3—6 (5)	4—7 (5)	7 ± 2	2—10	1—10	2—9
Amino acid or α amino N	19—34(27)	18—31(23)	40 ± 7	20—43	20—50	36—48
Unidentified[b]	18—25(21)	20—29(23)	19	15—30	15—30	10

[a] Range of values followed by average in parentheses.
[b] Total hydrolyzable N minus (ammonium + hexosamine + amino acid) N.

Table 32
THE CARBON-NITROGEN RATIO OF SOME ORGANIC MATERIALS

Material	C:N ratio[a]
Microbial tissue	8:1
Soil humus	10:1
Sweet clover (young)	12:1
Alfalfa hay	13:1
Tomato leaves	13:1
Bean stems	15:1
Manure (rotted)	20:1
Clover residues	23:1
Green rye	36:1
Cane trash	50:1
Corn stalks	60:1
Oak leaves	65:1
Oat straw	80:1
Timothy	80:1
Pine needles	225:1
Mature rye straw	350:1
Saw dust	400:1

[a] Values are approximate; in any particular material the ratio may vary considerably.

Carbon-Nitrogen Ratio

This ratio defines the relative quantities of these two elements in fresh organic materials, humus, or the soil. The C:N ratio is important because competition for available N from decomposing plant residues between microorganisms and growing plants quite often is a determining factor in plant growth. Table 32 presents C:N ratios for various organic materials. Figure 15 and Table 33[220] show the C and N contents, and C:N ratios for various soil orders by depth and soil horizons. Considerable variability exists among different soils. Parsons and Tinsley,[221] Campbell,[23] and Rodrigues[222] provide C:N data for many soils in different countries. Table 34 by Swift[223] summarizes data on C, N, S, and mineralized N and S for pasture and arable soils.

When organic materials with a C:N ratio greater than 30 are added to soils, immobilization of soil N occurs during initial decomposition of the residue. If the C:N ratio is between 20:1 and 30:1, neither immobilization nor release of mineral N may occur. When the C:N ratio is less than 17, mineral N is generally released early in the decomposition process. Figure 17[224] illustrates what happens during decomposition of fresh organic material. Initially there is a rise in the number of heterotrophic organisms producing CO_2. Organic residues with a wide C:N ratio produce a net immobilization of N (shaded in Figure 18). The C:N ratio narrows with further decay and microbial populations diminish due to decreased food supply, thus establishing a new equilibrium. Release of mineral N (crosshatched area under the top curve) can occur after the new equilibrium is reached. Depending on the amount and type of organic residues added, the organic matter level in the soil may increase. Soil moisture, pH, temperature, type, and quantity of organic material, and supply of utilizable N all affect the decomposition cycle.

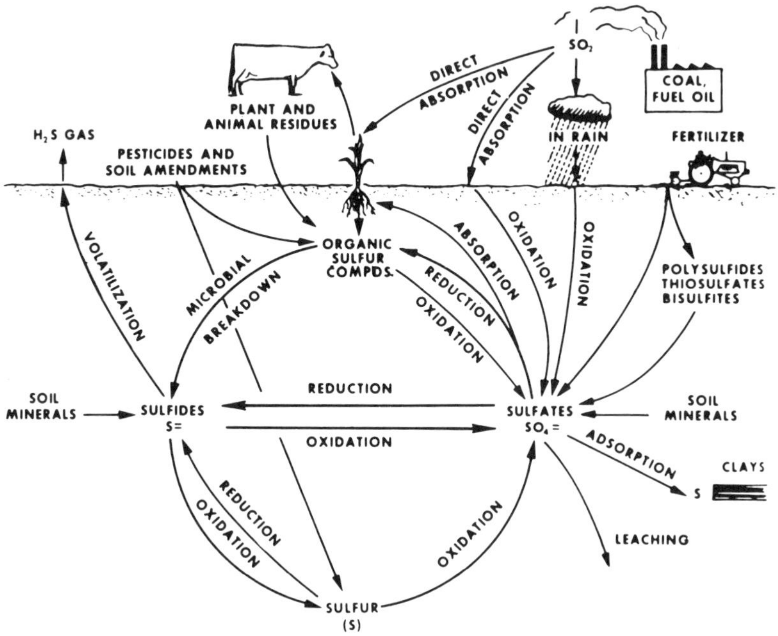

FIGURE 17. Sulfur cycle. (From Brady, N. C., *The Nature and Properties of Soils,* 8th ed., MacMillan, New York, 1969. With permission.)

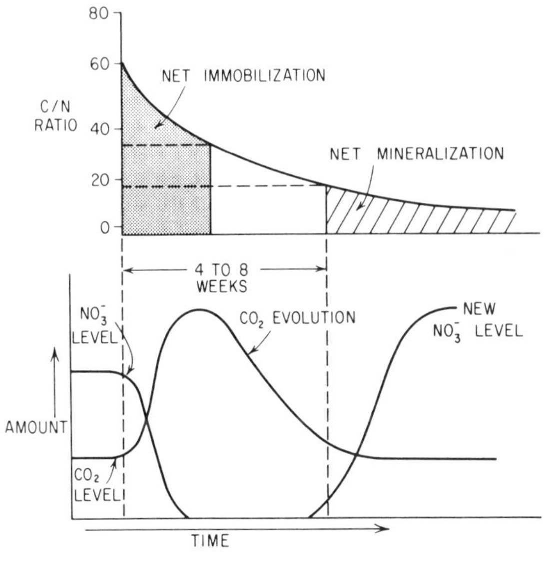

FIGURE 18. Changes in NO_3 and CO_2 levels during the decomposition of low nitrogen crop residues (Reproduced from Tisdale, S. L. and Nelson, W. L., *Soil Fertility and Fertilizers,* Tisdale, S. L. and Nelson, W. L., Eds., MacMillan, New York, 1975, 130. With permission.)

Table 33
DISTRIBUTION OF C:N RATIOS IN HORIZON SAMPLES OF SOME PACIFIC NORTHWEST AND NORTH CENTRAL U.S. SOILS

	Number of horizon samples in various ranges of C:N ratios									
	Below 5.5		Between 5.5 and 8.4		Between 8.5 and 11.4		Between 11.5 and 14.4		Above 14.4	
Soil horizon	a	b	a	b	a	b	a	b	a	b
A	0 (0)	0 (0)	3 (1)	0 (1)	8 (12)	7 (7)	16 (4)	18 (9)	3 (0)	5 (0)
B	0 (1)	0 (1)	8(14)	2 (11)	10 (4)	8 (7)	2 (0)	9 (0)	1 (0)	2 (0)
C	3 (3)	1 (0)	5 (4)	2 (5)	7 (0)	10 (1)	0 (1)	2 (1)	0 (0)	0 (0)
Total	3 (4)	1 (1)	16 (19)	4 (17)	25 (16)	25 (15)	18 (4)	29 (10)	4 (0)	7 (0)
Combined total	7	2	35	21	41	40	22	39	4	7

Note: a = organic C:total N ratio; b = organic C:organic N ratio. Figures in parentheses are comparative values for some north central U.S. soils.

From Young, J. L., *Soil Sci.*, 93, 397, 1962. With permission.

Table 34
C, N, S CONTENTS AND NET AMOUNTS OF N AND S RELEASED BY MINERALIZATION FROM SELECTED SOILS[223]

	Total (%)					Mineralized	
	C	N	S	C:N	C:S	Total NO_3^- + NH_4^+ (μg N/g soil)	S (μg S/g soil)
Pasture soils	5.0	0.38	0.060	13.2	85	111	6.3
	±0.3	±0.01	±0.004	±1.0	±8	±10	±0.9
Arable soils	2.1	0.17	0.034	12.8	63	55	3.1
	±0.1	±0.01	±0.003	±0.4	±4	±4	±1.0
All soils	3.6	0.27	0.047	13.0	74	83	4.7
	±0.4	±0.03	±0.004	±0.5	±5	±9	±0.8

From Swift, R. S., *Soil Organic Matter Studies*, Proc. Symp. FAO/IAEA, 1976, Braunschweig, Vol. 1, 1977, 275.

REFERENCES

1. **Allison, F. E.,** *Soil Organic Matter and Its Role in Crop Production,* Elsevier, Amsterdam, 1973.
2. **Nielson, G. A. and Hole, F. D.,** A study of the natural processes of incorporation of organic matter into soil in the University of Wisconsin Arboretum, *Wis. Acad. Sci.,* 52, 213, 1963.
3. **Foth, H. D.,** *Fundamentals of Soil Science,* 6th ed., John Wiley & Sons, New York, 1978.
4. **Brady, N. C.,** *The Nature and Properties of Soils,* 8th ed., MacMillan, New York, 1974.
5. **Larson, W. E., Holt, R. F., and Carlson, C. W.,** Residues of soil conservation, in *Crop Management Systems,* Special Publication No. 31, Oschwald, W. R., Ed., American Society of Agronomy, Madison, Wis., 1978, chap. 1.
6. **Jenny, H.,** *Factors of Soil Formation,* McGraw-Hill, New York, 1941, 281.
7. **Broadbent, F. E. and Bartholomew, W. V.,** The effect of quantity of plant material added to soil on its rate of decomposition, *Soil Sci. Soc. Am. Proc.,* 13, 271, 1948.
8. **Pinck, L. A. and Allison, F. E.,** Maintenance of soil organic matter. III. Influence of green manures on the release of native soil carbon, *Soil Sci.,* 71, 67, 1951.
9. **Jenkinson, D. S.,** Decomposition of labelled plant material in soil, in *Experimental Pedology,* Butterworths, London, 1964, 199.
10. **Sauerbeck, D. and Gonzalez, M. A.,** Field decomposition of carbon-14-labeled plant residues in various soils of the Federal Republic of Germany and Costa Rica, in *Soil Organic Matter Studies,* Proc. Symp. FAO/IAEA, Sept. 1976, Braunschweig, Germany, 1, 1977, 159.
11. **Paul, E. A. and McLaren, A. D.,** Biochemistry of the soil subsystem, in *Soil Biochemistry,* Vol. 3, Paul, E. A. and McLaren, A. D., Eds., Marcel Dekker, New York, 1975, 1.
12. **Sauerbeck, D. R. and Gonzalez, M. A.,** Decomposition of ^{14}C-labelled plant residues in different soils and climates, in *Abstracts for Commission Papers,* 11th Congress, International Society of Soil Science, Canada, 1978, 15.
13. **Jenkinson, D. S.,** Studies on the decomposition of C^{14} labeled organic matter in soil, *Soil Sci.,* 111, 64, 1971.
14. **Jenkinson, D. S.,** Studies on the decomposition of plant material in soil. IV. The effect of rate of addition, *J. Soil Sci.,* 28, 417, 1977.
15. **Jenkinson, D. S. and Ayanaba, A.,** Decomposition of carbon-14 labeled plant material under tropical conditions, *Soil Sci. Soc. Am. J.,* 41, 912, 1977.
16. **Jenkinson, D. S. and Rayner, J. H.,** The turnover of soil organic matter in some of the Rothamsted classical experiments, *Soil Sci.,* 123, 298, 1977.
17. **Herman, W. A., McGill, W. B., and Dormaar, J. F.,** Effects of initial chemical composition on decomposition of roots of three grass species, *Can. J. Soil Sci.,* 57, 205, 1977.
18. **Pinck, L. A., Allison, F. E., and Sherman, M. S.,** Maintenance of soil organic matter. II. Losses of carbon and nitrogen from young and mature plant materials during decomposition in soil, *Soil Sci.,* 69, 391, 1950.
19. **Hunt, H. W.,** A simulation model for decomposition in grasslands, *Ecology,* 58, 469, 1977.
20. **Salter, R. M. and Green, T. C.,** Factors affecting the accumulation and loss of nitrogen and organic carbon in cropped soils, *J. Am. Soc. Agron.,* 25, 622, 1933.
21. **Russell, J. S.,** Estimation of the time factor in soil organic matter equilibration under pastures, *Trans. Int. Soil Conf.,* Comm. IV and V, Sect. Soil Sci. N.Z., 1962, 191.
22. **Woodruff, C. M.,** Estimating the nitrogen delivery of soil from the organic matter determination as reflected by Sanborn Field, *Soil Sci. Soc. Am. Proc.,* 14, 208, 1949.
23. **Campbell, C. A.,** Soil organic carbon, nitrogen, and fertility, in *Soil Organic Matter,* Schnitzer, M. and Khan, S. U., Eds., Elsevier, New York, 1978, chap. 5.
24. **Lucas, R. E., Holtman, J. B., and Connor, L. J.,** Soil carbon dynamics and cropping practices, in *Agriculture and Energy,* Academic Press, New York, 1977, 333.
25. **Jenkinson, D. S.,** The turnover of organic matter in soil, *Report of the FOA/IAEA Tech. Meeting,* Pergamon Press, New York, 1963.
26. **Jenkinson, D. S.,** Studies on the decomposition of plant material, *J. Soil Sci.,* 16, 104, 1965.
27. **Lucas, R. E. and Vitosh, M. L.,** Soil organic matter dynamics, Research Report 358, Mich. State Univ. Agric. Exp. Stn., 1978.
28. **Larson, W. E., Clapp, C. E., Pierre, W. H., and Morachan, Y. B.,** Effect of increasing amounts of organic residues on continuous corn: organic carbon, nitrogen, phosphorous, and sulphur, *Agron. J.,* 64, 204, 1972.
29. Agriculture statistics, U.S. Department of Agriculture, Washington, D.C., 1975.
30. **Kowalenko, C. G.,** Organic nitrogen, phosphorus, and sulfur in soils, in *Soil Organic Matter,* Schnitzer, M. and Khan, S. U., Eds., Elsevier, Amsterdam, 1978, chap. 3.
31. **Wanke, D. D. and Barber, S. A.,** Root development and nutrient uptake by corn grown in solution culture, *Agric. J.,* 66, 514, 1974.

32. Our Land and Its Care, 5th ed., National Plant Food Institute, Washington, D.C., 1967.
33. **Jamison, V. C.,** Pertinent factors governing the availability of soil moisture to plants, *Soil Sci.,* 81, 459, 1956.
34. **Calhoun, F. G., Hammond, L. C., and Caldwell, R. E.,** Influence of particle size and organic matter on water retention in selected Florida soils, *Soil Crop Sci. Soc. Florida,* 32, 111, 1972.
35. **Broadbent, F. E.,** The soil organic fraction, in *Advances in Agronomy,* 5, 1953, 153.
36. **Mortensen, J. L.,** Complexing of metals by soil organic matter, *Soil Sci. Soc. Amer. Proc.,* 27, 179, 1963.
37. **Helling, C. S., Chesters, G., and Corey, R. B.,** Contribution of organic matter and clay to soil cation-exchange capacity as affected by the pH of saturating cation, *Soil Sci. Soc. Amer. Proc.,* 28, 517, 1964.
38. **Van Dijk, H.,** *The Use of Isotopes in Soil Organic Matter Studies,* Pergamon, New York, 1966, 129.
39. **Kamprath, E. J. and Welch, C. D.,** Retention and cation-exchange properties of organic matter in coastal plain soils, *Soil Sci. Soc. Amer. Proc.,* 26, 263, 1962.
40. **Yuan, T. L., Gammon, N., Jr., and Leighty, R. G.,** Relative concentration of organic and clay fractions to cation exchange capacity of sandy soils from several soil groups, *Soil Sci.,* 104, 123, 1967.
41. **Broadbent, F. E.,** Basic problems in organic matter transformations, *Soil Sci.,* 79, 107, 1955.
42. **Schnitzer, M.,** Humic substances: chemistry and reactions, in *Soil Organic Matter,* Schnitzer, M. and Khan, S. U., Eds., Elsevier, Amsterdam, 1978, chap. 1.
43. **Van Dijk, H.,** Cation binding of humic acids, *Geoderma,* 5, 53, 1971.
44. **Schnitzer, M. and Skinner, S. I. M.,** Organo-metallic interactions in soils. 4. Carboxyl and hydroxyl groups in organic matter and metal retention, *Soil Sci.,* 99, 278, 1965.
45. **Stevenson, F. J.,** Binding of metal ions by humic acid, in *Environmental Biogeochemistry,* Nriagu, J. O., Ed., Ann Arbor Science, Ann Arbor, Mich., 1976, chap. 33.
46. **Stevenson, F. J.,** Stability constants of Cu^{2+}, Pb^{2+}, and Cd^{2+} complexes with humic acids, *Soil Sci. Soc. Am. J.,* 40, 665, 1976.
47. **Schnitzer, M. and Hansen, E. H.,** Organo-metallic interactions in soils. VIII. An evaluation of methods for the determination of stability constants of metal-fulvic acid complexes, *Soil Sci.,* 109, 333, 1970.
48. **Stevenson, F. J.,** Nature of divalent transition metal complexes of humic acids as revealed by a modified poteniometric titration method, *Soil Sci.,* 123, 10, 1977.
49. **Coleman, N. T., McClung, A. C., and Moore, D. P.,** Formation constants for Cu(II)-peat complexes, *Science,* 123, 330, 1956.
50. **Courpron, C.,** Dêtermination des constants de stabilitê des complexes organo-mêtalliques des sols, *Ann. Agron.,* 18, 623, 1967.
51. **Schnitzer, M.,** Reactions between fulvic acid, a soil humic compound, and inorganic soil constituents, *Soil Sci. Soc. Am. Proc.,* 33, 75, 1969.
52. **Geering, H. R. and Hodgson, J. F.,** Micronutrient cation complexes in soil solution. III. Characterization of soil solution ligands and their complexes with Zn^{2+} and Cu^{2+}, *Soil Sci. Soc. Am. Proc.,* 33, 54, 1969.
53. **Hines, F. L. and Barber, S. A.,** Chelating ability of soil organic matter, *Soil Sci. Soc. Am. Proc.,* 21, 368, 1957.
54. **Miller, M. H. and Ohlrogge, A. J.,** Water soluble chelating agents in organic materials. I. Characterization of chelating agents and their reactions with trace metals in soils, *Soil Sci. Soc. Am. Proc.,* 22, 225, 1958.
55. **Randhawa, N. S. and Broadbent, F. E.,** Soil organic matter-metal complexes. VI. Stability constants of zinc-humic acid complexes at different pH values, *Soil Sci.,* 99, 362, 1965.
56. **Zende, N. A. and Raman, K. V.,** Organic matter - metal ion complexes in forrested and cultivated soils, *J. Indian Soc. Soil Sci.,* 3, 278, 1978.
57. **Matsuda, K. and Ito, S.,** Adsorption strength of zinc for soil humus. III. Relationship between stability constants of zinc-humic and -fulvic acid complexes, and the degree of humification, *Soil Sci. Plant Nutr.,* 16, 1, 1970.
58. **Ardakani, M. S. and Stevenson, F. J.,** A modified ion-exchange technique for the determination of stability constants of metal-soil organic matter complexes, *Soil Sci. Soc. Am. Proc.,* 36, 884, 1972.
59. **Mortenson, J. L.,** Partial extraction of organic matter, in *Methods of Soil Analyses,* Part 2, Black, C. A., Ed., American Society of Agronomy, Madison, Wisc., 1965.
60. **Stevenson, F. J.,** Gross chemical fractionation of organic matter, in *Methods of Soil Analyses,* Part 2, Black, C. A., Ed., American Society of Agronomy, Madison, Wisc., 1965.
61. **Bremner, J. M. and Lees, H.,** Studies on soil organic matter. II. The extraction of organic matter from soil by neutral reagents, *J. Agric. Sci.,* 39, 274, 1949.
62. **Bremner, J. M.,** Some soil organic matter problems, *Soil Fert.,* 19, 115, 1956.

63. **Choudri, M. B. and Stevenson, F. J.,** Chemical and physicochemical properties of soil colloids. III. Extraction of organic matter from soils, *Soil Sci. Soc. Am. Proc.,* 21, 508, 1957.
64. **Bremner, J. M.,** The amino-acid composition of the protein material in soil, *Biochem. J.,* 47, 538, 1950.
65. **Wolcott, A. R.,** Perspective on organic matter and nitrogen in soils. Outlines for a special course, *Soil Biochem.,* CCS 881, Michigan State Univ., E. Lansing, 1978.
66. **Mortensen, J. L. and Himes, F. L.,** Soil organic matter, in *Chemistry of the Soil,* Bear, F. E., Ed., Reinhold, New York, 1965, chap. 5.
67. **Loeppert, R. H.,** High Pressure Liquid Chromatography and Chemical Characterization of Extractable Soil Organic Matter, Ph.D. thesis, University of Florida, Gainesville, 1976.
68. **Hayes, M. H. B. and Swift, R. S.,** The chemistry of soil organic colloids, in *The Chemistry of Soil Constituents,* Greenland, D. J. and Hayes, M. H. B., Eds., John Wiley & Sons, New York, 1978, chap. 3.
69. **Loeppert, R. H. and Volk, B. G.,** Nature of organic matter extracted from Terra Ceia muck with selected solvents, *Soil Crop Sci. Soc. Fla. Proc.,* 33, 160, 1974.
70. **Demetriou, J. A., Macias, F. M., McArthur, M. J., and Beattie, J. M.,** Gel filtration chromatography of fluorescent phenolic and heterocyclic compounds, *J. Chromatogr.,* 34, 342, 1968.
71. **Posner, A. M.,** Importance of electrolyte in the determination of molecular weight by Sephadex gel filtration with special reference to humic acid, *Nature,* 198, 1161, 1963.
72. **Forss, K. G. and Stenlund, B. G.,** The influence of charged groups in gel permeation chromatography on polyelectrolytes, *J. Polym. Sci.,* 42, 951, 1975.
73. **Swift, R. S. and Posner, A. M.,** Gel chromatography of humic acid, *J. Soil Sci.,* 22, 237, 1971.
74. **Loeppert, R. H. and Volk, B. G.,** Applications of high-pressure liquid chromatography to studies of extractable soil organic matter, *Soil Organic Matter Studies,* Proc. IAEA-FAO Agrochemica Symposium (Braunschweig, 1976), 2, 213, 1977.
75. **Schnitzer, M. and Khan, S. U.,** *Humic Substances in the Environment,* Marcel Dekker, New York, 1972.
76. **Forsyth, W. G. C.,** Studies on the more soluble complexes of soil organic matter. I. A method of fractionation, *Biochem. J.,* 41, 176, 1947.
77. **MacDonald, A. M. G.,** The oxygen flask method, in *Advances in Analytical Chemistry and Instrumentation — 4,* Reilley, C. N., Ed., 1965, 75.
78. **Bremner, J. M.,** Organic forms of nitrogen, in *Methods of Soil Analysis,* Part 2, Black, C. A., Ed., American Society of Agronomy Madison, Wis. 1965, 1238.
79. **Allison, L. E.,** Organic carbon, in *Methods of Soil Analysis,* Part 2, Black, C. A., Ed., American Society of Agronomy, Madison, Wis., 1965, 1367.
80. **Martin, F. E., Dubach, P., Mehta, N. C., and Deuel, H.,** Bestimmung für funktionellen Gruppen von Huminstaffen, *Z. Pflanzenernähr, Düng. Bodenk.,* 103, 1963, 29.
81. **Brooks, J. D., Durie, R. A., and Sternhell, S.,** Chemistry of brown coals. I. Oxygen-containing functional groups in Victorian brown coals, *Aust. J. Appl. Sci.,* 8, 206, 1957.
82. **Wright, J. R. and Schnitzer, M.,** Oxygen containing functional groups in the organic matter of a podzol soil, *Nature,* 184, 1462, 1959.
83. **Schnitzer, M. and Gupta, U. C.,** Determination of acidity in soil organic matter, *Soil Sci. Soc. Am. Proc.,* 29, 274, 1965.
84. **Hubacher, M. H.,** Determination of carboxy groups in aromatic acids, *Anal. Chem.,* 21, 945, 1949.
85. **Wright, J. R. and Schnitzer, M.,** Oxygen containing functional groups in the organic matter of the Ao and Bh horizon of a podzol, *Trans. 7th Int. Congr. Soil Sci.,* Madison, Wis., 2, 120, 1960.
86. **Schnitzer, M. and Skinner, S. I. M.,** The carboxyl group in soil organic matter preparations, *Soil Sci. Soc. Am. Proc.,* 29, 400, 1965.
87. **Flaig, W., Scheffer, F., and Klamroth, B.,** Zur Kenntnis der Huminsäuren. VIII. Zur Charakterisierung der Huminsäuren des Bodens, *Z. Pflanzenernähr Düng. Bodenk.,* 7, 33, 1955.
88. **Schnitzer, M. and Skinner, S. I. M.,** A polarographic method for the determination of carbonyl groups in soil humic compounds, *Soil Sci.,* 101, 120, 1966.
89. **Vasilyevskaya, N. A., Glebko, L. I., and Maximov, O. B.,** Determination of humic acid quinoid groups, *Sov. Soil Sci.,* 4, 63, 1971.
90. **Berger, St. and Rieker, A.,** Identification and determination of quinones, in *The Chemistry of the Quinonoid Compounds,* Part 1, Patali, S., Ed., John Wiley & Sons, New York, 1974, 163.
91. **Steelink, C., Reid, T., and Tollin, G.,** On the nature of the free-radical moiety in lignin, *J. Am. Chem. Soc.,* 85, 4048, 1963.
92. **Van Krevelen, D. W.,** *Coal,* Elsevier, Amsterdam, 1961.
93. **Steyermark, A.,** *Quantitative Organic Microanalysis,* 2nd ed., Academic Press, N.Y., 1961.
94. **Stevenson, F. J. and Goh, K. M.,** Infrared spectra of humic acids and related substances, *Geochim. Cosmochim. Acta,* 35, 471, 1971.

95. **Flaig, W., Deutelspacher, H., and Rietz, E.,** Chemical composition and physical properties of humic substances, in *Soil Components,* Vol. 1, Gieseking, J. E., Ed., Springer-Verlag, New York, 1975.
96. **Mortenson, J. L. and Himes, F. L.,** Soil organic matter, in *Chemistry of the Soil,* Bear, F. E., Ed., Reinhold, New York, 1965.
97. **Stevenson, F. J. and Goh, K. M.,** Infrared spectra of humic acids: elimination of interference due to hygroscopic moisture and changes accompanying heating with KBr, *Soil Sci.,* 117, 34, 1974.
98. **Berkheiser, V. E.,** personal communication, 1980.
99. **Schnitzer, M.,** Characterization of humic constituents by spectroscopy, in *Soil Biochemistry,* Vol. 2, McLaren, A. D. and Skujins, J., Eds., Marcel Dekker, New York, 1971.
100. **Stuermer, D. H. and Payne, J. R.,** Investigations of sea water and terrestrial humic substances with carbon-13 and proton nuclear magnetic resonance, *Geochim. Cosmochim. Acta,* 40, 1109, 1976.
101. **Lentz, H., Ludemann, H., and Ziechmann, W.,** Proton resonance spectra of humic acid from the solum of a podzol, *Geoderma,* 18, 325, 1977.
102. **Sciacovelli, O., Senesi, N., Solinas, V., Testini, C.,** Spectroscopic studies of soil organic fractions. I. IR and NMR spectra, *Soil Biol. Biochem.,* 9, 287, 1977.
103. **Ruggiero, O., Sciacovelli, O., Testini, C., Interesse, F. S.,** Spectroscopic studies on soil organic fractions. II. IR and ^{1}H-NMR spectra of methylated and unmethylated fulvic acids, *Geochim. Cosmochim. Acta,* 42, 411, 1978.
104. **Loeppert, R. H. and Volk, B. G.,** ^{1}H-NMR spectroscopy of soils humic and fulvic acids, *Soil Sci. Am. J.,* in review.
105. **Villa, F., Lentz, H., and Ludemann, H.,** FT-^{13}C nuclear magnetic spectra of natural humic substances, *Biochem. Biophys. Res. Commun.,* 72, 1063, 1976.
106. **Grant, D.,** Chemical structure of humic structures, *Nature,* 270, 709, 1977.
107. **Wilson, M. A. and Goh, A. M.,** Proton-decoupled pulse Fourier-transform ^{13}C magnetic resonance of soil organic matter, *J. Soil Sci.,* 28, 645, 1977.
108. **Ogner, G.,** The ^{13}C nuclear magnetic resonance spectrum of methylated humic acid, *Soil Biol. Biochem.,* 11, 105, 1979.
109. **Ward, S. G.,** Coal — its constitution and utilisation as a chemical and as a raw material, *J. Inst. Fuel,* 21, 67, 1947.
110. **Schnitzer, M.,** Recent findings on the characterization of humic substances extracted from soils from widely differing climatic zones, in *Soil Organic Matter Structures,* Proc. IAEA-FAO-Agrochemica Symp., 2, 117, 1977.
111. **Gottlieb, S. and Hendricks, S. B.,** Soil organic matter as related to newer concepts of lignin chemistry, *Soil Sci. Soc. Am. Proc.,* 10, 117, 1945.
112. **Morrison, R. I.,** The alkaline nitrobenzene oxidation of soil organic matter, *J. Soil Sci.,* 9, 130, 1958.
113. **Morrison, R. I.,** Products of the alkaline nitrobenzene oxidation of soil organic matter, *J. Soil Sci.,* 14, 201, 1963.
114. **Wildung, R. E., Chesters, G., and Behmer, D. E.,** Alkaline nitrobenzene oxidation of plant lignins and soil humic colloids, *Plant Soil,* 32, 221, 1970.
115. **Wallis, A. F. A.,** Solvolysis by acids and bases, in *Lignins,* Sarkanen, K. V. and Ludwig, C. H., Eds., John Wiley & Sons, New York, 1971, 345.
116. **Chang, H. M. and Allan, G. G.,** Oxidation, in *Lignins,* Sarkanen, K. V. and Ludwig, C. H., Eds., John Wiley & Sons, New York, 1971, 433.
117. **Griffith, S. M. and Schnitzer, M.,** The alkaline cupric oxide oxidation of humic and fulvic acids extracted from tropical volcanic soils, *Soil Sci.,* 122, 191, 1976.
118. **Dence, C. W.,** Halogenation and nitration, in *Lignins,* Sarkanen, K. V. and Ludwig, C. H., Eds., John Wiley & Sons, New York, 1971, 373.
119. **Hayashi, T. and Nagai, T.,** On the components of soil humic acids. VIII. The oxidative decomposition by alkaline potassium permanganate and nitric acid, *Soil Plant Food,* 6, 170, 1961.
120. **Hansen, E. H. and Schnitzer, M.,** Nitric acid oxidation of Danish illuvial organic matter, *Soil Sci. Soc. Am. Proc.,* 31, 79, 1967.
121. **Yokakawa, C., Watanabe, Y., Kajiama, S., and Takiegami, Y.,** Studies on the chemical structure of coal, I. Stepwise oxidative degradation of coal, *Fuel,* 41, 209, 1962.
122. **Burges, N. A., Hurst, H. M., and Walken, B.,** The phenolic constituents of humic acid and their relation to the lignin of plant cover, *Geochim. Cosmochim. Acta,* 28, 1547, 1964.
123. **Stevenson, F. J. and Mendez, J.,** Reductive cleavage products of soil humic acids, *Soil Sci.,* 103, 383, 1967.
124. **Dormaar, J. F.,** Reductive cleavage of humic acid of chernozem soils, *Plant Soil,* 31, 182, 1969.
125. **Schnitzer, M. and Ortiz de Serra, M. I.,** The chemical degradation of a humic acid, *Can. J. Chem.,* 51, 1554, 1973.
126. **Schnitzer, M. and Ortiz de Serra, M. I.,** The sodium-amalgam reduction of soil and fungal humic substances, *Geoderma,* 9, 119, 1973.

127. **Martin, J. P., Haider, K., and Saiz-Jimenez, C.**, Sodium amalgam reductive degradation of fungal and model phenolic polymers, soil humic acids and simple phenolic compounds, *Soil Sci. Soc. Am. Proc.*, 38, 760, 1974.
128. **Piper, T. J. and Posner, A. M.**, Sodium amalgam reduction of humic acid. I. Evaluation of the method. II. Application of the method, *Soil Biol. Biochem.*, 4, 513, 1972.
129. **Cheshire, M. V., Cranwell, P. A., Falshaw, C. P., Floyd, A. J., and Haworth, R. D.**, Humic acid. II. Structure of humic acids, *Tetrahedron*, 23, 1669, 1967.
130. **Cheshire, M. V., Cranwell, P. A., and Haworth, R. D.**, Humic acid. III, *Tetrahedron*, 24, 5155, 1968.
131. **Hansen, E. H. and Schnitzer, M.**, Zinc-dust distillation of soil humic compounds, *Fuel*, 48, 41, 1969.
132. **Hansen, E. H. and Schnitzer, M.**, Zinc-dust distillation and fusion of a soil humic acid, *Soil Sci. Soc. Am. Proc.*, 33, 29, 1969.
133. **Felbeck, G. T.**, Studies on high pressure hydrogenolysis of organic matter from a muck soil, *Soil Sci. Soc. Am. Proc.*, 29, 48, 1965.
134. **Felbeck, G. T.**, Normal alkanes in muck soil organic matter hydrogenolysis products, *Int. Soil Sci. Soc.*, Trans. II and IV Comm., Aberdeen, 1966, 11, 1967.
135. **Maximov, O. B. and Krasaskaya, N. P.**, Action of metallic sodium on humic acids in liquid ammonia, *Geoderma*, 18, 227, 1977.
136. **Jackson, M. P., Swift, R. S., Posner, A. M., and Knox, J. R.**, Phenolic degradation of humic acid, *Soil Sci.*, 114, 75, 1972.
137. **Swift, R. S.**, Physico-chemical Studies on Soil Organic Matter, Ph.D. thesis, University of Birmingham, England, 1968.
138. **Craggs, J. D., Hayes, M. H. B., and Stacey, M.**, Sodium sulphide reactions with humic acid and model compounds, *Trans. 10th Int. Congr. Soil Sci., Moscow*, 2, 318, 1974.
139. **Hayes, M. H. B., Stacey, M., and Swift, R. S.**, Degradation of humic acid in a sodium sulphide solution, *Fuel*, 51, 211, 1972.
140. **Schnitzer, M. and Skinner, S. I. M.**, The peracetic acid oxidation of humic substances, *Soil Sci.*, 118, 322, 1974.
141. **Schnitzer, M. and Skinner, S. I. M.**, The low temperature oxidation of humic substances, *Can. J. Chem.*, 52, 1072, 1974.
142. **Ross, S. D., Finkelstein, M., and Rudd, E. J.**, *Anoidic Oxidation*, Academic Press, New York, 1975.
143. **Bühler, C. and Simon, W.**, Curie point pyrolysis gas chromatography, *J. Chromatogr. Sci.*, 8, 323, 1970.
144. **Simmonds, P. G., Shulman, G. P., and Stembridge, C. H.**, Organic analysis by pyrolysis-gas chromatography-mass spectrometry. A candidate experiment for the biological exploration of Mars, *J. Chromatogr. Sci.*, 7, 36, 1969.
145. **Wershaw, R. L. and Bohner, G. E.**, Pyrolysis of humic and fulvic acids, *Geochim. Cosmochim. Acta*, 33, 757, 1969.
146. **Martin, F.**, Pyrolysis gas chromatography of humic substances from different origin, *Z. Pflanzenern. Bodenk.*, 465, 407, 1975.
147. **Nagar, B. R., Waight, E. S., Neuzelaar, H. L. C., and Kistemaker, P. G.**, Studies on the structure and origin of soil humic acids by Curie point pyrolysis in direct combination with low voltage ionization mass spectrometry, *Plant Soil*, 43, 681, 1975.
148. **Martin, F., Saiz-Jimenez, C., and Cert, A.**, Pyrolysis-gas chromatography-mass spectrometry of soil humic fractions. I. The low boiling point compounds, *Soil Sci. Soc. Am. J.*, 41, 1114, 1977.
149. **Haider, K., Nagar, B. R., Saiz, C., Meuzelaar, H. L. C., and Martin, J. P.**, Studies on soil humic compounds, fungal melanins, and model polymers by pyrolysis mass spectrometry, in *Soil Organic Matter Studies*, Proc. IAEA-FAO-Agrochemica Symp., 2, 213, 1977.
150. **Meuzelaar, H. L. C., Haider, K., Nagar, B. R., and Martin, J. P.**, Comparative studies of pyrolysis-mass spectra of melanins, model phenolic polymers, and humic acids, *Geoderma*, 17, 239, 1977.
151. **Hayes, M. H. B., Swift, R. S., Wardel, R. E., and Brown, J. K.**, Humic materials from an organic soil: a comparison of extractants and of properties of extracts, *Geoderma*, 13, 231, 1975.
152. **Loeppert, R. H. and Volk, B. G.**, Nature of organic matter extracted from Terra Ceia muck with selected solvents, *Soil Crop Sci. Soc. Fla. Proc.*, 33, 160, 1973.
153. **Haider, K. and Martin, J. P.**, Decomposition of specifically carbon-14-labeled benzoic and cinnamic acid derivatives in soil, *Soil Sci. Soc. Am. Proc.*, 39, 657, 1975.
154. **Martin, J. P. and Haider, K.**, Decomposition of specifically carbon-14-labelled ferullic acid: free and linked into model humic-acid type polymers, *Soil Sci. Soc. Am. J.*, 40, 377, 1976.
155. **Martin, J. P. and Haider, K.**, Decomposition in soil of specifically ^{14}C-labeled DHP and corn stalk lignins, polymers, and coniferyl alcohols, in *Soil Organic Matter Studies*, Proc. IAEA-FAO-Agrochemica Symp., 2, 23, 1977.
156. **Haider, K., Martin, J. P., and Rietz, E.**, Decomposition in soil of ^{14}C-labeled coumaryl alcohols: free and linked into dehydropolymer and plant lignins, *Soil Sci. Soc. Am. J.*, 41, 556, 1977.

157. **Flaig, W.,** Organic compounds in soil, *Soil Sci.,* 111, 19, 1971.
158. **Lowe, L. E.,** Carbohydrates in soil, in *Soil Organic Matter,* Schnitzer, M. and Khan, S. U., Eds., Elsevier, Amsterdam, 1978, 65.
159. **Mehta, N. C., Dubach, P., and Deuel, H.,** Carbohydrates in the soil, *Adv. Carbohydr. Chem.,* 16, 1961, 335.
160. **Gupta, U. C.,** Carbohydrates, in *Soil Biochemistry,* Vol. 1, McLaren, A. D. and Peterson, G. H., Eds., Marcel Dekker, New York, 1967, 91.
161. **Swincer, G. D., Oades, J. M., and Greenland, D. J.,** Extraction, characterization, and significance of soil polysaccharides, *Adv. Agron.,* 21, 1969, 195.
162. **Finch, P., Hayes, M. H. B., and Stacey, M.,** The biochemistry of soil polysaccharides in *Soil Biochemistry,* Vol. 2, McLaren, A. D. and Skujins, J. J., Eds., Marcel Dekker, New York, 1971, 254.
163. **Greenland, D. J. and Oades, J. M.,** Saccharides, in *Soil Components,* Vol. 1, Gieseking, J. E., Ed., Springer Verlag, Berlin, 1975, 213.
164. **Gupta, U. C. and Sowden, F. J.,** Studies on methods for the determination of sugars and uronic acids in soils, *Can. J. Soil Sci.,* 45, 237, 1965.
165. **Cheshire, M. V. and Mundie, C. M.,** The hydrolytic extraction of carbohydrates from soil by sulphuric acid, *J. Soil Sci.,* 17, 372, 1966.
166. **Parsons, J. W. and Tinsley, J.,** Chemical studies of polysaccharide material in soils and composts based on extraction with anhydrous formic acid, *Soil Sci.,* 92, 46, 1961.
167. **Swincer, G. D., Oades, J. M., and Greenland, D. J.,** Studies on soil polysaccharides. I. The isolation of polysaccharides from soil. II. The composition and properties of polysaccharides in soils under pasture and under fallow wheat rotation, *Aust. J. Soil Res.,* 6, 211, 1968.
168. **Clapp, C. E.,** High Molecular Weight Water-Soluble Muck. Isolation and Determination of Constituent Sugars of a Borate Complex Forming Polysaccharide, Employing Electrophoretic Techniques, Ph.D. thesis, Cornell University, Ithaca, New York, 1957.
169. **Lee, Y. C. and Montgomery, R.,** Separations with molecular sieves, in *Methods in Carbohydrate Chemistry 5,* Whistler, R. L., Ed., 1965, 28.
170. **Churms, S. C.,** Gel chromatography of carbohydrates, *Adv. Carbohydr. Res.,* 25, 13, 1970.
171. **Clapp, C. E. and Davis, R. J.,** Properties of extracellular polysaccharides from *Rhizobium, Soil Biol. Biochem.,* 2, 109, 1970.
172. **Neukom, H. and Kuendig, W.,** Fractionation on diethylaminoethylcellulose columns: fractionation of neutral and acidic polysaccharides by ion-exchange column chromatography on diethylaminoethyl (DEAE)-cellulose, in *Methods in Carbohydrate Chemistry 5,* Whistler, R. L., Ed., 1965, 14.
173. **Parfitt, R. L. and Greenland, D. J.,** Adsorption of polysaccharides by montmorillonite, *Soil Sci. Soc. Am. Proc.,* 34, 862, 1970.
174. **Martin, J. P. and Aldrich, D. G.,** Influence of soil exchangeable cation ratios on the aggregating effects of natural and synthetic soil conditions, *Soil Sci. Soc. Am. Proc.,* 19, 50, 1955.
175. **Clapp, C. E., Davis, R. J., and Waugaman, S. H.,** The effect of rhizobial polysaccharides on aggregate stability, *Soil Sci. Soc. Am. Proc.,* 26, 466, 1962.
176. **Martin, J. P. and Richards, S. J.,** Influence of copper, zinc, iron, and aluminum salts of some microbial and plant polysaccharides on aggregation and hydraulic conductivity of Ramona sandy loam, *Soil Sci. Soc. Am. Proc.,* 33, 421, 1969.
177. **Greenland, D. J., Lindstrom, G. R., and Quirk, J. P.,** Organic materials which stabilize natural soil aggregates, *Soil Sci. Soc. Am. Proc.,* 26, 366, 1962.
178. **Acton, C. J., Paul, E. A., and Rennie, D. A.,** Measurement of the polysaccharide content of soils, *Can. J. Soil Sci.,* 43, 141, 1963.
179. **Webber, L. R.,** Soil polysaccharides and aggregation in crop sequences, *Soil Sci. Soc. Am. Proc.,* 29, 39, 1965.
180. **Clapp, C. E. and Emerson, W. W.,** The effect of periodate oxidation on the strength of soil crumbs. I. Qualitative studies. II. Quantitative studies, *Soil Sci. Soc. Am. Proc.,* 29, 127, 1965.
181. **Stefanson, R. C.,** Effect of periodate and pyrophosphate on the seasonal changes in aggregate stabilization, *Aust. J. Soil Res.,* 9, 33, 1971.
182. **Mehta, N. C., Streuli, H., Müller, M., and Deueul, H.,** Role of polysaccharides in soil aggregation, *J. Sci. Food Agric.,* 11, 40, 1960.
183. **Stefanson, R. C.,** Factors determining seasonal changes in the stabilities of soil aggregates, *Trans. 9th Congr. Int. Soil Sci. Soc.,* 2, 395, 1968.
184. **Martin, J. P.,** Decomposition and binding action of polysaccharides in soil, *Soil Biol. Biochem.,* 3, 33, 1971.
185. **Cheshire, M. V., Greaves, M. P., and Mundie, C. M.,** Decomposition of soil polysaccharides, *J. Soil Sci.,* 25, 483, 1974.
186. **Lynch, D. L. and Cotnoir, L. J.,** The influence of clay minerals on the breakdown of certain organic substrates, *Soil Sci. Soc. Am. Proc.,* 20, 367, 1956.

187. **Tan, K. H. and McCreery, R. A.**, The infrared identification of a humopolysaccharide ester in soil humic acid, *Soil Sci. Pl. Anal.*, 1, 75, 1970.
188. **Bradley, D. B. and Sieling, D. H.**, Effect of organic anions and sugars on phosphate precipitation by iron and aluminum as influenced by pH, *Soil Sci.*, 76, 175, 1953.
189. **Schnitzer, M. and Delong, W. A.**, Investigations on the mobilization and transport of iron in forested soils. II. The nature of the reaction of leaf extracts and leachates of iron, *Soil Sci. Soc. Am. Proc.*, 19, 363, 1955.
190. **Bloomfield, C.**, The deflocculation of kaolinite by aqueous leaf extracts, the role of certain constituents of the extracts, *Trans. 6th Int. Congr. Soil Sci.*, 27, 1956.
191. **Dormaar, J. F.**, Seasonal pattern of water-soluble constituents from leaves of Populus X 'Northwest' (Hort.), *J. Soil Sci.*, 21, 105, 1970.
192. **Forsyth, W. G. C.**, Studies on the more soluble complexes of soil organic matter. II. The composition of the soluble polysaccharide fraction, *Biochem. J.*, 46, 141, 1950.
193. **Whistler, R. L. and Kirby, K. W.**, Composition and behavior of soil polysaccharides, *J. Am. Chem. Soc.*, 78, 1755, 1956.
194. **Gupta, V. C., Sowden, F. J., and Stobbe, P. C.**, The characterization of carbohydrate constituents from different soil profiles, *Soil Sci. Soc. Am. Proc.*, 27, 380, 1963.
195. **Whitehead, D. C., Buchan, H., and Hastley, R. D.**, Components of soil organic matter under grass and arable cropping, *Soil Biol. Biochem.*, 7, 65, 1975.
196. **Guidi, G., Petruzzelli, G., and Sequi, P.**, Characterization of amino acid and carbohydrate components in fulvic acid, *Can. J. Soil Sci.*, 56, 159, 1976.
197. **Lowe, L. E.**, Amino acid distribution in forest humus layers in British Columbia, *Soil Sci. Soc. Am. Proc.*, 37, 569, 1973.
198. **Freney, J. R., Stevenson, F. J., and Beavers, A. H.**, Sulfur-containing amino acids in soil hydrolysates, *Soil Sci.*, 114, 468, 1972.
199. **Finlayson, A. J. and MacKenzie, S. L.**, A rapid method for methionine determination in plant materials, *Anal. Biochem.*, 70, 397, 1976.
200. **Bremner, J. M., Studies on soil organic matter.** I. The chemical nature of soil organic nitrogen, *J. Agric. Sci.*, 39, 183, 1949.
201. **Sowden, F. J.**, Nature of the amino acid compounds of soil. II. Amino acids and peptides produced by partial hydrolysis, *Soil Sci.*, 102, 264, 1966.
202. **Cheng, C. N.**, Extracting and desalting amino acids from soils and sediments: evaluation of methods, *Soil Biol. Biochem.*, 7, 319, 1975.
203. **Bremner, J. M.**, *Soil Biochemistry*, McLaren, A. D. and Peterson, G. H., Eds., Marcel Dekker, New York, 1967, chap. 2.
204. **Stevenson, F. J.**, Effect of some long-term rotations on the amino acid composition of the soil, *Soil Sci. Soc. Am. Proc.*, 20, 204, 1956.
205. **Sowden, F. J., Griffith, S. M., and Schnitzer, M.**, The distribution of nitrogen in some highly organic tropical volcanic soils, *Soil Biol. Biochem,*. 8, 55, 1976.
206. **Keeney, D. R. and Bremner, J. M.**, Effect of cultivation on the nitrogen distribution in soils, *Soil Sci. Soc. Am. Proc.*, 28, 653, 1964.
207. **Piper, T. J. and Posner, A. M.**, On the amino acids found in humic acids, *Soil Sci.*, 106, 188, 1968.
208. **Griffith, S. M., Sowden, F. J., and Schnitzer, M.**, The alkaline hydrolysis of acid-resistant soil and humic acid residues, *Soil Biol. Biochem.*, 8, 529, 1976.
209. **Freney, J. R.**, The extraction and partial characterization of nonhydrolysable nitrogen in soil, *Trans. 9th Int. Congr. Soil Sci.*, 3, 531, 1968.
210. **Biederbeck, V. O. and Paul, E. A.**, Fractionation of soil humate with phenolic solvents and purification of the nitrogen-rich portion with polyvinylpyrrolidone, *Soil Sci.*, 115, 357, 1973.
211. **Sequi, P., Guidi, G., and Petruzzelli, G.**, Distribution of amino acid and carbohydrate compounds in fulvic and fractionated on polyamide, *Can. J. Soil Sci.*, 55, 439, 1975.
212. **Felbeck, G. T., Jr.**, Structural hypotheses of soil humic acid, *Soil Sci.*, 111, 42, 1971.
213. **Schnitzer, M., Sowden, F. J., and Ivarson, K. C.**, Humic acid reaction with amino acids, *Soil Biol. Biochem.*, 6, 401, 1974.
214. **Mortland, M. M.**, Clay-organic complexes and interactions, *Adv. Agron.*, 22, 75, 1970.
215. **Chichester, F. W.**, Nitrogen in soil organo-mineral sedimentation fractions, *Soil Sci.*, 107, 356, 1969.
216. **Swift, R. S. and Posner, A. M.**, The distribution and extraction of soil nitrogen as a function of soil particle size, *Soil Biol. Biochem.*, 4, 181, 1972.
217. **Anderson, D. W., Paul, E. A., and St. Arnaud, R. J.**, Extraction and characterization of humus with reference to clay-associated humus, *Can. J. Soil Sci.*, 54, 317, 1974.
218. **Watson, J. R. and Parsons, J. W.**, Studies of soil organo-mineral factors. I. Isolation by ultrasonic dispersion, *J. Soil Sci.*, 25, 1, 1974.
219. **Calhoun, F. G., Carlisle, V. W., Caldwell, R. E., Zelazny, L. W., Hammond, L. C., and Breland, H. L.**, Characterization data for selected Florida soils, *Soil Sci.*, Res. Report 74-1, 1974.

220. **Young, J. L.**, Inorganic soil nitrogen and carbon nitrogen ratios of some Pacific Northwest soils, *Soil Sci.*, 93, 397, 1962.
221. **Parsons, J. W. and Tinsley, J.**, Nitrogenous substances, in *Soil Components*, Vol. 1, Gieseking, J. E., Ed., Springer-Verlag, New York, 1975, chap. 3.
222. **Rodrigues, G.**, Fixed ammonia in tropical soils, *J. Soil Sci.*, 5, 264, 1954.
223. **Swift, R. S.**, Mineralization of nitrogen and sulphur from soil organic matter, in *Soil Organic Matter Studies*, Proc. Symp. FAO/IAEA, 1976, Braunschweig, Vol. 1, 1977, 275.
224. **Tisdale, S. L. and Nelson, W. L.**, Soil and fertilizer nitrogen, in *Soil Fertility and Fertilizers*, Tisdale, S. L. and Nelson, W. L., Eds., MacMillan, New York, 1975, 130.
225. **Paul, E. A. and Van Veen, J. A.**, The use of tracers to determine the dynamic nature of organic matter, in *Symposia Papers*, 11th Congress, International Society of Soil Science, June 1978, 61.
226. **Buckman, H. O. and Brady, N. C.**, *The Nature and Properties of Soils*, 7th ed., MacMillan, New York, 1969.

SOIL FERTILITY, FERTILIZERS, AND PLANT NUTRITION

Robert D. Munson

INTRODUCTION

For ages man has attempted to increase his food supply to gain more control over his destiny. Food supply and soil productivity are directly related to soil fertility, fertilizer use, and plant nutrition. Early Greeks and Romans observed differences among soils and management practices. By the early 1800s scientific studies were underway to determine which elements plants required and how yields varied with the nutrient supply. Von Liebig, known as the father of agricultural chemistry, directed a great deal of thought and study to the fertility of soils. Gilbert and Lawes in England conducted some of the first systematic experiments using manufactured fertilizers for improving and maintaining soil fertility. Some of those early studies are still underway today at the Rothamsted Experimental Station.

Soil fertility is a general term directly related to the ability of the soil to supply the elements essential for crop growth, development and maturation for high yields. The degree of soil fertility is related to the capacity of soils to supply the essential elements at the rates and in the amounts required to produce high-yielding, high-quality crops on a sustained basis.

The soil system is very complex as one can conclude from Chapters 2, 3, 4, 5, 8, and 9. These physical, chemical, and biological factors interact in the presence of climate to influence the availability, uptake, and utilization of essential nutrients or elements and hence the fertility of soil. Man's objective in studying soil fertility is to determine action that can be taken to improve the nutrient availability and make soils more productive. In one situation that might mean draining the soil to improve it as a medium in which roots can grow, while in another it may be determining how much fertilizer containing various essential elements must be added to supply a crop with adequate nutrition for optimum production in a given environment.

Many excellent books have been written on soil fertility, fertilizers and plant nutrition, and their interrelated aspects. There are general books on soils and soil-plant relationships as related to soil fertility.[1-4] Soil fertility is is intimately linked to fertilizer use.[5-9]

Specific books or journals on the individual essential elements required by plants as related to soils, crops, and fertilizer use have been published on phosphorus,[10] potassium,[11-13] magnesium,[14-16] and calcium.[17] The benefits of liming in soil fertility have been covered in a book edited by Pearson and Adams.[18] The micronutrient role in soil fertility and fertilizer use has been discussed in depth in books edited by Mortvedt et al.[19] and Nicholas and Egan.[20] In-depth discussions of mineral and plant nutrition are included in books edited or authored by Stewart,[21] Arnon,[22] and Mengel and Kirkby.[23]

For the rèader searching for more detail than is possible in this chapter, the above will be helpful. However, readers must be mindful that a steady flow of new research and published journal articles continues to broaden the scope and documentation on soil fertility, fertilizer, and plant nutrition on the soils of the world.

ESSENTIAL ELEMENTS FOR PLANT GROWTH

The chemical elements essential for plant growth and crop production are shown in Table 1. The test for the essentiality of a given element is whether or not the plant life cycle can be completed in its absence. For most crops 16 chemical elements are essential. In most soil fertility literature elements and nutrients are used interchangeably.

Table 1
CHEMICAL ELEMENTS ESSENTIAL FOR MOST PLANTS OR CROPS, SYMBOL, AND FORM UTILIZED

Major or macronutrients	Secondary nutrients	Trace or micronutrients
Carbon (C), CO_2	Calcium (Ca), Ca^{2+}	Boron (B), $H_2BO_3^-$
Oxygen (O), CO_2	Magnesium (Mg), Mg^{2+}	Chlorine (Cl), Cl^-, Cl_2
Hydrogen (H), H_2O	Sulfur (S), SO_4^{2-}, SO_2	Cobalt (Co), Co^{2+}
Nitrogen (N), NH_4^+, NO_3^-, N_2 (Legumes)		Copper (Cu), Cu^{2+}
		Iron (Fe), Fe^{2+}
Phosphorus (P), $H_2PO_4^-$, HPO_4^{2-}		Manganese (Mn), Mn^{2+}
		Molybdenum (Mo), MoO_4^{2-}
Potassium (K), K^+		Zinc (Zn), Zn^{2+}

These nutrients are often referred to as major or macro, secondary, and trace or micronutrients, based on the relative amounts and their role in the nutrition of the crop. Each essential element has a specific role in the physiology and nutrition of the crop, but they all function in concert to produce optimum yields. About 95% of most crops are composed of C, H, and O which come from the carbon dioxide (CO_2) of the air and water (H_2O) from the soil. It should be remembered soil moisture is extremely important in the uptake and utilization of all the essential elements. The other 13 or 14 elements are primarily taken up in ionic form from the soil solution,[24] although S and Cl can be absorbed from the air as SO_2 and Cl_2. Legumes, of course, are capable of fixing N_2 through the symbiotic relationship of *Rhizobia* in root nodules. While other macronutrients are dwarfed by C, H, and O, their relative requirements are much greater than the secondary or micronutrients.

Sodium (Na), silicon (Si), cobalt (Co) and vanadium (V) at times may produce yield increases, but they are not essential elements by normal definition. Sodium shows a partial substitution for K and has a complimentary effect on yield for crops such as sugar beets and celery. Cobalt, on the other hand, increases the yield of legumes because of its role in the formation of vitamin B_{12} needed by nitrogen fixing *Rhizobia* in root nodules.[25] It is not required by nonleguminous plants.

Fertilizers are substances containing essential elements,[6-9,26] that are applied to soils to improve the productivity of those found to be deficient in their native state or following cropping. They, also, are applied annually or periodically to maintain yields and keep soils in a high state of productivity or fertility. They are usually inorganic or organic salts, but may be added as organic crop residues, animal manures, or sewage sludge. Predicting the amount of fertilizer or nutrients to add for a specific crop on a specific soil is determined by chemical soil tests, crop requirement, past management, past yields, and crop response results and environmental conditions.[27]

The Role of Essential Elements and Deficiency Symptoms

Each essential element has an individual and collective role in nutrition[28] during the growth, development, and maturation of a crop. Elements are absorbed primarily in inorganic ionic form.[24] Various elements are taken up and utilized at different rates and amounts, interacting metabolically to influence the biochemical and biophysical characteristics of plants and the final yield. When an essential element is or becomes limiting or stressed, biochemical and biophysical changes occur in plants which produce rather specific visible changes or deficiency symptoms[29] for various crops depending upon their age or stage of growth. These are often referred to as "hunger signs". The greater the degree of deficiency, the greater the yield loss.

The basic objective is to use these elements wisely so that no nutrient stress occurs to reduce CO_2 fixation through photosynthesis. Thus the highest rate of photosynthesis will be achieved to produce the highest possible edible, usable, or marketable yield of a quality product. It is in this context that control of the essential elements is studied.

The role of the essential mineral elements and the symptoms produced in the developing crops when they are deficient will be discussed.

Nitrogen

Nitrogen is the major element in amino and nucleic acids, the buiding blocks of protein, enzymes, and genetic cellular material. It, along with properly formed chlorophyll when other essential elements are adequate, contributes to the dark green color of leaves, vigor, and productivity of plants and crops.

Crops without adequate N are stunted, light green in color and unproductive. In large monocotyledonous plants (grasses), such as corn or sugar cane, a V-shaped tip firing or yellowing occurs that extends up the mid-rib of the leaves. The symptoms show up on the lower leaves first, because N is very mobile, being translocated to the meristematic tissue and rapidly expanding new leaves. With serious deficiency these lower leaves dry up and die prematurely.

In dicotyledonous plants, leaves develop a pale light green color. Plants are stunted. Lower leaves dry up and drop off the plant as the N is mobilized and transported to newly developing tissue. Yields of forage or forage seed, grain, or fruit are seriously reduced by N deficiency.

Phosphorus

Phosphorus is a key portion of RNA and DNA molecules containing the coding for protein formation and cellular genetic information.

Phosphorus is intimately tied to plant energy reactions of photosynthesis and respiration through phosphorylation and conversion of ATP, which are so necessary for capturing and recycling of energy within plants.

Phosphorus is bound and stored as phospholipids in cellular membranes. Also it is laid down and stored in seeds as phytin. Stored P in seed is the energy source for its germination and seedling development for the next crop or generation.

With low P concentrations, plants are stunted and leaves may have a yellow, bronze, or purple color. Yields are reduced and maturity is often delayed.

Potassium

Potassium is the major univalent cation required by plants. Concentrations are very high in meristematic tissue where cells are rapidly dividing and growing. It has many roles in plant nutrition involving maintenance of electro-neutrality, photosynthesis, enzyme activation, osmotic regulation, and photosynthate transport and storage.[30,31] For example, K has been shown to influence the formation and photosynthetic activity of chloroplasts at the subcellular level in leaves. It also influences the opening and closing of stomates and the size of the stomate opening by concentrating in the guard cells when leaves are exposed to light.[32,33] Thus, it affects the quantity and rate of CO_2 entering the leaf for photosynthesis.

Potassium has been shown to activate over 60 enzymes,[34] including such important ones as pyruvic kinase and nitrate reductase. It is closely tied to the production of ATP and the plant's "energy system". In addition, it is related to phloem loading affecting the rate of photosynthate transport, as well as the deposition of starch in seed through starch synthetase.

Plants deficient in K appear chlorotic, with marginal scorching on the leaves of grasses. Deficiencies first appear on the lower leaves, because K is mobile and it is redistributed to metabolically active tissue. Carbohydrates within the young crop may be low, with early leaf drop of lower leaves, followed by bud, flower, and pod or fruit drop, or unfilled grain, reducing yields. Root development is reduced, reducing exploitation of the soil for moisture and other nutrients.

Calcium

The role of Ca involves enzyme activation and inhibition, membrane structure and activity, cell wall structure, and interaction with other elements.[17,35] Calcium concentrations are usually high in areas of rapid cell division or high metabolic activity. Calcium is essential for root activity and growth and functions to influence the selectivity of ions in nutrient uptake. Growing points and leaves become malformed, inactive, and die with increasing Ca deficiency. In cell wall structure Ca stabilizes cells, being laid down as Ca pectate in the middle lamella.

Calcium also interacts with other elements. In some cases, it can have a sparing effect on micronutrients, while in others it may induce the need for increased micronutrient applications. Calcium mobility in some crops has been shown to be related to Cu, with Ca deficiency symptoms being related to low Cu.

Magnesium

Magnesium occupies the center of the chlorophyll molecule and is directly related to photosynthesis. It also activates enzymes, especially those associated with energy transfer. The enzymes activated are associated with photosynthesis, glycolysis, tricarboxylic acid cycle, respiration, and sulfate reduction.[36] Magnesium concentration also influences the proper formation of cellular organells, as well as carbohydrate metabolism and protein synthesis.

Magnesium needs and uptake are interrelated with K, Ca, and P. The source of N can have an effect on Mg uptake. Ammonium ion (NH_4^+) can have a depressive effect on Mg uptake unless the soil Mg is high. Heavy applications of K may decrease Mg uptake. Also, high concentrations of naturally occurring Al in acid soils depress Mg uptake and utilization.

With Mg deficiency, monocots exhibit a light green color with striping. Dicotyledonous plants exhibit a chlorosis, followed by a purpling and necrosis. Unless deficiencies are serious, yield losses are usually not as great as one would anticipate from the symptoms observed. Magnesium in seeds of crops is laid down with P in the formation of phytin.

Sulfur

Sulfur plays a major role in protein synthesis through the amino acids cystine and methionine. When plants are deficient in S, they are light green to yellow in color, particularly the new leaves that emerge. The older leaves may be fairly dark green. Sulfur does not appear to be mobilized from older tissue and transported to meristematic tissue to the degree that elements such as N and K are.

Sulfur is a constituent of biotin and thiamin, important vitamins and components in plant metabolism. It is also part of Coenzyme A and Acetyl CoA. Acetyl CoA is intimately involved in fatty acid metabolism.

Without adequate S plants are stunted. Legumes that depend on symbiotically fixed N do not grow properly because of low N fixation and utilization under S deficiency. It therefore follows that the protein concentrations of forages and the protein and oil concentrations of seed or grain would be reduced.

Boron[37]

Boron deficiency results in the disruption of cell division and growth and ultimate death of meristematic tissue or growing tips. Root growth is also reduced. New leaves become malformed and chlorotic, grading from yellow to purplish with increasing deficiency. Crops with severe deficiency do not flower and therefore cannot produce seed.

Grasses such as corn show white stripes due to coalescence of chlorotic spots that start between the veins. The length of the stripes are from 0.5 to 2 in. The chlorotic areas may exude a clear gelatinous substance. Internodes remain short and the new leaves remain whorled.[38]

Boron is relatively immobile and when uptake is decreased and stress occurs, it is not redistributed from the older tissue. Therefore, the lower leaves remain green, but those on the new shoots or tips are deficient. These conditions are particularly prevalent during periods of moisture stress, which reduced the release of boron from organic matter, as well as its rate of uptake.

Boron is incorporated into the structure of the cell wall. It contributes to the germination and stability of pollen tubes during the pollination process. It also influences the development of the vascular tissue of plants. When B is deficient, sugar translocation does not occur. This effect, as suggested by Mengel and Kirkby,[39] may be due to the lack of a physiological sink for sugars under B deficiency. Unless photosynthate is transported from leaves to sinks, it accumulates and the photosynthetic capacity of leaves shuts down.

Chlorine

Chlorine deficiency causes younger leaves of plants to be chlorotic and produce an overall wilted appearance, indicating that water uptake and/or transpiration may be influenced. Chloride, because of its hydrophilic characteristics, increases the osmotic pressure of cells.[40]

Metabolically chloride is involved in photosynthesis. Chloroplasts contain rather high concentrations of chloride, which is related to the photosystem II in the evolution of oxygen.

Cobalt

Cobalt is not required by all crops. It is, however, required by legumes, such as soybeans and alfalfa, that depend on nodulation and N-fixation for their source of N. Cobalt influences the formation of vitamin B_{12} which is essential for the formation of the leghemoglobin associated with the N fixation process.[25] Cobalt-deficient legumes develop the symptoms of nitrogen deficient plants.[41] Cobalt deficiency can be overcome by providing Co or by providing the legume crop with N.

Copper

Copper deficiency in grasses such as corn shows up initially in young leaves, which become yellow and stunted. Tissue along the tips and edges of the leaves becomes necrotic and dead, similar to the effect observed with K deficiency.

In small grains new leaves are light yellow and plants are stunted. Later these growing tip leaves become twisted, dead, and broken. Grain production is nonexistent or drastically reduced with copper deficiency.

Copper has been shown to be related to Ca transport to the growing tips of plants. This may indicate that Cu is associated with the formation of conductive xylem tissue needed to enable Ca to reach the growing tips and maintain the integrity of the rapidly dividing and growing cells.

Most of a plant's Cu is associated with the organelles, such as the chloroplasts. Copper is required for plastocyanin, a chloroplast protein. It is also essential for the formation of several oxidase enzymes, including cytochrome oxidase, ascrobate oxidase, and polyphenol oxidase.[42,43]

Iron

Grasses that are Fe deficient show an interveinal chlorosis, especially in younger leaves. In severe deficiency the leaves are very bleached and stunted, and yields are seriously reduced.

With dicotyledonous plants, such as soybeans, the new leaves at the top of the plant are the ones that have a general chlorosis, while those on the lower portion of the plant may be quite green under mild Fe deficiency. The reason for this condition is that Fe is quite immobile and is not readily transported from the older tissue to the developing leaves under Fe stress. A continuous supply is needed to keep the deficiency from showing.

Iron is important in chlorophyll production and photosynthetic functioning through Ferredoxin. It also functions in the activity of enzymes, namely catalase, peroxidase, and cytochrome oxidase. Only a small portion of the total plant iron is active, while the bulk of it is stored as a Fe-P protein, phytoferritin.[44]

Manganese

Leaves of monocots that are Mn deficient are light green and develop an alternate interveinal striping. Deficiency in dicots shows up on the new leaves, with a chlorosis or bleaching between the veins, but with the veins remaining green. Both groups are stunted and unproductive when Mn is deficient.[45]

Chloroplasts in Mn deficient plants are not properly formed, nor do they function properly. Manganese is essential for the photosystem II or oxygen evolution phase of photosynthesis, as well as electron transport.[43]

Molybdenum

Molybdenum deficiency resembles nitrogen deficiency because of its direct link with N metabolism through the nitrate reductase enzyme in nonlegume plants and the nitrogenase enzyme in N-fixing legumes. Therefore, plants develop a pale or light yellowish-green color, with small malformed twisted new leaves. On crops such as cauliflower, only the mid-rib of the leaves develops under deficiency. This is sometimes referred to as "whiptail". On crops such as tomatoes, the symptoms appear as a mottling, with a curling of the older leaves. The chlorotic areas initially appear puffy, but then become necrotic and dead.[45,46]

Flower formation is decreased and seed may fail to develop following pollination, reducing yields, and viable seed production. Adequate levels of Mo in large seeded legumes, such as peas or soybeans, may be sufficient to produce satisfactory crop yields on soils that would be considered deficient in Mo.

Zinc

Zinc deficiency in large grasses, such as corn, stunts growth. New leaves emerging from the whorl may be nearly white, giving rise to a symptom sometimes referred to as "white bud". Older leaves under severe deficiency may show a creamy bleached color on either side of the mid-rib, while the mid-rib itself and the outer edges of the leaves remain green. With time, the light chlorotic area may turn purplish, become necrotic and show signs of tearing because of the thinness of the tissue.

In dicots, leaves are pale green, yellow, and mottled. Later the mottling takes on a rusted appearance. This is particularly true in soybeans. Another effect is a rosetting of leaves and the production of light green very small leaves, sometimes referred to as "little leaf". Buds do not form and develop properly. Therefore, yields are markedly reduced.

The chlorosis produced by Zn deficiency is related to abnormal formation and development of the grana in chloroplasts, reducing photosynthesis. The deficiency also reduces RNA production and development needed for cell division and growth. Without adequate Zn, indolacetic acid production appears to be reduced, which would reduce expansion growth of cells in leaves and may be part of the reason for the "little leaf" symptom in some crops.[44,47]

Zinc deficiency is one of the more common micronutrient deficiencies.[48,49]

NUTRIENT-CROP RELATIONSHIPS

Optimum Nutrient Concentrations

Each of the essential elements has an optimum concentration which under given conditions will produce the maximum or most profitable yield of a crop. In the past, "critical levels" or concentrations of the various elements were determined by supplying increasing amounts of an essential element, assuming others to be present in optimum amounts, and relating the percent yield of the crop produced to the concentration found in a specific plant part sample at a particular physiological stage of development. The "critical level" was defined as the concentration accepted as that producing 90% of the optimum (or 100%) yield level.[50] With that definition one accepts a 10% reduction in growth rate or yield and a concentration below that which produces the optimum growth rate or maximum yield.[51] The real question relates to the interrelationship of nutrients and what happens to economic yields if one accepts a reduction of 10% in yield associated with the "critical concentration" for each of the 13 elements involved in mineral nutrition.[52]

A model curve showing the percent growth per unit time with the deficient, critical, optimum, and toxic or imbalance concentrations is shown in Figure 1. Nitrogen, P, K, S, Ca, and Mg concentrations are usually expressed in percent and those of the micronutrients are expressed in parts per million on a dry matter basis.

Yield-Fertilizer Nutrient Yield Curves

As a deficient nutrient is supplied to a soil the crop responds, increasing the rate of growth and yield of the crop. This is particularly true for annual crops. The shape of the curve depends on the degree of the deficiency. A model of a typical yield nutrient response on a very deficient soil is shown in Figure 2. If a soil contains very high concentrations of the available nutrient, adding more in the fertilizer may produce little, if any, response or may, in fact, reduce yields.

On a deficient irrigated sandy soil, increasing N rates produce nearly linear and large responses to the first rates of N, but then as the limits of production for that system or some other factor became limiting, the response decreases and levels off (Figure 3). This curve is not unlike a 10-year average N-corn yield response curves from finer textured soils of humid areas, such as Ohio, and the rate of the nutrient applied and its availability to the crop.

On fertile, productive soils containing adequate amounts of a nutrient, further additions of that nutrient in applied fertilizers will do little to increase yields. The lack of corn response to N over a 6-year period on a fertile virgin soil is shown in Figure 4. This soil had the capacity to mineralize adequate N from the soil organic matter to meet the needs of the crop. Note the extremes in yield differences from the highest to the lowest yield among the years included in the 6-year average. Factors other than N were having a much greater effect on corn yield.

Soil testing or chemical analyses can be used to predict soils that are deficient in

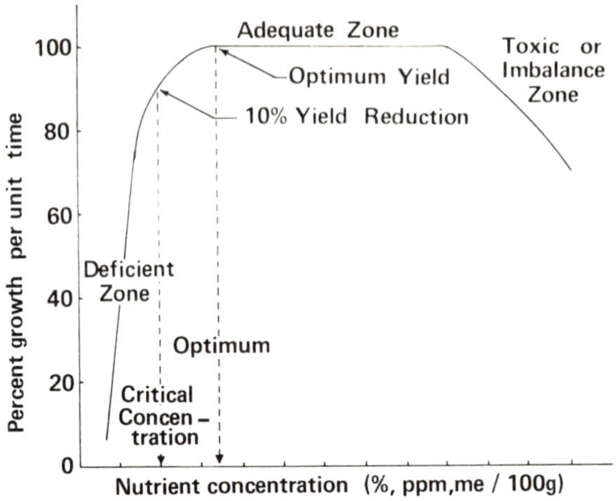

FIGURE 1. Model curve of the percentage growth of a crop with respect to deficient, critical, optimum, or toxic concentrations of an essential element in a specific plant part sampled at a given physiological stage of development.[50]

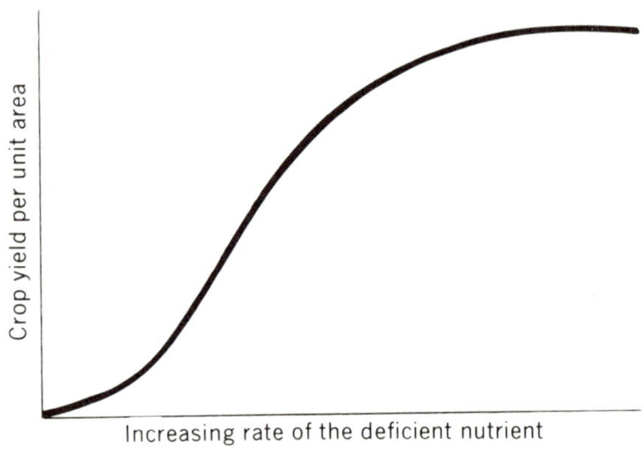

FIGURE 2. Model of a crop yield nutrient response curve when increasing amounts of a fertilizer nutrient are applied to a very deficient soil.

nutrients and that have a high probability of producing a profitable response to a nutrient or nutrients applied in fertilizers. Experimental results are used to define the soil test levels and fertilizer nutrients and rates that will be profitable to use.[27,55] Relations are seldom perfect and the availability of soil nutrients can vary widely from season to season and year to year. The proper interpretation and continual updating of research information is extremely important in effective recommendations based on soil tests.[56,57]

Experience indicates that soil tests are a great improvement over guessing, which would be the alternative. They provide a means of monitoring soil fertility as one of the diagnostic techniques of modern farming.

Yield-Nutrient Concentrations and Interrelationships of Nutrients

When increasing amounts of a nutrient are added to a crop on a deficient or responsive soil, as the rate of the nutrient increases, the concentration of the nutrient in the crop or a part of the crop sampled increases, accompanied by an increase in yield as long as the nutrient is the major limiting factor. The relationship between the yield of

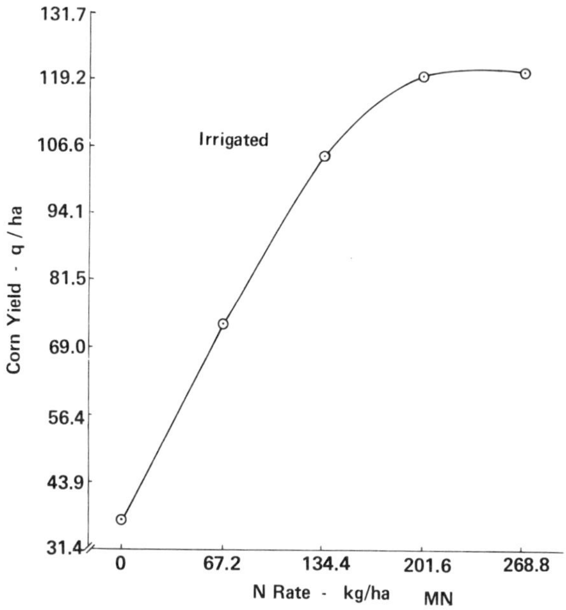

FIGURE 3. Applied fertilizer nutrients on deficient soils can produce large yield increases, the response being proportional to the deficiency as in this example of corn response to N.[53]

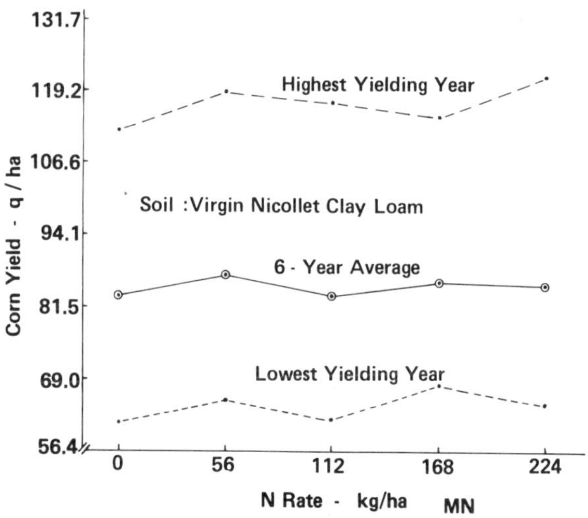

FIGURE 4. Additions of a nonlimiting nutrient as fertilizer produce little yield response, as shown in this lack of corn response to N on a fertile virgin soil.[54]

corn and the concentration of the corn leaf K sampled at silking is shown in Figure 5. With other nutrients in adequate supply, there is an optimum concentration for K photosynthesis,[59] as there is for each of the elements.

When the availability of a single nutrient is increased, there is usually a shift in the concentration of other elements. This may be caused by increased dry matter accumulation and thus a dilution of elements that are not being added in fertilizers. Using the corn leaf K concentrations and relating them to the concentrations of leaf Ca and Mg, one can understand the interrelationships that exist within the crop as the availability

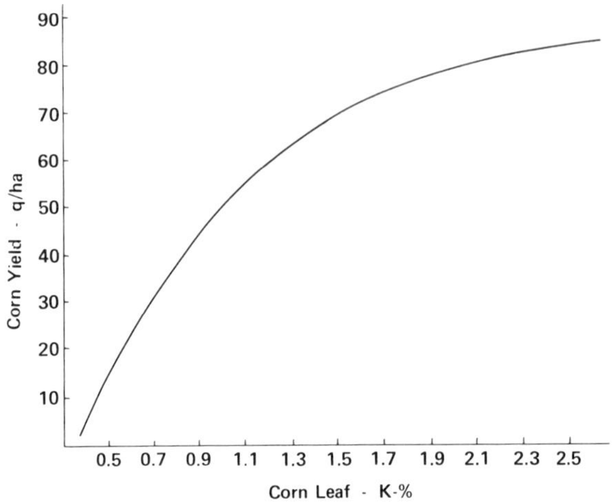

FIGURE 5. The relationship of corn grain yield to the percent K in the corn leaf.[58]

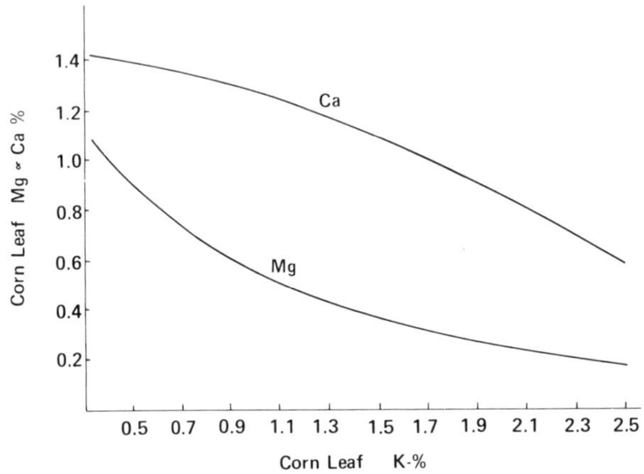

FIGURE 6. The interrelationships of corn leaf Ca and Mg to increasing corn leaf K concentrations.[58]

of a deficient nutrient is increased (Figure 6). As long as the nutrient concentrations are moving toward the optimum level, yields will be increased.

Crop Responses to More Than One Nutrient

When two or more nutrients are limiting yields, the addition of one or the other nutrient may slightly increase yields, but then decrease yield as higher rates are applied. However, if the two nutrients are applied together, based on the degree of the deficiency in the soil and the ability of the crop to utilize them, yields can be markedly increased due to their interaction or synergistic effect.

Early studies in Iowa provided a classic example of the value of applying two nutrients together to increase yields.[60] When N or phosphate was applied alone, there was little yield response of corn to the treatment. However, when they were applied together the interaction was tremendous, nearly quadrupling yields above the best yield when only one nutrient was applied (Figure 7).

From a practical standpoint, soil test levels can be increased for elements such as P and K, and variable rates of N applied. In a study on corn in Wisconsin in which

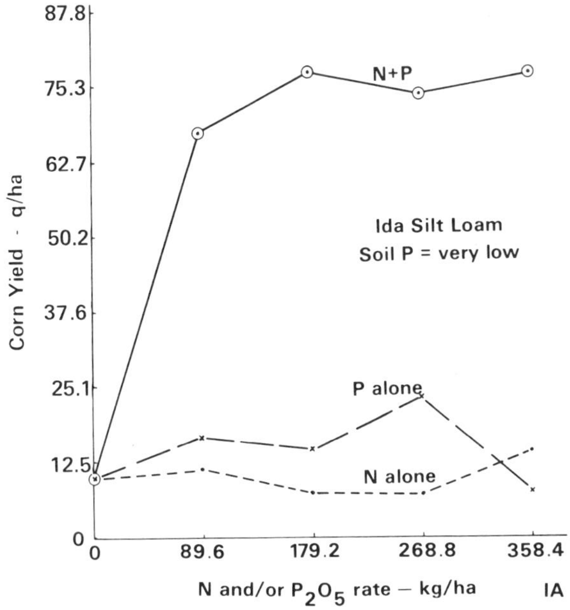

FIGURE 7. Adding one nutrient when another is also limiting can be ineffective. Adding increasing rates of both nutrients, as in this example, can produce a positive interaction or synergistic effect, greatly increasing yields.[60]

variable potassium soil test levels were established, using potash fertilizer and variable rates of N, a response surface was obtained[61] (Figure 8). With a soil K test of 112 kg/ha (50 ppm) with no applied N, the corn yield was 53.3 q/ha. However, with increasing K soil test and N rate the maximum yield was 123.6 q/ha. The optimum soil K level was 349 kg/ha (155 ppm) and the N rate for maximum yield was 263 kg N/ha.

The maximum economic yield was 120.4 q/ha and was obtained with 204 kg N/ha with the optimum soil K test of 349 kg/ha. These researchers estimated that 6.9 kg K_2O/ha were required to increase the soil test K level by 1 kg/ha on this particular prairie soil. Once the optimum K level was reached, 95 kg K_2O/ha/year would be required to maintain a 125 q/ha (200 bushel/acre) yield level.

Associated with such yield response curves are changes in the concentrations of nutrients. Sampling time, as well as the plant part sampled, is important in showing relationships with yields. Early samplings may indicate that concentrations in the untreated control plants are adequate. However, later samplings after soil supplies of an element are depleted will clearly indicate decreased concentrations, while plants receiving adequate amounts of the nutrient in question will have higher concentrations.

Fertilizer Rates and Improved Technology

When improved technology or cultural practices are introduced, yield potentials are often increased, increasing the amount of fertilizer or nutrient that can profitably be used (Figure 9). With such improvements, the total profit increases as the rate of fertilizer increases. These improved practices could be such items as timely planting, improved plant population, improved hybrid or variety, tillage to conserve soil moisture, or improved weed and insect control.

If improved cultural practices, such as increased plant population, are used without adequate nutrient applications, negative responses may be obtained (Figure 10). However, if optimum rates of nutrients are applied with the improved cultural practice,

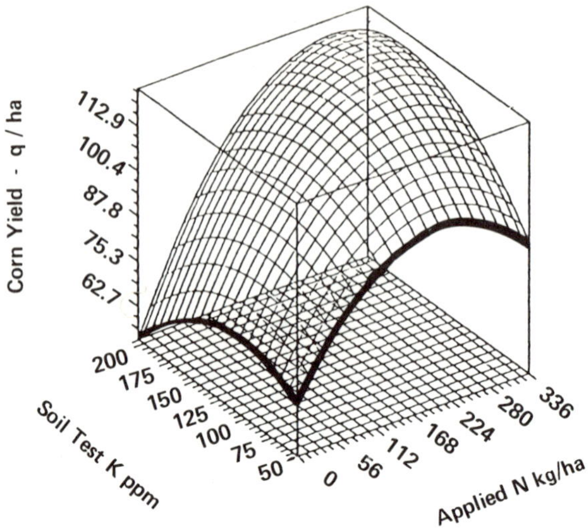

FIGURE 8. Optimum soil test levels with optimum rates of fertilizer nutrients can be used to establish maximum yields of a crop or maximum economic yields and rates of nutrients.[61]

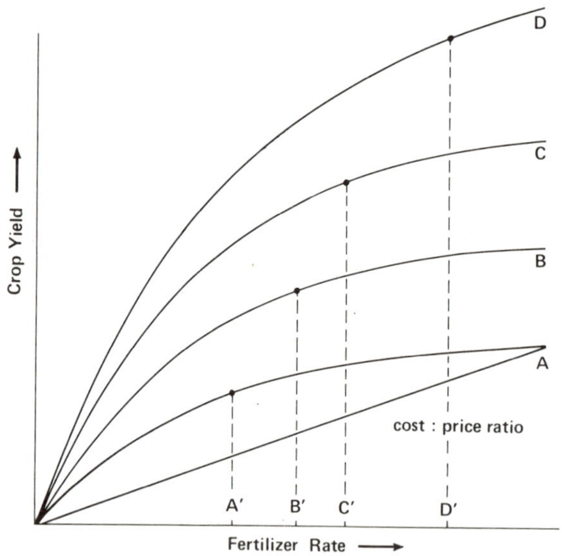

FIGURE 9. As new technology or improved cultural practices are introduced, the optimum rate of fertilizer increases, increasing both yield and profits.

the response shows increasing or linear returns, as in the case with N P K and the higher populations.[62]

Yield Levels

As yield levels are increased with improved management and fertilization, the rate of uptake of nutrients, as well as the total uptake, is increased. For example, recent studies on corn indicate how the levels of N, P, and K changed as the corn grain yield level increased from 69 to 145 q/ha (Table 2).[62] While the total amounts of nutrients increased with grain yield level, the amounts of nutrient per unit of grain produced

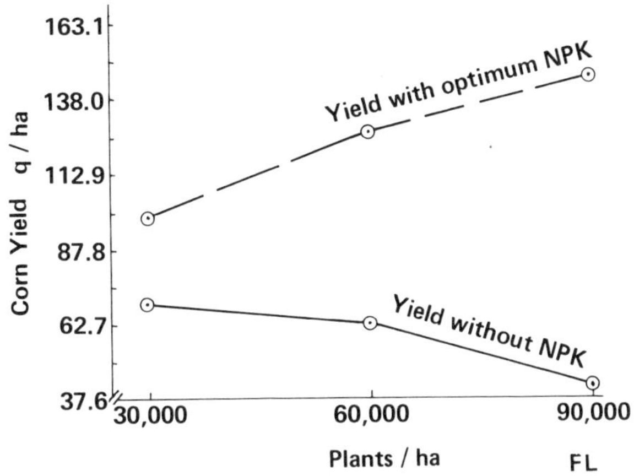

FIGURE 10. Improved cultural practices without adequate nutrients can be unproductive. Response with optimum rates of N, P, K may be linear.[62]

Table 2
INCREASE IN GRAIN YIELD OF CORN WITH TOTAL UPTAKE OF N, P, AND K IN CROP 87 DAYS AFTER PLANTING

Corn yield level		Nutrient uptake, excluding roots		
		N	P	K
(q/ha)	(bu/A)	(kg/ha)		
69	110	92	24	131
100	159	162	34	185
127	202	204	43	250
145	231	239	45	290

From Rhoads, F. M. and Stanley, R. L., Jr., *Soil Crop Sci. Soc. Fla.*, 38, 78, 1978. With permission.

were surprisingly uniform. For the higher yield levels, it varied from 1.61 to 1.65 kg N/q, from 0.31 to 0.34 kg P/q, and from 1.85 to 2.0 kg K/q for the total uptake and grain yields achieved.

Similar results for soybean yield levels indicate how the total nutrient content changed and increased with bean yield level and total dry matter production (Table 3).[63] Soybeans are capable of fixing considerable N symbiotically. Availability of nutrients other than N can have a marked influence on soybean ability to fix N_2 and, hence, their total uptakes are mutually but not proportionally related.

For forage crops (such as alfalfa) which produce increased yields through multiple cuttings taken every 30 to 35 days where the climate, soil moisture, and adequate nutrients are available, yield levels and nutrient uptake increase together. Once optimum concentrations are known, the amount of nutrient removed per ton of dry matter or hay can be determined and related to yield (Table 4). These data were based on the average of 15 studies which produced an average yield of 15.2 T/ha, with a range of 7.8 to 24.1 T/ha.

If other elements are present and maintained in optimum amounts, most of the N needed by legumes can be fixed from atmospheric N_2.[64] With increased costs of energy

Table 3
INCREASE IN NUTRIENT UPTAKE AND YIELD LEVEL IN SOYBEANS

Soybean yield level		Total dry matter	Maximum nutrient uptake excluding roots				
			N	P	K	Ca	Mg
(kg/ha)	(bu/A)	(kg/ha)	(kg/ha)				
2351	35	8,549	148.8	16.1	121.3	63.2	30.0
3360	50	10,204	287.8	23.3	175.0	54.9	21.5
5376	80	13,885	460.5	37.8	232.5	80.5	35.8

From Henderson, J. B. and Kamprath, E. J., *N.C. Agric. Exp. Sta. Tech. Bull.*, No. 197, 1970. With permission.

Table 4
INCREASE IN NUTRIENT UPTAKE AND YIELD LEVEL IN ALFALFA

Tons	Nutrient (kg per unit)										
/ha	N	P	K	Ca	Mg	S	Fe	Cu	Zn	Mn	B	Mo
1	29.1	2.49	23.0	12.5	2.6	2.4	0.16	0.006	0.020	0.051	0.026	0.004
11.2	326	28	258	140	29	27	1.8	0.07	0.22	0.57	0.29	0.05
22.4	652	56	516	280	58	54	3.6	0.14	0.44	1.14	0.58	0.10

used in the manufacture of fertilizer N, it is expected that in some areas of the world the use of N-fixing crops as a source of N will again increase.

The impact of fertilizer use on crop yield levels can be evaluated in a number of ways. One way is to look at the trend in the yield of a crop or crops over time and relate it to the trend in fertilizer use in a county or state. This has recently been done for the state of Illinois, which has within its boundaries some of the most fertile soils in the world, but nevertheless soils that produce marked crop responses when fertilizer nutrients are applied with excellent management. The trend in corn yields from 1866 to 1979 is shown in Figure 11 and the trend in N, P_2O_5 and K_2O use in short tons from 1930 to 1978 are shown in Figure 12.[65] While one cannot attribute all of the yield increases over time to increased fertilizer use, the yield trends in Illinois and the U.S. are definitely associated to increased emphasis on soil fertility, including proper lime and fertilizer use. Prior to 1940, during the period of "organic" farming, yields were very low. The takeoff point for yields coincides quite closely to the summation of the nitrogen, phosphate, and potash use curves.

From 1866 to 1937, the increases in Illinois corn yields were negligible, starting at about 18.8 q/ha. Fertilizer use was essentially nil, except for some use of rock phosphate along with a liming program. Following the advent of hybrid corn, which increased the yield potential of corn, and the end of World War II, fertilizer nutrients became increasingly available. Both yields and fertilizer use increased rapidly. In 1979, the U.S. Department of Agriculture reported that the average corn yield in Illinois had reached 80.3 q/ha (128 bushel/acre) and total N, P_2O_5, and K_2O usage for the state was up to 822,033; 487,123; and 712,422 tons, respectively.

Von Peter has reported on the research of Couston of the Food and Agricultural Organization (FAO) showing that the crop value index of 41 countries is highly related to the kg of fertilizer use per hectare.[66] Similar studies on the yield levels of cereal grain per unit area in countries of the world are highly related to the kg of N, P_2O_5 and K_2O used per hectare.[67]

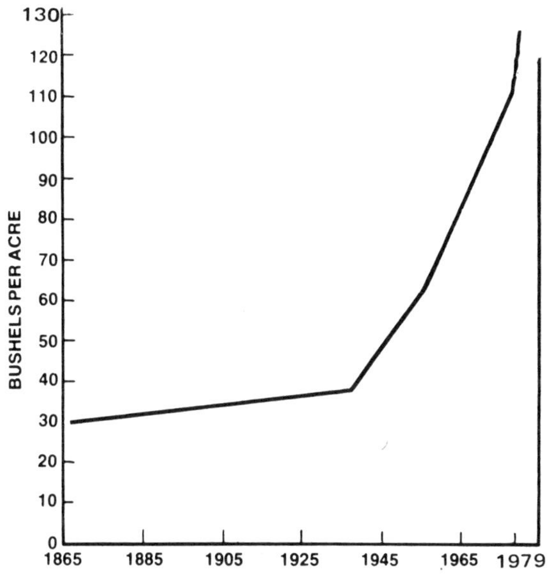

FIGURE 11. Trend in average Illinois corn yields from 1866 to 1979.[65]

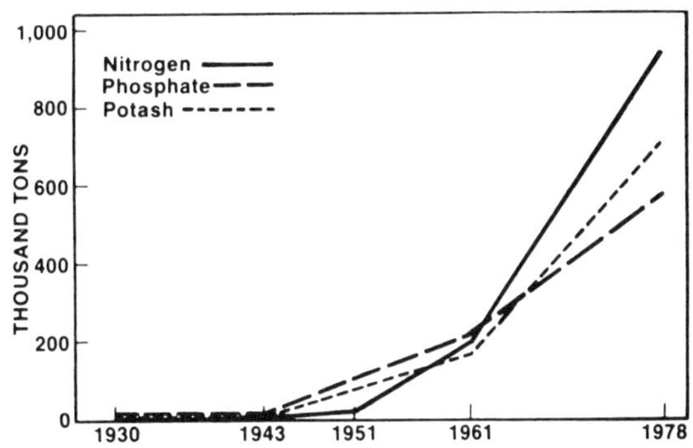

FIGURE 12. Trend in nitrogen, phosphate, and potash use in Illinois from 1930 to 1980.[65]

TILLAGE PRACTICES CAN CHANGE FERTILITY AND FERTILIZER NEEDS

In the U.S. there is a shift towards reduced tillage systems. These are sometimes referred to as minimum or conservation tillage systems. In the cooler regions of temperate zones or in areas of higher elevation, changing the tillage system can increase the soil test level of nutrients required, as well as change the yield level that can be achieved. This is particularly true for potassium (Figure 13). Plowed, till-planted corn was more responsive to increased soil K levels established through fertilization, than was till-planted corn that had not been plowed.[68] Some research indicates that on minimally tilled soils, corn yields can be equal to conventionally tilled soils when properly fertilized.

Evidence also indicates that the use of row fertilizer on soils testing low and medium is a very effective means of overcoming the differential yield levels between plowed and no-till systems (Figure 14). At high soil potassium levels, the row fertilizer contin-

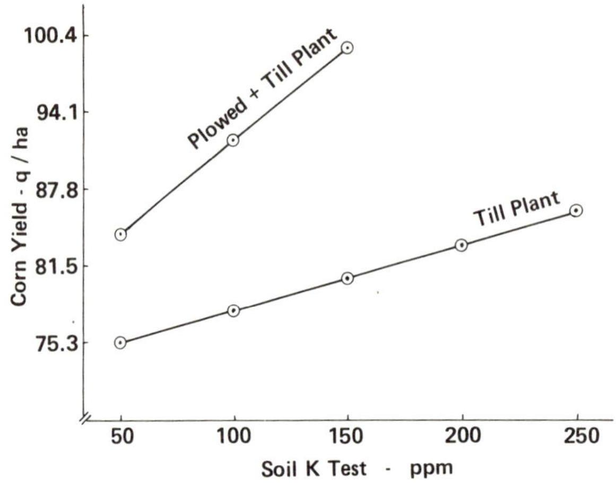

FIGURE 13. Under conditions of reduced tillage, higher soil K levels are required to meet the needs of corn.[68]

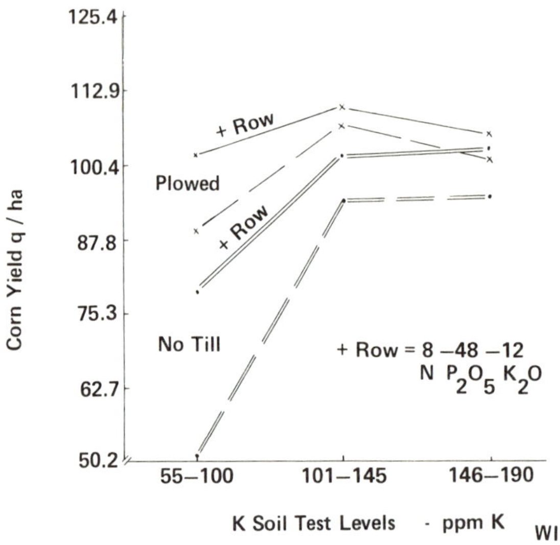

FIGURE 14. Higher soil test levels and row fertilizer use are more important in minimum or reduced tillage systems.[69]

ued to show benefits on the no-till system. Because of data such as these, it is recommended that soil test levels be increased to the medium-to-high range before shifting to minimum or reduced tillage systems. Part of the problem involves a poor distribution of the nutrients in the soil profile with reduced tillage.[70] Therefore, periodic plowing to redistribute nutrients within the soil is also suggested.

INCREASING FERTILIZER NUTRIENT EFFICIENCY

There are many ways of increasing nutrient efficiency. Application of the very best technical and management information available is essential. Proper seedbed preparation; selection of a disease-resistant or tolerant variety or hybrid with a high-yield potential; planting at the optimum date, depth, plant population, and row width; proper weed and pest control; and supplemental irrigation are all practices that have potential for increasing the use and response to native soil nutrients and applied fertil-

Table 5
EFFECT OF ROW AND BROADCAST POTASH ON CORN YIELDS ON EDINA SILT LOAM TESTS
(78.4 kg K/ha)

Row K[a] (K_2O kg/ha)		Corn yields (kg/ha) Plowed-down K rates (kg K_2O/ha)		
		0	67.2	268.8
0		7,589	9,910	10,725
33.6		9,659	10,412	10,725
	Increase	2,070	502	0

[a] Average of 4 K sources.

From Webb, J. R., *Annual Progress Report — 1971 for Outlying Experimental Farm 9,* Iowa Agric. Exp. Sta., Ames, 1972, 2. With permission.

izer nutrients. On many soils banding or row-applying the fertilizer itself can increase the response to a given amount of nutrient. Very often the row- or band-applied P and/or K may double the effectiveness of broadcast fertilizer for lower rates. That is particularly true on low-testing soils that have the capacity to "fix" or adsorb nutrients. However, the most profitable yields often result from a combination of row plus broadcast nutrients.

Results from combinations of row and plowed-down K are shown in Table 5 for a low K soil.[71] The 33.6 kg K_2O row applied produced nearly the same yield as double that rate broadcast and plowed down. The most profitable yield was produced by 67.2 kg K_2O plowed down with 33.6 kg applied in the row. As indicated by these results, high soil test levels or high broadcast rates usually reduce the response to row-applied nutrients.

Row or banded fertilizers may induce deficiencies on soils marginally deficient in a nutrient. This has been observed with P and Zn[72] (Figure 15). When P was banded, Zn deficiency was much more serious (left of the sign) than where an equal amount was broadcast (right of the sign). However, when Zn was applied (background) at 11.2 kg/ha, the yields of the banded P were slightly superior.

In other cases, row or banded fertilizer may increase the availability of soil Zn or Mn because of its acidifying effect on the soil. Banded anhydrous ammonia, sources of phosphate and sulfur can drastically change the pH of the soil and around the fertilizer band, increasing the solubility and availability of elements. Randall and his colleagues in Wisconsin have clearly demonstrated that row-applied monoammonium and diammonium phosphate increased the availability of soil Mn to soybeans in soils that were Mn deficient.[74] Boawn and his colleagues in Washington have shown how increasing N rates increase the availability of soil Zn to crops.[75] Therefore, applied nutrients have both direct and indirect effects on the nutrition of crops.

NUTRIENT UPTAKE BY HIGH YIELDING CROPS

The environmental conditions must be favorable, the management practices must be in place, and adequate nutrients must be available in order to produce high yields.

The total uptake of nutrients for a 15.99 ton/ha corn grain crop contained the following nutrients (kg/ha) in grain and stover:[76]

N	P_2O_5	K_2O	Mg	S	B	Cu	Fe	Mn	Zn
365	235	410	52	31	0.14	0.10	6.0	1.1	2.3

From the micronutrient standpoint, the Fe and Mn appear unduly high, perhaps because of irrigation water. Usually, the total micronutrient content of very high-yielding crops is less than 3.5 kg/ha. That does not make them any less important if they are deficient or unavailable to the developing crop. However, the content of the macro and secondary nutrients on an elemental basis is about 940 kg/ha.

The nutrient uptake for a range of high-yielding crops is given in Table 6. It should be remembered that these should be used cautiously. Many soils of the world simply do not have the productive capacity to produce such yields, nor do native crops have the genetic capability for these yields. The soils vary in their native or natural state of fertility, and need to be measured experimentally and related to fertilizer responses before recommendations are made. In most cases, however, soils properly fertilized are capable of producing far greater yields than most believe.

THE FUTURE

Fertilizer use in the world must expand in order to meet the food requirements of growing populations. The economic, social, and environmental advantages of proper fertilizer use in the context of world problems should make it a high priority item. Little stability can be achieved until people are properly fed, clothed, and sheltered. As crop yields are increased, a greater percentage of the yield will be due to fertilizer nutrients, because if nutrients are not present, high yields cannot be produced.

FIGURE 15. The effect of banded and broadcast P without Zn (foreground) and with adequate Zn (background).[73]

Table 6
THE N, P_2O_5, K_2O, Mg, AND S UPTAKE OF HIGH YIELDING CROPS

Crop	Yield per hectare plant part	N (kg/ha)	P_2O_5 (kg/ha)	K_2O (kg/ha)	Mg (kg/ha)	S (kg/ha)
Grain crops						
Barley	5,376 kg grain	123	45	39	9	11
	Straw	45	17	129	10	11
	Total	168	62	168	19	22
Buckwheat	1,613 kg grain	34	17	11	—	6
	Straw	13	6	28	—	3
	Total	47	23	39	—	9
Corn	12,544 kg grain	168	97	64	20	17
	Stover	130	30	234	53	20
	Total	298	127	298	73	37
Oats	3,584 kg	90	28	22	6	9
	Straw	39	17	140	17	12
	Total	129	45	162	23	21
Rice	7,840 kg grain	86	52	31	9	6
	Straw	39	16	134	7	8
	Total	125	68	165	16	14
Sorghum	8,960 kg grain	134	67	34	16	25
	Stover	146	34	190	34	18
	Total	280	101	224	50	43
Wheat	5,376 kg grain	106	49	30	13	6
	Straw	47	11	151	13	17
	Total	153	60	181	26	23
Oil crops						
Flax	1,344 kg seed	45	19	17	—	3
	Straw	16	6	34	—	3
	Total	61	25	51	—	6
Oil Palm	24,640 kg	73	28	111	21	—
	Fronds, stems	120	55	189	40	—
	Total	193	83	300	61	—
Peanuts	4,480 kg nuts	157	25	39	4	11
	Vines	112	19	168	22	12
	Total	269	44	207	28	22
Rapeseed	1,960 kg seed	74	36	18	—	13
	Straw	44	16	75	—	10
	Total	118	52	97	—	23
Soybeans	4,032 kg beans	269	54	94	19	13
	Stover	94	18	65	11	15
	Total	363	72	159	30	28
Sunflower	3,920 kg seed	140	67	44	13	7
	Stover	57	11	100	34	11
	Total	197	78	144	47	18
Fiber crops						
Cotton	1,680 kg lint	105	43	49	12	8
	Stalks	96	28	92	27	26
	Total	201	71	141	39	34
Pulp Wood	98 cords	168	34	90	22	—
(Slash pine)	Bark, branches	212	8	67	34	—
	Total	380	42	157	56	—
Sugar						
Sugar Cane	224 T cane	179	101	375	45	60
	Trash	224	74	308	67	36
	Total	403	175	683	112	96
Sugar Beets	67 T	140	17	280	30	11
	Tops	146	28	336	59	39
	Total	286	45	616	89	50

Table 6 (continued)
THE N, P_2O_5, K_2O, Mg, AND S UPTAKE OF HIGH YIELDING CROPS

Crop	Yield per hectare plant part	kg/ha N	P_2O_5	K_2O	Mg	S
Silage crops						
Corn silage	72 T	298	128	298	73	37
Forage sorghum	18 T	222	75	320	39	20
Oat silage	18 T	123	34	263	22	17
Sorghum-Sudan	18 T	357	137	523	53	—
Forage crops						
Alfalfa	22 T	672	134	672	59	57
Big Bluestem	7 T	71	57	118	13	—
Birdsfoot Trefoil	9 T	215	94	305	36	—
Bromegrass	11 T	246	73	353	11	22
Clover-grass	13 T	336	101	403	34	34
Guineagrass	26 T	323	113	488	111	52
Hybrid Bermudagrass	22 T	560	157	470	56	56
Indiangrass	7 T	67	57	81	10	—
Lespedeza	7 T	168	56	168	28	22
Little Bluestem	7 T	74	57	97	13	—
Napiergrass	28 T	340	165	678	71	84
Orchard grass	13 T	336	112	420	28	39
Pangola grass	26 T	335	121	482	75	52
Paragrass	27 T	345	110	515	88	46
Pensacola Bahiagrass	16 T	339	97	271	39	30
Ryegrass	11 T	241	95	269	45	—
Switchgrass	7 T	77	67	128	17	—
Tall fescue	7.8 T	151	73	207	15	—
Timothy	9 T	168	62	280	11	18
Turf Grass						
Bluegrass (turf)	7 T	224	62	202	22	28
Bentgrass (turf)	6 T	291	74	164	15	11
Bermudagrass (turf)	9 T	252	45	179	22	17
Vegetable crops						
Bell peppers	20,160 kg	81	24	99	8	—
	Vines	73	35	144	40	—
	Total	154	59	243	48	—
Cabbage	78,400 kg	157	39	143	10	72
	Stems and leaves	146	31	136	30	—
	Total	303	70	279	40	72
Celery	168 T	286	146	762	—	—
	Roots	28	39	78	—	—
	Total	314	185	840	—	—
Cucumbers	22 T	45	16	74	4	—
	Vines	56	16	121	24	—
	Total	101	32	195	28	—
Lettuce	44.8 T heads	101	34	207	—	—
Onions	67.2 T	202	90	179	20	41
Peas	2,800 kg	103	27	35	7	7
	Vines	81	12	83	13	4
	Total	184	39	118	20	11
Potatoes	56 T	194	82	315	17	17
	Vines	108	19	297	40	8
	Total	302	101	612	57	25
Snap beans	9 T	78	24	86	9	—
	Vines	76	13	96	10	—
	Total	154	37	182	19	—

Table 6 (continued)
THE N, P_2O_5, K_2O, Mg, AND S UPTAKE OF HIGH YIELDING CROPS

Crop	Yield per hectare plant part	kg/ha N	P_2O_5	K_2O	Mg	S
Sweet corn	10,080 kg	90	25	58	7	6
	Stover	67	28	94	16	7
	Total	157	53	152	23	13
Sweet potato	33.6 T	82	44	189	9	—
	Vines	93	34	161	11	—
	Total	175	78	350	20	—
Table beets	56 T	190	34	235	34	15
	Tops	213	15	414	83	31
	Total	403	49	649	117	46
Tomatoes	67,200 kg	112	26	242	9	24
	Vines	90	28	134	22	22
	Total	202	54	376	31	46
Fruits						
Apples	28 T	22	9	56	2	—
	Leaves and stems	90	43	146	25	—
	Total	112	52	202	27	—
Cantaloupe	19,600 kg	44	18	91	7	—
	Vines	29	6	40	7	—
	Total	73	24	131	14	—
Grapes	27 T	74	26	134	11	—
	Vines	40	13	40	9	—
	Total	114	39	174	20	—
Oranges	605 q	101	26	181	11	8
	Leaves and stems	196	36	188	31	24
	Total	297	62	369	42	32
Peaches	322 q	35	10	65	12	—
	Leaves and stems	60	30	55	10	—
	Total	95	40	120	22	—
Tropical crops						
Bananas	1,200 plants	448	448	1,680	175	—
Cocoa	1,008 kg beans	20	8	12	2	—
	Husks, stems, and leaves	446	113	809	131	—
	Total	446	131	821	133	—
Coconuts	1200 nuts	108	35	231	15	9
Oil palm	24,640 kg	73	28	111	21	—
	Fronds and stems	120	55	189	40	—
	Total	193	83	300	61	—
Pineapple	399.8 q	171	140	668	72	16
Rubber	2,262 kg	19	11	19	4	—
	Leaves and stems	65	30	104	15	—
	Total	84	41	123	19	—
Sugar cane	244 T	179	101	375	45	60
	Trash	224	74	308	67	36
	Total	403	175	683	112	96
Tobacco						
Flue cured	3,360 kg	95	17	174	17	13
	Stalks, etc.	46	12	114	10	8
	Total	141	29	288	27	21
Burley	4,480 kg	165	15	166	22	19
	Stalks, etc.	160	27	194	15	8
	Total	325	42	360	37	27

From Agerton, R. and Martin, S., Eds., *Better Crops with Plant Food,* 63, 4, 1979. With permission.

REFERENCES

1. Stefferud, A., Ed., *Soil* The Yearbook of Agriculture, U.S. Department of Agriculture, Washington, D.C., 1957.
2. Black, C. A., *Soil-Plant Relationships,* John Wiley & Sons, New York, 1968.
3. Brady, N. C., *The Nature and Properties of Soils,* 8th ed., Macmillan, New York, 1974.
4. Russell, E. W., *Soil Conditions and Plant Growth,* 10th ed., Longman, London, 1973.
5. Millar, C. E., *Soil Fertility,* John Wiley & Sons, New York, 1955.
6. McVickar, M. H., Bridger, G. L., and Nelson, L. B., Eds., *Fertilizer Technology and Usage,* Soil Science Society of America, Madison, Wis., 1963.
7. Nelson, L. B., Ed., *Changing Patterns in Fertilizer Use,* Soil Science Society of America, Madison, Wis., 1968.
8. Olson, R. A., Ed., *Fertilizer Technology and Use,* 2nd ed., Soil Science Society of America, Madison, Wis., 1971.
9. Tisdale, S. L. and Nelson, W. L., *Soil Fertility and Fertilizers,* 3rd ed., Macmillan, New York, 1975.
10. Khasawneh, F. E., Sample, E. C., and Kamprath, E. J., Eds., *The Role of Phosphorus in Agriculture,* Soil Science Society of America, Madison, Wis., 1980.
11. Kilmer, V. J., Younts, S. E., and Brady, N. E., Eds., *The Role of Potassium in Agriculture,* Soil Science Society of America, Madison, Wis., 1968.
12. Sekhon, G. S., Ed., *Potassium in Soils and Crops,* Potash Research Institute of India, New Delhi,1978.
13. Potassium for Agriculture, a Situation Analysis, Potash and Phosphate Institute, Atlanta, 1980.
14. Jacob, A. *Magnesium: The Fifth Major Nutrient,* Staples Press, London, 1958.
15. Anderson, G. C., Jencks, E. M., and Horvath, D. J., Eds., Magnesium and Agriculture, West Virginia University, Morgantown, 1959.
16. National Advisory Service, Soil Potassium and Magnesium, Ministry of Agriculture, Fisheries, and Food, Tech. Bull. 14, Her Majesty's Stationery Office, London, 1967.
17. Jones, J. B., Eds., International Symposium on Calcium Nutrition on Economic Crops, *Commun. Soil Sci., Plant Analy.,* 10 (1 and 2), 1979.
18. Pearson, R. W. and Adams, F., Eds., *Soil Acidity and Liming,* American Society of Agronomy, Madison, Wis., 1967.
19. Mortvedt, J. J., Giordano, P. M., and Lindsay, W. L., Eds., *Micronutrients in Agriculture,* Soil Science Society of America, Madison, Wis., 1972.
20. Nicholas, D. J. D. and Egan, A. R., Eds., *Trace Elements in Soil-Plant-Animal Systems,* Academic Press, New York, 1975.
21. Stewart, F. D., Ed., *Inorganic Nutrition of Plants, Plant Physiology, Vol. III,* Academic Press, New York, 1963.
22. Arnon, I., *Mineral Nutrition of Maize,* International Potash Institute, Bern, Switzerland, 1975.
23. Mengel, K. and Kirkby, E. A., *Principles of Plant Nutrition,* International Potash Institute, Bern, Switzerland, 1978.
24. Mengel, K., Ion uptake and translocation, in *The Plant Root and Its Environment,* Carlson, E. W., Ed., University Press of Virginia, Charlottesville, 1974, 83.
25. Ahmed, S. and Evans, H. J., Cobalt: a Micronutrient element for the growth of soybean plants under symbiotic conditions, *Soil Science,* 90, 205, 1960.
26. Hignett, T. P., Ed., *Fertilizer Manual,* International Fertilizer Development Center, Muscle Shoals, Ala., and United Nations Industrial Development Organization, United Nations, New York, 1980.
27. Hanway, J. J., Experimental methods for correlating and calibrating soil tests, in *Soil Testing and Plant Analysis,* Walsh, L. M. and Beaton, J. D., Eds., Soil Science Society of America, Madison, Wis., 1973, 55.
28. Nason, A. and McElroy, W. D., Modes of action of the essential mineral elements, in *Inorganic Nutrition of Plants, Plant Physiology Vol. III,* Stewart, F. C., Ed., Academic Press, New York, 1963, 451.
29. Sprague, H. B., Ed., *Hunger Signs in Crops,* David McKay, New York, 1964.
30. Potassium in Biochemistry and Physiology, International Potash Institute, Bern, Switzerland, 1971.
31. Wallingford, W., Functions of potassium in plants, in *Potassium for Agriculture, a Situation Analysis,* Potash & Phosphate Institute, Atlanta, 1980, 10.
32. Humble, G. D. and Raschke, R., Stomatal opening quantitatively related to K transport, *Plant Physiol.,* 48, 447, 1971.
33. Fischer, R. A., The role of potassium in stomatal opening, in *Potash Review,* Subj. 3, Biology 44th Suite, International Potash Institute, Bern, Switzerland, 1972.
34. Evans, H. J. and Wildes, R. A., Potassium and its role in enzyme activation, in *Potassium in Biochemistry and Physiology,* International Potash Institute, Bern, Switzerland, 1971, 13.

35. **Foy, C. D.,** Effects of soil calcium on plant growth, in *The Plant Root and Its Environment,* University Press of Virginia, Charlottesville, 1974, 565.
36. **Gauch, H. G. and Krauss, R. W.,** Roles of magnesium in plants, in *Magnesium and Agriculture,* West Virginia University, Morgantown, 1959, 39.
37. **Jackson, J. F. and Chapman, K. S. R.,** The role of boron in plants, in *Trace Elements in Soil-Plant-Animal Systems,* Nicholas, D. J. D. and Eagan, A. R., Eds., Academic Press, New York, 1975, 213.
38. **Berger, K. C., Heikkinen, T., and Zube, E.,** Boron deficiency, a cause of blank stalks and barren ears in corn, *Soil Sci. Soc. of Am.,* 21, 629, 1957.
39. **Mengel, K. and Kirkby, E. A.,** Boron, in *Principles of Plant Nutrition,* International Potash Institute, Bern, Switzerland, 1978, 483.
40. **Mengel, K. and Kirkby, E. A.,** Chlorine, in *Principles of Plant Nutrition,* International Potash Institute, Bern, Switzerland, 1978, 495.
41. **Ahmed, S. and Evans, H. J.,** The essentiality of cobalt for soybean plants grown under symbiotic conditions, Proc. Natl. Acad. Sci., 47, 24, 1961.
42. **Mengel, K., and Kirkby, E. A.,** Copper, in *Principles of Plant Nutrition,* International Potash Institute, Bern, Switzerland, 1978, 463.
43. **Boardman, N. K.,** Trace elements in photosynthesis, in *Trace Elements in Soil-Plant-Animal Systems,* Nicholas, D. J. D. and Egan, A. R., Eds., Academic Press, New York, 1975, 199.
44. **Nicholas, D. J. D.,**The functions of trace elements in plants, in *Trace Elements in Soil-Plant-Animal Systems,* Nicholas, D. J. D. and Egan, A. R., Eds., Academic Press, New York, 1975, 181.
45. **Hewitt, E. J.,** The essential nutrient elements requirements and interactions in plants, in *Inorganic Nutrition of Plants, Plant Physiology III,* Stewart, F. C., Ed., Academic Press, New York, 1963, 137.
46. **Evans, H. J.,** Role of molybdenum in plant nutrition, *Soil Sci.,* 81, 199, 1956.
47. **Price, C. A., Clark, H. E., and Funkhouser, E. A.,** Functions of micronutrients in plants, in *Micronutrients in Agriculture,* Soil Science Society of America, Madison, Wis., 1972, 231.
48. **Berger, K. C. and Pratt, P. F.,** Advances in secondary and micronutrient fertilization, in *Fertilizer Technology and Usage,* Soil Science Society of America, Madison, Wis., 1963, 287.
49. **Loneragan, J. F.,** The availability and absorption of trace elements in soil-plant systems and their relationship to movement and concentrations of trace elements in plants, in *Trace Elements in Soil-Plant-Animal Systems,* Nicholas, D. J. D. and Egan, A. R., Eds., Academic Press, New York, 1975, 109.
50. **Ulrich, A.,** Plant tissue analysis: plant analysis as a guide in fertilizing crops, in *Soil and Plant-Tissue Testing in California,* Calif. Ext. Bull. 1879, 1976, 1.
51. **Munson, R. D. and Nelson, W. L.,** Principles and practices in plant analysis, in *Soil Testing and Plant Analysis,* Walsh, L. M. and Beaton, J. D., Eds., Soil Science Society of America, Madison, Wis., 1973, 223.
52. **Munson, R. D.,** Let's take a new look at plant nutrition concepts, *Fert. Solutions Mag.,* 22, 86, 1978.
53. **Malzer, G. L.,** Personal communication, 1979.
54. **Fenster, W. E., Overdahl, C. J., Randall, G. W., and Schoper, R. P.,** Effect of Nitrogen Fertilizer on Corn Yield and Soil Nitrates, Minn. Agric. Exp. Sta., Misc. Rep., 153, 1978.
55. **Viets, F. G., Jr. and Hanway, J. J.,** How to determine nutrient needs, in *Soil,* Stefferud, A., Ed., U.S. Department of Agriculture, Washington, D.C., 1957, 172.
56. **Barber, S. A.,** The changing philosophy of soil test interpretation, in *Soil Testing and Plant Analysis,* Walsh, L. M. and Beaton, J. D., Eds., Soil Science Society of America, Madison, Wis., 1973, 201.
57. **McLean, E. O.,** Fertilizer and lime recommendations based on soil tests: good, but could be better?, *Commun. Soil Sci. Plant Anal.,* 8, 441, 1977.
58. **Loue, A.,** Studies about the Nutrition and Fertilization of Maize Conducted in the Pan (Atlantic Pyrenees) region from 1953 to 1968, Agronomic Services of Commercial Society of Potash and Nitrogen, Mulhouse, France.
59. **Koch, D. W. and Estes, G. O.,** Influence of K stress on growth, stomatal behavior, and CO_2 assimilation in corn, *Crop Science,* 15, 697, 1975.
60. **Heady, E. O., Pesek, J. T., and Brown, W. G.,** Crop Response Surfaces and Economic optima in Fertilizer Use, Iowa Agric. Exp. Sta., Res. Bull., 424, 1955.
61. **Moncrief, J. F., Walsh, L. M., and Schulte, E. E.,** Crop production and soil fertility: response surface approach to yield maximization, *Better Crops with Plant Food,* 63, 12, 1979/80.
62. **Rhoads, F. M. and Stanley, R. L., Jr.,** Effect of population and fertility on nutrient uptake and yield components of irrigated corn, *Soil and Crop Sci. Soc. Fla.,* 38, 78, 1978.
63. **Henderson, J. B. and Kamprath, E. J.,** Nutrient and Dry Matter Accumulation by Soybeans, North Carolina Agric. Exp. Sta. Tech. Bull. No. 197, 1970.
64. **Duke, S. H., Collins, M., and Soberalske, R. M.,** Effect of potassium fertilization on nitrogen fixation and nodule enzymes of nitrogen metabolism in alfalfa, *Crop Sci.,* 20, 213, 1980.

65. **Walker, W. M., Swanson, E. R., and Carmer, S. G.,** No yield plateau at present for corn and soybeans in Illinois, *Ill. Res.,* 22, 9, 1980.
66. **von Peter, A.,** Fertilizer Requirements in Developing Countries, Fertiliser Society of London, Proceedings No. 188, 27, 1980.
67. **Hignett, T. P., Ed.,** Role of Fertilizer in Agriculture, *Fertilizer Manual,* International Fertilizer Development Center, Muscle Shoals, Ala., 20, 1980.
68. **Moncrief, J. F., Walsh, L. M., and Schulte, E. E.,** Nutrient requirements of corn as influenced by tillage, in *Proc. 1979 Fert., Aglime Pest Manage. Conf.,* College of Agricultural and Life Sciences, University of Wisconsin, Madison, 62, 1979.
69. **Schulte, E. E.,** Effectiveness of row fertilizer in Wisconsin, in *Proc. 1980 Fert., Aglime, Pest Managem. Conf.,* College of Agricultural and Life Sciences, University of Wisconsin, Madison, 113, 1980.
70. **Randall, G. W.,** Fertilization practices for conservation tillage, in *32nd Annu. Fert. Agric. Chem. Dealers Conf.,* University Extension, Iowa State University, Ames, EC-1498K, 1980.
71. **Webb, J. R.,** Source, rate, and method of applying potassium fertilizers for corn, in *Annual Progress Report — 1971 for Outlying Experimental Farm 9,* Iowa Agric. Exp. Sta., Ames, 1972, 2.
72. **Adriano, D. C. and Murphy, L. S.,** Effects of ammonium polyphosphates on yield and chemical composition of irrigated corn, *Agron. J.,* 62, 561, 1970.
73. **Murphy, L. S.,** Personal communication.
74. **Randall, G. W., Schulte, E. E., and Corey, R. B.,** Soil Mn availability to soybeans as affected by mono and diammonium phosphate, *Agron. J.,* 67, 705, 1975.
75. **Boawn, L. C., Viets, F. G., Jr., Crawford, C. L., and Nelson, J. C.,** Effect of nitrogen carrier, nitrogen rate, zinc rate and soil pH on zinc uptake by sorghum, potatoes, and sugar beets, *Soil Sci.,* 90, 329, 1960.
76. **Dowdy, R. H.,** Personal communication.
77. **Agerton, B. and Martin, S., Eds.,** Uptake by plant parts and total at harvest, in *Better Crops with Plant Food,* 63, 4, 1979.

FERTILIZER SOURCES AND COMPOSITION

G. L. Terman

Commercial chemical fertilizers are produced largely to supply nutrients in plant-available forms for crop production. The terms "fertilizers," "plant nutrients," and "plant foods" are used more or less interchangeably in the fertilizer industry. Minerals, composts, stable manure, sewage sludge, etc., low in available nutrient content are usually termed soil amendments, rather than commercial fertilizers.

Sixteen nutrients are now considered to be essential for plant growth. Of these, 13 nutrients are supplied largely in fertilizers to supplement quantities present in soils. These are commonly divided into primary macronutrients — nitrogen (N), phosphorus (P), and potassium (K); secondary macronutrients — calcium (Ca), magnesium (Mg), and sulfur (S); and micronutrients — trace, or minor elements — boron (B), chlorine (Cl), copper (Cu), iron (Fe), manganese (Mn), molybdenum (Mo), and zinc (Zn). Sodium (Na), cobalt (Co), vanadium (V), and silicon (Si) also often give beneficial effects on plant growth, and may be essential for growth of some plants. Other essential macronutrients include carbon (C), hydrogen (H), and oxygen (O), but these are supplied from air and water, not primarily from fertilizers.

Methods for chemical analyses of fertilizers are given by the Association of Official Analytical Chemists (AOAC).[1] Available P in fertilizers in relation to crop response was summarized by Terman, Hoffman, and Wright[2] and by others. The principles of agronomic evaluation of nutrients in fertilizers have been discussed by Terman, Bouldin, and Webb,[3] by Terman and Engelstad,[4] and by others. Tisdale and Nelson[5] and Buckman and Brady[6] are among the most useful and better known textbooks covering fertilizer-soil-crop relationships.

FERTILIZER EXPRESSIONS AND CONVERSIONS

In the U.S., contents of the macronutrients (NPK) expressed as %N-%P_2O_5 (pentoxide, commonly termed phosphate)-%K_2O (oxide, commonly termed potash) content in this order are termed the fertilizer grade. Thus 100 kg of 13-13-13 contains 13 kg each of N, P_2O_5, and K_2O. Each grade or formulation is usually composed of several fertilizer salts. Other nutrients were formerly expressed also as the oxides, but now are expressed largely as the elements. This is also true for N, P, and K in many other countries.

In the U.S., fertilizer regulations have been incorporated into individual state (as opposed to Federal) fertilizer control laws. Each law requires guarantees of the minimum percentages of each of the three primary nutrients — total N, available P_2O_5, and soluble K_2O — and other constituents if they are listed on the fertilizer label or tag. The development and publication of standardized methods of analysis necessary for the successful operation of fertilizer control laws is a function of AOAC.[1] Some states specify a choice of AOAC or state methods in their fertilizer laws.

Element-oxide conversions for P and K are: $P \times 2.291 = P_2O_5$; $P_2O_5 \times 0.436 = P$; $K \times 1.205 = K_2O$; and $K_2O \times 0.830 = K$.

In U.S. fertilizer usage the term "unit" is 20 lb of N, P_2O_5, or K_2O, or 1%/ton of fertilizer. Thus, a ton of 10-10-10 fertilizer contains 10 units each of N, P_2O_5, and K_2O.

Chemical fertilizers may be produced and/or used as solids, as solutions, or as suspensions. Large amounts of N are applied as liquid anhydrous ammonia, which becomes gaseous as it is injected into the soil.

English metric conversions are: 1 lb = 0.454 kg; 1 kg = 2.205 lb; 1 English short ton (2000 lb) = 0.907 metric ton; 1 metric ton = 1.102 English ton.

PRIMARY SOURCES OF NUTRIENTS

Nitrogen

Most fertilizer N is initially produced as anhydrous ammonia (NH_3) by combining N_2 from air (78% N) with H_2 from petroleum products by the Haber-Bosch process (high temperature and pressure, together with use of catalysts). About 35,000 to 40,000 ft_3 (9,900 to 11,000 m^3) of natural gas is required to produce 1 ton of NH_3.

Leguminous plants also fix atmospheric N_2 biologically in root nodules containing *Rhizobium* or other species of microorganisms. Nonlegumes, such as corn, wheat, and cotton, cannot use N directly from the atmosphere. Small amounts of chemical N are obtained from natural nitrate deposits, such as Chilean nitrate and guano. Various organic residues and wastes also supply N. These sources are discussed in another chapter. Nitrogen and S both occur in mineral, atmospheric, and organic forms. Other nutrients occur largely in mineral forms; they are also components of plants.

Phosphorus

The first chemical P fertilizer was made well over a century ago by acidulating bones with H_2SO_4. Various natural phosphate rock deposits are now the chief source of fertilizer P (or phosphate raw material). The more soluble ores may be ground finely and used effectively in some situations for direct application. Most phosphate rock is beneficiated and then acidulated to produce more soluble P fertilizers.

Potassium

Most fertilizer K (or potash) is mined from natural subsurface solid salt deposits, is recovered by solution mining followed by evaporation, or by evaporation of saline surface waters.

Calcium and Magnesium

The source of most Ca and Mg is natural lime rock formations of calcite ($CaCO_3$) or dolomite [$CaMg(CO_3)_2$]. Phosphate rocks also contain considerable Ca as free lime or as apatite. Some Mg is recovered by evaporation of sea water or other brines. Liming materials are summarized in another chapter.

Sulfur

Most elemental S is obtained by use of hot water injected into subsurface deposits to melt the S, followed by recovery as a solid after cooling at the surface. Much of the S is oxidized to H_2SO_4 for use in producing acidulated phosphates (about 50% of S production) or for other uses. Considerable S is also recovered as sulfate byproducts of the smelting of pyritic (FeS_2) ores. Research is underway to evaluate the potential for recovery of SO_2 from stack gases.

Micronutrients

Micronutrients are recovered by smelting of various natural ores, followed by further processing as determined by each final product. Boron is obtained largely from natural desert borax deposits.

NITROGEN FERTILIZER PROCESSES (TABLE 1)

As mentioned previously, *anhydrous NH_3* made from N_2 and H_2 is the basic compound from which most other N fertilizers are produced, largely as granular products. Ammonia or *aqua ammonia* (NH_4OH) may also be injected into the soil or irrigation water for use by crops.

Anhydrous NH_3 is oxidized to produce *HNO_3*, which can be reacted with more NH_3 to produce *ammonium nitrate* (AN,NH_4NO_3). Ground limestone ($CaCO_3$) is added to AN in some countries to produce *ammonium nitrate limestone* (ANL). *Ammonium sulfate* (AS) is produced by reacting NH_3 and H_2SO_4. It is also made as a coke-oven byproduct. Ammonium sulfate or NH_3 plus H_2SO_4 can be added to AN to produce *ammonium nitrate sulfates* (ANS). *Calcium nitrate* [$Ca(NO_3)_2 \cdot 4H_2O$] has been produced by reacting HNO_3 with $CaCO_3$, but most of it for fertilizer use is a byproduct from nitric phosphate manufacture. *Ammonium chloride* is usually produced by reacting NH_3 with byproduct HCl (largely for fertilizer use in Japan). *Urea* produced by reacting NH_3 and CO_2 became the leading world N fertilizer in 1975. Ammonium nitrate still exceeded urea production in the U.S. in 1977 by a substantial margin. Urea and AN are produced in pellet form as prills or granules. Large amounts of both AN and urea are used as solutions. *Sulfur-coated urea* (SCU) is a product developed by TVA by coating granular urea with molten elemental S, wax, sealant, and surface conditioner dust. It is a slow-release product for which initial and later dissolution rates can be varied widely by the nature and amount of S coating and sealant used. *Urea formaldehyde* (UF) is another slow-release but expensive N source produced by reaction of urea and formaldehyde. It has been produced for several years for use in specialty fertilizers. Its solubility can also be varied widely by temperature and other conditions during reaction. Other slow-release types of N produced in small amounts include oxamide (32% N), crotonylidene diurea (28% N), dicyandiamide (42% N), and thiourea (37% N).

Calcium cyanamid ($CaCN_2$) is produced by reacting heated calcium carbide with N. The carbide is formed by reacting $CaCO_3$ with carbon from coke in an electric furnace. Calcium cyanamid now has very limited use as fertilizer.

The most common *N solutions* are mixtures of urea, NH_4NO_3, and water (28 to 32% N). Other concentrations are also produced, including higher analysis suspensions.

Use of *Chilean nitrate* ($NaNO_3$) has declined from a major source to a minor source of N since development of the Haber-Bosch process for fixing atmospheric N.

PHOSPHATE FERTILIZER PROCESSES (TABLE 1)

Phosphate rock (PR) is the chief raw material source of P fertilizers, and H_2SO_4 is a primary reactant for acidulation. In production of *ordinary superphosphate* (OSP) an amount of H_2SO_4 is added to PR necessary to solubilize the P. The resulting product (16 to 20% P_2O_5) contains largely $Ca(H_2PO_4)_2 \cdot H_2O$ and gypsum ($CaSO_4 \cdot 2H_2O$).

An excess of H_2SO_4 is reacted with PR to produce *H_3PO_4* (wet process). This acid is then used widely to produce other phosphate fertilizers. *Triple superphosphate* (TSP) is produced by reacting H_3PO_4 and PR. Either OSP or TSP may be treated with NH_3 to produce ammoniated superphosphates, usually as a component of mixed fertilizers. Ammoniation of superphosphates tends to decrease water solubility of the P.

Phosphate fertilizers were formerly produced by TVA from elemental P. The *elemental P* was produced by reacting PR, carbon, and silica in electric furnaces. Cost of electricity is now too high for production of elemental P for use in fertilizers. *Cal-*

Table 1
COMMERCIAL AND EXPERIMENTAL FERTILIZERS — MANUFACTURE AND ANALYSES[3,5]

Fertilizer and common abbreviations	Typical grades ($N\text{-}P_2O_5\text{-}K_2O$)	Water solubility of N, P or K[a]	Main reactants or process used	Major fertilizer compounds present	Macronutrient content (%)					
					N	P	K	Ca	Mg	S
N Sources		N								
Ammonia	82-0-0	100	$N_2 + H_2$	NH_3	82.0	—	—	—	—	—
Anhydrous aqua	20-0-0	100	$NH_3 + H_2O$	NH_3	15—30	—	—	—	—	—
Ammonium chloride	28-0-0	100	$NH_3 + HCl$	NH_4Cl	28.0	—	—	—	—	—
Ammonium nitrate (AN)	33.5-0-0	100	$NH_4 + NHO_3$	NH_4NO_3	33—34	—	—	—	—	—
Ammonium nitrate limestone (ANL)	20.5-0-0	100	NH_4NO_3 + lime	NH_4NO_3	20.5	—	—	8.2	4.4	0.4
Ammonium nitrate sulfate (ANS)	30-0-0	100	$NH_3 + HNO_3 + H_2SO_4$	H_4NO_3, $(NH_4)_2SO_4$	30.0	—	—	—	—	12.0
Ammonium sulfate (AS)	21-0-0	100	$NH_3 + H_2SO_4$	$(NH_4)_2SO_4$	21.0	—	—	0.3	—	24.0
Calcium cyanamid	22-0-0	>90	$CaC_2 + N_2$	$CaCN_2$	22.0	—	—	40.0	0.1	0.3
Calcium nitrate	15.5-0-0	100	$CaCO_3 + HNO_3$	$Ca(NO_3)_2$	15.5	—	—	19.0	1.5	—
Nitrogen solutions	32-0-0	100	NH_4NO_3 + urea	NH_4NO_3, $CO(NH_2)_2$	21—36	—	—	—	—	—
Sodium nitrate	16-0-0	100	Beneficiation	$NaNO_3$	16.0	—	0.2	0.1	0.1	0.1
Chilean synthetic	16-0-0	100	$Na_2CO_3 + HNO_3$	$NaNO_3$	16.0	—	—	—	—	—
Urea	46-0-0	100	$NH_3 + CO_2$	$CO(NH_2)_2$	46.0	—	—	—	—	—
Urea, S-coated (SCU)	35-0-0	v	$CO(NH_2)_2 + S$	$CO(NH_2)_2$, S	35.0	—	—	—	—	7—10
Urea-form (UF)	38-0-0	v	$CO(NH_2)_2 + CH_20$	$CO(NH_2)_2$: CH_2O	38.0	—	—	—	—	—
P Sources		P								
Bone meal	3-25-0	< 1	Grinding + steam		2—4	9—12	0.2	20—25	0.4	0.1
Phosphate rocks (PR)	0-32-0	< 1	Beneficiation	Apatites	—	11—17	—	33—36	—	—
Phosphoric acid										
Wet process	0-55-0	100	PR + H_2SO_4	H_3PO_4	—	22—26	—	—	—	—
Furnace grade	0-55-0	100	Elemental P + H_2O	H_3PO_4	—	22—26	—	—	—	—
Super acid	0-78-0	100	Elemental P + H_2O	H_3PO_4, poly P acids	—	32—34	—	—	—	—
Superphosphates										
Ordinary (OSP)	0-20-0	85	PR + H_2SO_4	$Ca(H_2PO_4)_2 \cdot H_2O$, $CaSO_4 \cdot \frac{1}{2}H_2O$	—	7—9	0.2	20	0.2	12.0
Triple (TSP)	0-45-0	87	PR + H_3PO_4	$Ca(H_2PO_4)_2 \cdot 2H_2O$	—	18—22	0.3	14	0.3	1.4
Concentrated (CSP)	0-48-0	90	PR + H_3PO_4 (furnace grade)	$Ca(H_2PO_4)_2 \cdot H_2O$	—	21	—	12—14	—	<1.0
Dicalcium phosphate										
HCl process	0-40-0	4	$H_3PO_4 + CaCO_3$	$CaHPO_4 \cdot 2H_2O$, $CaHPO_4$	—	17	—	28	—	—
Furnace H_3PO_4	0-48-0	3	$H_3PO_4 + CaCO_3$	$CaHPO_4$, $CaHPO_4 \cdot 2H_2O$	—	21	—	19	—	—

Calcium metaphosphate (CMP)	0-62-0	5	PR + P (heat)	Vitreous Ca polyphosphate, $Ca_2P_2O_7$	—	27	—	19	—	—
Fused tricalcium phosphate (FTP)	0-62-0	<2	PR + heat	$Ca_3(PO_4)_2$	—	12	—	20	—	—
Rhenania phosphate	0-33-0	<2	PR + Na_2CO_3 + SiO_2 + heat	Silico phosphates	—	14	—	30	0.3	—
Serpentine phosphate	0-22-0	<2	PR + serpentine	Ca, Mg silico phosphates	—	9	—	20	8	—
Basic slag	0-9-0	<2	Iron ore + lime + heat	Ca silico carnotite	—	3.5	—	29	3	0.3
Colloidal "clay" phosphate	0-22-0	<1	Tailings from PR beneficiation	Apatites, Al, phosphates	—	9	—	—	—	—
N, NK, & PK Sources										
Muriate of potash	0-0-60	100	Beneficiation	KCl	—	—	50—52	0.1	0.1	—
Sulfate of potash	0-0-50	100	KCl + H_2SO_4, beneficiation	K_2SO_4	—	—	40—43	—	0.1	18
K, Mg sulfate (sulpomag)	0-0-23	100	Beneficiation	$K_2SO_4 \cdot 2MgSO_4$	—	—	18	0.1	11	18
Nitrate of potash	13-0-44	100	KCl + HNO_3 or $NaNO_3$	KNO_3	13	—	37	0.3	0.3	0.2
K polyphosphate (KPP)	0-59-39	5	KCl + H_3PO_4	K polyphosphates	—	26	32	—	—	—
NP Sources										
Ammoniated OSP	4-14-0	35	OSP + NH_3	$NH_4H_2PO_4$, $CaHPO_4$, apatite	4	6	—	16	0.3	10
Ammoniated CSP	5-47-0	50	CSP + NH_3	$NH_4H_2PO_4$, $CaHPO_4$	5	20	—	12—14	—	<1
Ammoniated TSP		50	TSP + NH_3	$(NH_4)_2HPO_4$, apatite	9	—	—	—	—	—
Ammonium phosphate nitrate (APN)	30-10-0)	100	H_3PO_4 + NH_3 + HNO_3	$NH_4H_2PO_4$,	30	4	—	—	—	—
	25-25-0)	100		NH_4PO_3	25	11	—	—	—	—
	28-14-0)	100		$(NH_4)_2HPO_4$	28	6	—	—	—	—
Monoammonium phosphate (MAP)	11-48-0	>90	NH_3 + HF_3PO_4	$NH_4H_2PO_3$	11	21	—	1.5	0.2	4.5
Ammonium phosphate sulfate (APS)	13-39-0)	>90	H_3PO_4 + NH_3 + H_2SO_4	$NH_4H_2PO_4$	13	17	—	—	—	2
	16-20-0)	>90		$(NH_4)_2HPO_4$	16	9	—	—	—	15
	16-48-0)	>90		$(NH_4)_2SO_4$	16	21	—	—	—	2
Diammonium phosphate (DAP)	21-53-0)	100	H_3PO_4 + NH_3	$(NH_4)_2HPO_4$	21	23	—	—	—	—
	18-55-0)	>95			18	20	—	—	—	—
Ammonium polyphosphate (APP)	15-60-0)	100	Poly P acid + NH_3	$NH_4H_2PO_4$	15	26	—	—	—	—
	12-55-0)	>90		$(NH_4)_3HP_2O_7$	12	24	—	—	—	—
Urea ammonium phosphate (UAP)	34-17-0)	100	$CO(NH_2)_2$ +) $(NH_4)_2HPO_4$)	$CO(NH_2)_2$, $(NH_4)_2HPO_4$	34	7	—	—	—	—
	29-29-0)	100			29	13	—	—	—	—
	25-35-0)	100			25	15	—	—	—	—
Urea ammonium polyphosphate (UAPP)	36-18-0)	100	$CO(NH_2)_2$ + poly P acids	$CO(NH_2)_2$,	36	8	—	—	—	—
	30-30-0)	100		$(NH_4)_3HP_2O_7$,	30	13	—	—	—	—
	22-44-0)	100		$NH_4H_2PO_4$	22	19	—	—	—	—

Table 1 (continued)
COMMERCIAL AND EXPERIMENTAL FERTILIZERS — MANUFACTURE AND ANALYSES[3,5]

Fertilizer and common abbreviations	Typical grades (N-P_2O_5-K_2O)	Water solubility of N, P or K[a]	Main reactants or process used	Major fertilizer compounds present	Macronutrient content (%)					
					N	P	K	Ca	Mg	S
Urea phosphate (UP)	18-45-0	100	$CO(NH_2)_2 + H_3PO_4$	$CO(NH_2)_2 \cdot H_3PO_4$	18	20	—	—	—	—
Nitric phosphate (NP)	20-20-0	40	PR + HNO_3(+ H_3PO_4 + H_2SO_4)	$CaHPO_4$, $Ca(NO_3)_2$, $NH_4H_2PO_4$, apatite	20	9	—	—	—	—

[a] v = variable.

cium metaphosphate was also formerly made by TVA by burning elemental P with PR in a furnace. Polyphosphoric acid, the basis for production of fluid fertilizers, is produced by concentration of wet process H_3PO_4. Higher energy costs have largely eliminated heat treatment processes for phosphate fertilizer production in the U.S.

A wide range of solid and fluid ammonium phosphates is made by reacting NH_3 and H_3PO_4 alone or plus polyphosphoric acids. *Monoammonium phosphate* (MAP) is usually produced by reacting wet process H_3PO_4 with NH_3. Twice the amount of NH_3 per unit of P is used to produce *diammonium phosphate* (DAP). HNO_3 is also added to produce *ammonium phosphate nitrates* (APN) and H_2SO_4 or solid AS to produce *ammonium phosphate sulfates* (APS), but production is low.

Ammonium polyphosphate (APP) is now produced by reacting NH_3 with concentrated wet process H_3PO_4 in pipe reactors to produce granular grades such as 11-55-0. This product is used in bulk blending and in formulating suspension fertilizers.

Urea is being increasingly used in formulating NP and NPK fertilizers. The usual process is to react NH_3 and wet process H_3PO_4 or concentrated H_3PO_4 (poly acids) and add urea as a melt or as a solid. Grades such as 29-29-0 and 19-19-19 are called *urea ammonium phosphates* (UAP).

Urea phosphate is an adduct $[CO(NH_2)_2 \cdot H_3PO_4]$ formed by reaction of urea and H_3PO_4. It is used both as fertilizer and as an animal feed supplement.

Nitric phosphates are produced by acidulating PR with HNO_3 or mixtures of HNO_3, H_3PO_4, or H_2SO_4 to give a wide range of grades and P solubilities. As a result of environmental problems and costs, the trend in recent years has been away from nitric phosphates to ammonium phosphate and urea ammonium phosphate fertilizers.

Other P-containing fertilizers produced in limited amounts include *metal ammonium phosphates* (ex., $MgNH_4PO_4 \cdot H_2O$, 9% N). These sources have been used largely as slow-release sources of N and P.

POTASSIUM FERTILIZER PROCESSES (TABLE 1)

The chief K fertilizer is *potassium chloride* (KCl, or muriate of potash, from muriatic acid, HCl). Both solid and solution-recrystallization mining methods are used for separation and recovery of component salts. *Sulpomag* (langbenite, $K_2SO_4 \cdot 2MgSO_4$) is recovered largely from salt deposits near Carlsbad, New Mexico.

Potassium sulfate (K_2SO_4) is produced largely by reaction of KCl with sulfate salts or H_2SO_4. Consequently, its unit cost is greater than that of KCl.

Potassium nitrate (KNO_3) is produced by reacting KCl with HNO_3 or $NaNO_3$. Unit cost of the K in KNO_3 is thus also higher than in KCl, but is used for crops such as tobacco, for which Cl may have adverse effects.

Potassium polyphosphate has been produced by TVA and others by reaction of KCl and H_3PO_4 in a furnace. Recent high energy costs and problems with disposal of by-product HCl have resulted in production being discontinued.

Other K fertilizers produced to a limited extent include K_2CO_3 (56% K), $KHCO_3$ (39% K), manure salts (about 20% K), kainite ($KCl \cdot MgSO_4 \cdot 3H_2O$, 12 to 17% K), and the K orthophosphates (KH_2PO_4 and K_2HPO_4).

SALT INDEX OF FERTILIZERS (TABLE 2)

Most fertilizers contain soluble salts and appreciable concentrations of these salts in contact with seeds can reduce germination and injure plant roots. Salt index is usually calculated as the ratio of the increase in osmotic pressure produced by a fertilizer to the increase produced by an equal weight of $NaNO_3$, taken as 100. Soluble N and K

Table 2
SALT INDICES OF SELECTED FERTILIZERS[5,7,8]

Fertilizer	Fertilizer grade	Salt index: On a material weight basis	Salt index: Per unit (20 lb) of plant nutrients
N sources			
$NaNO_3$	16.5-0-0	100	6.06
Anhydrous NH_3	82-0-0	47	0.57
NH_4NO_3	33.5-0-0	100	3.00
$(NH_4)_2SO_4$	21-0-0	69	3.25
$Ca(NO_3)_2$	15.5-0-0	65	4.20
$CaCN_2$	21-0-0	31	1.48
Urea	46-0-0	75	1.62
UAN solution	40-0-0	78	1.93
Organic N	5—13-0-0	4	0.3 to 0.7
P sources			
OSP	0-20-0	8	0.39
CSP, TSP	0-45-0	10	0.22
K sources			
KC1	0-0-60	116	1.94
K_2SO_4	0-0-50	46	0.85
$K_2SO_4 \cdot 2MgSO_4$	0-0-23	43	1.97
KNO_3	0-0-47	74	1.22
NP, NK, PK, or NPK sources			
MAP	11-48-0	30	0.41
DAP	18-46-0	34	0.46
KNO_3	13-0-44	74	1.22
APS	16-20-0	45	—
NPK	5-10-10	38	—
	10-20-20	57	—
	15-15-15	71	—

compounds have much higher indices than do phosphate compounds. Salt indices of mixed fertilizers are estimated from the amounts of various components.

Free ammonia released from urea or ammonium (NH_4−N) compounds in neutral to alkaline soils and nitrite formed from NH_4−N are often more injurious if placed near germinating seeds and seedling roots than is soluble salt content. Urea and DAP are the most subject to release of free NH_3, and thus should not be applied near (closer than 2 to 5 cm) germinating seeds. The same is true of fertilizers having high salt indices, especially in soil low in moisture content or in saline soils.

ACIDITY OR BASICITY OF FERTILIZERS (TABLE 3)

W. H. Pierre developed a method in 1933 for determining the net acid- or base-forming effects of fertilizers applied to soil. He assumed that (1) Cl, S, one-third of the P, and half the N content contributed to soil acidification, that is, decreased soil pH; and (2) that the basic cations (Ca, Mg, K, and Na) contributed to soil basicity, that is, increased soil pH. He also assumed that half the applied N was absorbed by plants as HNO_3 and half as a salt, such as $Ca(NO_3)_2$. He calculated that 1.8 kg of $CaCO_3$ would be required to neutralize the acidity resulting from each kg of applied N. This method, with some modifications, has been adopted by AOAC.[1]

That the N content of fertilizers is indeed important in contributing to soil acidity

Table 3
EQUIVALENT ACIDITY (A) OR BASICITY (B) OF SELECTED FERTILIZERS[a] [5,8]

Fertilizer	Grade	Acidic or basic	$CaCO_3$ (lb) Per unit (20 lb) of N	Per 100 lb of fertilizer
Anhydrous NH_3	82-0-0	A	36	146
NH_4NO_3	33.5-0-0	A	36	60
$(NH_4)_2SO_4$	20.5-0-0	A	107	110
Urea	46-0-0	A	36	84
N solution	32-0-0	A	36	57
Organic sources	2 to 13-0-0	A	30—50	7—20
$Ca(NO_3)_2$	16-0-0	B	22	20
$NaNO_3$	16-0-0	B	36	29
$CaCN_2$	22-0-0	B	57	63
KNO_3	13-0-44	B	40	26
MAP	11-48-0	A	36	20
DAP	18-46-0	A	36	32
Superphosphates	0-16 to 45-0	B	—	5—12
KCl or K_2SO_4	0-0-50 to 60	—	—	0

[a] AOAC method, equivalent pounds of $CaCO_3$/ton of fertilizer:
Equivalent acidity due to N = % N × 35.7
Basicity or acidity of ignited residue = net ml of 0.5 *N* NaOH; titration × 50 (1 g of fertilizer sample);
Basicity due to citrate insoluble P_2O_5 = % citrate insoluble P_2O_5 × 28.2
Net = equivalent acidity or basicity

is emphasized by the usually high correlation between nitrate-N and basic cations leached from soils and the accompanying lower soil pH.

SOLID FERTILIZERS

Solid fertilizers may be produced as powders, semigranular materials, granules, or prills. *Granular fertilizers* have obvious advantages in regard to ease of blending, handling, and uniform application by means of fertilizer drills and bulk spreading. Most fertilizers now used are granular or prilled. Bulk densities of some commercial granular products are given in Table 4.

Size of granules is important in blending operations, since dissimilar sizes tend to segregate (Figure 1) during bulk blending and transportation. Density is a much less important factor in fertilizer segregation.

Hygroscopicity of fertilizer salts is their tendency to absorb moisture from the air if its vapor pressure exceeds that of a saturated solution of the salt. Table 5 gives the relative humidities of air at which a salt or salt mixture begins to absorb moisture at 86°F (30°C).

Fertilizers which are very hygroscopic are usually bagged in moisture-tight plastic bags for handling and transport. However, hygroscopic fertilizers can be bulk handled with controlled humidity, covering of surface of piles in storage, and reasonable protection in transport.

Less hygroscopic fertilizers may be either bagged or handled in bulk. Bulk handling and spreading have increased rapidly in the U.S. (Figure 2), with a consequent decline in bagging. Numbers of bulk blend fertilizer plants have also increased greatly (Figure 3) to about 5200 in 1976.

Table 4
BULK DENSITIES OF COMMERCIAL FERTILIZERS[16]

Product	Grade	Average bulk density (lb/ft³)
Urea		
Prilled	46-0-0	45
Granular	46-0-0	46
Ammonium nitrate		
Prilled		
High density	33.5-0-0	58
Low density	33.5-0-0	45
Granular	33.5-0-0	46
Urea ammonium phosphate	28-28-0	48
	34-17-0	
Urea ammonium sulfate	40-0-0	45
Ammonium sulfate	21-0-0	60
Sulfur-coated urea	36-0-0	45
Diammonium phosphate	18-46-0	55
Monoammonium phosphate	11-52-0	56
Normal superphosphate	0-19-0	68
Triple superphosphate	0-45-0	66
Potassium chloride		
Granular	0-0-60	67
Coarse	0-0-60	68

FLUID FERTILIZERS

Fluid (nonsolid) fertilizers include solutions (all solids dissolved) and suspensions (solids only partly dissolved). Use of fluid fertilizers has been growing steadily in recent years (Figure 2).

Fluid fertilizers have certain advantages over solid fertilizers: (1) They can be pumped to reduce labor, handling, and application costs; (2) they can be easily blended, and the ingredients do not segregate during handling; (3) they serve as a base with which micronutrients, pesticides, etc. can be easily blended for more uniform application to soils; (4) equipment for mixing and mixing costs may be lower than for solid fertilizers. A disadvantage of fluid fertilizers is that the water component dilutes the plant nutrients and adds to transportation costs. As a result, fluid fertilizer mixing plants serve only local areas. Numbers of such plants (Figure 4) have been increasing rapidly in the U.S., to a total of about 3200 in 1976.

Nonpressure N solutions usually contain AN or AN + urea. Low pressure solutions contain aqua ammonia (NH_4OH) alone, NH_3 + H_2O + AN, or NH_4OH + AN + urea. Properties of some common solutions are shown in Table 6.

Fluids containing both N and P, or N, P, and K are usually based on ammonium phosphates. Solubility-N/P_2O_5 ratio relationships for saturated ammonium orthophosphates at 32° F are shown in Figure 5. Solubility-pH and corrosive properties are shown in Figure 6. The highest solubility point (eutectic) at 32°F is an N-P_2O_5 concentration of about 9-28-0. This increases to about 11-34-0 at 86°F, and to 14-44-0 at 174°F. The usual grades made, however, are 8-24-0, 11-33-0, and 14-42-0.

Most recent advances in use of fertilizer solutions have been based on ammonium polyphosphate, made by ammoniation of polyphosphoric acids, rather than on ammonium orthophosphates. Solubility of ammonium polyphosphates is higher, so that

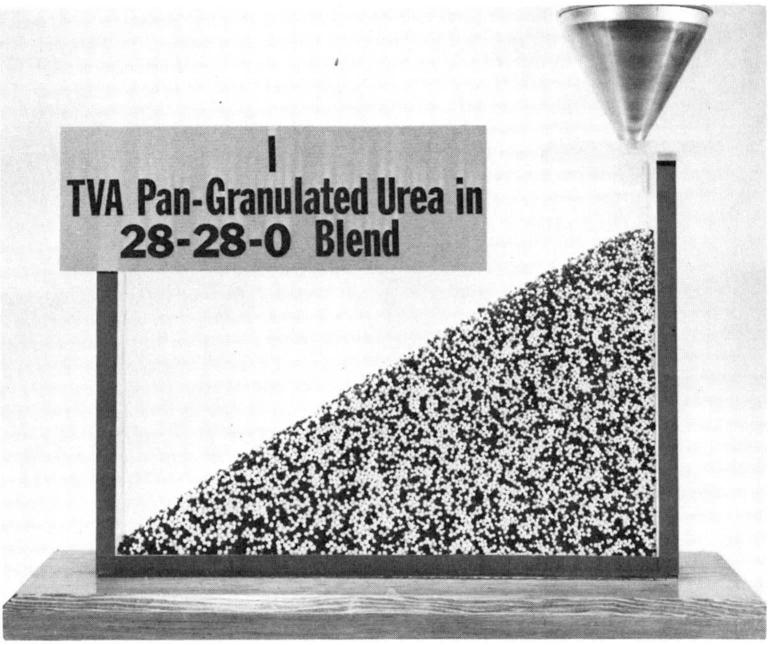

1A

1B

FIGURE 1. A uniform mixing of similar sized DAP (dark granules) and pan-granulated urea (light granules); (B) segregation resulting from mixing DAP and smaller urea prills. (From Hoffmeister, G., TVA, Muscle Shoals, Ala., unpublished material.)

Table 5
HYGROSCOPICITY OF SOME PURE FERTILIZER COMPOUNDS AND MIXTURES

Fertilizer salt	Critical relative humidity at which the salt or its mixture with other salts begins to absorb moisture at 86°F[a]												
Calcium nitrate $Ca(NO_3)_2 \cdot 4H_2O$													47
Ammonium nitrate NH_4NO_3												59	24[b]
Sodium nitrate $NaNO_3$											72	46[b]	38[b]
Urea $CO(NH_2)_2$										73	46[b]	18[b]	67[c]
Sodium chloride $NaCl$									75	53[b]	68[b]	[d]	
Ammonium chloride NH_4Cl								77	69[b]	58[b]	42—52[e]	51[b]	
Ammonium sulfate $(NH_4)_2SO_4$							79	71[b]		56[b]	[d]	62[b]	
Potassium chloride KCl						84	[d]	74[b]	72[b]	60[b]	[d]	[d]	[e]
Potassium nitrate KNO_3					91	79[b]	62—69[e,f]	55—68[e,f]	61—67[e]	65[b]	65[b]	60[b,f]	31[b]
Ammonium phosphate $NH_4H_2PO_4$				92	60[e,f]	[d]	76[b]	74[b]		65[b]	49—64[e]	58[b]	[d]
Potassium phosphate KH_2PO_4			93	91[b]	90[b]	83[b]	[d]	73[e,f]		70[b]	63[e]	[d]	[d]
Calcium phosphate $Ca(H_2PO_4)_2$		94	91[b]	89[b]	44—88[e]	78[e]	[d]	74[e]		65[b]	36—68[e]	26—53[e]	46[b]
Potassium sulfate K_2SO_4	96	[d]	94[b]	79[e,f]	88[b]	83[b]	81[b]	71[e,f]		72[b]	[d]	[d]	[d]
Calcium sulfate $CaSO_4$			93—94[e]	77—92[e]	30—92[e]								

[a] The first value is the approximate critical relative humidity of the salt alone. Read down or across to find the critical relative humidity of its mixture with most of the other salts listed.
[b] Nonreacting salt pairs. Data for solution saturated with both salts.
[c] The double salt, $Ca(NO_3)_2 \cdot 4CO(NH_2)_2$.
[d] Unstable salt pair of reciprocal salt pairs. See stable salt pair.
[e] Stable salt pair of reciprocal salt pairs.
[f] Solid phases include solid solutions.

From *Farm Chemicals Handbook*, Meister, Willoughby, Ohio 1978. With permission.

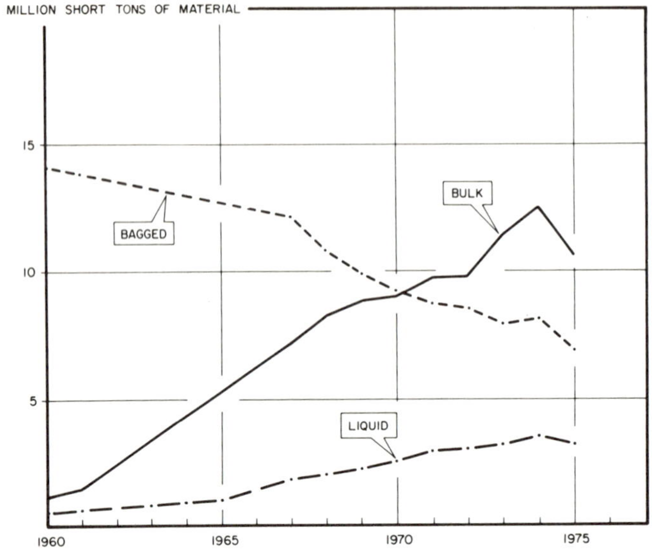

FIGURE 2. Consumption of bagged, bulk, and fluid fertilizers in the U.S. (From Harre, E. A., Goodson, M. N., and Bridges, J. D., Fertilizer Trends, 1976, TVA Bull. Y-111, Muscle Shoals, Ala., 1977.)

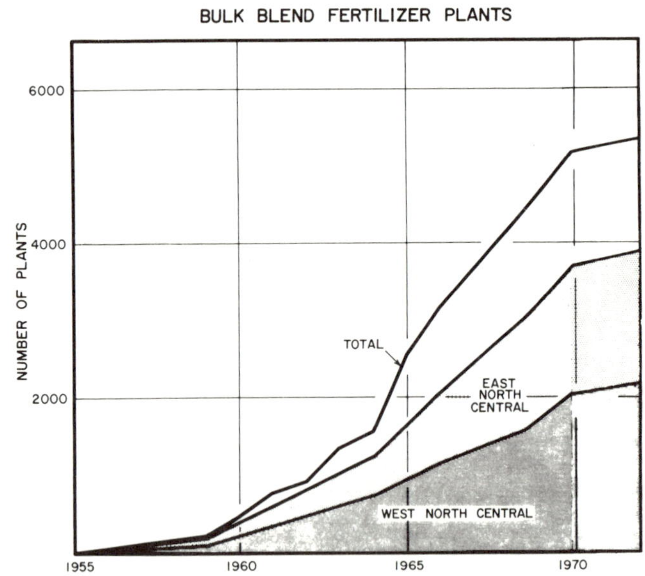

FIGURE 3. Increase in numbers of bulk blend fertilizer plants in the U.S. (From Fertilizer Trends.[20])

solutions such as 10-34-0 or 11-37-0 made from furnace-grade H_3PO_4 are stable at 32°F. Solubility diagrams for ammonium polyphosphates made from wet process acids (more impurities) show eutectics at 39% (N + P_2O_5) for zero polyphosphate content, 46% for 45% polyphosphate content (% of total P), and 48% for 70% polyphosphate content. The N/P_2O_5 ratios at the eutectics are between 0.30 and 0.32 (TVA data), and the pH between 5.8 and 6.3. In addition, most micronutrients are also more soluble in polyphosphate solutions. Higher grades than these, especially if made from wet

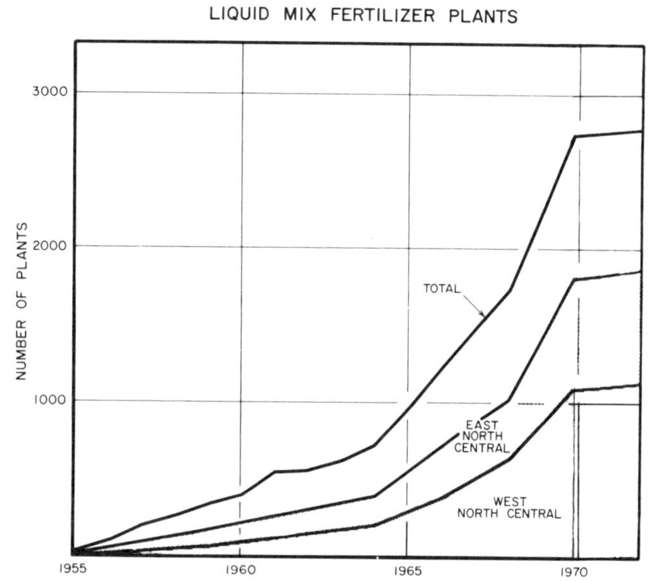

FIGURE 4. Increase in numbers of liquid mix fertilizer plants in the U.S. (From Harre, E. A., Fertilizer Trends, 1973, TVA Bull. Y-77, Muscle Shoals, Ala., 1974.)

process H_3PO_4 and including KCl are usually suspensions. Some of the more popular suspension grades are shown in Table 7.

WORLD FERTILIZER RESERVES

Nitrogen

Elemental N_2 in the atmosphere (78% of the air) used to produce anhydrous ammonia is unlimited in terms of needs for fertilizers. Supplies of natural gas, used primarily as the source of hydrogen (H_2) to combine with N_2 to produce anhydrous ammonia, however, are increasingly higher in cost. The same is true for petroleum derivatives sometimes used instead of natural gas. Coal may be used more extensively to produce H_2 if natural gas supplies become critical. Investment and production costs will be substantially higher.

Phosphates

Phosphate rock (PR) ores, used primarily as the source of P fertilizers, occur in many countries. Deposits vary widely in phosphate content (usually expressed as P_2O_5 in percent), or bone phosphate of lime (BPL = $\%P_2O_5 \times 2.185$). Table 8 shows recent estimates of the higher grades (32% P_2O_5 or higher) of PR of sedimentary origin (about 84% of the world total). Some of these ores are satisfactory for grinding for direct application for crop production, as well as for acidulation to produce more soluble phosphates. Table 9 shows similar estimates for PR of igneous origin (about 12% of the world total). Igneous PR is less reactive than many PRs of sedimentary origin and is suitable for crop production only after acidulation.

Reactivity of PR is commonly determined by its dissolution in neutral ammonium citrate. Relative absolute citrate solubilities (solubility based on content of apatite in the PR) of various PRs are shown in Figures 7 and 8. Some of the more reactive PRs of sedimentary origin occur in North Carolina and North Africa.

Other types of phosphate occurrence such as guano and aluminum phosphates con-

Table 6
CHEMICAL COMPOSITION AND PHYSICAL PROPERTIES OF NONPRESSURE AND LOW PRESSURE LIQUID NITROGEN MATERIALS

Suggested product designation	Composition: Ammonia (%)	Composition: Ammonium nitrate (%)	Composition: Urea (%)	Composition: Water (%)	Nitrogen Content: Ammonia N: Free (%)	Nitrogen Content: Ammonia N: Combined (%)	Nitrate N (%)	Urea N (%)	Total N (%)	Salting out Temperature (°F)	Specific Gravity at 60°F	Weight of solution per gal. at 60°F (lb)	Weight of nitrogen per gal. at 60°F (lb)	Volume per 100 lbs. of nitrogen at 60°F (gal)	Vapor pressure: 60°F (psi)	Vapor pressure: 104 °F (psi)
Aqueous solutions of nitrogen salts																
160(0-46-0)	—	45.7	—	54.3	—	8.0	8.0	—	16.0	11	1.207	10.07	1.61	62.1		
170(0-49-0)	—	48.6	—	51.4	—	8.5	8.5	—	17.0	16	1.222	10.19	1.73	57.8		
175(0-50-0)	—	50.0	—	50.0	—	8.8	8.8	—	17.6	22	1.231	10.27	1.81	55.2		
180(0-51-0)	—	51.4	—	48.6	—	9.0	9.0	—	18.0	23	1.235	10.30	1.85	54.1		
190(0-54-0)	—	54.3	—	45.7	—	9.5	9.5	—	19.0	33	1.253	10.45	1.99	50.3		
200(0-57-0)	—	57.2	—	42.8	—	10.0	10.0	—	20.0	41	1.264	10.54	2.11	47.4		
210(0-60-0)	—	60.0	—	40.0	—	10.5	10.5	—	21.0	49	1.286	10.73	2.25	44.4		
228(0-65-0)	—	65.0	—	35.0	—	11.4	11.4	—	22.8	65	1.307	10.90	2.49	40.2		
245(0-70-0)	—	70.0	—	30.0	—	12.3	12.3	—	24.6	86	1.346	11.23	2.76	36.2		
280(0-39-31)	—	39.3	30.6	30.1	—	6.9	6.9	14.3	28.1	−1	1.279	10.67	3.00	33.3		
280(0-40-30)	—	40.1	30.0	29.9	—	7.0	7.0	14.0	28.0	−1	1.283	10.70	3.00	33.3		
280(0-80-0)	—	80.0	—	20.0	—	14.0	14.0	—	28.0	136	1.360	11.34	3.18	31.4		
290(0-83-0)	—	82.9	—	17.1	—	14.5	14.5	—	29.0	154	1.380	11.51	3.34	29.9		
300(0-42-33)	—	42.2	32.7	25.1	—	7.4	7.4	15.3	30.1	15	1.301	10.85	3.27	30.6		
315(0-0-68)	—	—	67.5	29.5	—	—	—	31.5	31.5	135	1.170	9.76	3.07	32.6		
320(0-44-35)	—	44.3	35.4	20.3	—	7.8	7.8	16.5	32.1	32	1.327	11.07	3.55	28.2		
320(0-45-35)	—	45.0	35.0	20.0	—	7.9	7.9	16.3	32.1	32	1.325	11.05	3.55	28.2		
338(0-0-72)	—	—	72.5	24.5	—	—	—	33.8	33.8	155	1.18	9.84	3.33	30.0		
376(0-0-80)	—	—	80.5	16.5	—	—	—	37.6	37.6	190	1.19	9.92	3.73	26.8		

								Aqua ammonia								
206(25-0-0)	25.0	—	—	75.0	20.6	—	—	—	20.6	−103	0.911	7.60	1.56	64.1	−8	2
242(29-0-0)	29.4	—	—	70.6	24.2	—	—	—	24.2	—	0.897	7.48	1.84	54.3	—	10
								Ammonia-ammonium nitrate solution								
318(8-72-0)	8.0	72.0	—	20.0	6.6	12.6	12.6	—	31.8	89	1.257	10.48	3.33	30.0	—	1
334(10-72-0)	10.0	72.0	—	18.0	8.2	12.6	12.6	—	33.4	85	1.230	10.26	3.43	29.2	—	4
340(19-54-0)	18.5	54.2	—	27.3	15.2	9.5	9.5	—	34.2	12	1.118	9.32	3.19	31.3	—	1
370(17-67-0)	16.6	66.8	—	16.6	13.7	11.7	11.7	—	37.1	50	1.184	9.87	3.66	27.3	−9	2
410(19-73-0)	19.0	72.5	—	8.5	15.6	12.7	12.7	—	41.0	61	1.194	9.96	4.08	24.5	−6	7
410(22-65-0)	22.2	65.0	—	12.8	18.3	11.4	11.4	—	41.1	21	1.138	9.49	3.90	25.6	−5	10
410(22-66-0)	21.9	65.7	—	12.4	18.0	11.5	11.5	—	41.0	21	1.139	9.50	3.90	25.6	−5	10
410(24-60-0)	24.3	60.0	—	15.7	20.0	10.5	10.5	—	41.0	1	1.105	9.22	3.78	26.5	—	13
410(26-55-0)	26.5	55.0	—	18.5	21.8	9.6	9.6	—	41.0	−23	1.077	8.98	3.68	27.2	—	16
410(26-56-0)	26.3	55.5	—	18.2	21.6	9.7	9.7	—	41.0	−25	1.078	8.99	3.69	27.1	−3	17
414(19-74-0)	19.0	74.0	—	7.0	15.6	13.0	13.0	—	41.6	64	1.186	9.89	4.11	24.3	−3	8
440(24-70-0)	23.8	69.8	—	6.4	19.6	12.2	12.2	—	44.0	25	1.147	9.57	4.21	23.8	−2	18
448(25-69-0)	25.0	69.0	—	6.0	20.6	12.1	12.1	—	44.8	6	1.124	9.37	4.20	23.8	0	17
								Ammonia-urea								
210(4-0-38)	4.3	—	37.5	58.2	3.5	—	—	17.5	21.0	33	1.085	9.05	1.90	52.6	—	—
								Ammonia-ammonium nitrate-urea solutions								
370(13-53-16)	13.3	53.4	15.9	9.4	10.9	9.3	9.3	7.4	36.9	36	1.235	10.30	3.80	26.3	0	—
405(16-70-6)	16.0	70.0	6.0	8.0	13.2	12.3	12.3	2.8	40.6	66	1.228	10.24	4.16	24.0	—	4
410(19-58-11)	19.0	58.0	11.0	12.0	15.6	10.2	10.2	5.1	41.1	7	1.162	9.69	3.98	25.1	−6	10
410(19-66-5)	19.0	66.0	5.0	10.0	15.6	11.6	11.6	2.3	41.1	44	1.182	9.86	4.05	24.7	—	8
414(19-66-6)	19.0	65.6	6.0	9.4	15.6	11.5	11.5	2.8	41.4	35	1.178	9.82	4.07	24.6	−7	11
420(20-66-6)	19.5	66.3	6.0	8.2	16.0	11.6	11.6	2.8	42.0	34	1.179	9.83	4.13	24.2	—	10
430(20-68-6)	20.0	68.0	6.0	6.0	16.4	11.9	11.9	2.8	43.0	33	1.176	9.81	4.22	23.7	−3	15
430(22-62-7)	21.8	62.4	7.0	8.8	17.9	10.9	10.9	3.3	43.0	12	1.165	9.72	4.18	23.9	—	15
440(22-66-6)	22.0	66.0	6.0	6.0	18.1	11.6	11.6	2.8	44.1	13	1.155	9.63	4.25	23.5	−2	19
440(24-56-10)	24.0	56.0	10.0	10.0	19.7	9.8	9.8	4.7	44.0	−13	1.119	9.33	4.11	24.3	—	19

From *Liquid Fertilizer Handbook*, National Fertilizer Solutions Assoc., Peoria, Ill. With permission.

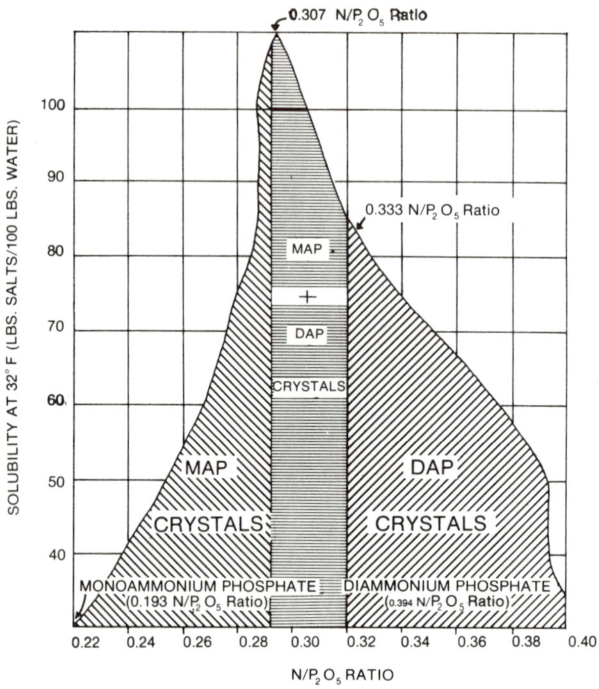

FIGURE 5. Solubility relationships of saturated ammonium orthophosphate solutions at 32°F.[16]

stitute only about 4% of the world total. If lower grade PR is included in the reserve estimates, variable tonnages many times those shown in Tables 8 and 9 can be obtained. Such estimates indicate adequate phosphate supplies for several hundred years, although at higher recovery costs as higher analysis reserves are depleted. Phosphates in lower analysis reserves must be concentrated (beneficiated) for satisfactory acidulation and fertilizer production.

Potash

Potash reserves occur in many countries as soluble brines or ores. Chemical and mineral composition is shown in Table 10. Reserves are quite extensive, as shown in Table 11. Supplies are adequate for the foreseeable future.

TRENDS IN FERTILIZER USE

World consumption of fertilizer N, P, and K has increased in recent years (Figure 9) with use of N increasing more rapidly than use of P and K.

For the U.S., increases in tonnages of all fertilizers and mixed fertilizers since 1930 are given in Table 12, together with contents of N + P_2O_5 + K_2O and other data. Use of N fertilizers for direct application (Figure 10) has been increasing more rapidly than use in mixtures. This has resulted from direct use of anhydrous ammonia and N solutions. Anhydrous ammonia consumption (Figure 11) has been increasing faster than has that of N solutions, AN, or urea. Use of AN has leveled off or even decreased in recent years.

As for N and K, consumption of P dropped in 1975 (Figure 12), but increases have resumed since then. Use of ammonium phosphates (Figure 13) continues to increase,

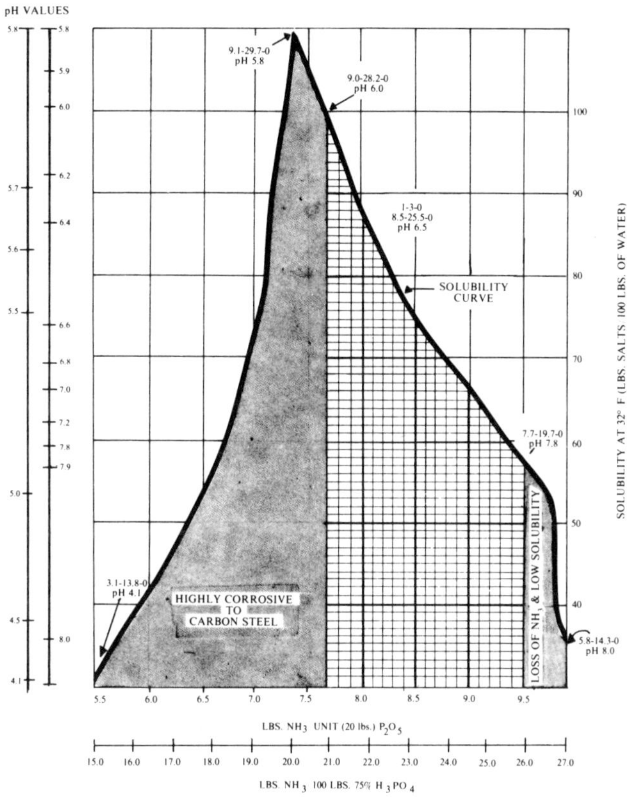

FIGURE 6. Solubility relationships of saturated ammonium orthophosphate solutions at 32°F in relation to pH and corrosive properties. (From Liquid Fertilizer Handbook, National Fertilizer Solutions Association, Peoria, Ill.)

while use of concentrated superphosphate has leveled off, and use of normal superphosphate has declined for the past 15 years. This decline in normal superphosphate consumption is due largely to higher transportation costs per unit of P contained and the lower analysis granular grades resulting from its use.

Total consumption of K fertilizers and use for direct application (Figure 14) have been increasing more than use in mixtures because of increases in bulk blending and spreading.

Table 7
POPULAR SUSPENSION FERTILIZER GRADES[16]

Fertilizer grade	Anhydrous NH_3 (82-0-0)	Aqua NH_3 (20-0-0)	UAN soluble (32-0-0)	11-33-0 base suspension	Solid APP (12-54-0)	KCl (0-0-62)	Clay	Water
	Ingredients, lb/ton of product							
11-33-0 base suspension								
A	—	367	—	—	1222	—	30	381
B	90	—	—	—	1222	—	30	658
Based on 11-33-0 base suspension								
4-12-24	—	—	—	728	—	775	20	477
6-18-18	—	—	—	1091	—	581	14	314
13-13-13	—	—	542	788	—	420	19	231
16-8-16	—	—	834	485	—	517	23	141
Based on UAN solution, H_3PO_4, and DAP				H_3PO_4 (0-54-0)	DAP (18-46-0)			
18-6-6	9	—	1000	67	183	194	30	517
16-8-8	11	—	834	89	243	258	30	535
12-12-12	16	—	503	133	365	387	30	566
8-16-16	21	—	172	178	487	516	30	596
6-18-18	26	—	—	200	548	581	30	615
8-24-8	35	—	—	267	730	258	30	680
10-30-0	44	—	—	334	913	—	30	679

Table 8
RESERVES, RESOURCES AND PRODUCTION OF MARINE PHOSPHORITE (IN METRIC TONS OF MATERIAL CONTAINING AT LEAST 30% P_2O_5)

Location	Reserves	Resources	Production 1974 1/
North Africa			
Algeria	500×10^6	600×10^6	0.7×10^6
Morocco	$5{,}000 \times 10^6$	$35{,}000 \times 10^6$	19.7×10^6
Tunisia	500×10^6	800×10^6	3.8×10^6
West Africa			
Angola	20×10^6	100×10^6	
Senegal (Taiba)	100×10^6	$1{,}000 \times 10^6$	0.4×10^6 (Al)
(Thies)	90×10^6	$2{,}000 \times 10^6$	1.5×10^6 (Ca)
Spanish Sahara	$1{,}600 \times 10^6$	$15{,}000 \times 10^6$	2.4×10^6
Togo	100×10^6	200×10^6	2.6×10^6
Middle East			
UAR (Egypt)	800×10^6	$2{,}000 \times 10^6$	0.5×10^6
Iran	30×10^6	100×10^6	
Iraq	60×10^6	600×10^6	
Israel	100×10^6	200×10^6	0.9×10^6
Jordan	100×10^6	200×10^6	1.6×10^6
Saudi Arabia		$1{,}000 \times 10^6$	
Syria	400×10^6	400×10^6	
Turkey		300×10^6	
Europe			
U.S.S.R. (Kazakhstan)	250×10^6	250×10^6	
(Karatau)	700×10^6	700×10^6	11.5×10^6
(Aldan, Yakut)	500×10^6	$2{,}000 \times 10^6$	
Other (France)			
(Belgium)			
(Germany)	15×10^6	30×10^6	0.1×10^6
Asia			
Australia	500×10^6	$1{,}500 \times 10^6$	Tr.
China	100×10^6	$1{,}000 \times 10^6$	3.0×10^6
India	70×10^6	200×10^6	0.1×10^6
Mongolia	250×10^6	700×10^6	2/
North Vietnam		400×10^6	1.2×10^6
Pakistan	100×10^6	150×10^6	
North America			
Mexico (Baja)		$1{,}000 \times 10^6$	
(Zacatecas)		140×10^6	0.07×10^6
U.S. (Eastern)	$1{,}600 \times 10^6$	$6{,}000 \times 10^6$ }	
(Western)	$6{,}000 \times 10^6$	$7{,}000 \times 10^6$ }	41.4×10^6
South America			
Brazil (Bambui)	200×10^6	500×10^6	
(Olinda)		20×10^6	
Colombia		600×10^6	Tr.
Peru (Sechura)		$6{,}100 \times 10^6$	
Venezuela	20×10^6	20×10^6	0.1×10^6
Totals	$19{,}705 \times 10^6$	$87{,}810 \times 10^6$	92.1×10^6
			(83.8(%) of world total)

From Cathcart, J. B., in *Role of Phosphorus in Agriculture*, Soil Sci. Soc. Am., Madison, Wisc., 1979. With permission.

Table 9
RESERVES, RESOURCES, AND PRODUCTION OF APATITE OF IGNEOUS ORIGIN (IN METRIC TONS OF MATERIAL CONTAINING AT LEAST 30% P_2O_5)

Location	Reserves	Resources	Production 1974
Kola Peninsula, U.S.S.R.	400×10^6	400×10^6	11.0×10^6
Palabora, South Africa	100×10^6	1300×10^6	1.4×10^6
Brazil (Carbonatites)			
Araxa	50×10^6	50×10^6	
Jacupiranga	12×10^6		
Catalao	75×10^6	75×10^6	0.2×10^6
Tapira	100×10^6	150×10^6	
Others		500×10^6	
Eastern Uganda	40×10^6	160×10^6	0.02×10^6
Finland	50×10^6	100×10^6	
North Korea	5×10^6	30×10^6	0.4×10^6
Southern Rhodesia			
Dorowa	10×10^6	10×10^6	0.1×10^6
Others	10×10^6	30×10^6	
Totals	852×10^6	2805×10^6	13.1×10^6 (11.9% of total world production)

From Cathcart, J. B., *Role of Phosphorus in Agriculture* Soil Sci. Soc. Am., Madison, Wisc., 1979. With permission.

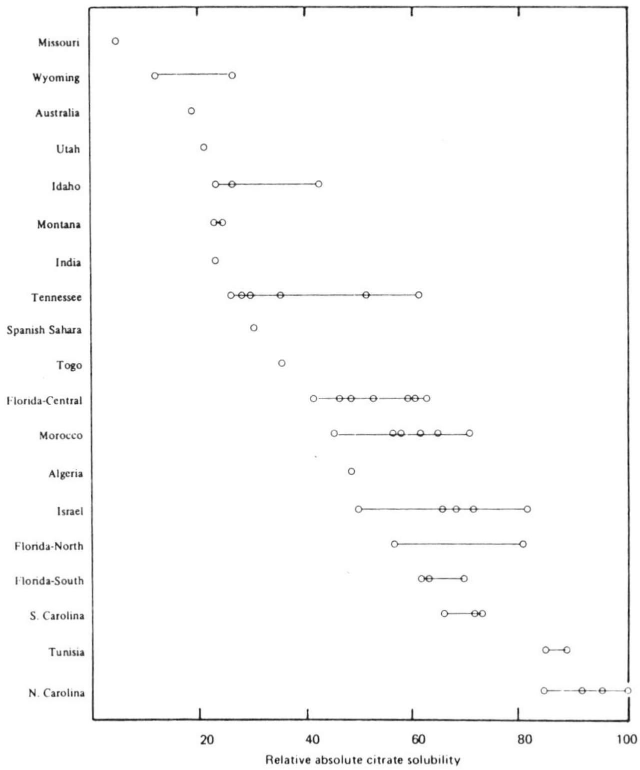

FIGURE 7. Citrate solubilities of phosphate rocks (North Carolina PR = 100). (From Lehr, J. R. and McClellan, G. H., TVA Bull. Y-43, Muscle Shoals, Ala., 1972.)

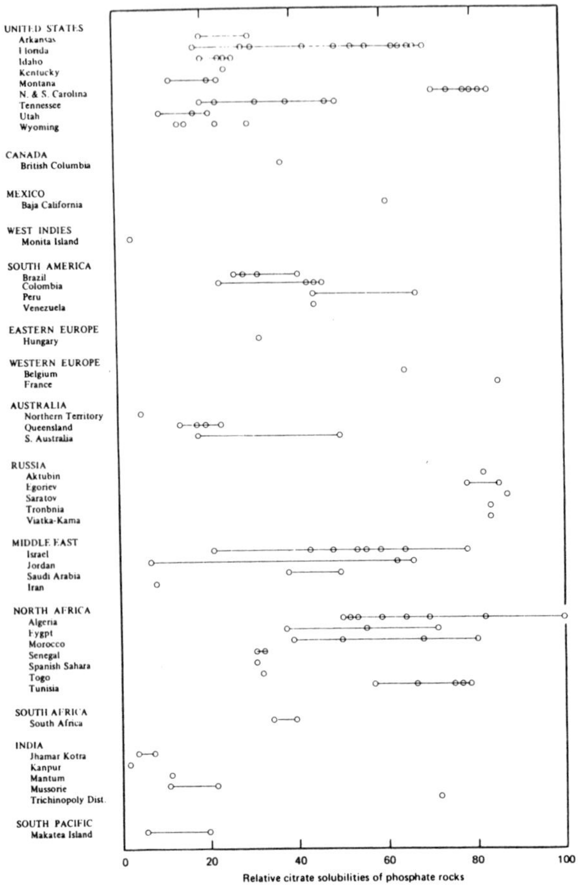

FIGURE 8. Citrate solubilities of phosphate rocks (Algerian PR = 100). (From Lehr, J. R., and McClellan, G. H., TVA Bull. Y-43, Muscle Shoals, Ala., 1972.)

Table 10
COMMON POTASH MINERALS, ORES, AND PRODUCTS

Name	Formula	K_2O (%)	K (%)	Remarks
	MINERALS			
Chlorides				
Sylvite	KCl	63.17	52.44	Principal ore mineral
Carnallite	$KCl \cdot MgCl_2 \cdot 6H_2O$	16.95	14.07	Ore mineral and contaminant
Kainite	$4KCl \cdot 4MgSO_4 \cdot 11H_2O$	19.26	15.99	Important ore mineral
Halite	$NaCl$	0	0	Principal ore contaminant
Sulfates				
Polyhalite	$K_2SO_4 \cdot MgSO_4 \cdot 2CaSO_4 \cdot 2H_2O$	15.62	12.97	Ore contaminant
Langbeinite	$K_2SO_4 \cdot 2MgSO_4$	22.69	18.84	Important ore mineral
Leonite	$K_2SO_4 \cdot MgSO_4 \cdot 4H_2O$	25.68	21.32	Ore contaminant
Schoenite	$K_2SO_4 \cdot MgSO_4 \cdot 6H_2O$	23.39	19.42	Rare
Glaserite	$3K_2SO_4 \cdot Na_2SO_4$	42.51	35.29	Rare
Syngenite	$K_2SO_4 \cdot CaSO_4 \cdot H_2O$	28.68	23.81	Rare
Alunite	$K_2Al_6(OH)_{12}(SO_4)_4$	11.4	9.46	
Kieserite	$MgSO \cdot H_2O$	0	0	Common ore contaminant
Anhydrite	$CaSO_4$	0	0	Common ore contaminant
Nitrates				
Niter	KNO_3	46.5	38.60	
	ORES			
Sylvinite[a]	$KCl + NaCl$	10—35	8—29	Canada, U.S., U.S.S.R.
Hartsalz	$KCl + NaCl + CaSO_4 (MgSO_4 \cdot H_2O)$	10—20	8—17	Germany
Carnallitite[b]	$KCl \cdot MgCl_2 \cdot 6H_2O + NaCl$	10—16	8—14	Germany, Congo, Brazil
Langbeinite[c]	$K_2SO_4 \cdot 2MgSO_4 + NaCl$	7—12	5—10	U.S., U.S.S.R.
Mischsalz	Hartsalz + Carnallite	8—20	6—17	Germany
Kainite	$4KCl \cdot 4MgSO_4 \cdot 11H_2O + NaCl$	13—18	10—14	Italy, Ethiopia
Niter (Caliche)	$KNO_3 + NaNO_3 + NaSO_4 + NaCl$	0.6—1.9	0.5—1.5	Chile

Table 10 (continued)
COMMON POTASH MINERALS, ORES, AND PRODUCTS

Name	Formula	K_2O (%)	K (%)	Remarks
PRODUCTS				
Potassium chloride	KCl	63.17	52.44	Principal potassium product
Potassium sulfate	K_2SO_4	54.05	44.87	Artificial product
Potassium-magnesium sulfate (langbeinite)	$K_2SO_4 \cdot 2MgSO_4$	22.69	18.84	Natural product (New Mexico)
Manure salts	KCl + NaCl	40—60	33—50	U.S.S.R. and East Germany
Niter	KNO_3	46.5	38.6	
Potassium nitrate	$KNO_3 + NaNO_3$	10—14	8—11	

[a] May contain one or more sulfate minerals or carnallite.
[b] May contain sylvite.
[c] May contain sylvite and leonite, kainite, or other sulfate minerals.

From Kilmer, V. J., *The Role of Potassium in Agriculture,* American Society of Agronomy, Madison, Wisc., 1968. With permission.

Table 11
WORLD POTASSIUM RESERVES

Country	Mineralogical ore types	Grade % K_2O	Depth (ft)	Recoverable metric K_2O tons (millions): Total tons	Recoverable metric K_2O tons (millions): Breakdown
North America					
Canada				18,000	
Saskatchewan, Alberta and Manitoba	Sylvinite and Carnallitite	14—32	3,000—8,000		5,000—13,000
Nova Scotia	Carnallitite	5	3,850—4,000	Potential	

Arctic Islands	ND[a]				Potential	225—360
U.S.						
New Mexico	Sylvinite	10—22	600—2,000			
	Langbeinite	7—12	600—2,000			
Utah	Sylvinite	20—25	3,000			
	Brine	0.5	Surface		90—136	
	Brine	0.7	Lake bed			
Arizona	Sylvinite	14—16	1,800			
North Dakota	Sylvinite	NA[b]	5,500—12,500		Potential	
California	Brine	3.0	Lake bed			
Nevada	Brine	1.2	500—1,000			
Mexico	Carnallitite and Sylvinite	NA	200—2,000		Potential	
South America				16		
Brazil	Carnallitite and Sylvinite (?)	8—16	1,500			5
Chile	Nitrates	1.0	Surface			NA
Peru	Brine	2.0	Shallow			11
Europe					5,723—9,400	
Czechoslovakia					Potential	
Denmark	Sylvinite and Carnallitite	10—20	700—3,000			6
England	Sylvinite	15—30	3,200—4,500			13—45
	Shale	11	Surface			10
France	Sylvinite	15—25	1,400—3,400			180—225
East Germany	Sylvinite	13—17	1,000—3,200			3,600—5,400
	Carnallitite	12	1,000—3,200			
West Germany	Sylvinite	15—20	1,300—3,500			1,800—3,600
	Hartsalz	9—15	1,300—3,500			
	Carnallitite	9—15	1,300—3,500			
Greece					Potential	
Italy	Kainite	12—18	1,000—2,600			23
	Carnallitite	11—13	1,000—2,600			
	Sylvinite	18—19	1,000—2,600			
Netherlands					Potential	
Poland	Carnallitite	8	600—900			8
Rumania					Potential	
Spain	Sylvinite	15—25	1,000—4,000			63
	Carnallitite	14	1,000—4,000			20

Table 11 (continued)
WORLD POTASSIUM RESERVES

Country	Mineralogical ore types	Grade % K_2O	Depth (ft)	Recoverable metric K_2O tons (millions)	
				Total tons	Breakdown
Russia				24,000	
West Ukraine	Hartsalz	16	250—1,000		
(Stebnik-Kalush)	Sylvinite	8—30	250—1,000		
	Kainite	10	250—1,000		
	Langbeinite	NA	250—1,000		
Byelorussia	Sylvinite	9—30	1,000—3,100		
(Starobin-Soligorsk)					
Urals	Sylvinite	15—30	220—700		
(Solikamsk-Brezhniki)	Carnallitite	11—17	220—700		
Siberia					
Irkutsk	Sylvinite	NA	2,300—3,400		
Krasnoyarsk	Sylvinite, Kainite and Langbeinite	NA	4,000—4,600		
Yakutsk	NA	NA	NA		
Africa				141—150	
Angola					Potential
Congo (Brazzaville)	Sylvinite	15—20	850—1,300		27
	Carnallitite	15	850—1,300		30
Egypt					Potential
Ethiopia	Sylvinite	20—30	80—400		18—27
	Kainite	NA			
Gabon	Carnallitite and Sylvinite				Potential
Libya	Brine and Salts	1—4	Shallow		23
Mali	Sulfates				Potential
Morocco	Sylvinite	10—11	1,600—2,600		23
	Carnallitite	11—13	1,600—2,600		
Niger and Nigeria	Brine	2.4—6.0	NA		Potential
South Africa	Nitrates	NA	Surface		NA
Tunisia	Brine	NA	Surface		NA

Asia				544	
China	NA	NA	NA		NA
Israel and Jordan	Brine	0.7	Surface		544
Pakistan	Brine	4.2	4,000—8,000		Potential
	Sylvinite	NA	NA		Potential
	Carnallitite	NA	NA		Potential
	Langbeinite	NA	NA		Potential
Thailand					Potential
Oceania					
Australia	Brine	3.6	Surface		Potential
	Bedded salt				Potential
Total					48,649—52,470

[a] ND = none discovered.
[b] NA = not available.

From Kilmer, V. J., *The Role of Potassium in Agriculture,* American Society of Agronomy, Madison, Wisc., 1968. With permission.

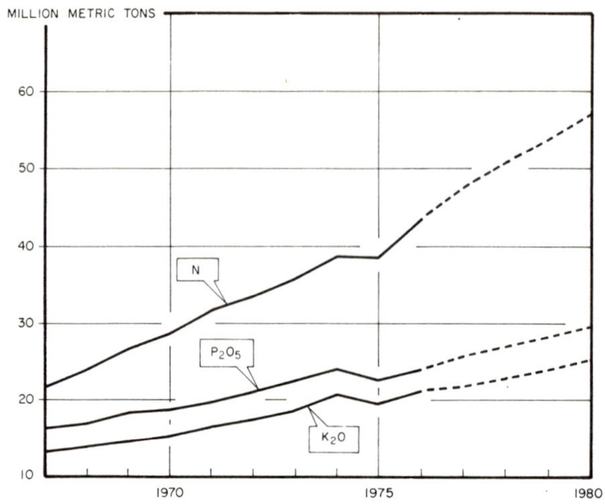

FIGURE 9. World plant nutrient consumption. (From Harre, E. A. and Shields, J. T., Proc. Natl. Fert. Solutions Assoc., St. Louis, 1977, 12.)

Table 12
U.S. FERTILIZER CONSUMPTION AND COMPOSITION[a]

	All fertilizers							Mixed fertilizers				
			Primary nutrients						Primary nutrient content (% available)			
Year	Total material (1000 tons)	Total primary nutrients (1000 tons)	Concentration (%)	N (1000 tons)	Avail. P_2O_5 (1000 tons)	K_2O (1000 tons)	Proportion of all fert. (%)	Total material (1000 tons)	Total	N	P_2O_5	K_2O
1930	8,425	1,523	18.08	377	793	354	67	5,616	17.90	3.10	9.80	5.00
1940	8,656	1,768	20.42	420	912	435	64	5,513	19.90	3.76	9.64	6.50
1950—51	20,991	4,727	23.23	1,237	2,110	1,380	67	13,978	24.19	4.18	11.03	8.98
1952—53	23,414	5,646	25.05	1,637	2,271	1,738	67	15,722	25.84	4.63	11.34	9.87
1954—55	22,726	6,119	27.90	1,961	2,284	1,875	68	15,348	27.90	5.24	11.86	10.80
1955—56	22,193	6,055	28.29	1,933	2,247	1,875	67	14,776	28.67	5.39	12.08	11.20
1956—57	22,709	6,377	29.30	2,135	2,305	1,937	65	14,703	29.53	5.74	12.36	11.43
1957—58	22,516	6,512	30.18	2,284	2,293	1,935	64	14,353	30.22	5.96	12.53	11.73
1958—59	25,308	7,435	30.87	2,672	2,572	2,191	64	16,069	30.79	6.22	12.66	11.91
1959—60	24,877	7,464	31.76	2,738	2,572	2,153	63	15,650	31.55	6.50	12.99	12.06
1960—61	25,567	7,844	32.18	3,031	2,645	2,169	62	15,735	31.93	6.81	13.15	11.97
1961—62	26,615	8,448	33.38	3,370	2,807	2,271	61	16,205	32.95	7.08	13.70	12.17
1962—63	28,844	9,505	34.71	3,929	3,073	2,503	59	17,157	34.30	7.36	14.42	12.52
1963—64	30,681	10,460	35.82	4,353	3,378	2,730	59	18,093	35.24	7.61	14.95	12.68
1964—65	31,836	10,985	36.16	4,639	3,512	2,835	59	18,559	36.13	7.92	15.50	12.71
1965—66	34,532	12,445	37.62	5,326	3,897	3,221	57	19,447	36.53	8.10	15.82	12.61
1966—67	37,082	13,973	39.15	6,027	4,305	3,642	57	21,132	37.70	8.27	16.43	13.00
1967—68	38,743	15,033	40.49	6,788	4,453	3,793	55	21,324	38.54	8.70	17.13	12.71
1968—69	38,944	15,513	41.22	6,959	4,662	3,892	55	21,222	39.02	8.95	17.53	12.54
1969—70	39,589	16,069	41.96	7,459	4,574	4,036	53	20,961	39.65	9.25	17.69	12.71
1970—71	41,118	17,168	43.03	8,134	4,803	4,231	52	21,513	40.92	9.59	18.33	13.00
1971—72	41,206	17,213	43.14	8,022	4,864	4,327	52	21,511	41.42	9.91	18.58	12.93
1972—73	43,288	18,029	43.11	8,295	5,085	4,649	52	22,547	41.76	10.21	18.79	12.76
1973—74	47,094	19,338	43.01	9,157	5,099	5,083	51	24,067	40.51	10.10	17.75	12.66
1974—75	42,484	17,561	43.25	8,601	4,507	4,453	49	20,647	40.66	10.17	17.99	12.50
1975—76	49,189	20,849	44.46	10,412	5,228	5,210	47	22,958	42.34	10.53	19.29	12.52

Table 12 (continued)
U.S. FERTILIZER CONSUMPTION AND COMPOSITION[a]

Year	All fertilizers							Mixed fertilizers				
	Total material (1000 tons)	Total primary nutrients (1000 tons)	Primary nutrients				Proportion of all fert. (%)	Total material (1000 tons)	Primary nutrient content (% available)			
			Concentration (%)	N (1000 tons)	Avail. P_2O_5 (1000 tons)	K_2O (1000 tons)			Total	N	P_2O_5	K_2O
1976—77	51,624	22,111	45.03	10,647	5,630	5,834	47	24,099	43.25	10.59	19.88	12.78
1977—78	47,587	20,600	45.07	9,973	5,099	5,527	46	22,121	43.16	10.73	19.64	12.79

[a] Statistics for 1930 to 1940 are from *Agricultural Statistics*, U.S. Department of Agriculture, 1952. Data for 1950 to 1951 through 1977 to 1978 are from annual reports of the U.S. Department of Agriculture, *Consumption of Commercial Fertilizers and Primary Plant Nutrients in the United States;* statistics for 1976 to 1977 and 1977 to 1978 are from the Annual Report issued November 1978.

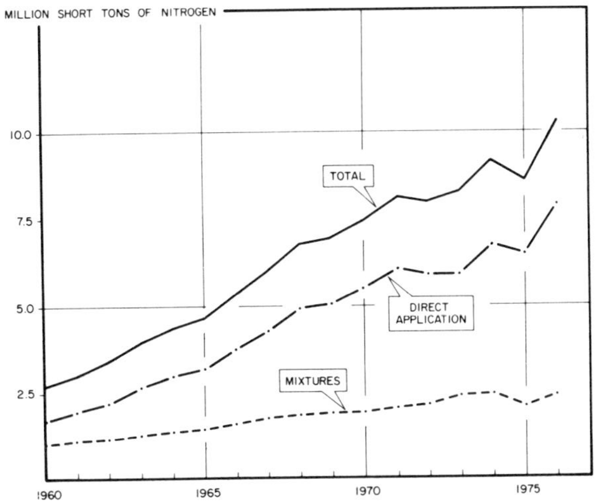

FIGURE 10. Nitrogen fertilizer consumption in the U.S. (From Harre, E. A., Goodson, M. N., and Bridges, J. D., Fertilizer Trends, 1976, TVA Bull. Y-111, Muscle Shoals, Ala., 1977.)

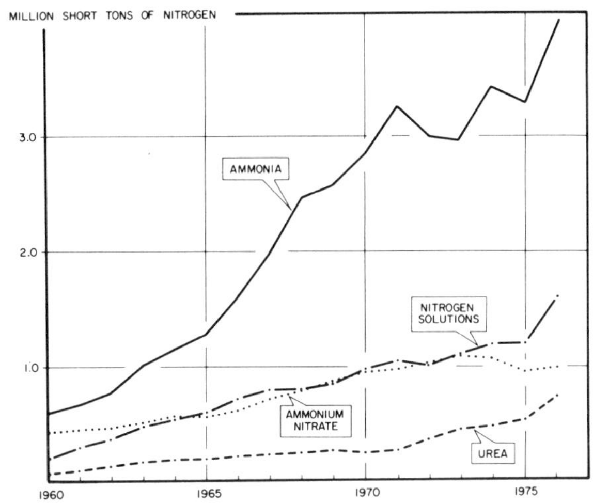

FIGURE 11. Nitrogen fertilizer material consumption in the U.S. (From Harre, E. A., Goodson, M. N., and Bridges, J. D., Fertilizer Trends, 1976, TVA Bull. Y-111, Muscle Shoals, Ala., 1977.)

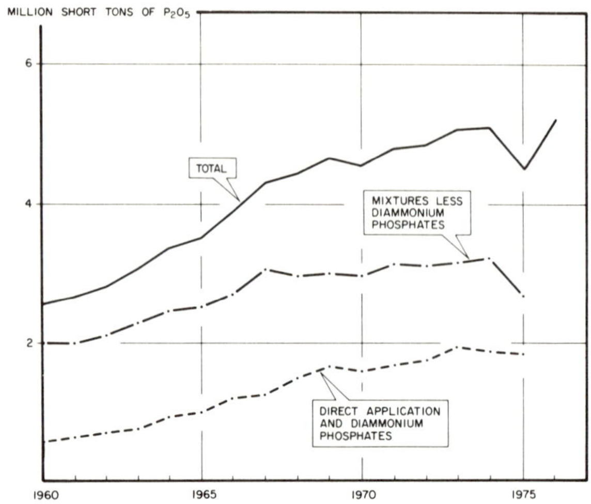

FIGURE 12. Phosphate fertilizer consumption in the U.S. (From Harre, E. A., Goodson, M. N., and Bridges, J. D., Fertilizer Trends, 1976, TVA Bull. Y-111, Muscle Shoals, Ala., 1977.)

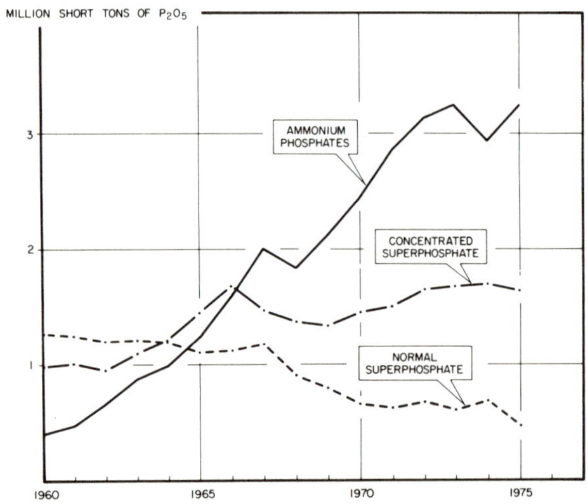

FIGURE 13. Production of ammonium phosphates and superphosphates in the U.S. (From Harre, E. A., Goodson, M. N., and Bridges, J. D., Fertilizer Trends, 1976, TVA Bull. Y-111, Muscle Shoals, Ala. 1977).

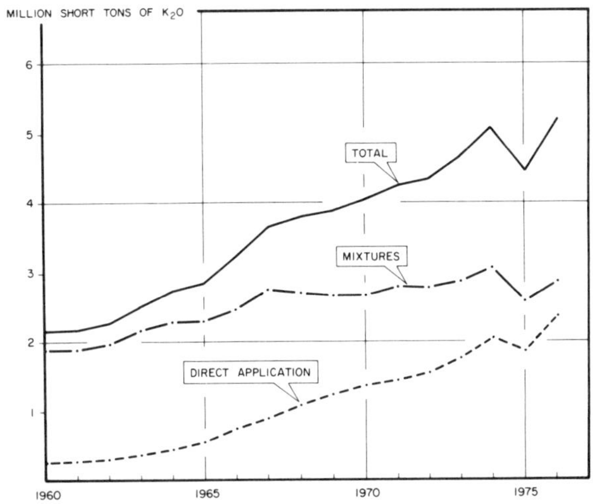

FIGURE 14. Potash fertilizer consumption in the U.S. (From Harre, E. A., Goodson, M. N., and Bridges, J. D., Fertilizer Trends, 1976, TVA Bull. Y-111, Muscle Shoals, Ala., 1977.)

REFERENCES

1. Official Methods of Analysis, 12th ed., Association of Official Analytical Chemists, Washington, D.C., 1975.
2. **Terman, G. L., Hoffman, W. M., and Wright, B. C.,** Crop response to fertilizers in relation to content of "available" phosphorus, *Advan. Agron.,* 16, 59, 1964.
3. **Terman, G. L., Bouldin, D. R., and Webb, J. R.,** Evaluation of fertilizers by biological methods, *Advan. Agron.,* 14, 265, 1962.
4. **Terman, G. L. and Engelstad, O. P.,** Agronomic Evaluation of Fertilizers — Principles and Practices, TVA Bull. 21, Revised, 1976.
5. **Tisdale, S. L. and Nelson, W. L.,** *Soil Fertility and Fertilizers,* 2nd ed., Macmillan, New York, 1966.
6. **Buckman, H. O. and Brady, N. C.,** *The Nature and Properties of Soils,* 7th ed., Macmillan, New York, 1969.
7. **Hardesty, J. O.,** Watch that salt content, *Farm Chem.,* 14(10), 36, 1977.
8. *Farm Chemicals Handbook,* Meister, Willoughby, Ohio, 1978.
9. **Harre, E. A., Goodson, M. N., and Bridges, J. D.,** Fertilizer Trends, 1976, TVA Bull. Y-111, Muscle Shoals, Ala., 1977.
10. **Harre, E. A.,** Fertilizer Trends, 1973, TVA Bull. Y-77, Muscle Shoals, Ala., 1974.
11. *Liquid Fertilizer Handbook,* National Fertilizer Solutions Assoc., Peoria.
12. **Kamprath, E. G., Khasowneh, F. E., and Sample, E. C.,** *The Role of Phosphorus in Agriculture,* Soil Science Society of America, Madison, Wisc., 1979.
13. **Lehr, J. R. and McClellan, G. H.,** A Revised Laboratory Scale for Evaluating Phosphate Rocks for Direct Application, TVA Bull, Y-43, Muscle Shoals, Ala., 1972.
14. **Kilmer, V. J., Younts, S. E., and Brady, N. C.,** *The Role of Potassium in Agriculture,* American Society of Agronomy, Madison, Wisc., 1968.
15. **Harre, E. A. and Shields, J. T.,** NPK — World Production, *Proc. Natl. Fert. Solutions Assoc.,* St. Louis, 1977, 12.
16. **Balay, H. V.,** National Fertilizer Development Center, TVA, Muscle Shoals, Ala., unpublished data.

LIME, LIME MATERIALS, AND OTHER SOIL AMENDMENTS*

Donald R. Christenson

INTRODUCTION

Application of amendments to soils has been practiced for many centuries. Writings dating to 2500 B.C. indicate that some form of amendments must have been responsible for good yields in the Tigris and Euphrates river valleys.[1] Whether this was due to amendments as we now know them is not clear. Romans used and classified liming materials, learning about the practice from the Greeks.

Limestone usage in the U.S. increased rapidly after the initiation of Federal Agricultural Conservation Programs in 1936 (Table 1). Peak usage occurred in the late 1940s, followed by a reduction in usage during the 1950s and early 1960s. Another increase and decline occurred in the later 1960s and early 1970s. Usage again increased from 1974 to 1976. The reasons for this varied usage of limestone are not clear. These data suggest that between 2.5 and 3.5 times as much agricultural limestone could be used in the U.S.

DEFINITIONS

Agricultural liming material — This is any material whose Ca and Mg compounds are capable of neutralizing soil acidity.[3]

Calcite — Pure calcite is $CaCO_3$ and contains 40.0% Ca.[2]

Calcitic lime and dolomitic lime — These terms are descriptive referring to the relative calcium and magnesium content. The dividing line between calcitic and dolomitic lime is not clearly defined.[4]

Calcium carbonate equivalent — This is the acid-neutralizing capacity of a material expressed as a weight percent of $CaCO_3$.[2] It may also be expressed on a volume basis. For example, if a cubic yard of a material has a calcium carbonate equivalent of 1240 lb, this means that a cubic yard of the material will neutralize the same amount of acidity as 1240 pounds of $CaCO_3$.[4]

Calcium oxide — This is formed by heating $CaCO_3$ to drive off CO_2 leaving CaO. It is also known as quicklime, unslaked lime, burnt lime, lump lime, stone lime, caustic lime, or builders lime. It does not occur in nature.

Dolomite — Pure dolomite is $CaMg(CO_3)_2$ and contains 21.6% Ca and 13.1% Mg.[2]

Gypsum — This is a hydrated form of $CaSO_4$. Since it does not correct acidity, it is not a liming material.[3]

Hydrated lime — This is formed by adding sufficient water to CaO to form $Ca(OH)_2$. It is also called *slaked lime* or *quick lime*.[3]

Lime — Chemically lime is CaO. The term is broadly used in agriculture to define any material containing Ca or Ca and Mg in forms capable of correcting soil acidity.[3]

Lime requirement — This is the quantity of lime required to bring an acid soil to a predetermined pH or degree of acidity.[3]

Marl — This is a soft $CaCO_3$ deposit formed when lime-charged water flows into low lying areas with a resultant loss of CO_2 and precipitation of $CaCO_3$.[2]

Neutralizing value — See calcium carbonate equivalent.

Refuse lime — This is liming material from sugar factories, paper mills, acetylene

* Contribution from the Michigan Agricultural Experiment Station, East Lansing, Michigan 48824, as Journal Article Number 8753.

Table 1
AGRICULTURAL LIMESTONE USAGE IN THE U.S. AS COMPARED TO NEEDS FOR SELECTED YEARS FROM 1929 THROUGH 1976[a]

Year	Annual usage (Tons)	Use vs. needs (%)
1929	3,907,000	4.86
1932	1,810,627	2.25
1935	3,504,595	4.36
1938	7,859,225	9.77
1941	15,916,431	19.78
1944	24,567,884	30.54
1947	30,282,902	37.64
1950	29,842,145	37.09
1953	20,669,096	25.61
1956	22,021,393	27.53
1959	22,725,969	28.41
1962	23,616,235	29.52
1965	28,075,297	35.01
1968	30,535,501	35.05
1971	23,679,002	26.93
1974	27,806,918	29.42
1975	31,319,779	33.09
1976	38,146,914	39.99

[a] Data from National Limestone Institute, Inc.

plants, rock wool plants, water softening plants and lime from lead mines.[2] It is also called *waste lime*.[3]

Slag — This is any of a series of byproducts from furnace operations. Blast furnace slag is a byproduct of pig iron production. Open hearth slag is a byproduct from furnaces in the manufacture of steel. Basic slag is formed when pig iron from highly phosphatic iron ores is converted into steel.[2]

COMPOSITION

Chichilo and Whittaker[5] and the *Dictionary of Plant Foods*[6] presented analyses of numerous limestone sources. These data are summarized in Tables 2 and 3 and show the wide variation of constituents one can expect in various sources of liming materials. Chichilo and Whittaker indicated that some of the cations other than Ca and Mg were present as silicates.

The presence of heavy metals in limestone sources has been reported by several workers. Warren and Delavault[7] reported values of 3 to 65 ppm Pb on 112 limestone sources from Canada. Murray et al.[8] reported values up to 10 ppm Cr in 20 of 25 limestone samples analyzed. Keller et al.[9] reported a wide variation between layers in a quarry in Missouri. The ranges were 1 to 5 ppm Ni, 10 to 50 ppm V, 50 to 3000 ppm Cr, and 3 to 20 ppm Pb. No cadmium was detected. Data on heavy metal loading rates for various soils are presented in a later section, "Other Soil Amendments,."

Table 2
CHEMICAL COMPOSITION OF 194 SAMPLES OF AGRICULTURAL LIMESTONE[5]

Element	Range (%)	Average (%)
Ca	17.98—39.76	30.21
Mg	0.04—12.86	4.92
Si	0.03—15.53	2.36
Al	0.01—2.15	0.45
Fe	0.01—3.11	0.43
K	<0.0001—1.80	0.23
S	<0.01—1.35	0.11
Na	<0.001—0.15	0.03
	ppm	ppm
Mn	20—3000	330
F	<10—1410	230
P	10—5660	210
Zn	<1—425	31
V	<1—106	11
B	<1—21	4
Cu	<0.3—89	2.7
Mo	<0.1—92	1.1
Co	<1—6	1

Table 3
ANALYSES OF VARIOUS SOURCES OF LIMESTONE[6]

Constituent	Number of samples	Range (%)	Mean (%)
P_2O_5	593	0.0—2.59	0.11
K_2O	511	tr—1.07	0.31
Ca	1892	21.53—39.83	31.70
Mg	1569	tr—14.33	3.43
C	843	8.79—12.98	11.15
S	98	0.0—1.17	0.04
Si	918	0.0—14.88	2.63
		ppm	ppm
Cu	17	4—88	42
Mn	430	tr—7900	4900
Zn	25	1—460	150
B	3	0—160	30

Barber[2] discussed the use of slag materials as liming sources. When slags are sufficiently finely ground, their effectiveness is equal or nearly equal to the carbonates and oxides of Ca and Mg when applied in equivalent amounts. Analyses of slag samples are given in Table 4.

Marl is a soft $CaCO_3$ deposit which is dredged and spread when wet. Even though

Table 4
RANGE IN CHEMICAL COMPOSITION OF SLAGS

Element	Blast furnace (%)	Open hearth (%)	Basic (%)
Ca	26—32	16	27—42
Mg	2—7	5	1—5
Si	15—20	8	1.4—6
Al	5.4—8.5	1.5	1—2
Fe	0.2—2	24	4—17
Mn	0.2—1.2	10	0.6—4
S	1—3	0.1	0.2—11
P	<1	1	5—10

Reproduced from Barber, S. A., *Soil Acidity and Liming*, Pearson, R. W. and Adams F., Eds., American Society of Agronomy, Madison, Wis., 1967, chap 3. With permission.

Table 5
ANALYSES OF 27 MARL SAMPLES SUBMITTED TO THE MICHIGAN STATE UNIVERSITY SOIL TESTING LABORATORY

Analyses	Average	Range
$MgCO_3$ (%)	3.06	1.40—5.18
Neutralizing value[a] (%)	84.2	32.2—102.9
Calcium carbonate equivalent (lb/yd^3)[b]	1075	175—1811

[a] Neutralizing value expressed in terms of calcium carbonate.
[b] Expressed on a volume basis.

it is finely divided, it does not react any more quickly than some other materials due to its lumpy consistency. Table 5 shows analyses of some marl samples submitted to the Michigan State University Soil Testing Laboratory from 1973 to 1977. These values are comparable to those reported by Shickluna et al.[10] As is the case with other materials, there is considerable variability among samples. Shickluna et al. reported as much as 85% variation in $MgCO_3$ content in the same marl deposit over a depth of 15 ft.

AMOUNTS NEEDED

Historically, lime needs have been based on previous experience, soil type, pH, soil texture, and/or interactions among these factors. One example is given in Table 6 relating soil pH and texture to lime needs. In general, these methods were not satisfactory, resulting in underliming a majority of the time but overliming on occasion.

Research has shown that soil pH, total acidity, texture, organic matter, crop to be grown, weathering stage of soil, and depth of plowing influence amount of lime required. The contribution of various soil factors to lime requirement was the subject of several investigations;[11-13] these effects are summarized in Table 7. When the percent of clay alone is considered, Ross et al.[11] accounted for 63% of the variability in lime requirement, while Keeney and Corey[12] found poor correlation between percent of clay

Table 6
TONS OF LIMESTONE REQUIRED TO RAISE THE pH OF A 6⅔-IN. PLOW LAYER OF DIFFERENT TEXTURAL GROUPS TO pH 6.5[4]

	Tons of lime recommended[a]			
Texture of plow layer	4.5—4.9 pH	5.0—5.4 pH	5.5—5.9 pH	6.0—6.4 pH
Clay and silty clay	6	5	4	2.5
Clay loams and loams	5	4	3	2
Sandy loams	4	3	2.5	1.5
Loamy sands	3	2.5	2	1
Sands	2.5	2	1.5	0.5

[a] Material of which 25% passes a 100-mesh sieve and has a neutralizing value of 90%.

Table 7
RELATIONSHIP (r) BETWEEN VARIOUS SOIL FACTORS AND LIME REQUIREMENT AS DETERMINED BY INCUBATION OF SOILS AND RAISING SOIL pH TO A PREDETERMINED VALUE

Soil factor	Ross et al[11] (r)[a]	Keeney/ Corey[12] (r)	Pionke et al.[13] (r)
Percent clay	0.794[b]	0.189	—
Percent organic matter	0.949[c]	0.620[c]	—
$\Delta pH^{d} \times$ % organic matter	0.927[c]	0.884[c]	0.887[c]
Cation exchange capacity	0.546[c]	0.956[c]	—
pH (H_2O)	ns	−0.389	—
Exchangeable H	0.877[c]	—	0.816[c,e]
Exchangeable Al	—	0.086	—
SMP buffer[13]	0.944[c]	0.949[c]	0.904[c]

[a] Simple correlation coefficient.
[b] 5% level of significance.
[c] 1% level of significance.
[d] 6.5 − soil pH
[e] % base unsaturation.

and lime requirement. Ross et al. pointed out that it is easier to explain a good correlation between lime requirement and percent of clay than a poor correlation. Soil pH and exchangeable Al both correlated poorly. Organic matter, cation exchange, and exchangeable hydrogen singly related somewhat better to lime requirement, but the degree was not consistent among studies. A pH change — organic matter function correlated well in three studies, accounting for 78 to 85% of total variability. Pionke et al.[13] developed the following equation, accounting for 89.5% of the variability in lime requirement:

$$\text{Lime requirement} = 0.47 + 0.97\,\Delta pH(KCl) \times \%\ \text{Organic matter} + 0.03\,\Delta pH(KCl) \times \%\ \text{clay} + 1.73\ \text{exchangeable Al} + 0.53\ \text{nonexchangeable acidic Al}$$

where ΔpH is 6.5 − soil pH measured in KCl. Their ranking of soil properties as predictors of $CaCO_3$-equilibrated lime requirement followed the order: pH-dependent sites on organic matter > nonexchangeable acidic Al > exchangeable Al > pH-dependent sites on clay.

The SMP buffer[14] was also tested and accounted for as much variability in incubated lime requirement as any soil property singly. The buffer solution contains para-nitrophenol, triethanolamine, K_2CrO_4, Ca-acetate and $CaCl_2$ which is adjusted to pH 7.5. A pH depression of the buffer-soil suspension is correlated with lime requirement. The authors[14] suggested that the buffer system appeared to be adequate for indicating lime needs regardless of extractable (soluble) Al present. However, in further studies McLean et al.[15] concluded that SMP buffer was well adapted for soils requiring more than 4000 lb lime/acre, having a pH less than 5.8, containing less than 10% organic matter and having appreciable quantities of soluble Al. More recently McLean et al.[16] offered an "improved buffer" which appears promising for soils over a wide range of soluble Al levels. Additional investigations are needed to verify these findings.

In recent years a controversy has existed over whether to lime soils to some predetermined pH or to lime for the reduction of toxic concentrations of Al and Mn. The evidence suggests that highly weathered soils (Ultisols, Oxisols, Latosols) should be limed to reduce toxic levels, while less weathered soils (Alfisols, Mollisols) should be limed to a predetermined pH.

Liming Alfisols and Mollisols to pH 6.0 to 6.5 is the recommended practice for four reasons. First, nutrient availability and balance are more easily maintained in this range. Second, the cost of lime is usually reasonable, adding less than $10/acre/year. This cost is usually recovered from increased yields. Third, use of ammoniated phosphates or other ammonium N sources hastens the development of soil acidity. Fourth, it has been suggested that liming to pH 6.5 reduces uptake and damage of heavy metals such as nickel and cadmium.[17]

Liming for the reduction of toxicities appears to be the accepted practice on more highly weathered soils (Ultisols, Oxisols) for several reasons. First, yields are reduced when soils are overlimed. Second, nutrient availability is reduced when the resulting soil pH is too high. Third, excessive Ca disperses soil aggregates causing soil physical problems. McLean[18] summarized phenomena relating to liming various soils in the acid soil continuum (Table 8).

Numerous articles have been published on the effect of toxic concentrations of Al and Mn on plant growth. Critical levels of each are difficult to tabulate due to differences in soils and differential effects of crop varieties. The reader is referred to several articles, books, and reviews[19-32] for further information on this subject.

pH PREFERENCES OF PLANTS

Wheeler et al.,[33] Hartwell and Damon,[34] and Spurway[35] recognized that different plant species varied in susceptibility to the effects of soil acidity. Some crops such as alfalfa and sweet clover require slightly acid to neutral pH levels (Table 9). Others such as blueberries and cranberries require strongly acid soils. Some special cases exist, as in Irish potatoes, where low pH values are required for control of scab or other diseases.

pH-NUTRIENT AVAILABILITY

Truog[36] discussed availability of nutrients as a function of soil pH. These basic ideas have not changed appreciably and are summarized in Figure 1. It is easy to see that

Table 8
SUMMARY OF VARIOUS PHENOMENA RELATED TO CROP RESPONSES TO LIMING IN THE ACID SOIL CONTINUUM[18]

Liming effect	Mollisols	Alfisols	Ultisols	Oxisols
Optimum pH range[a]	6.2—6.8	6.0—6.6	5.6—6.2	5.0—5.6
Crop response — rel. mag.[b]	Fair to good	Excellent	Excellent	Good to excellent
Cause of favorable lime response[c]				
Chemical	pH adjustment	pH adjustment	Al (and Mn) inactivation	Al, Mn inactivation
	Mobilized Ca	Al inactivation	Ca addition	Ca addition
	N fixation and metabolism	Mobilized Ca N fixation and metabolism		Slowed weathering
Physical	Improved aggreg. (Flocculation and OM coatings)	Slightly improved aggreg.	None when limed @ opt. pH	None when limed @ opt. pH
Adverse effects of overliming[c]				
Chemical	Hardly any	Zn or Mn def.	Zn, Mn (Cu, B) def.	Zn, Mn, Cu, B def.
		Mo toxicity	Mo toxicity	Mo toxicity
			Excess Ca	Excess Ca
Physical	Hardly any	None	Dispersed aggreg.	Dispersed aggreg.

[a] Optimum pH naturally varies with crop and even with strains of the same crop, but these pH ranges seem to approximate those where most of the crops grown on these soils do best.
[b] Assuming pH is approximately 1 unit below optimum.
[c] Arranged in decreasing order in which it most likely occurs.

Table 9
OPTIMUM pH RANGES OF SELECTED PLANTS[35]

Field Crops	
Alfalfa	6.2—7.8
Barley	6.5—7.8
Bean, field	6.0—7.5
Beets, sugar	6.5—8.0
Bluegrass, Ky.	5.5—7.5
Clover, red	6.0—7.5
Clover, sweet	6.5—7.5
Clover, white	5.6—7.0
Corn	5.5—7.5
Flax	5.0—7.0
Oats	5.0—7.5
Pea, field	6.0—7.5
Peanut	5.3—6.6
Rice	5.0—6.5
Rye	5.0—7.0
Sorghum	5.5—7.5
Soybean	6.0—7.0
Sugar cane	6.0—8.0
Tobacco	5.5—7.5
Wheat	5.5—7.5
Vegetable Crops	
Asparagus	6.0—8.0
Beets, table	6.0—7.5
Broccoli	6.0—7.0
Cabbage	6.0—7.5
Carrot	5.5—7.0
Cauliflower	5.5—7.5
Celery	5.8—7.0
Cucumber	5.5—7.0
Lettuce	6.0—7.0
Muskmelon	6.0—7.0
Onion	5.8—7.0
Potato	4.8—6.5
Rhubarb	5.5—7.0
Spinach	6.0—7.5
Tomato	5.5—7.5
Flowers and Shrubs	
African violet	6.0—7.0
Almond, flowering	6.0—7.0
Alyssum	6.0—7.5
Azalea	4.5—5.0
Barberry, Japanese	6.0—7.5
Begonia	5.5—7.0
Burning bush	5.5—7.5
Calendula	5.5—7.0
Carnation	6.0—7.5
Chrysanthemum	6.0—7.5
Gardenia	5.0—6.0
Geranium	6.0—8.0
Holly, American	5.0—6.0
Ivy, Boston	6.0—8.0
Lilac	6.0—7.5
Lily, Easter	6.0—7.0
Magnolia	5.0—6.0
Orchid	4.0—5.0
Phlox	5.0—6.0
Poinsettia	6.0—7.0
Quince, flowering	6.0—7.0
Rhododendron	4.5—6.0
Rose, hybrid tea	5.5—7.0
Snapdragon	6.0—7.5
Snowball	6.5—7.5
Sweet William	6.0—7.5
Zinnia	5.5—7.5
Forest Plants	
Ash, white	6.0—7.5
Aspen, American	3.8—5.5
Beech	5.0—6.7
Birch, European (white)	4.5—6.0
Cedar, white	4.5—5.0
Club moss	4.5—5.0
Fir, balsam	5.0—6.0
Fir, Douglas	6.0—7.0
Heather	4.5—6.0
Hemlock	5.0—6.0
Larch, European	5.0—6.5
Maple, sugar	6.0—7.5
Moss, sphagnum	3.5—5.0
Oak, black	6.0—7.0
Oak, pin	5.0—6.5
Oak, white	5.0—6.5
Pine, jack	4.5—5.0
Pine, loblolly	5.0—6.0
Pine, red	5.0—6.0
Pine, white	4.5—6.0
Spruce, black	4.0—5.0
Spruce, Colorado	6.0—7.0
Spruce, white	5.0—6.0
Sycamore	6.0—7.5
Tamarack	5.0—6.5
Walnut, black	6.0—8.0
Yew, Japanese	6.0—7.0
Weeds	
Dandelion	5.5—7.0
Dodder	5.5—7.0
Foxtail	6.0—7.5
Goldenrod	5.0—7.5
Grass, crab	6.0—7.0
Grass, quack	5.5—6.5
Horse tail	4.5—6.0
Milkweed	4.0—5.0
Mustard, wild	6.0—8.0
Thistle, Canada	5.0—7.5
Fruits	
Apple	5.0—6.5
Apricot	6.0—7.0
Arbor Vitae	6.0—7.5
Blueberry, high bush	4.0—5.0
Cherry, sour	6.0—7.0
Cherry, sweet	6.0—7.5
Crab apple	6.0—7.5
Cranberry, large	4.2—5.0
Peach	6.0—7.5
Pineapple	5.0—6.0
Raspberry, red	5.5—7.0
Strawberry	5.0—6.5

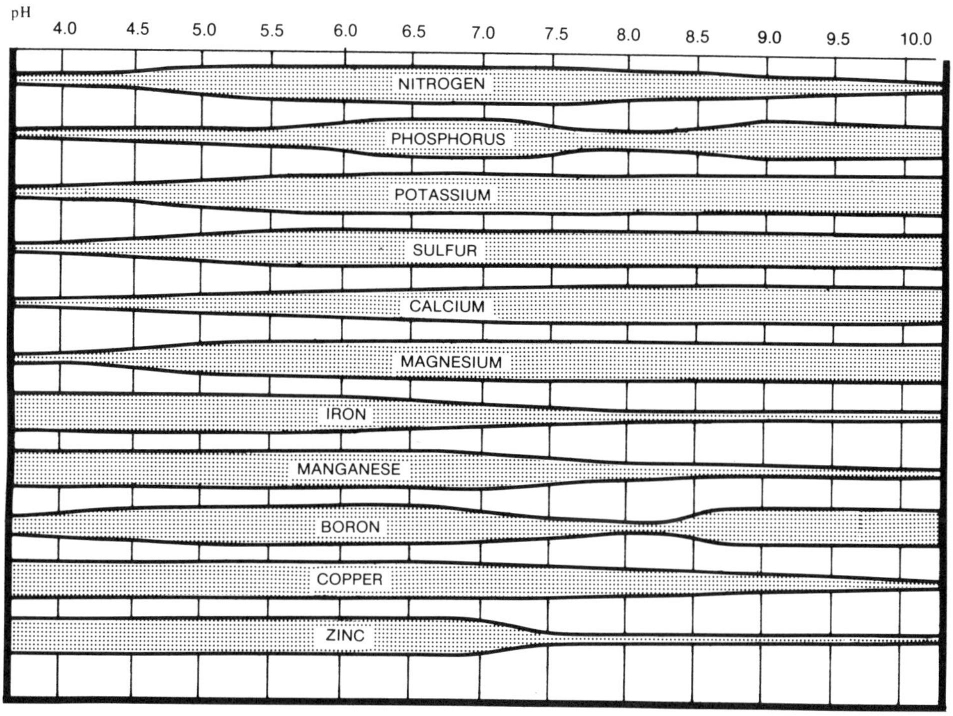

FIGURE 1. The general relation of pH to the availability of plant nutrients in the soil; the width of the bar indicates relative availability.

when all nutrients are considered, nutrient availability is at a maximum when soil pH is at 6.0 to 6.5. The reader is referred to several reviews[20,37,38] for further discussion concerning the effect of soil pH on nutrient availability.

ACID FORMING FERTILIZERS

Ammoniated phosphates and ammonium forms of nitrogen fertilizer contribute to soil acidity due to the microbial transformation of NH^{+}_{4} to NO^{-}_{3}. Wolcott et al.[39] has shown the relative acidifying effects of nitrogen fertilizers to be in the following order: $(NH_4)_2SO_4 > NH_4NO_3 \simeq NH_3 \simeq$ Urea. They also found that the residual basicity from $Ca(NO_3)_2$ and $NaNO_3$ was approximately equal to the acidifying potential of climate and management.

Table 10 presents commonly accepted values for the acidifying effects of various fertilizers. Variations from these theoretical values are attributed to differential uptake of ions by plants, denitrification, ammonia volatilization, and differential retention of ions by soils.[39,40]

Black[41] suggests that of common fertilizer materials only superphosphate is inherently strongly acid. However, even though the solution around the particle is extremely acid (pH 1 to 2), the effect is only temporary. The long-term influence does not exist because of the tendency of phosphate to react with hydrous oxides of aluminum and iron, releasing hydroxyl ions.

Table 10
ACIDIFYING EFFECTS OF VARIOUS FERTILIZERS[6]

Fertilizer material	Pure lime required to neutralize acidifying effect (lb lime/1b N)
Nitrogen Sources	
Anhydrous ammonia (83% N)	1.8
Aqua ammonia (21% N)	1.8
Ammonium nitrate (33% N)	1.8
Mono ammonium phosphate (11-48-0)	5.9
Ammonium sulfate (21% N)	5.5
Diammonium phosphate (18-46-0)	3.6
Urea (45% N)	1.9
Nitrogen solutions (28—32% N)	1.8
Urea — form (38% N)	1.8
Phosphorus Sources	
None except ammoniated phosphates are acidifying	
Potassium Sources	
None are acidifying	

Table 11
AVERAGE AMOUNTS (LB/TON) OF VARIOUS NUTRIENTS IN MANURES FROM DIFFERENT FARM ANIMALS[45,46]

Nutrient	Diary cattle	Beef cattle	Poultry	Swine	Sheep
Nitrogen (N)	10.0	14.0	25.0	10.0	28.0
Phosphorus (P)	2.0	4.0	11.0	2.8	4.2
Potassium (K)	8.0	9.0	10.0	7.6	20.0
Sulfur (S)	1.5	1.7	3.2	2.7	1.8
Calcium (Ca)	5.0	2.4	36.0	11.4	11.7
Iron (Fe)	0.1	0.1	2.3	0.6	0.3
Magnesium (Mg)	2.0	2.0	6.0	1.6	3.7
Boron (B)	0.01	0.03	0.01	0.09	0.02
Copper (Cu)	0.01	0.01	0.01	0.04	0.01
Manganese (Mn)	0.03	0.01	0.18	0.04	0.02
Zinc (Zn)	0.04	0.03	0.01	0.12	0.05
% moisture	85	85	72	82	77

OTHER SOIL AMENDMENTS

Animal Manures

The value of animal manures in crop production has long been recognized. It provides nutrients and organic matter to the soil which aids in maintaining yields. Vitosh et al.[42] reported that a yearly application of 67.2 metric tons of fresh manure per hectare (30 tons/acre) increased the organic matter content of a Metea sandy loam soil approximately 0.1% per year. However, the most favorable rate was 22.4 metric tons/hectare (10 tons/acre). Wallingford et al.[43] reported that excessive rates of manure would result in soluble salt accumulation and depressed yield. Annual rates ranging from 13 to 30 dry metric tons/hectare produced nearly maximum corn silage yields without causing excessive salt accumulation in the soil. (These rates correspond to 29 to 68 wet tons/acre). Lund et al.[44] showed that dairy cattle manure rates of 22.5 to 45 metric tons/hectare produced top yields of millet and rye. When rates exceeded 90 metric tons, the forage was high in nitrate and had K/(Ca + Mg) ratios that could be detrimental to animal health.

There is considerable variation in nutrient content of manure among animals, feed, and housing conditions. Generally about 75% of the nitrogen, 80% of the phosphorus, and 90% of the potassium are recovered in the excrement. However, only about half of the nutrients in manure are recovered in crop production due to losses from leaching, volatilization, and sorption by the soil. Some average composition values of various manures are given in Table 11.

Accumulation of manure in loafing barns or sheds will be kept moist and packed by animal feet. Added bedding absorbs liquid which reduces losses of nitrogen and organic matter due to decay or fermentation. Manure exposed to outdoor weather will lose significant amounts of nutrients and organic matter. Under such conditions, deep, well-packed piles with straight sides reduces fermentation and leaching losses. Tight-walled pits or ponds reduce losses for both solid and slurry (<8% solids) systems.

Plant Residues

Nutrient concentrations in plant residues vary with stage of growth. Values given in Table 12 represent nutrient levels of more mature plant tissue. These levels vary from one reporting source to another.

Sewage Sludge and Wastewaters

Other sources of plant nutrients for field crops are municipal sewage sludges and wastewater effluents. Sludges can be a very beneficial soil amendment, providing nutrients and organic matter. Sludges vary in moisture content from a slurry of less than 1% solids to a "dried" cake with greater than 40% solids. Excessive concentrations of metals like copper, zinc, nickel, chromium, and others can be a problem if present in excessive concentrations. Total elemental concentrations for over 200 sludge samples are given in Table 13.

Wastewaters come from two main sources: industrial wastewaters and municipal sources. The latter may include industrial wastes. Three estimates of chemical concentrations of secondary effluents are given in Table 14. Industrial wastewaters contain varying quantities of chemicals, each somewhat unique to the particular industry.

Soil loading values were presented by Galloway and Jacobs[51] (Table 15). It should be noted that these values are a maximum for the lifetime of the soil. These loading rates make an effort to take into account various soil properties which affect metal availability.

Gypsum and Sulfur

These two materials are used as both fertilizers and soil amendments. In the former case gypsum is used to supply either sulfur or calcium. Elemental sulfur is used to correct sulfur deficiencies; however, it must be applied well before it is needed, since microbial oxidation is required before it can be utilized by higher plants.

Sodic Soils

A soil in which the sodium-adsorption ratio of the saturation extract is 15 or more is considered to be a sodic soil. These soils contain sufficient sodium to reduce plant growth due to dispersion of soil colloidial materials, resulting in reduced air and water movement. Gypsum and sulfur are among the materials which can be applied to amend the soil. This results in the replacement of sodium on the exchange complex which is subsequently leached. For each milliequivalent of sodium which must be replaced, 1800 lb $CaSO_4$ or 320 lb sulfur per acre 6 in. of soil are required.[57] Other materials may also be used (Table 16).

Table 12
APPROXIMATE NUTRIENT CONCENTRATION (%) IN VARIOUS CROPS

Crop	N	P	K	Ca	Mg	S	Ref.
Forages							
Alfalfa hay, all analyses	2.45	0.24	1.97	1.47	0.31	0.29	47
	—	0.28	2.25	2.12	0.33	0.29	48
Bermuda grass hay (coastal)	1.47	0.19	1.42	0.37	0.15	—	47
Bluegrass hay, all analyses	1.66	0.27	1.67	0.40	0.19	0.12	47
Clover hay, all analyses	1.94	0.20	1.65	1.28	0.37	0.14	47
	—	0.27	2.26	1.45	0.35	0.22	48
Cowpea hay	1.97	0.30	1.66	1.38	0.37	0.32	47
	—	0.32	2.80	1.52	0.41	0.35	48
Grass hay, all analyses	1.12	0.21	1.20	0.48	0.16	0.15	47
Johnsongrass hay, all analyses	1.04	0.26	1.22	0.87	0.31	—	47
Lespedeza hay, all analyses	2.10	0.19	0.94	1.00	0.19	—	47
	—	0.20	1.02	1.10	0.22	—	48
Millet hay (Japanese)	1.33	0.16	2.10	0.20	0.20	0.14	47
	—	0.22	3.20	0.96	0.41	—	48
Milo, stover	0.51	0.11	—	0.58	—	—	47
Milo above ground, all analyses	—	0.33	3.06	0.72	—	—	48
Oat hay	1.31	0.19	0.83	0.21	0.16	—	47
	—	0.23	1.15	0.35	0.25	0.08	48
Orchard grass hay	1.30	0.18	1.92	0.27	0.21	0.23	47
Peanut hay, with nuts	2.14	0.15	0.85	1.13	0.37	—	47
Soybean forage, all analyses	2.43	0.07	0.22	0.26	0.13	—	47
	—	0.26	1.24	1.12	0.50	0.29	48
Sweet clover, first year hay	2.64	0.26	1.57	1.37	0.10	0.13	47
Sweet clover, all analyses	—	0.14	1.38	1.26	0.34	0.45	48
Timothy, in bloom	1.30	0.08	0.54	0.08	0.04	0.04	47
Timothy hay, all analyses	—	0.21	1.78	0.38	0.14	0.14	48
Grain Crops							
Barley (grain)	2.03	0.40	0.49	0.06	0.13	0.15	47
	—	0.41	0.55	0.06	0.14	0.17	48
Barley (straw)	0.59	0.10	1.33	0.33	0.07	0.14	47
	—	0.12	1.48	0.35	0.08	0.15	48
Corn (grain)	1.39	0.27	0.29	0.02	0.11	—	47
	—	0.43	0.40	0.02	0.16	0.14	48
Corn (stover)	0.94	0.09	1.49	0.54	0.41	0.15	47
	—	0.15	0.74	0.32	0.24	0.17	48
Flax (straw)	1.15	0.10	1.62	0.67	0.29	0.25	47
Flax (grain)	3.84	0.52	0.79	0.22	0.23	0.03	47
	—	0.59	0.63	0.28	0.43	0.06	48
Millet (straw)	0.61	—	1.44	0.08	—	—	47
Millet (grain)							
Foxtail	1.94	0.20	0.31	—	—	—	47
Proso	1.90	0.30	0.43	0.05	0.16	—	47
Japanese	1.70	0.44	0.33	—	—	—	47
Oats (grain)	1.92	0.33	0.43	0.09	0.16	0.21	47
	—	0.34	0.48	0.10	0.16	0.19	48
Oats (straw)	0.66	0.09	2.00	0.24	0.18	0.18	47
	—	0.11	1.51	0.21	0.24	0.25	48
Rice (grain)	1.26	0.32	0.34	0.08	0.14	—	47
	—	0.36	0.38	0.09	0.16	—	48
Rice (straw)	0.62	0.07	1.22	0.19	0.10	—	47
	—	0.13	1.64	0.30	0.11	—	48

Table 12 (continued)
APPROXIMATE NUTRIENT CONCENTRATION (%) IN VARIOUS CROPS

Crop	N	P	K	Ca	Mg	S	Ref.
Rye (grain)	2.02	0.33	0.47	0.10	0.12	0.16	47
	—	0.37	0.53	0.12	0.14	0.18	48
Rye (straw)	0.56	0.09	0.90	0.26	0.07	0.10	47
	—	0.10	0.97	0.28	0.07	0.11	48
Sorghum (grain)	1.73	0.32	0.37	0.02	—	—	47
Sorghum (fodder)	0.99	0.14	1.29	0.34	0.31	—	47
(straw)	—	0.13	1.29	0.16	—	—	48
Sunflower (seed, whole)	2.69	0.52	0.66	0.17	—	—	47
Sunflower (stems, leaves, receptacle)	0.69	0.14	2.37	1.34	0.71	0.18	49
Wheat (grain)	2.11	0.39	0.42	0.04	0.14	0.20	47
	—	0.40	0.48	0.05	0.17	0.18	48
Wheat (straw)	0.62	0.07	0.62	0.15	0.11	0.17	47
	—	0.08	0.85	0.23	0.10	0.17	48
Other Crops							
Beans, dry (straw)	0.98	0.13	1.02	1.67	0.12	0.10	47
	1.18	0.06	0.92	1.48	0.44	—	50
	—	0.11	1.22	1.71	0.13	0.11	48
Beans, dry (seed)	3.66	0.45	1.27	0.15	0.17	0.23	47
	—	0.63	1.41	0.17	0.19	0.26	48
Cotton (seed, whole)	3.70	0.68	1.11	0.14	0.32	0.24	47
	—	0.75	1.20	0.15	0.35	0.26	48
Cotton (burs)	—	0.10	3.50	0.67	0.26	—	48
(lint)	—	0.05	0.59	0.13	0.09	0.05	48
Peanuts (nuts)	4.86	0.44	0.54	0.06	0.18	0.25	47
Peas (seed)	—	0.57	1.41	0.19	0.18	0.13	48
(straw)	0.98	0.10	1.08	—	—	—	47
Potatoes (tubers)	0.35	0.05	0.48	0.01	0.03	0.02	47
	—	0.25	2.28	0.05	0.13	0.12	48
Potatoes (vines)	—	0.26	1.62	4.12	1.15	0.68	48
Soybeans (seeds)	6.06	0.59	1.50	0.25	0.28	0.22	47
	—	0.78	1.83	0.24	0.31	0.24	48
Soybeans (straw)	0.62	0.05	0.53	—	—	—	47
Sugarbeet (roots)	0.26	0.04	0.25	0.04	0.03	0.01	47
	—	0.11	0.64	0.22	0.16	0.07	48
Sugarbeet (leaves)	—	0.22	3.28	1.14	0.51	0.48	48
Sugarcane (whole plant)	0.16	0.04	0.37	—	—	—	47
Sugarcane (leaves & tops)	0.21	0.07	0.76	0.09	—	—	47
Sugarcane (tops)	—	0.17	1.83	0.21	0.14	—	48
Sugarcane (stalks)	—	0.07	0.56	0.07	0.06	0.06	48
Tomatoes (fruit)	—	0.55	4.80	0.24	0.30	0.32	48

Table 13
TOTAL ELEMENTAL COMPOSITION DRY WEIGHT BASIS OF OVER 200 SEWAGE SLUDGE SAMPLES FROM 8 STATES

	Concentration (%)		
Component	Minimum	Maximum	Median
Organic C	6.5	48.0	30.4
Inorganic C	0.3	543.0	1.4
Total N	<0.1	17.6	3.3
NH_4-N	<0.1	6.7	1.0
NO_3-N	<0.1	0.5	<0.1
Total P	<0.1	14.3	2.3
Inorganic P	<0.1	2.4	1.6
Total S	0.6	1.5	1.1
Ca	0.10	25.0	3.9
Fe	<0.10	15.3	1.1
Al	0.10	13.5	0.4
Na	0.01	3.1	0.2
K	0.02	2.6	0.3
Mg	0.03	2.0	0.4
	Concentration (ppm)		
Zn	101	27,800	1,740
Cu	84	10,400	850
Ni	2	3,515	82
Cr	10	99,000	890
Mn	18	7,100	260
Cd	3	3,410	16
Pb	13	19,730	500
Hg	<1	10,600	5
Co	1	18	4
Mo	5	39	30
Ba	21	8,980	162
As	6	230	10
B	4	757	33

From Dowdy, R. H., Larson, R. E., and Epstein, E., *Proc. Conf. Land Appl. Waste Mater.*, Soil Conservation Society of America, Ankeny, Iowa, 1976, 140. With permission.

Table 14
THREE ESTIMATES OF THE PROPERTIES OF SECONDARY EFFLUENTS

Constituent	Concentration (mg/ℓ)			
	Range[54]	Typical value or range[55]	Typical value[56]	Range[56]
Suspended solids	13—62	25	—	—
Biochemical oxygen demand	13—75	25	—	—
Chemical oxygen demand	50—160	70	—	—
Nitrogen, total	—	20	25	15—40
Phosphorus, total	7—10	10	10	0.5—40
Trace metals				
Cadmium	—	0.015	<0.005	<0.005—6.4
Chromium	0.01—2.5	0.02—0.14	0.025	<0.05—6.8
Cobalt	—	—	<0.05	<0.05—0.05
Copper	0.10—1.4	0.07—0.14	0.10	<0.02—5.9
Iron	0.10—3.0	0.10—4.3	—	—
Lead	0.01—1.0	0.01—0.03	0.05	<0.02—6.0
Manganese	—	0.20	—	—
Mercury	<0.005	0.01	0.001	<0.0001—0.125
Nickel	0.02—2.0	0.03—0.20	0.02	<0.02—6.0
Zinc	0.10—1.10	0.20—0.44	0.15	<0.02—20
Other Parameters				
Boron	0.5—1.0	1.0	—	—
Calcium	1—40	20	—	—
Magnesium	1—10	17	—	—
Potassium	7—10	14	—	—
Sodium	40—100	50	—	—
Chloride	40—100	45	—	—
Sulfate	12—52	—	—	—
Oil	0—10	—	—	—
Phenol	0—1	—	—	—

From Elliott, L. F. and Stevenson, F. J., *Soils for Management of Organic Wastes and Waste Waters*, American Society of Agronomy, Madison, Wis., 1977, 56. With permission.

Table 15
TOTAL AMOUNT OF SLUDGE METALS ALLOWED ON AGRICULTURAL LAND[51]

Metal	Soil cation exchange capacity (me/100g)[a] <5	5—15	>15
	Maximum metal load (lb/acre)		
Lead	500	1000	2000
Zinc	250	500	1,000
Copper	125	250	500
Nickel	50	100	200
Cadmium	5	10	20

[a] Determined by the pH 7 ammonium acetate procedure.

Table 16
AMOUNT OF OTHER MATERIALS TO EQUAL 1 LB OF SULFUR AS A SOIL AMENDMENT[57]

Amendment	Equivalent to 1 lb of sulfur (lb)
Lime-sulfur solution (CaS_2)	4.17
Sulfuric acid	3.06
Gypsum ($CaSO_4 \cdot 2H_2O$)	5.38
Iron sulfate ($FeSO_4 \cdot 7H_2O$)	8.69
Aluminum sulfate ($Al_2(SO_4)_3 \cdot 18H_2O$)	6.94

Table 17
QUANTITY OF SULFUR TO REDUCE SOIL pH FOR A DEPTH OF SEVEN INCHES[58]

Desired pH change	Sulfur (lb/acre) Sands	Loamy sands	Sandy loams	Loams	Clay loams
7.0—6.0	300	400	500	700	1100
6.0—5.0	800	1000	1200	1400	1400
5.0—4.0	800	1000	1200	1400	1800

Acidifying Soils

On occasion it is desirable to acidify soils for some special purpose. Generally it is not economically feasible to reduce soil pH over large land areas, but it may be for small areas. The amount of sulfur required will vary among soils, but some general figures for pH reduction are given in Table 17. The materials listed in Table 16 may also be used except gypsum which will not acidify soils to a large extent.

REFERENCES

1. **Tisdale, S. L. and Nelson, W. L.**, *Soil Fertility and Fertilizers*, 3rd ed., Macmillan, New York, 1975, chap. 1.
2. **Barber, S. A.**, Liming materials and practices, in *Soil Acidity and Liming*, Pearson, R. W. and Adams F., Eds., American Society of Agronomy, Madison, Wis., 1967, chap. 3.
3. **Whittaker, C. W., Anderson, M. S., and Reitemeir, R. F.**, Liming Soils — an Aid to Better Farming, Farmers Bulletin No. 2124, U.S. Department of Agriculture, Washington, D.C., 1966.
4. **Christenson, D. R. and Doll, E. C.**, Lime for Michigan Soils, Michigan State Univ. Coop. Ext. Serv. Ext. Bull. 471, 1973.
5. **Chichilo, P. and Whittaker, C. W.**, Trace elements in agricultural limestones of the United States, *Agron. J.*, 53, 139, 1961.
6. *Dictionary of Plant Foods*, Meister, Willoughby, Ohio, 1965, C-179.
7. **Warren, H. V. and Delavault, R. E.**, The lead, copper, zinc and molybdenum contents of some limestones and related rocks in southern Ontario, *Econ. Geol.*, 56, 1265, 1961.
8. **Murray, J. A., Fischer, H. C., and Rolnick, L. S.**, Shrinkage of high calcium limestones during burning, *J. Am. Ceram. Soc.*, 37, 323, 1954.
9. **Keller, W., Klemme, A., and Pickett, E.**, Survey of chemical composition of rock layers in an agricultural limestone, *Econ. Geol.*, 45, 461, 1950.
10. **Shickluna, J. C., Longnecker, E. D., and Porter, J. A.**, A rapid chemical determination of the magnesium carbonate content of marl coupled with a survey of this material in Michigan marl deposits, *Mich. Agr. Exp. Sta. Quar. Bull.*, 41, 198, 1958.
11. **Ross, G. J., Lawton, K., and Ellis, B. G.**, Lime requirement related to physical and chemical properties of nine Michigan soils, *Soil Sci. Soc. Am. Proc.*, 28, 209, 1964.
12. **Keeney, D. R. and Corey, R. B.**, Factors affecting lime requirements of Wisconsin soils, *Soil Sci. Soc. Am. Proc.*, 27, 277, 1963.
13. **Pionke, H. B., Corey, R. B., and Schulte, E. E.**, Contributions of soil factors to lime requirements and lime requirement tests, *Soil Sci. Soc. Am. Proc.*, 32, 113, 1968.
14. **Shoemaker, H. E., McLean, E. O., and Pratt, P. F.**, Buffer methods for determining lime requirement of soils with appreciable amounts of extractable aluminum, *Soil Sci. Soc. Am. Proc.*, 25, 274, 1961.
15. **McLean, E. O., Dumford, S. W., and Coronel, F.**, A comparison of several methods of determining lime requirements of soils, *Soil Sci. Soc. Am. Proc.*, 30, 26, 1966.
16. **McLean, E. O., Eckert, D. J., Reddy, G. Y., and Trierweiler, J. F.**, An improved SMP soil lime requirement method incorporating double-buffer and quick-test, *Soil Sci. Soc. Am. Proc.*, 42, 311, 1978.
17. **Leeper, G. W.**, *Managing the Heavy Metals on Land*, Marcel Dekker, New York, 1978.
18. **McLean, E. O.**, Potentially beneficial effects from liming: chemical and physical, *Soil Crop Sci. Soc. Fla.*, 31, 189, 1971.
19. **Adams, F. and Pearson, R. W.**, Crop response to lime in the southern United States and Puerto Rico, in *Soil Acidity and Liming*, Pearson, R. W. and F. Adams, Eds., American Society Agronomy, Madison, Wis., 1967, chap. 4.
20. **Abruna-Rodriguez, F., Vincente-Chandler, J., Pearson, R. W., and Silva, S.**, Crop response to soil acidity factors in Ultisols and Oxisols. I. Tobacco, *Soil Sci. Soc. Am. Proc.*, 34, 629, 1970.
21. **Evans, C. E. and Kamprath, E. J.**, Lime response as related to percent Al saturation, solution Al and organic matter content, *Soil Sci. Soc. Am. Proc.*, 34, 893, 1970.
22. **Kamprath, E. J.**, Exchangeable aluminum as a criterion for liming leached mineral soils, *Soil Sci. Soc. Am. Proc.*, 34, 252, 1970.
23. **Reeve, N. G. and Sumner, M. E.**, Lime requirements of Natal Oxisols based on exchangeable aluminum, *Soil Sci. Soc. Am. Proc.*, 34, 595, 1970.
24. **Martini, J. A., Kockhann, R. A., Siqueira, O. J., and Barkert, C. M.**, Response of soybeans to liming as related to soil acidity, Al and Mn toxicities and P in some Oxisols of Brazil, *Soil Sci. Soc. Am. Proc.*, 38, 616, 1974.
25. **Mendez, J. and Kamprath E. J.**, Liming of Latosols and the effect on phosphorus response, *Soil Sci. Soc. Am. Proc.*, 42, 86, 1978.
26. **Foy, C. D. and Brown, J. C.**, Toxic factors in acid soils. II. Differential aluminum tolerance of plant species, *Soil Sci. Soc. Am. Proc.*, 28, 27, 1964.
27. **Martini, J. A., Kockhann, R. A., Gomes, E. P., and Langer, F.**, Response of wheat cultivars to liming in some high Al Oxisols of Rio Grande do Sul, Brazil, *Agron. J.*, 69, 612, 1977.
28. **Sartain, J. B. and Kamprath, E. J.**, Effect of soil Al saturation on nutrient concentration of soybean tops, roots and nodules, *Agron. J.*, 69, 843, 1977.

29. **MacLeod, L. B. and Jackson, L. P.,** Aluminum tolerance of two barley varieties in nutrient solution, peat and soil culture, *Agron. J.,* 59, 359, 1967.
30. **Foy, C. D., Lafever, H. N., Schwartz, J. W., and Fleming, A. L.,** Aluminum tolerance of wheat cultivars related to region of origin, *Agron. J.,* 66, 751, 1974.
31. **Foy, C. D.,** Effects of aluminum on plant growth, in *The Plant Root and Its Environment,* Carson, E. W., Ed., University Press of Virginia, Charlottesville, 1974, chap. 20.
32. **Wright, M. J.,** Plant Adaptation to Mineral Stress in Problem Soils, Special Publication, Cornell Agricultural Experimental Station, Ithaca, N.Y., 1976.
33. **Wheeler, H. J., Tucker, G. M., and Hartwell, B. L.,** On the neutralizing action of lime, *R. I. Agric. Exp. Sta. Annu. Rep.,* 9, 294, 1896.
34. **Hartwell, B. L. and Damon, S. C.,** A Field Comparison of Hydrated Lime with Limestone of Different Degrees of Fineness, *R. I. Agric. Exp. Stn. Bull.,* 180, 1919.
35. **Spurway, C. H.,** Soil Reaction Preferences of Plants, *Mich. Agric. Exp. Sta. Spec. Bull.,* 306, 1941.
36. **Truog, E.,** Lime in relation to availability of plant nutrients, *Soil Sci.,* 65, 1, 1948.
37. **Russell, E. W.,** *Soil Conditions and Plant Growth,* John Wiley & Sons, 1962, chap. 26 to 28.
38. **Kamprath, E. J. and Foy, C. D.,** Lime-fertilizer interactions in acid soils, in *Fertilizer Technology and Use,* Olson, R. A., Army, T. J., and Kilmer, V. J., Eds., Soil Science Society America, Madison, Wis., 1971, chap 5.
39. **Wolcott, A. R., Foth, H. D., Davis, J. F., and Schikluna, J. C.,** Nitrogen Carriers. I. Soil effects, *Soil Sci. Soc. Am. Proc.,* 29, 405, 1965.
40. **Jolley, V. D. and Pierre, W. H.,** Soil acidity from long term use of nitrogen fertilizer and its relationship to recovery of the nitrogen, *Soil Sci. Soc. Am. Proc.,* 41, 368, 1977.
41. **Black, C. A.,** *Soil-Plant Relationships,* John Wiley, & Sons, New York, 1968.
42. **Vitosh, M. L., Davis, J. F., and Knezek, B. D.,** Long term effects of manure, fertilizer and plow depth on chemical properties of soils and nutrient movement in a monoculture system, *J. Environ. Qual.,* 2, 296, 1973.
43. **Wallingford, G. W., Murphy, L. S., Powers, W. L., and Manges, H. L.,** Disposal of beef-feedlot manure: effects of residual and yearly applications on corn and soil chemical properties, *J. Environ. Qual.,* 4, 526, 1975.
44. **Lund, Z. F., Doss, B. D., and Lowry, F. E.,** Dairy cattle manure — its effect on rye and millet forage yield and quality, *J. Environ. Qual.,* 4, 195, 1975.
45. **Walsh, L. M. and Hensler, R. F.,** Manage Manure for Its Value, Wisconsin Extension Circular 1971, 550.
46. **Miller, H. F., Ross, I. J., and Thomas, G. W.,** Farm Manure: Production-Value-Use, ID-19, University Kentucky Cooperative Extension Service, 1976.
47. **Morrison, F. B.,** *Feeds and Feeding,* 22nd ed., Morrison, Ithaca, New York, 1956, 1000.
48. **Beeson, K. C.,** The Mineral Composition of Crops with Particular Reference to the Soils in Which They Were Grown, U.S. Department of Agriculture Miscellaneous Publication No. 369, 1941.
49. **Robinson, R. G.,** Sunflower date of planting and chemical composition at various growth stages, *Agron. J.,* 62, 665, 1970.
50. **Janatawat, S.,** The Residual Effects of Zinc Fertilizers on the Growth and Nutrient Uptake of the White Pea Bean (*Phaseolus vulgaris* var. Sanilac) and the Red Kideny Bean (*Phaseolus vulgaris* var. Charlevoix), M.S. thesis, Michigan State University, East Lansing, 1969.
51. **Galloway, H. M. and Jacobs, L. W.,** Sewage sludges-characteristics and management, in *Utilizing Municipal Sewage Wastewaters and Sludges on Land for Agricultural Production,* Jacobs, L. W., Ed., North Central Reg. Ext. Pub. No. 52, 1977, 3.
52. **Dowdy R. H., Larson, R. E., and Epstein, E.,** Sewage sludge and effluent use in agriculture, *Proc. Conf. Land Appl. Waste Mater,* Soil Conservation Society of America, Ankeny, Iowa, 1976, 138.
53. **Elliott, L. F. and Stevenson, F. J.,** *Soils for Management of Organic Wastes and Waste Waters,* American Society Agronomy, Madison, Wis., 1977, 56.
54. **Buzzell, T.,** Secondary Treatment Processes, in Wastewater Management by Disposal on the Land, Cold Regions Res. and Eng. Lab., Corps of Engineers, U.S. Army, Special Rep. 171, 1972, 35.
55. **Driver, C. H., Hrutfiord, B. F., Spyridakis, D. E., Welch, E. B., and Wooldridge, D. D.,** Assessment of the Effectiveness and Effects of Land Disposal Methodologies of Wastewater Management, Corps of Engineers, U.S. Army, Wastewater Mgmt. Rep. 72-1, 1972.
56. **Menzies, J. D. and Chaney, R. L.,** Waste Characteristics in Factors Involved in Land Application of Agricultural and Municipal Wastes, National Program Staff, Soil, Water and Air Sciences, Agricultural Research Service, U.S. Department of Agriculture, Beltsville, Md., 1974.
57. **Richards, L. A.,** Diagnosis and Improvement of Saline and Alkali Soils, USDA Agricultural Handbook No. 60, U.S. Department of Agriculture, Washington, D.C., 1954.
58. **Christenson, D. R., Lucas, R. E., and Doll, E. C.,** Fertilizer Recommendations for Michigan Vegetables and Field Crops, Mich. State Univ. Coop. Ext. Serv. Ext. Bull. E-550, Michigan State University, East Lansing, 1972.

SOIL AND WATER MANAGEMENT AND CONSERVATION

Jerry V. Mannering

SEVERITY OF THE EROSION PROBLEM AND ITS GEOGRAPHIC LOCATION

Soil erosion is the wearing away of the land surface by water, wind, ice, and other geological agents.[1] Erosion of the soil by water is particularly serious since it not only reduces soil productivity through the loss of fertile topsoil, it also degrades water quality when sediment and associated nutrients are transported to water courses.[2]

Erosion is a natural process of soil formation (geologic erosion) and is responsible for much of the land form in the world today. However, it is also greatly influenced by man's activities (accelerated erosion). Accelerated erosion results when man makes major changes in land use without replacing the natural barriers to erosion with effective conservation practices.

Present day economic factors are increasing the seriousness of erosion problems in the U.S. as well as in other parts of the world. In the developing countries problems are arising because of pressures to produce more food, which in certain cases results in over-cropping of fragile lands. In the developed countries many of the proven conservation practices, such as contouring, strip cropping, terracing, sod waterways, and small fields, are not compatible with the highly efficient large tillage, planting, and harvesting equipment presently being used. Also, the high cost of land and large capital investments associated with modern farming technology require maximum short-term economic returns, resulting in increased acreages of more erosive row-crops at the expense of soil-conserving, grass-legume-based rotations.

The U.S. Conservation Needs Inventory (CNI) of 1965,[3] found susceptibility to erosion to be the most widespread conservation problem limiting the land capability. It was found to be the dominant problem on about half or 738 million acres of the area included in the national inventory. Although serious erosion occurs on rangeland, pasture land, forest land and from nonagricultural uses, such as construction areas, road banks, and stream banks,[4] perhaps the most serious threat to society comes from the erosion of cropland. The CNI study[5,6] showed 277 million acres, or more than 60% of the cropland to need conservation treatment. It further showed that about 20% of 438 million acres of cropland in the U.S. were averaging more than 8 tons of soil loss per acre per year, 50% between 3 and 8 tons and 30% less than 3 tons. A maximum annual soil loss of 5 tons/acre/year has been suggested[7] as permissible if soil productivity is to be maintained. Excessive erosion also results in serious sedimentation problems. A recent estimate[8] attributes about half of the U.S. 4 billion tons annual sediment production to cropland sources.

The geographical location of erosion problems in the contiguous 48 states of the U.S. is shown in Figures 1 through 5. Figure 1 shows the 156 major land resource areas.[57] Figure 2 is the estimate of average annual potential direct runoff (that portion of the precipitation that appears in surface streams).[58] Figure 3 shows croplands as percentages of total land resource areas (see Figure 1).[59] Figure 4 estimates croplands on which erosion is the dominant limitation for their agricultural use as percentages of total land resource areas.[60] Figure 5 shows the relative potential contribution of cropland to water-shed sediment yields.[61]

These figures show in a very general way the relationships of soils, land use, and climate to soil erosion and sedimentation. Effects of these factors on the soil erosion process will be explained in more detail in the next section.

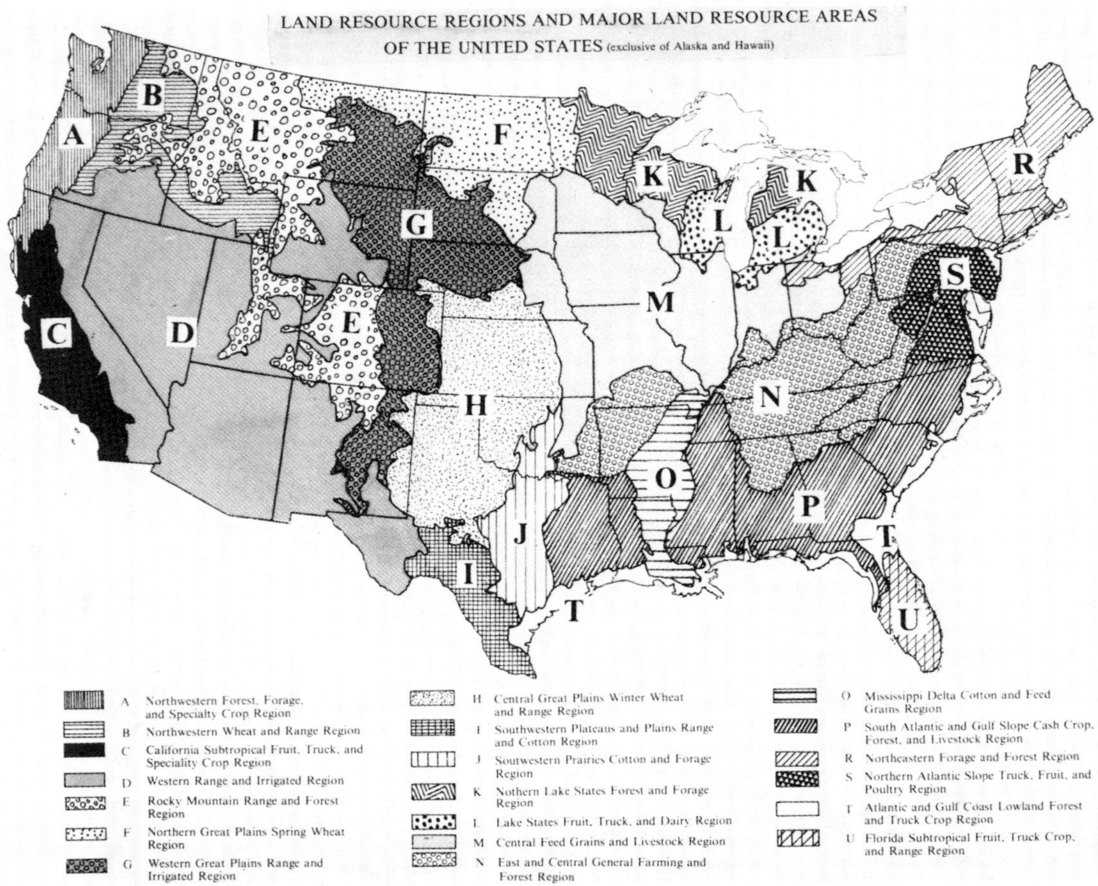

FIGURE 1. Land resource regions and major land resource areas of the U.S. (From Wischmeier, W. H., Control of Water Pollution from Cropland, Agric. Res. Serv., U.S. Department of Agriculture and Environmental Protection Agency, 1(3), 6, 1975.)

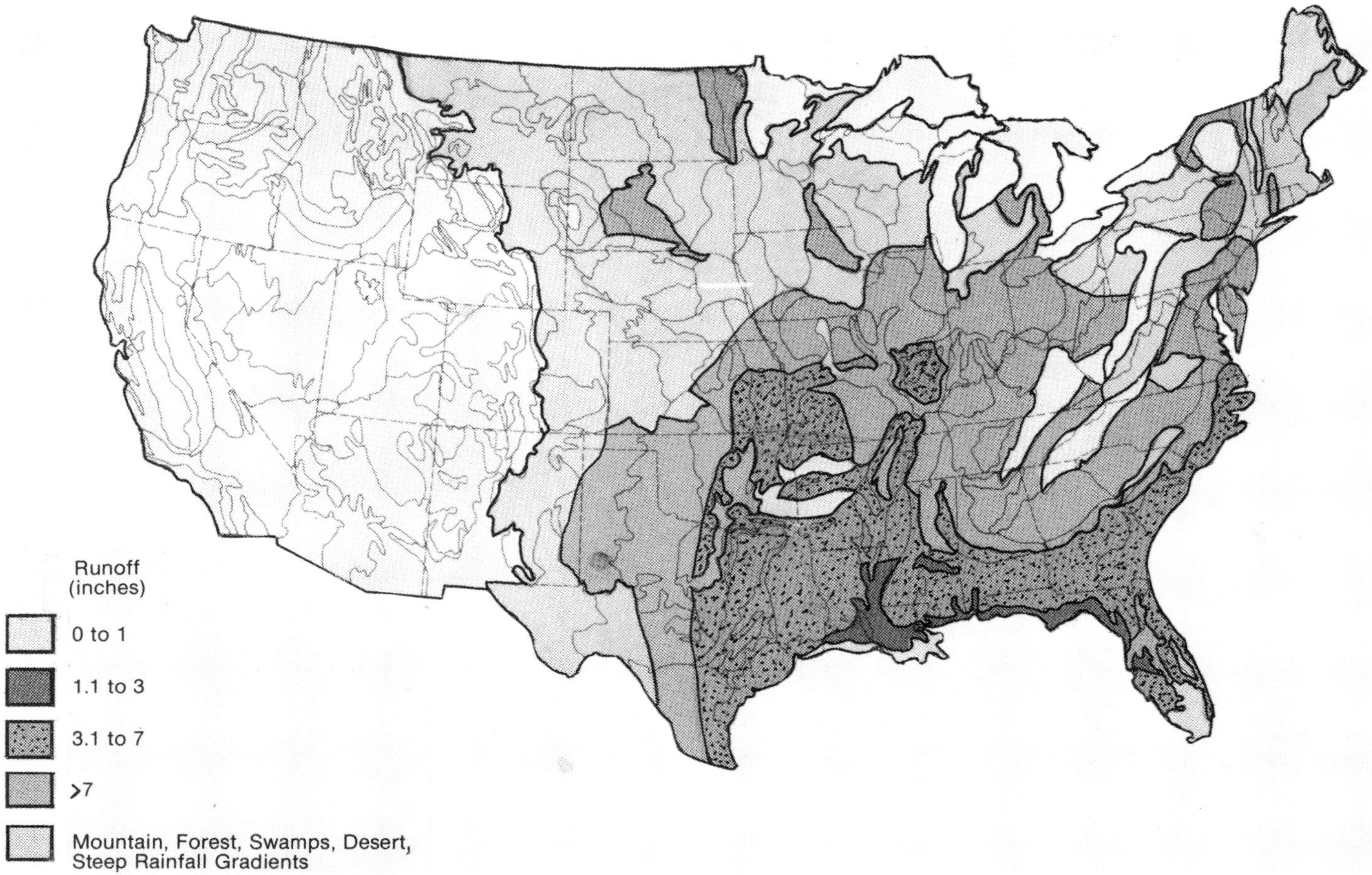

FIGURE 2. Average annual potential direct runoff. (From Wischmeier, W. H., Control of Water Pollution from Cropland, Agric. Res. Serv., U.S. Department of Agriculture and Environmental Protection Agency, 1(3), 8, 1975.)

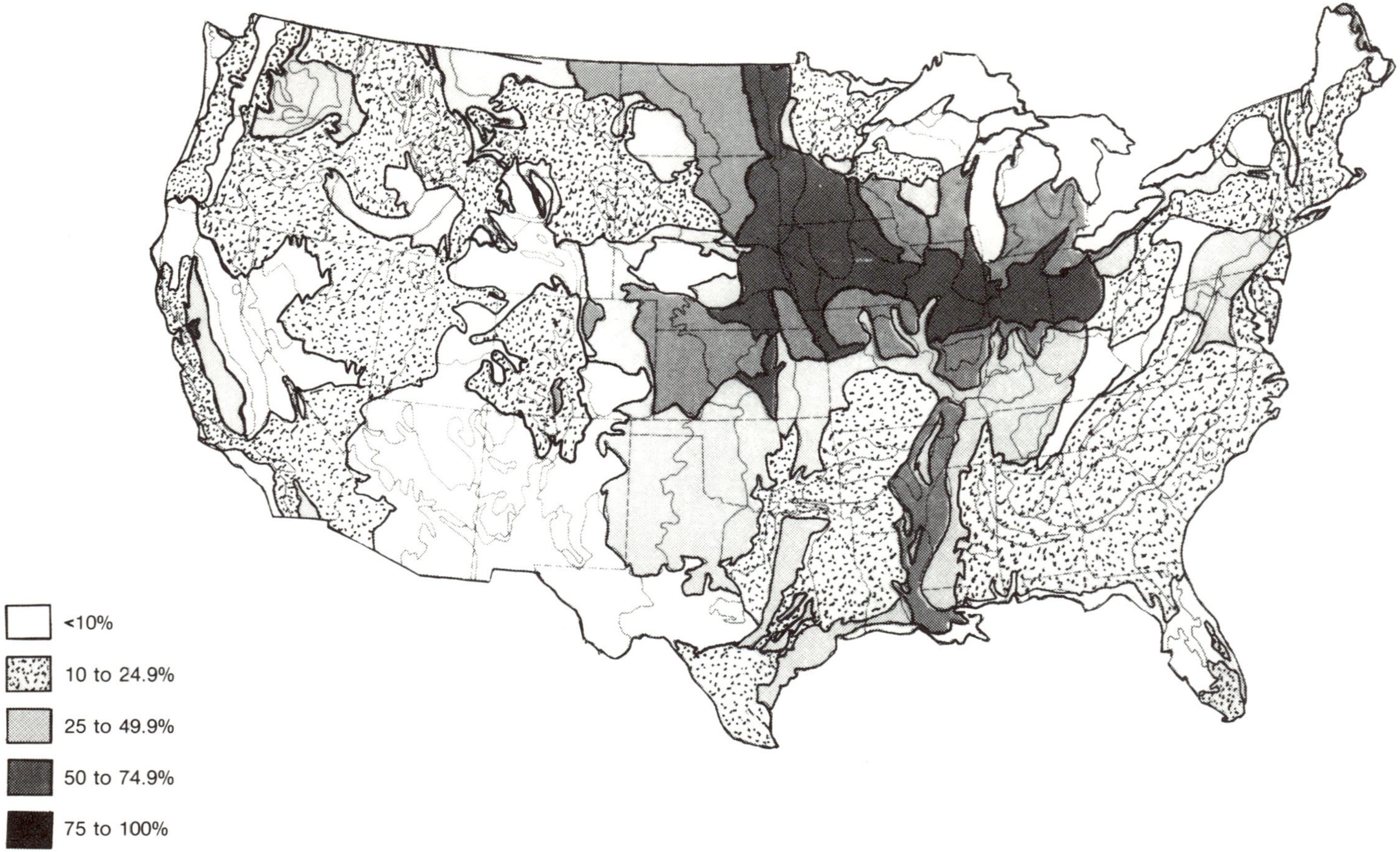

FIGURE 3. Croplands as percentages of total land resource areas. (From Wischmeier, W. H., Control of Water Pollution from Cropland, Agric. Res. Serv., U.S. Department of Agriculture and Environmental Protection Agency, 1(3), 12, 1975.)

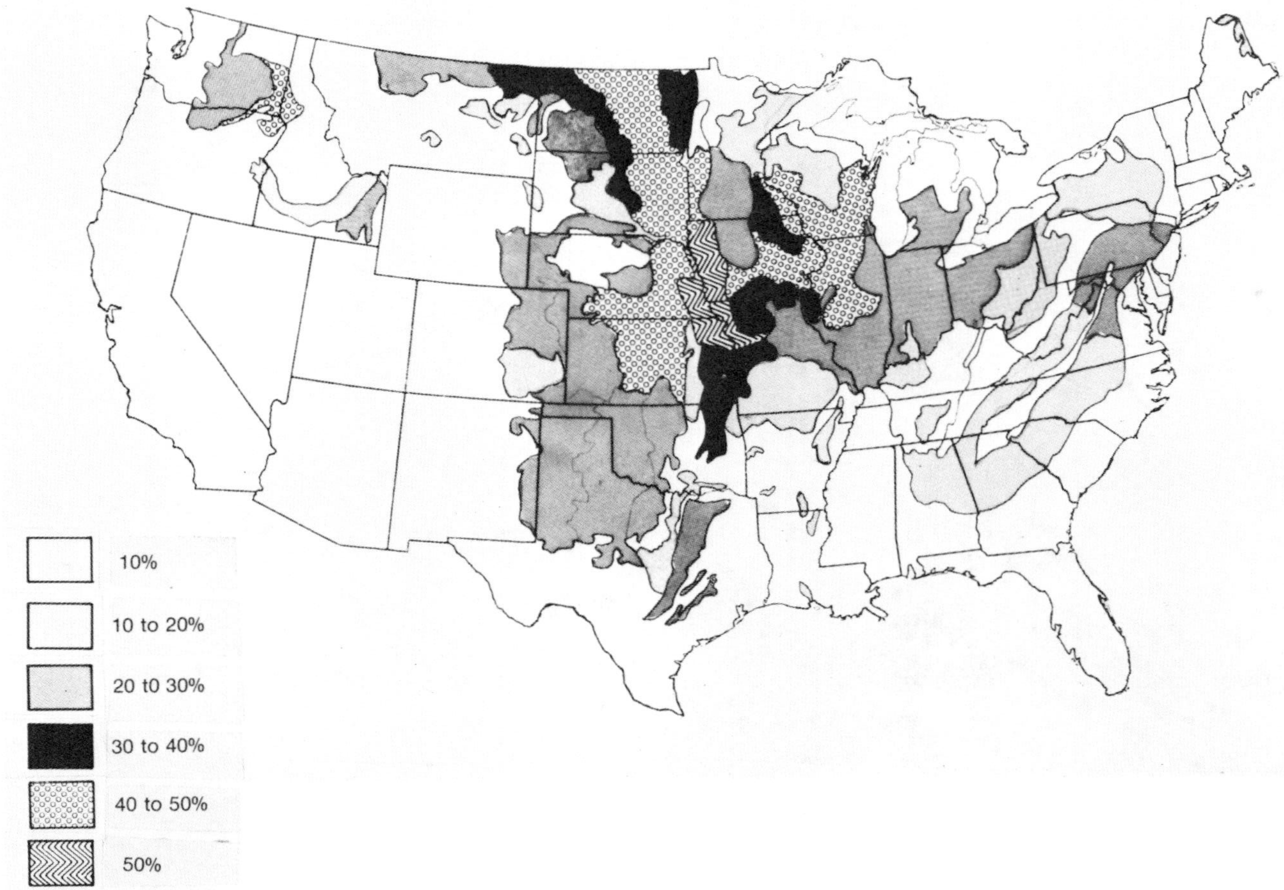

FIGURE 4. Croplands on which erosion is the dominant limitation for their agricultural use as percentages of total land resource areas. (From Wischmeier, W. H., Control of Water Pollution from Cropland, Agric. Res. Serv., U.S. Department of Agriculture and Environmental Protection Agency, 1(3), 13, 1975.)

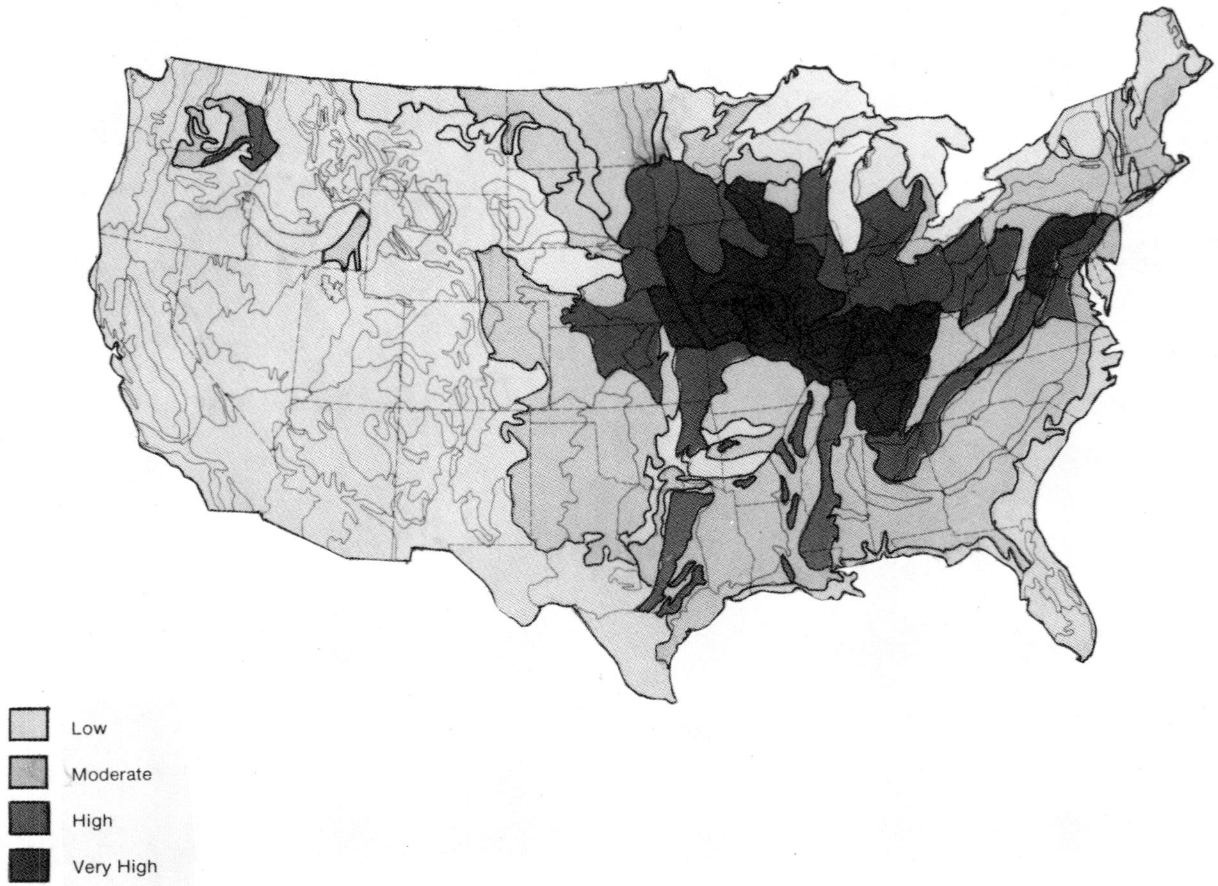

FIGURE 5. Relative potential contribution of cropland to watershed sediment yields. (From Wischmeier, W. H., Control of Water Pollution from Cropland, Agric. Res. Serv., U.S. Department of Agriculture and Environmental Protection Agency, 1(3), 15, 1975.)

FACTORS AFFECTING SOIL EROSION BY WATER

Soil erosion is a process of detachment and transportation of soil materials by erosive agents.[9] It is a mechanical process that requires energy. It can take the form of sheet erosion,[10] raindrop splash and shallow-overland-flow-transport of some of the detached particles; rill erosion[1] where numerous shallow channels are formed from runoff; gully erosion[1] where water concentrates in narrow channels and, over short periods, removes the soil from a narrow area to depths of 1 to 2 ft or more; and mass movement[11] on gully banks, where gravitational and seepage forces cause movement of the soil from a higher to a lower elevation. The form or forms of erosion predominating in a particular region will depend upon climate, topography, vegetation, soils, and the human factor.[12]

We will discuss these factors individually in the form and order that they appear in the "Universal Soil Loss Prediction Equation".[13] This empirical equation is in the form A = RKLSCP where A is the computed annual soil loss per unit area, R is the rainfall factor, K is the soil erodibility factor, L is the slope-length factor, S is the slope-steepness factor, C is the cropping-management factor, and P is the erosion control practice factor.

Climate

The principal influence of climate is the type, amount, and temporal distribution of the rainfall. The erosive potential of rainfall depends on the raindrop fall velocities, size distribution, and total mass at impact. The kinetic energy of falling raindrops is tremendous, for example, one 30-min thunderstorm in the Midwest can have an impact energy of 1000 ft-ton/acre.[14] Wischmeier and Smith[13] account for differences in rainfall intensity-duration-frequency in different regions of the country in their isoerodent map (Figure 6).[62] R is the average number of erosion-index units in a year of rainfall and is computed as the product of the total kinetic energy of a rainstorm, times its maximum 30-min intensity.[15] This product was statistically determined to be highly correlated with soil erosion. It includes not only the influence of raindrop splash, but also runoff on soil erosion. The R (climatic) factor does not account for soil losses due to snow melt and wind. Where these are significant, they need to be treated separately.

Not only is total annual kinetic energy important to the erosion process; temporal distribution of rainfall kinetic energy within a year is also important. If, for example, most of the high-intensity storms in a region occur at a time when most of the cropland is bare, serious erosion will result. If on the other hand, cropland is well protected by vegetative cover during periods of intense rainfall, little serious erosion will result. This distribution of the erosion potential is included in assessing the erodibility of crops and cultural practices in the different regions.[13] Figure 7[63] illustrates how dramatically this distribution can vary from one region to another.

Soils

Some soils erode more readily than others, even when rainfall, slope, vegetative cover, and management are the same. This has been referred to as soil erodibility and is represented by "K" in the universal soil loss prediction equation. It refers to the inherent susceptibility of soil to erosion by rainfall and runoff. Soil properties that influence erodibility are those that affect infiltration and permeability and those that resist dispersion, splashing, abrasion, and transport forces of rainfall and runoff. Numerous soil parameters have been found to be significantly related to soil erodibility.[16] However, Wischmeier et al.[17] found that by knowing only soil particle size distribution,

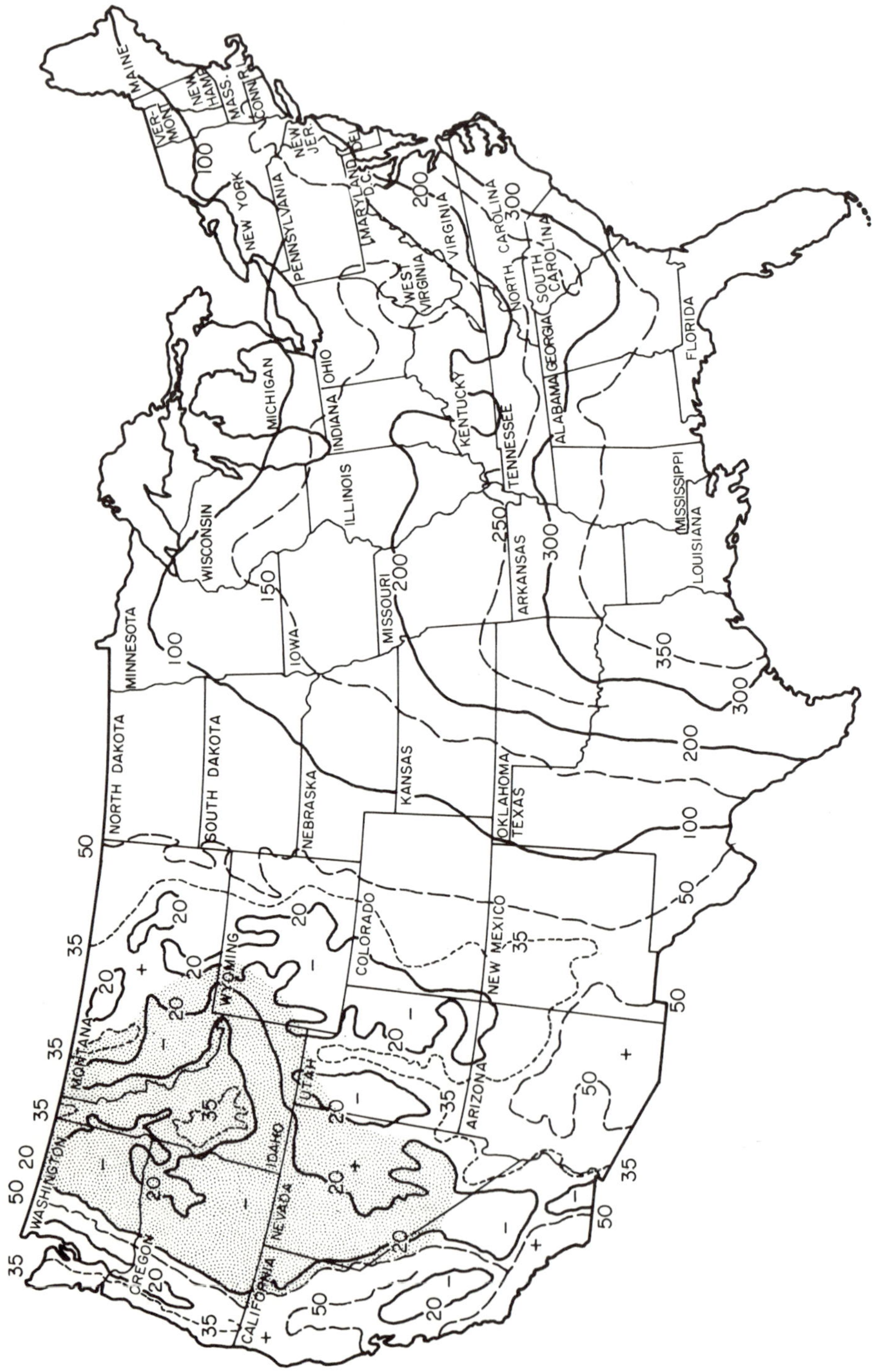

FIGURE 6. Average annual values of the rainfall-erosivity factor, R. (From Wischmeier, W. H., Control of Water Pollution from Cropland, Agric. Res. Serv., U.S. Department of Agriculture and Environmental Protection Agency, 1(3), 17, 1975.)

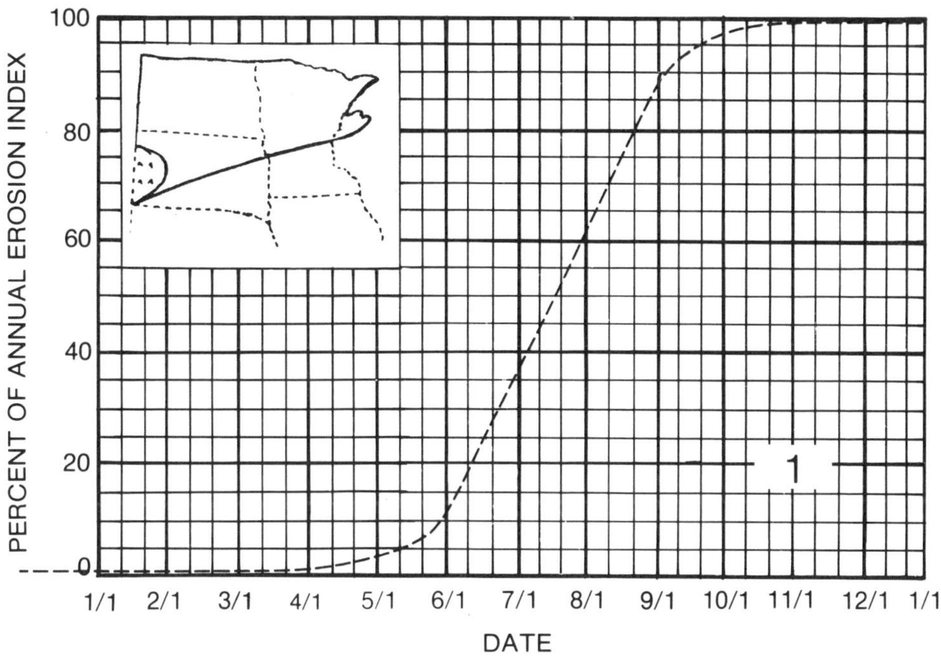

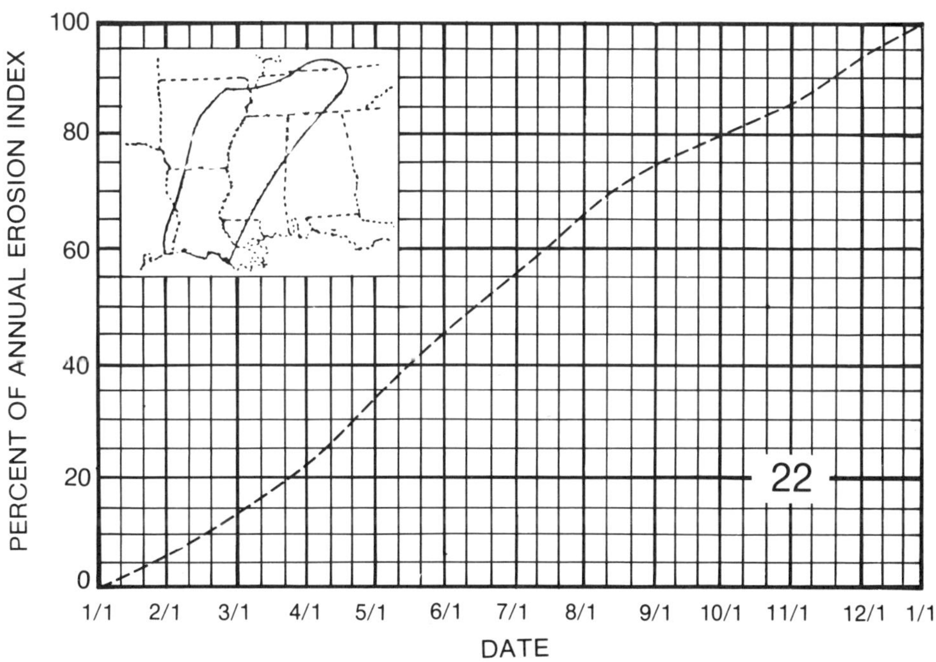

FIGURE 7. Erosion-index distribution curves 1 and 22: The Dakotas and parts of Minnesota, parts of Tennessee, Arkansas, Louisiana, Mississippi, and Alabama. (From Wischmeier, W. H. and Smith, D. D., *U.S. Dep. Agric. Agric. Handb. 282,* 18, 28, 1965.)

soil organic matter content, a soil structure index, and profile permeability class, he could estimate an erodibility factor suitable for predicting erosion in the field. This "so-called" soil erodibility nomograph is shown in Figure 8.[64] Table 1 indicates the

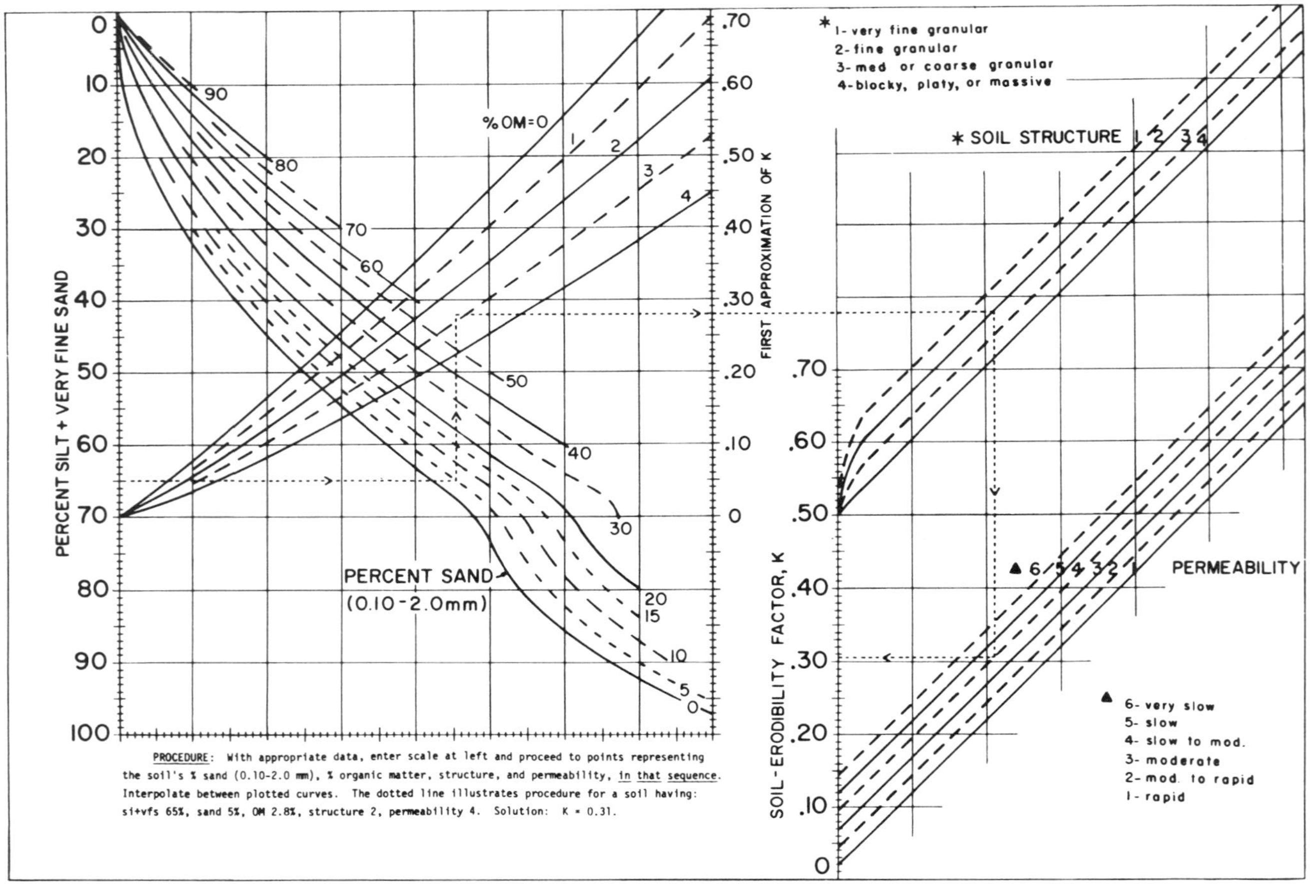

FIGURE 8. Nomograph for determining soil-erodibility factor, K, for U.S. mainland soils. (From Wischmeier, W. H., Control of Water Pollution from Cropland, Agric. Res. Serv., U.S. Department of Agriculture and Environmental Protection Agency, 1(3), 19, 1975.)

Table 1
INDICATIONS OF THE GENERAL MAGNITUDE OF THE SOIL-ERODIBILITY FACTOR (K)[a]

Texture class	Organic matter content <0.5% (K)	2% (K)	4% (K)
Sand	0.05	0.03	0.02
Fine sand	0.16	0.14	0.10
Very fine sand	0.42	0.36	0.28
Loamy sand	0.12	0.10	0.08
Loamy fine sand	0.24	0.20	0.16
Loamy very fine sand	0.44	0.38	0.30
Sandy loam	0.27	0.24	0.19
Fine sandy loam	0.35	0.30	0.24
Very fine sandy loam	0.47	0.41	0.33
Loam	0.38	0.34	0.29
Silt loam	0.48	0.42	0.33
Silt	0.60	0.52	0.42
Sandy clay loam	0.27	0.25	0.21
Clay loam	0.28	0.25	0.21
Silty clay loam	0.37	0.32	0.26
Sandy clay	0.14	0.13	0.12
Silty clay	0.25	0.23	0.19
Clay		0.13—0.29	

[a] The values shown are estimated averages of broad ranges of specific-soil values. When a texture is near the borderline of two texture classes, use the average of the two K values. For specific soils, use of Figure 11 or Soil Conservation Service K-value tables will provide much greater accuracy.

From Wischmeier, W. H., Agric. Res. Serv., U.S. Dept. Agriculture, 1(3), 20, 1975.

general magnitude of soil erodibility factors (K) across a broad range of textures and different levels of organic matter.[65]

Slope

Length, steepness, and shape are components of slope that have significant effects on runoff and soil erosion. Slope length is defined as the distance from the point of origin of overland flow to the point where runoff enters a well-defined channel or the steepness of the slope decreases enough that deposition begins.[13] The effect of slope length is primarily due to greater accumulation and more channelization of runoff as the slope becomes longer.

Slope steepness affects both runoff and soil loss. Runoff tends to increase linearly with increases in slope on small grain and curvilinearly on rowcrop land.[18] Soil loss increases more rapidly than runoff as the slope steepens. The combined effect of length and steepness for uniform slopes is given in Table 2.[66] These values which are used to represent the LS (length and degree of slope) variables in the universal soil loss prediction equation were derived by the formula

Table 2
VALUES OF THE EROSION EQUATION TOPOGRAPHIC FACTOR, LS, FOR SPECIFIED COMBINATIONS OF SLOPE LENGTH AND STEEPNESS[a]

Slope	Slope length (ft)											
(%)	25	50	75	100	150	200	300	400	500	600	800	1000
0.5	0.07	0.08	0.09	0.10	0.11	0.12	0.14	0.15	0.16	0.17	0.19	0.20
1	0.09	0.10	0.12	0.13	0.15	0.16	0.18	0.20	0.21	0.22	0.24	0.26
2	0.13	0.16	0.19	0.20	0.23	0.25	0.28	0.31	0.33	0.34	0.38	0.40
3	0.19	0.23	0.26	0.29	0.33	0.35	0.40	0.44	0.47	0.49	0.54	0.57
4	0.23	0.30	0.36	0.40	0.47	0.53	0.62	0.70	0.76	0.82	0.92	1.0
5	0.27	0.38	0.46	0.54	0.66	0.76	0.93	1.1	1.2	1.3	1.5	1.7
6	0.34	0.48	0.58	0.67	0.82	0.95	1.2	1.4	1.5	1.7	1.9	2.1
8	0.50	0.70	0.86	0.99	1.2	1.4	1.7	2.0	2.2	2.4	2.8	3.1
10	0.69	0.97	1.2	1.4	1.7	1.9	2.4	2.7	3.1	3.4	3.9	4.3
12	0.90	1.3	1.6	1.8	2.2	2.6	3.1	3.6	4.0	4.4	5.1	5.7
14	1.2	1.6	2.0	2.3	2.8	3.3	4.0	4.6	5.1	5.6	6.5	7.3
16	1.4	2.0	2.5	2.8	3.5	4.0	4.9	5.7	6.4	7.0	8.0	9.0
18	1.7	2.4	3.0	3.4	4.2	4.9	6.0	6.9	7.7	8.4	9.7	11.0
20	2.0	2.9	3.5	4.1	5.0	5.8	7.1	8.2	9.1	10.0	12.0	13.0
25	3.0	4.2	5.1	5.9	7.2	8.3	10.0	12.0	13.0	14.0	17.0	19.0
30	4.0	5.6	6.9	8.0	9.7	11.0	14.0	16.0	18.0	20.0	23.0	25.0
40	6.3	9.0	11.0	13.0	16.0	18.0	22.0	25.0	28.0	31.0	—	—
50	8.9	13.0	15.0	18.0	22.0	25.0	31.0	—	—	—	—	—
60	12.0	16.0	20.0	23.0	28.0	—	—	—	—	—	—	—

[a] Values given for slopes longer than 300 ft or steeper than 18% are extrapolations beyond the range of the research data and therefore, less certain than the others. Adjustments for irregularity of slope are discussed in Volume II.

From Wischmeier, W. H., Agric. Res. Serv., U.S. Dept. Agric., 1(3), 22, 1975.

$$LS = \left(\frac{\Lambda}{72.6}\right)^{m} \frac{430 \sin^2 \theta + 30 \sin \theta + 0.43}{6.574} \qquad (1)$$

where Λ = slope length in feet and m = 0.5 if the slope is steeper than 4%, 0.4 for 4% slopes and 0.3 for slopes of 3% or less; and θ = the angle of slope.[19] The relationship is also shown in graphic form in Figure 9.[67]

Shape of the slope (uniform, convex, and concave) is handled by a recently published formula[20] which breaks complex slopes into segments and through a technique of averaging computes a more accurate factor.

Cover and Management

The effects of cropping and management variables cannot be evaluated independently when assessing their influence on erosion because of the many interactions involved.[19-21] Almost any crop can be grown continuously or in rotation. The sequence within a system can be varied. The tillage system used can have a major influence on erosion. Crop residues can be removed, left on the surface, incorporated near the surface, or plowed under. Seedbeds can be left rough or they can be left smooth. The effectiveness of the crop residues depends on the amount. This in turn depends on rainfall distribution, fertility level, and other management decisions made by the farmer. Similarly the residual influence of meadow-sod plowed under before row crops depends on the type and quality of the meadow and length of elapsed time since the sod was turned under. The canopy protection of crops depends on type of vegetation,

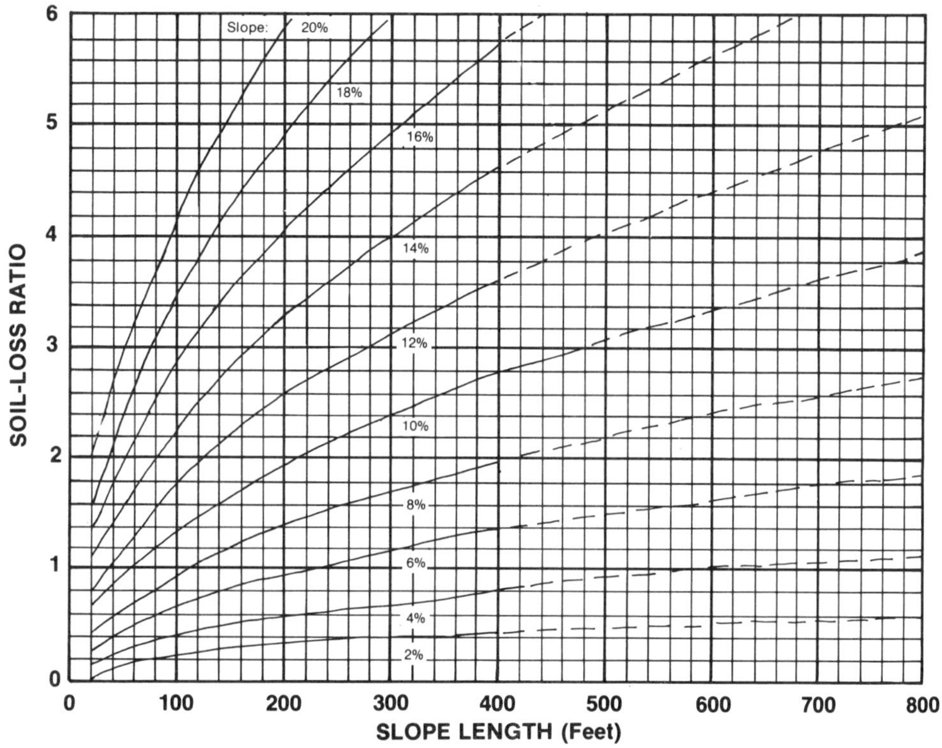

FIGURE 9. Slope effect chart (topographic factor, LS). (From Wischmeier, W. H. and Smith, D. D., Predicting runoff-erosion losses from cropland east of the Rocky Mountains, *U.S. Dep. Agric. Handb.* 181, 8, 1965.)

stand, quality of growth, and varies greatly in different seasonal periods. Different combinations of these variables will have different effects on soil loss.

This complex problem is handled in the Universal Soil Loss Prediction Equation by defining a factor C, which is the ratio of soil loss from land cropped under specified conditions to the corresponding loss from clean-tilled continuous fallow.[13] In order to account for the interaction between amount of cover and type of management, the cropping year or rotation cycle is divided into a series of crop stage periods[13] defined so that within each period, cover and management are approximately constant. The fraction of the annual erosion index (R) that normally occurs in that locality during each of the crop stages is determined. It is then multiplied by the soil loss ratio for that period (ratio of existing cover and management to fallow soil). The sum of products obtained for the crop stage periods in any 1 year is the C value for that particular crop in that system. Table 3 illustrates the range in C values for a variety of crops and management practices.[68] Note particularly the major influence of conservation tillage in reducing the value of C.

Erosion Control Measures

In addition to the influence of cover and management described above, other practices can have significant influence on soil erosion. These are sometimes called supporting practices and the principal ones include contour tillage, contour strip cropping, and terrace systems. They are designed to slow the runoff water and thus reduce the amount of soil it can carry. Contour tillage is the practice of tillage and planting on the contour. Contour strip cropping is the practice in which contour strips of sod are

Table 3
GENERALIZED VALUES OF THE COVER AND MANAGEMENT FACTOR, C, IN THE 37 STATES EAST OF THE ROCKY MOUNTAINS[a]

Line no.	Crop, rotation, and management[c]	Productivity level[b] (C value) High	Mod.
Base value: continuous fallow, tilled up and down slope		1.00	1.00
Corn			
1	C, RdR, fall TP, conv (1)	0.54	0.62
2	C, RdR, spring TP, conv (1)	0.50	0.59
3	C, RdL, fall TP, conv (1)	0.42	0.52
4	C, RdR, seeding, spring TP, conv (1)	0.40	0.49
5	C, RdL, standing, spring TP, conv (1)	0.38	0.48
6	C, fall shred stalks, spring TP, conv (1)	0.35	0.44
7	C(silage)-W(RdL, fall TP) (2)	0.31	0.35
8	C, RdL, fall chisel, spring disk, 40—30% rc (1)	0.24	0.30
9	C(silage), W seeding, no-till pl in c-k W (1)	0.20	0.24
10	C(RdL)-W(RdL, spring TP) (2)	0.20	0.28
11	C, fall shred stalks, chisel pl, 40—30% rc (1)	0.19	0.26
12	C—C—C—W—M, RdL, TP for C, disk for W (5)	0.17	0.23
13	C, RdL, strip till row zones, 55—40% rc (1)	0.16	0.24
14	C—C—C—W—M—M, RdL TP for C, disk for W (6)	0.14	0.20
15	C—C—W—M, RdL, TP for C, disk for W (4)	0.12	0.17
16	C, fall shred, no-till pl, 70—50% rc (1)	0.11	0.18
17	C—C—W—M—M, RdL TP for C, disk for W (5)	0.087	0.14
18	C—C—C—W—M, RdL, no-till pl 2nd & 3rd C(5)	0.076	0.13
19	C—C—W—M, RdL, no-till pl 2nd C (4)	0.068	0.11
20	C, no-till pl in c-k wheat, 90—70% rc (1)	0.062	0.14
21	C—C—C—W—M—M, no-till pl 2nd & 3rd C (6)	0.061	0.11
22	C—W—M, RdL, TP for C, disk for W (3)	0.055	0.095
23	C—C—W—M—M, RdL, no-till pl 2nd C (5)	0.051	0.094
24	C—W—M—M, RdL, TP for C, disk for W (4)	0.039	0.074
25	C—W—M—M—M, RdL, TP for C, disk for W (5)	0.032	0.061
26	C, no-till pl in c-k sod, 95—80% rc (1)	0.017	0.053
Cotton[d]			
27	Cot, conv (Western Plains) (1)	0.42	0.49
28	Cot. conv (South) (1)	0.34	0.40
Meadow			
29	Grass and legume mix	0.004	0.01
30	Alfalfa, lespedeza or Sericia	0.020	
31	Sweet clover	0.025	
Sorghum, grain (Western Plains)[d]			
32	RdL, spring TP, conv (1)	0.43	0.53
33	No-till pl in shredded 70—50% rc	0.11	0.18
Soybeans[d]			
34	B, RdL, spring TP, conv (1)	0.48	0.54
35	C—B, TP annually, conv (2)	0.43	0.51
36	B, no-till pl	0.22	0.28
37	C—B, no-till pl, fall shred C stalks (2)	0.18	0.22
Wheat			
38	W—F, fall TP after W (2)	0.38	
39	W—F, stubble mulch, 500 lb rc (2)	0.32	
40	W—F, stubble mulch, 1000 lb rc (2)	0.21	
41	Spring W, RdL, Sept. TP, conv (N & S Dak) (1)	0.23	
42	Winter W, RdL, Aug. TP, conv (Kans) (1)	0.19	
43	Spring W, stubble mulch, 750 lb rc (1)	0.15	

Table 3 (continued)
GENERALIZED VALUES OF THE COVER AND MANAGEMENT FACTOR, C, IN THE 37 STATES EAST OF THE ROCKY MOUNTAINS[a]

Line no.	Crop, rotation, and management[c]	Productivity level[b] (C value) High	Mod.
44	Spring W, stubble mulch, 1250 lb rc (1)	0.12	
45	Winter W, stubble mulch, 750 lb rc (1)	0.11	
46	Winter W, stubble mulch, 1250 lb rc (1)	0.10	
47	W—M, conv (2)	0.054	
48	W—M—M, conv (3)	0.026	
49	W—M—M—M, conv (4)	0.021	

Note: Abbreviations defined:
B — soybeans
C — corn
c-k — chemically killed
conv — conventional
cot — cotton
F — fallow
M — grass & legume hay
pl — plant
W — wheat
wc — winter cover
lbs rc — pounds of crop residue per acre remaining on surface after new crop seeding
% rc — percentage of soil surface covered by residue mulch after new crop seeding
70—50% rc — 70% cover for C values in first column; 50% for second column
RdR — residues (corn stover, straw, etc.) removed or burned
RdL — all residues left on field (on surface or incorporated)
TP — turn plowed (upper 5 or more in. of soil inverted, covering residues)

[a] This table is for illustrative purposes only and is not a complete list of cropping systems or potential practices. Values of C differ with rainfall pattern and planting dates. These generalized values show approximately the relative erosion-reducing effectiveness of various crop systems, but locationally derived C values should be used for conservation planning at the field level. Tables of local values are available from the Soil Conservation Service.

[b] High level is exemplified by long-term yield averages greater than 75 bu. corn or 3 tons grass-and-legume hay; or cotton management that regularly provides good stands and growth.

[c] Numbers in parentheses indicate number of years in the rotation cycle. No. (1) designates a continuous one-crop system

[d] Grain sorghum, soybeans, or cotton may be substituted for corn in lines 12, 14, 15, 17—19, 21—25 to estimate C values for sod-based rotations.

From Wischmeier, W. H. and Smith, D. D., *U.S. Dep. Agric. Agric Handb.*, 282, 23, 1965.

alternated with strips of row crop or small grain. Terracing is the practice of constructing earthen barriers on the contour so as to intercept runoff and either remove it by a channel with slight grade so as not to create serious erosion or to contain it until it soaks into the soil. The supporting practice factor is identified in the Universal Soil Loss Prediction Equation as P and is defined as the ratio of soil loss with contouring, strip cropping, or terracing to that with straight row, up-and-downhill culture.[13] Table 4[69] shows the influence of these practices on soil loss over several slope ranges.

Table 4
VALUES OF SUPPORT-PRACTICE FACTOR, P

Practice	Land slope (%)				
	1.1—2	2.1—7	7.1—12	12.1—18	18.1—24
			(Factor P)		
Contouring (P_c)	0.60	0.50	0.60	0.80	0.90
Contour strip cropping (P_{sc})					
R—R—M—M[a]	0.30	0.25	0.30	0.40	0.45
R—W—M—M	0.30	0.25	0.30	0.40	0.45
R—R—W—M	0.45	0.38	0.45	0.60	0.68
R—W	0.52	0.44	0.52	0.70	0.90
R—O	0.60	0.50	0.60	0.80	0.90
Contour listing or ridge planting (P_{cl})	0.30	0.25	0.30	0.40	0.45
Contour terracing (P_t)[b]	$0.6/\sqrt{n}$[c]	$0.5/\sqrt{n}$	$0.6/\sqrt{n}$	$0.8/\sqrt{n}$	$0.9/\sqrt{n}$
No support practice	1.0	1.0	1.0	1.0	1.0

[a] R = rowcrop, W = fall-seeded grain, O = spring-seeded grain, M = meadow. The crops are grown in rotation and so arranged on the field that rowcrop strips are always separated by a meadow or winter-grain strip.

[b] These P_t values estimate the amount of soil eroded to the terrace channels and are used for conservation planning. For prediction of off-field sediment, the P_t values are multiplied by 0.2

[c] n = number of approximately equal-length intervals into which the field slope is divided by the terraces. Tillage operations must be parallel to the terraces.

From Wischmeier, W. H., Control of Water Pollution from Cropland, Agric. Res. Serv., U.S. Department of Agriculture and Environmental Protection Agency, 1(3), 27, 1975.

Concept of Soil Conservation Programs

Soil conservation planning for cultivated land is based on predicting annual soil loss by the use of the Universal Soil Loss Prediction Equation.[13] This methodical procedure, based on statistical analyses of thousands of plot-years of data from 47 locations in 24 states, enables land management planners to estimate erosion rates for a wide range of climatic, soil, slope, crop, and management conditions and to select the combinations best suited to each particular situation.

The philosophy of conservation planning is to keep annual soil loss within tolerance limits (T value). In recent years erosion controls to improve water quality and the environment have also been stressed. The basis for T values, which range from 2 to 5 tons/acre/year, is that soils can sustain this amount of loss without losing productive potential.[13,22] Since erosion in excess of renewal rates reduces soil depth (maintenance of adequate soil depth being essential for good plant growth), shallow soils are assigned lower T values than deep soils.

In practice, A in the equation $A = RKLSCP$ is set equal to T to obtain the maximum CP-value that will keep expected annual soil loss within prescribed tolerance limits. It is the C and P portions of the equation that man can modify to keep soil erosion within tolerance limits. Generally, several alternative conditions will meet this requirement, so the land-management planner along with the farmer can then select the most feasible. This approach forms the basis for conservation farm planning presently used by the Soil Conservation Service of the U.S. Department of Agriculture.

Erosion usually can be controlled by practices that minimize raindrop impact on the soil surface and weaken the erosive forces of the runoff by reducing its velocity and channelization. In many situations, erosion can be controlled with agronimic practices that improve crop residue management, cropping sequences, tillage and seeding methods, and timing of field operations. Generally, farming parallel to the field contours will further reduce erosion. However, in some situations (excessive slope length and runoff concentration) they must be supported by practices such as terraces, diversions, contour strip cropping, waterways, and control structures. Table 5[70] lists the principal types of erosion-control practices and some of their favorable and unfavorable features.

CURRENT RESEARCH PROGRAMS TO FURTHER IMPROVE SOIL LOSS PREDICTION AND CONTROL OF SOIL EROSION

Research is being continued to further refine the factors presently used in soil loss prediction. Through further field and laboratory testing and through modeling efforts, progress is being made in understanding the "mechanics" of the erosion process.[23-25] The overall erosion process is very complex and current researchers are attempting to separate the process into components that can be explained in a physical rather than empirical sense. For example, present models separate "rill" and "inter rill" erosion; they separate raindrop detachment of soil particles and transport functions of runoff; they can account for aggradation and deposition phases of sediment transport; they reflect the interactive behavior of the upland-stream channel system. Although these models are still quite complicated and are still in need of more accurate input parameters, they offer great potential.

Success has already been achieved in the development of a runoff equation that greatly improves precision of soil loss estimates when included in a soil-loss-predictive equation for single storm events. There is current research underway to determine the optimum treatment and placement of very limited residues and the optimum amount and type of associated tillage required to minimize erosion in the absence of adequate cover. Critical slope-length limits for effectiveness of partial mulch covers and favorable microtopographies provided by conservation tillage practices are being studied. Expanded studies to determine more accurately the soil physical and chemical factors affecting erodibility over a broader range of soils is underway. Climatic variables are being sought where the standard erosion-index of rainfall does not hold up. More accurate estimates of the influence of complex and extremely long, gentle slopes are being attempted. Finally, major emphasis is underway to understand better the relationship of soil erosion to sedimentation and its influence on water quality. These research efforts involve many preliminary and secondary investigations that are not enumerated.

Table 5
PRINCIPAL TYPES OF CROPLAND EROSION CONTROL PRACTICES AND THEIR HIGHLIGHTS

No.	Erosion control practice	Practice highlights	Ref.
E 1	No-till plant in prior-crop residues	Most effective in dormant grass or small grain; highly effective in crop residues; minimizes spring sediment surges and provides year-round control; reduces man, machine and fuel requirements; delays soil warming and drying; requires more pesticides and nitrogen; limits fertilizer- and pesticide placement options; some climatic and soil restrictions	26—31
E 2	Conservation tillage	Includes a variety of no-plow systems that retain some of the residues on the surface; more widely adaptable but somewhat less effective than E 1; advantages and disadvantages generally same as E 1 but to lesser degree	26, 28—36
E 3	Sod-based rotations	Good meadows lose virtually no soil and reduce erosion from succeeding crops; total soil loss greatly reduced but losses unequally distributed over rotation cycles:aid in control of some diseases and pests; more fertilizer-placement options; less realized income from hay years; grater potential transport of water soluble P; some climatic restrictions	13, 19, 21, 37—40
E 4	Meadowless rotations	Aid in disease and pest control; may provide more continuous soil protection than one-crop systems; much less effective than E 3	13, 19, 41
E 5	Winter cover crops	Reduce winter erosion where corn stover has been removed and after low-residue crops; provide good base for slot-planting next crop; usually no advantage over heavy cover of chopped stalk or straw; may reduce leaching of nitrate; water use by winter cover may reduce yield of cash crop	13, 19
E 6	Improved soil fertility	Can substantially reduce erosion hazards as well as increase crop yields	13, 19, 42, 43
E 7	Timing of field operations	Fall plowing facilitates more timely planting in wet springs, but it greatly increases winter and early spring erosion hazards; optimum timing of spring operations can reduce erosion and increase yields	19
E 8	Plow-plant systems	Rough, cloddy surface increases infiltration and reduces erosion; much les effective than E 1 and E 2 when long rain periods occur; seedling stands may be poor when moisture conditions are less than optimum. Mulch effect is lost by plowing	13, 19, 44, 45
E 9	Contouring	Can reduce average soil loss by 50% on moderate slopes, but less on steep slopes; loses effectiveness if rows break over; must be supported by terraces on long slopes; soil, climatic, and topographic limitations; not compatible with use of large farming equipment on many topographies. Does not affect fertilizer and pesticide rates	13, 19, 46—48
E 10	Graded rows	Similar to contouring but less susceptible to row breakovers	19, 47
E 11	Contour strip cropping	Rowcrop and hay in alternate 50- to 100-foot strips reduce soil loss to about 50% of that with the same rotation contoured only; fall seeded grain in lieu of meadow about half as effective; alternating corn and spring grain not effective; area must be suitable for across-slope farming and establishment of rotation meadows; favorable and unfavorable features similar to E 3 and E 9	13, 19, 46—48

Table 5 (continued)
PRINCIPAL TYPES OF CROPLAND EROSION CONTROL PRACTICES AND THEIR HIGHLIGHTS

No.	Erosion control practice	Practice highlights	Ref.
E 12	Terraces	Support contouring and agronomic practices by reducing effective slope length and runoff concentration; reduce erosion and conserve soil moisture; facilitate more intensive cropping; conventional gradient terraces often incompatible with use of large equipment, but new designs have alleviated this problem; substantial initial cost and some maintenance costs	13, 19, 46, 47, 49—52
E 13	Grassed outlets	Facilitate drainge of graded rows and terrace channels with minimal erosion; involve establishment and maintenance costs and may interfere with use of large implements	13, 19, 46, 48
E 14	Ridge planting	Earlier warming and drying of row zone; reduces erosion by concentrating runoff flow in mulch-covered furrows; most effective when rows are across slope	19, 29, 48, 53
E 15	Contour listing	Minimizes row breakover; can reduce annual soil loss by 50%; loses effectiveness with post-emergence corn cultivation; disadvantages same as E 9	13, 19, 47, 48, 54
E 16	Change in land use	Sometimes the only solution. Well-managed permanent grass or woodland effective where other control practices are inadequate; lost acreage can be compensated for by more intensive use of less erodible land	19, 47
E 17	Other practices	Contour furrows, diversions, subsurface drainage, land forming, closer row spacing, etc.	19, 47, 55, 56

From Wischmeier, W. H., Control of Water Pollution from Cropland, Agric. Res. Serv., U.S. Department of Agriculture and Environmental Protection Agency, 1(4), 63, 1975.

REFERENCES

1. *Resource Conservation Glossary,* Soil Conservation Society of America, Ankeny, Iowa, 1976, 52.
2. **Holt, R. F., Johnson, H. P., and McDowell, L. L.,** Surface water quality, in *Proc. Natl. Conserv. Tillage Conf.,* Soil Conserv. Soc. Am., Ankeny, Iowa, 1973, 141.
3. Soil and Water Conservation Needs — A National Inventory, *U.S. Department of Agriculture, Misc. Publ.,* 1965, 971.
4. **Kimberlin, L. W. and Moldenhauer, W. C.,** Predicting soil erosion, in *Proc. Natl. Symp. Soil Erosion and Sediment. by Water,* Am. Soc. Agric. Eng., St. Joseph, Mo., 1977, 31.
5. Basic Statistics — National Inventory of Soil and Water Conservation Needs, Stat. Bull. No. 461, U.S. Department of Agriculture, Washington, D.C., 1971.
6. **Dideriksen, R. I. and Grunewald, A. R.,** What is happening to our soils resources: the current inventory of rural land, in Proc. 29th Annu. Mtg. Soil Conserv. Soc. Am., Ankeny, Iowa, 1974, 15.
7. **Stamey, W. L. and Smith, R. M.,** A conservation definition of erosion tolerance, *Soil Sci.,* 97, 183, 1964.
8. **Wadleigh, C. H.,** Wastes in relation to agriculture and forestry, *U.S. Dept. Agric. Misc. Publ.,* 1065, 112, 1968.
9. **Ellison, W. D.,** Soil erosion studies. I, *Agric. Eng.,* 28, 145, 1947.
10. **Baur, A. J.,** Soil and water conservation glossary, *J. Soil Water Conserv.,* 7, 41, 1952.
11. **Beasley, R. P.,** Water erosion, in *Erosion and Sediment Pollution Control,* Iowa State University Press, Ames, 1972, chap. 2.
12. **Bauer, L. D.,** *Soil Physics,* John Wiley & Sons, New York, 370, 1940.

13. **Wischmeier, W. H. and Smith, D. D.**, Predicting rainfall erosion losses, from cropland east of the Rocky Mountains, *U.S. Dep. Agric. Handb.*, 282, 1965.
14. **Wischmeier, W. H. and Smith, D. D.**, Rainfall energy and its relationship to soil loss, *Trans. Am. Geophys. Unon*, 39, 285, 1958.
15. **Wischmeier, W. H.**, A rainfall erosion index for a universal soil-loss prediction equation, *Soil Sci. Soc. Am. Proc.*, 23, 246, 1959.
16. **Wischmeier, W. H. and Mannering, J. V.**, Relation of soil properties to its erodibility, *Soil Sci. Soc. Am, Proc.*, 33, 131, 1969.
17. **Wischmeier, W. H., Johnson, C. B., and Cross, B. V.**, A soil erodibility nomograph for farmland and construction sites, *J. Soil Water Conser.*, 26, 189, 1971.
18. **Wischmeier, W. H.**, Relation of field-plot runoff to management and physical factors, *Soil Sci. Soc. Am. Proc.*, 30, 272, 1966.
19. **Wischmeier, W. H.**, Cropland erosion and sedimentation, in *Control of Water Pollution from Cropland*, Agric. Res. Serv., U.S. Department of Agriculture, and Environmental Protection Agency, Washington, D.C., 2, 31, 1976.
20. **Foster, G. R. and Wischmeier, W. H.**, Evaluating irregular slopes for soil loss prediction, *Trans. Am. Soc. Agric. Eng.*, 17, 305, 1974.
21. **Mannering, J. V. and Fenster, C. M.**, Vegetative water erosion control for agricultural areas, in *Proc. Natl. Symp. Soil Erosion Sediment. by Water*, Am. Soc. Agric. Eng., St. Joseph, Mo., 1977, 91.
22. **Smith, R. M. and Stamey, W. L.**, Determining the range of tolerable erosion, *Soil Sci.*, 100, 414, 1965.
23. **Foster, G. R. Meyer, L. D., and Onstad, C. A.**, A runoff erosivity factor and variable slope length exponents for soil loss estimates, *Trans. Am. Soc. Agric. Eng.*, in press.
24. **Onstad, C. A. and Foster, G. R.**, Erosion modeling on a watershed, *Trans. Am. Soc. Agric. Eng.*, 18, 288, 1975.
25. **David, W. P. and Beer, C. E.**, Simulation of soil erosion, *Trans. Am. Soc. Agric. Eng.*, 18, 126, 1975.
26. **Wischmeier, W. H.**, Conservation tillage to control water erosion, in *Conserv. Tillage, Proc. Natl. Conf.*, Soil Conserv. Soc. Am., Ankeny, Iowa, 1973, 133.
27. **Harrold, L. L. and Edwards, W. M.**, A severe rainstorm test of no-till corn, *J. Soil Water Conserv.*, 27, 30, 1972.
28. **Oschwald, W. and Siemens, J.**, Conservation tillage: a perspective, *SM-30 Agron. Facts*, University of Illinois, Urbana, 1976.
29. **Griffith, D. R., Mannering, J. V., and Moldenhauer, W. C.**, Conservation tillage in the eastern Corn Belt, *J. Soil Water Conserv.*, 32, 20, 1977.
30. **Van Doren, D. M., Jr. and Triplett, G. B.**, Mulch and tillage relations in corn culture, *Soil Sci. Soc. Am. Proc.*, 37, 766, 1973.
31. **Jones, J. H., Moody, J. E., and Lillard, J. H.**, Effects of tillage, no tillage, and mulch on soil water and plant growth, *Agron. J.*, 61, 719, 1969.
32. **Witmuss, H. D., Triplett, G. B., Jr., and Grib, B. W.**, Concepts of conservation tillage systems using surface mulches, in *Proc. Natl. Conserv. Tillage Conf.*, Soil Conserv. Soc. Am., Ankeny, Iowa, 1973, 5.
33. **Adams, J. E.**, Influence of mulches on runoff, erosion, and soil moisture depletion, *Soil Sci. Soc. of Am. Proc.* 30, 110, 1966.
34. **Larson, W. E. and Gill, W. R.**, Soil physical parameters for designing new tillage systems, in *Proc. Natl. Conserv. Tillage Conf.*, Ankeny, Iowa, 1973, 13.
35. **Unger, P. W., Allen, R. R., and Wiese, A. F.**, Tillage and herbicides for surface residue maintenance, weed control, and water conservation, *J. Soil Water Conserv.*, 26, 147, 1971.
36. **Fenster, C. R.**, Conservation tillage in the Northern Plains, *J. Soil Water Conserv.*, 32, 37, 1977.
37. **Mannering, J. V., Johnson, L. C., Meyer, L. D., and Jones, B. A.**, The erosion-control, effectiveness of rotation meadows, *J. Soil Water Conserv.*, 19, 91, 1964.
38. **Mannering, J. V., Meyer, L. D., and Johnson, C. B.**, Effect of cropping intensity on erosion and infiltration, *Agron. J.*, 60, 206, 1968.
39. **Wischmeier, W. H.**, Erosion from corn after meadow depends on quality of sod crops, *Crops Soils*, 12, 25, 1960.
40. **Browning, G. M., Parish, C. L., and Glas, J. A.**, A method for determining the use and limitation of rotation and conservation practices in control of soil-erosion practices in Iowa, *Soil Sci. Soc. Am. Proc.*, 23, 246, 1947.
41. **Van Doren, D. M., Jr., Triplett, G. B., Jr., and Henry, J. E.**, Long-term influence of tillage, rotation, and soil on corn yield, *Ohio Report*, 80, 1975.

42. **Moldenhauer, W. C., Wischmeier, W. H., and Parker, D. T.,** The influence of crop management of runoff, erosion, and soil properties of a Marshall silty clay loam, *Soil Sci. Soc. Am. Proc.*, 31, 541, 1967.
43. **Wischmeier, W. H.,** Cropping-management factor evaluations for a universal soil-loss equation, *Soil Sci. Soc. Am. Proc.*, 24, 322, 1960.
44. **Mannering, J. V., Meyer, L. D., and Johnson, C. B.,** Infiltration and erosion as affected by minimum tillage for corn, *Soil Sci. Soc. Am. Proc.*, 30, 101, 1966.
45. **Harrold, L. L.,** The watershed hydrology of plow-plant corn, *J. Soil Water Conserv.*, 15, 183, 1960.
46. **Schwab, G. O., Frevert, R. K., Edminster, T. W., and Barnes, K. K.,** in *Soil Water Conserv. and Eng.*, 2nd ed., John Wiley & Sons, New York, 1966.
47. **Meyer, L. D. and Mannering, J. V.,** Tillage and land modification for water erosion control, in *Proc. Tillage for Greater Crop Prod.*, Am. Soc. of Agric. Eng., St. Joseph, Mo., 1967, 58.
48. **Beasley, R. P.,** Planning agricultural systems, in *Erosion and Sediment Pollution Control,* Iowa State University Press., Ames, 1972, chap. 11.
49. **Beasley, R. P.,** Terraces, in *Erosion and Sediment Pollution Control,* Iowa State University Press, Ames, 1972, chap. 7.
50. **Jacobson, P.,** Improvements in bench terraces, *Trans. Am. Soc. Agric. Eng.*, 11, No. 4, 1968.
51. **Phillips, R. L. and Beauchamp, K. W.,** Design, layout, construction, and maintenance of terraces, *Am. Soc. Agric. Eng., Yearbook,* 1972.
52. **Jeschke, J. L., Moe, R. C., and Carmak, W. J.,** Mechanical erosion control for agricultural areas, in *Proc. Natl. Symp. Soil Erosion Sediment. Water,* Am. Soc. Agric. Eng., St. Joseph, Mo., 1977, 107.
53. **Mannering, J. V. and Burwell, R. E.,** Tillage methods to reduce runoff and erosion in the Corn Belt, *Agric. Inf. Bull. 330,* U.S. Department of Agriculture, Washington, D.C., 14, 1968.
54. **Moldenhauer, W. C. and Wischmeier, W. H.,** Soil and water losses and infiltration rates on Ida silt loam as influenced by cropping systems, tillage practices, and rainfall characteristics, *Soil Sci. Soc. Am. Proc.*, 24, 409, 1960.
55. **Beasley, R. P.,** Cross-slope channels, diversions, and basins, in *Erosion and Sediment Pollution Control,* Iowa State University Press, Ames, 1972, chap. 8.
56. **Mannering, J. V. and Johnson, C. B.,** Effect of crop row spacing on erosion and infiltration, *Agron. J.*, 61, 902, 1969.
57. **Wischmeier, W. H.,** Control of Water Pollution from Cropland, Agric. Res. Serv., U.S. Department of Agriculture and Environmental Protection Agency, Washington, D.C., 1(3), 6, 1975.
58. **Wischmeier, W. H.,** Control of Water Pollution from Cropland, Agric. Res. Serv., U.S. Department of Agriculture and Environmental Protection Agency, Washington, D.C., 1(3), 8, 1975.
59. **Wischmeier, W. H.,** Control of Water Pollution from Cropland, Agric. Res. Serv., U.S. Department of Agriculture and Environmental Protection Agency, Washington, D.C., 1(3), 12, 1975.
60. **Wischmeier, W. H.,** Control of Water Pollution from Cropland, Agric. Res. Serv., U.S. Department of Agriculture and Environmental Protection Agency, Washington, D.C., 1(3), 13, 1975.
61. **Wischmeier, W. H.,** Control of Water Pollution from Cropland, Agric. Res. Serv., U.S. Department of Agriculture and Environmental Protection Agency, Washington, D.C., 1(3), 15, 1975.
62. **Wischmeier, W. H.,** Control of Water Pollution from Cropland, Agric. Res. Serv., U.S. Department of Agriculture and Environmental Protection Agency, Washington, D.C., 1(3), 17, 1975.
63. **Wischmeier, W. H. and Smith, D. D.,** Predicting rainfall-erosion losses from cropland, east of the Rocky Mountains, *U.S. Dep. Agric. Agric. Handb.*, 282, 18, 28, 1965.
64. **Wischmeier, W. H.,** Control of Water Pollution from Cropland, Agric. Res. Ser., U.S. Department of Agriculture and Environmental Protection Agency, Washington, D.C., 1(3), 19, 1975.
65. **Wischmeier, W. H.,** Control of Water Pollution from Cropland, Agric. Res. Serv., U.S. Department of Agriculture and Environmental Protection Agency, Washington, D.C., 1(3), 20, 1975.
66. **Wischmeier, W. H.,** Control of Water Pollution from Cropland, Agric. Res. Serv., U.S. Department of Agriculture and Environmental Protection Agency, Washington, D.C. 1(3), 22, 1975.
67. **Wischmeier, W. H. and Smith, D. D.,** Predicting runoff-erosion losses from cropland east of the Rocky Mountains, *U.S. Dep. Agric. Agric. Handb.*, 282, 8, 1965.
68. **Wischmeier, W. H.,** Control of Water Pollution from Cropland, Agric. Res. Ser., U.S. Department of Agriculture and Environmental Protection Agency, Washington, D.C., 1(3), 23, 1975.
69. **Wischmeier, W. H.,** Control of Water Pollution from Cropland, Agric. Res. Serv., U.S. Department of Agriculture and Environmental Protection Agency, Washington, D.C., 1(3), 27, 1975.
70. **Wischmeier, W. H.,** Control of Water Pollution from Cropland, Agric. Res. Serv. U.S. Department of Agriculture and Environmental Protection Agency, Washington, D.C. 1(4), 63, 1975.

SOIL AND WATER MANAGEMENT AND CONSERVATION: WIND EROSION*

E. L. Skidmore

THE WIND EROSION PROBLEM

Extensive aeolian deposits from past geologic eras give evidence that wind erosion is not a recent phenomenon. In recent years satellite photographs have revealed much about the origin and extent of duststorms.[1]

General areas most susceptible to wind erosion on agricultural land include much of North Africa and the Near East, parts of southern and eastern Asia, Australia, and southern South America, and the semiarid and arid portions of North America.[2] In addition, such agricultural areas as the Siberian Plain and others in the U.S.S.R. have a potential for wind erosion.

Wind erosion is the dominant problem on about 30 million ha of land in the U.S.[3] About 2 million ha are moderately to severely damaged each year. Wind erosion can be a problem whenever the following soil, vegetative, and climatic conditions exist: (1) the soil is loose, dry, and reasonably finely divided; (2) the soil surface is smooth and vegetative cover is absent or sparse; (3) the field is sufficiently large; and (4) the wind is strong enough to move soil. Those conditions more often prevail in semiarid areas when precipitation is inadequate, but they are sometimes present in subhumid and humid areas, especially on noncohesive soils.

Soil erosion by wind occurs when wind exerts enough force on the surface of the ground that the most easily detachable soil particles or sand grains dislodge and are transported by the wind. Bagnold[4] described the initial motion: "surface grains, previously at rest, began to be rolled along the surface by the direct pressure of the wind . . . gathered sufficient speed to start bouncing off the ground." Others[5,6] observed that as the "fluid threshold" was approached, some particles began to vibrate or rock back and forth. Erodible particles vibrated with increasing intensity as wind speed increased and then left the surface instantaneously as if ejected.

Particles 0.1 to 0.5 mm in diameter rise almost vertically, travel 10 to 15 times their height of rise, and then return to the surface with an angle of descent of about 6 to 12° from the horizontal.[7-9] On striking the surface, the particles either rebound and continue their movement by striking and then rebounding from the surface, which is called saltation, or they impart most of their energy by striking other particles, causing the particles struck to rise upward or roll along the surface.

The rolling or sliding of larger particles with energy from saltating particles is called creep. Bagnold[4] observed that at low wind speeds the grains move in jerks, a few millimeters at a time, but as the wind speed increases, the distance particles move increases and more particles are set in motion until, in high winds, the whole surface appears to be creeping forward.

Particles smaller than about 0.1 mm may enter suspension and be carried to great heights by eddies of the erosive winds. The impact of particles in saltation usually starts movement of these fine particles. Although most of the soil eroded by wind is moved by saltation and surface creep, that moved by suspension is the most spectacular and easily recognized from a distance. Bennett[10] estimated that a single dust storm

* Contribution from the Science and Education Administration, Agricultural Research, U.S. Department of Agriculture, in cooperation with the Kansas Agricultural Experiment Station, Department of Agronomy Contribution No. 80-185-R.

occurring on May 12, 1934 carried 272 million metric tons of topsoil out of the Great Plains. Hagen and Woodruff[11] estimated that eroding lands of the Great Plains contributed 220 and 70 million metric tons of dust per year to the atmosphere in the 1950s and 1960s, respectively.

Wind erosion sorts many soils. The fine and porous particles are removed, leaving behind the coarser and denser particles,[12-14] Of those removed, the coarsest particles usually end up in a soil drift and the remainder enter suspension to be transported, often great distances, before they are deposited.[15] Wind erosion sometimes virtually removes the entire surface soil.[12,16,17] This nonselective removal by wind is associated primarily with loess sorted and deposited from the atmosphere during past geologic eras.

Wind erosion physically removes from the field the most fertile portion of the soil and, therefore, lowers productivity.[18] Dust obscures visibility and pollutes the air, causes traffic hazards, fouls machinery, and imperils animal and human health.

Blowing soil fills road ditches, reduces seedling survival and growth, lowers the marketability of vegetable crops like asparagus, green beans, and lettuce, increases susceptibility of plants to certain types of stress including disease, and contributes to transmission of some plant pathogens.[19-21]

FACTORS AFFECTING SOIL EROSION BY WIND

Studies to understand the mechanics of the wind erosion process, to identify major factors influencing wind erosion, and to develop wind erosion control methods led to the development of a wind erosion equation.[7,22] The equation here is used as an outline to discuss the major factors affecting wind erosion. The general functional relationship between the independent variable, E (the potential average annual soil loss), and the equivalent variables or major factors is $E = f(I, K, C, L, V)$, where I is a soil erodibility index, K is a soil ridge-roughness factor, C is a climatic factor, L is field length along the prevailing wind erosion direction, and V is equivalent vegetative cover.

Soils

Soils vary greatly in the ease with which they are detached and transported by wind (erodibility). The noncohesive sandy soils have a very fragile structure; they are essentially single grains and, as such, are already detached. The particles of the finer-textured cohesive soils form into compound particles or aggregates of various size and stability.

Chepil[23] determined from wind tunnel studies that mineral soil aggregates larger than greater than 0.84 mm, as determined by dry sieving, is used to indicate erodibility of soil by wind. The erodibility of soils with different percentages of nonerodible fractions exceeding 0.84 mm is listed in Table 1.

Percentages of dry soil fractions greater than 0.84 mm are obtained by standard dry sieving in the field or the laboratory.[24] In practice, to avoid sampling in the field and sieving, soil erodibility is often estimated by grouping soils according to predominant soil textural class.

Actual erodibility of most soils is extremely dynamic and varies seasonally, annually, and with management practices. In a study on the effects of season on soil erodibility, Chepil[25] found that erodibility was higher in the spring than in the previous fall in all cases where the soil had been intermittently moist during the winter, but the increases were not of the same magnitude in all soils. The greatest increase in erodibility from fall to spring occurred in the finest-textured soils, the least in the coarsest. Sandy loam was highly erodible in both fall and spring. Clay was least erodible in the fall but was

Table 1
SOIL ERODIBILITY I FOR SOILS WITH DIFFERENT PERCENTAGES OF NONERODIBLE FRACTIONS AS DETERMINED BY STANDARD DRY SIEVING

Percentage of dry soil fractions > 0.84 mm

	Units (metric tons/ha)									
Tens	0	1	2	3	4	5	6	7	8	9
0	—	695	560	493	437	404	381	359	336	314
10	300	294	287	280	271	262	253	244	238	228
20	220	213	206	202	197	193	186	182	177	170
30	166	161	159	155	150	146	141	139	134	130
40	126	121	117	114	112	108	105	101	96	92
50	85	80	75	70	65	61	28	54	52	49
60	47	45	43	40	38	36	36	34	31	29
70	27	25	22	18	16	13	9	7	7	4
80	4	—	—	—	—	—	—	—	—	—

From Woodruff, N. P. and Siddoway, F. H., *Soil Sci. Soc. Am. Proc.*, 29, 602, 1965. With permission.

about as highly erodible as sandy loam in the spring. The intermediate-textured soils had an intermediate erodibility in both spring and fall.

Ridge Roughness

Forming soil having nonerodible clods and aggregates into ridges reduces erosion.[26] The experimentally derived relationship between relative quantity of eroded material and ridge roughness shows that ridging may reduce wind erosion up to 50%.[22]

The soil ridge-roughness factor K as influenced by ridge spacing and ridge height is given in Table 2. Ridge-spacing combinations that yield soil ridge-roughness factors of 0.5 and 0.6 approximate ridged surfaces; 0.7 and 0.8, semiridged; and 0.9 and 1.0, smooth surfaces. The Soil Conservation Service (SCS) evaluates fields as smooth, semi-ridged, or ridged and then assigns 1.0, 0.75, and 0.5, respectively, as soil ridge-roughness factors.[27]

Climate

Before wind erosion can occur, the wind must exert a shear stress on the ground surface that is greater than the forces tending to keep the particles at rest. When soil particles are loose and dry, the minimum or threshold wind speed required to initiate soil movement is about 5.0 m/sec at 30 cm height.[15,28,29] Several investigators[4,30,31] found that when wind speed was greater than that required to barely move the soil, the rate of soil movement was directly proportional to the friction velocity cubed. The friction velocity squared is directly proportional to the vertical flux of horizontal momentum or surface shear stress.

Water in the soil forms a cohesive bonding between particles. The force between the soil particles must be overcome by the force of the wind before erosion can occur. Chepil[32] found the resistance due to cohesion (γ) of the adsorbed water films was proportional to the square of the water content:

$$\gamma = 6\left(\frac{\text{soil water content}}{\text{soil water content at 15 bars suction}}\right)^2$$

Table 2
SOIL RIDGE ROUGHNESS FACTOR K

Ridge spacing (cm)	Ridge height (cm)														
	1	2	4	6	8	10	12	14	16	18	20	22	24	26	28	30
1	0.6	0.6	0.8													
4	0.9	0.6	0.6	0.8												
8	0.9	0.7	0.5	0.6	0.8											
12	1.0	0.8	0.5	0.5	0.7	0.8										
16	1.0	0.9	0.6	0.5	0.6	0.8										
20	1.0	0.9	0.6	0.5	0.5	0.6	0.8									
24	1.0	0.9	0.6	0.5	0.5	0.6	0.7	0.8								
28	1.0	0.9	0.7	0.5	0.5	0.5	0.7	0.8								
32	1.0	0.9	0.7	0.5	0.5	0.5	0.6	0.7	0.8							
36	1.0	0.9	0.7	0.6	0.5	0.5	0.6	0.7	0.8							
40	1.0	0.9	0.8	0.6	0.5	0.5	0.5	0.6	0.8							
44	1.0	0.9	0.8	0.6	0.5	0.5	0.5	0.6	0.7	0.8						
48	1.0	1.0	0.8	0.6	0.5	0.5	0.5	0.6	0.7	0.8						
52	1.0	1.0	0.8	0.6	0.5	0.5	0.5	0.5	0.6	0.8						
56	1.0	1.0	0.8	0.6	0.5	0.5	0.5	0.5	0.6	0.7	0.8					
60	1.0	1.0	0.8	0.6	0.6	0.5	0.5	0.5	0.6	0.7	0.8					
64	1.0	1.0	0.9	0.7	0.6	0.5	0.5	0.5	0.6	0.6	0.6	0.8				
68	1.0	1.0	0.9	0.7	0.6	0.5	0.5	0.5	0.5	0.6	0.7	0.8				
72	1.0	1.0	0.9	0.7	0.6	0.5	0.5	0.5	0.5	0.6	0.7	0.8				
76	1.0	1.0	0.9	0.7	0.6	0.5	0.5	0.5	0.5	0.6	0.7	0.8				
80	1.0	1.0	0.9	0.7	0.6	0.5	0.5	0.5	0.5	0.6	0.6	0.7	0.8			
84	1.0	1.0	0.9	0.7	0.6	0.5	0.5	0.5	0.5	0.6	0.6	0.7	0.8			
88	1.0	1.0	0.9	0.8	0.6	0.5	0.5	0.5	0.5	0.5	0.6	0.7	0.8			
92	1.0	1.0	0.9	0.8	0.6	0.5	0.5	0.5	0.5	0.5	0.6	0.7	0.8			
96	1.0	1.0	0.9	0.8	0.6	0.6	0.5	0.5	0.5	0.5	0.6	0.6	0.7	0.8		
100	1.0	1.0	0.9	0.8	0.6	0.6	0.5	0.5	0.5	0.5	0.6	0.6	0.7	0.8		
104	1.0	1.0	0.9	0.8	0.6	0.6	0.5	0.5	0.5	0.5	0.6	0.6	0.7	0.8		
108	1.0	1.0	0.9	0.8	0.6	0.6	0.5	0.5	0.5	0.5	0.5	0.6	0.7	0.8		
112	1.0	1.0	0.9	0.8	0.7	0.6	0.5	0.5	0.5	0.5	0.5	0.6	0.7	0.7	0.8	
116	1.0	1.0	0.9	0.8	0.7	0.6	0.5	0.5	0.5	0.5	0.5	0.6	0.6	0.7	0.8	
120	1.0	1.0	0.9	0.8	0.7	0.6	0.5	0.5	0.5	0.5	0.5	0.6	0.6	0.7	0.8	

On the basis that erodibility of a soil varies inversely with the equivalent moisture in the surface soil particles and is proportional to wind speed cubed, Chepil et al.[33] proposed a wind erosion climatic factor, C. For example, the wind erosion climatic factor relative to Garden City, Kan. is expressed as follows:

$$C = \frac{1}{2.9} \frac{U^3}{(P\text{-}E)^2}$$

where U is mean annual wind speed for a standard height of 9.1 m (30 ft), P-E is moisture index, and 2.9 is the approximate average value of $U^3/(P\text{-}E)^2$ for Garden City, Kan. The annual climatic factor for much of the U.S. is shown in Figure 1.

Field Width

Soil movement on an eroding field surrounded by stubble field strips begins with zero on the windward side and increases with distance downwind.[34] The cumulative rate of soil movement with distance away from the windward edge of eroding fields was the main cause of increasing abrasion and gradual decrease in surface roughness along the direction of wind.[35] Chepil[35] called the increase in rate of soil flow with distance downwind "avalanching".

If the fields are large enough, the rate of particle flow reaches a maximum that a wind of a given velocity can carry. Beyond that point the rate of flow remains essentially constant.[36]

Vegetative Cover

Vegetation, both living and as residue from harvested crops, protects the soil against wind erosion. Standing crop residues provide nonerodible elements that absorb much of the shear stress in the boundary layer. When vegetation and crop residues are sufficiently high and dense to prevent the intervening soil surface drag from exceeding the threshold drag, soil will not erode. Flattened stubble, though not so effective as standing, also protects the soil from wind erosion.[37]

Studies[37-39] to quantify specific properties of vegetative covers influencing wind erosion led to the relationship developed by Woodruff and Siddoway[22] showing the influence of an equivalent vegetative cover of small grain and sorghum stubble for various orientations and heights.

Research efforts have continued to evaluate the protective role of additional crops,[40,42] range grasses,[41] feedlot manure,[43] and the protective requirements of equivalent residue needed to control wind erosion.[44,45]

WIND EROSION CONSERVATION PROGRAMS

Soil conservation planning for wind erosion control is based on predicting potential average annual soil loss with the wind erosion equation.[22] When the potential soil loss is greater than the considered "tolerable soil loss," the equation can be used to determine the field conditions (soil cloddiness, roughness, vegetative cover, sheltering by barriers, or width and orientation of field) necessary to reduce potential erosion to an acceptable level.

The SCS has used the Woodruff and Siddoway[22] equation extensively to plan wind erosion control practices and to determine crop tolerance to wind erosion.[20,21,27,46] The equation also is a useful guide to wind erosion control principles.[47-49] Other uses of the equation include: (1) determining spacing for barriers in narrow strip-barrier systems,[50] (2) estimating fugitive dust emissions from agricultural and subdivision lands,[51,52] (3) predicting horizontal soil fluxes for comparison with vertical aerosol

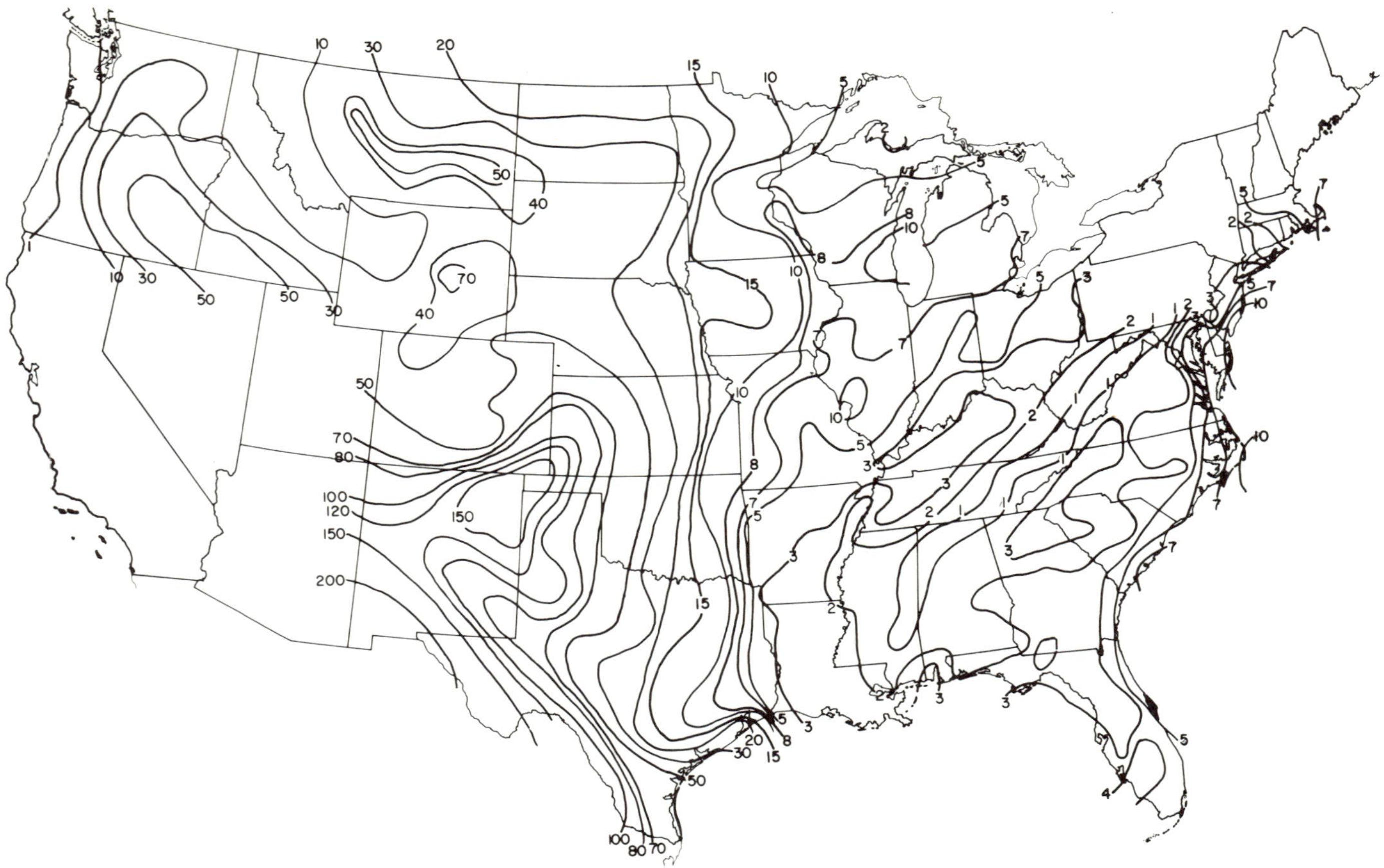

FIGURE 1. Annual climatic factor C. (Original drawing April 17, 1968, by D. V. Armbrust. Arkansas, Iowa, Kentucky, Louisiana, Tennessee, and West Virginia added November 24, 1971, by N. P. Woodruff.)

fluxes,[53] (4) estimating effects of wind erosion on productivity,[54,55] (5) evaluating stubble requirements in field strips to trap windblown soil,[56] and (6) delineating cropland where residues might be removed without exposing the soil to wind erosion.[45]

Those not familiar with the literature on conservation planning for wind erosion control may wish to consult the following references. These include reviews,[7,57] wind erosion equation,[22] information helpful in using the wind erosion equation,[58] use of computer to solve the wind erosion equation,[59,60] and a guide to wind erosion contol practices.[49]

Example

Suppose one wished to know the average annual potential soil loss from a field 400 m wide with 1000 kg/ha of flat wheat stubble in northwestern Kansas. The field has ridges 76 cm apart and 10 cm high.

Sieving showed that 25% of the soil aggregates were greater than 0.84 mm. Table 1 shows that soil erodibility I is 193 metric tons/ha. Table 2 shows that for 76-cm ridge spacing and 10-cm ridge height, the soil ridge-roughness factor is 0.5. Figure 1 shows that the climatic factor in northwestern Kansas is 50.

Table 3 shows that for Goodland (northwestern Kansas), the prevailing wind erosion direction (February and March) is 338° (NNW) with a preponderance of 2.5. Then from Table 4, wind erosion direction factor is 1.33. Field width of 400 m times wind erosion direction factor gives 530 m for median unsheltered distance across the field.

The procedure to determine potential average annual soil loss, $E = f(I, K, C, L, V)$, from those data follows. E_1, E_2, E_3, and E_4 are solutions at intermediate steps in solving the final answer.

1. $E_1 = I = 193$ metric tons/ha
2. Determine $E_2 = IK = (193$ metric tons/ha$)(0.5) = 96.5$ metric tons/ha
3. Determine $E_3 = IKC = (E_2)(C) = (96.5)(0.5) = 48.3$ metric tons/ha
4. Determine $E_4 = IKC(L)$ for $L = 530$ m. Use Table 5. Find $E_2 = 90$ in first column and follow the row right to column with heading of 610 m and read 1.90. Interpolate between 305 and 610 m for L and between 90 and 112 for E_2 to obtain 2.8 divisions for curve deviation. Find E_3 of approximately 48.3 at the bottom of Table 5. Then move right 2.8 scale divisions to obtain $E_4 = 37$ metric tons/ha.
5. Determine E. In the left column of Figure 2, find the row for equivalent flat small grain residue of 1000 kg/ha and move right to where E_4 or $IKCL = 37$; then move vertically down to $IKCLV = 8.4$ metric tons/ ha/year.

If in Step 4 residue is given as something different than flat small grain residue, convert to flat small grain residue using Figures 3 and 4 or variations thereof. Research is in progress to obtain the data needed to convert more kinds of vegetative material to equivalent flat small grain residue.

Other procedures are available to solve the wind erosion equation for potential average annual soil loss. These include graphical,[22,58] slide rule,[61] and computer.[59,60] Potential average annual soil loss is shown for a few combinations of variables in Table 6.

Table 3
PREVAILING WIND EROSION DIRECTION[a] AND PREPONDERANCE OF PREVAILING WIND EROSION FORCES IN THE PREVAILING WIND EROSION DIRECTION[58]

Location	Jan	Feb	Mar	Apr	May	Jun	Jul	Aug	Sep	Oct	Nov	Dec
Alaska												
Anchorage	22	22	0	180	180	180	180	180	180	180	180	0
	2.0	3.4	2.7	2.7	3.4	3.1	2.7	3.2	3.1	3.2	3.1	3.1
Fairbanks	45	225	45	225	45	247	247	247	247	247	225	45
	2.6	2.9	2.5	1.3	1.9	2.1	2.2	2.4	2.0	2.5	1.3	3.3
Arizona												
Ajo	180	180	337	225	225	203	225	180	202	180	180	180
	1.9	2.1	1.2	1.2	1.4	1.8	1.3	1.2	1.6	1.5	1.7	1.4
Douglas	201	203	225	247	248	225	337	315	90	90	90	180
	1.3	1.6	1.5	1.4	1.5	1.6	1.4	1.4	1.5	1.8	1.0	1.0
Kingman	225	203	203	225	225	225	225	225	225	225	225	0
	2.9	1.7	2.6	2.8	2.7	3.6	2.4	2.9	1.9	1.5	2.0	1.6
Phoenix	90	293	292	90	157	135	180	135	158	180	180	337
	1.5	1.4	1.6	1.7	1.4	1.5	1.5	1.7	1.7	1.6	1.0	1.4
Prescott	202	202	225	203	203	225	225	203	203	202	202	202
	1.7	1.4	1.7	2.0	2.7	2.3	1.9	1.6	2.0	2.1	2.2	1.5
Tucson	113	292	292	90	248	292	113	113	113	113	113	113
	2.6	1.7	1.6	1.4	1.7	1.2	1.5	1.6	2.1	2.3	3.4	2.6
Yuma	0	0	337	293	135	157	157	157	158	337	0	0
	2.6	2.1	1.5	1.5	1.7	2.2	3.0	2.5	1.9	1.4	2.4	2.7
California												
Arcata	135	315	315	315	315	315	315	315	315	336	135	135
	2.3	2.8	3.3	3.9	4.4	6.6	6.9	5.0	3.6	2.1	3.1	2.5

[a] Clockwise from the north through 360°.

Location	Jan	Feb	Mar	Apr	May	Jun	Jul	Aug	Sep	Oct	Nov	Dec
Bishop	0	0	0	0	0	180	158	180	180	180	0	0
	4.1	4.0	3.2	2.6	2.0	2.4	2.1	2.3	2.5	3.6	5.7	3.9
Blythe	0	0	0	225	180	225	180	180	180	203	0	337
	2.3	2.4	2.2	2.3	1.4	2.1	1.9	3.9	1.8	1.9	4.2	1.9
China Lake	203	225	225	225	225	225	225	203	225	225	225	225
	1.6	1.7	1.7	1.6	1.8	2.0	2.1	2.4	2.4	1.9	1.6	1.7

El Centro	248	248	248	248	248	270	270	270	270	270	270	248
	1.6	2.0	2.9	3.6	3.8	3.7	2.1	2.3	2.6	4.1	1.6	1.4
Fort Bragg	113	315	315	315	337	315	338	337	337	135	113	135
	2.3	2.3	2.3	4.3	4.1	4.5	3.9	3.7	4.5	3.6	1.7	1.6
Fresno	157	315	315	315	315	293	293	293	315	158	135	315
	2.7	2.9	3.1	2.6	2.2	4.6	4.2	3.1	2.4	1.9	2.0	2.0
Marysville	157	135	315	135	135	135	135	157	315	135	315	338
	3.3	4.3	3.0	3.2	6.4	4.6	3.6	3.2	4.9	4.6	9.3	1.9
Merced	135	135	337	337	337	337	337	337	337	337	315	135
	3.8	3.5	3.3	4.2	3.3	3.6	4.0	4.0	3.6	3.6	3.5	4.1
Palmdale	247	225	247	225	225	225	225	225	225	225	248	225
	1.8	2.1	1.6	2.0	2.4	4.2	5.7	5.0	4.3	2.4	1.9	1.7
Riverside	23	338	293	293	293	293	315	293	293	293	67	69
	1.2	1.2	1.6	2.0	2.5	2.8	3.6	3.7	3.1	2.3	1.1	1.4
Salinas	135	135	135	293	292	292	292	315	315	315	135	135
	6.8	4.5	3.5	2.8	3.0	3.3	3.4	3.2	4.2	3.0	4.2	5.3
San Diego	180	180	180	202	248	203	293	248	293	203	180	337
	1.5	1.5	1.4	1.3	1.1	1.2	1.3	1.1	1.5	1.4	1.4	1.3
San Miguel I	157	157	337	337	315	315	293	315	315	293	294	337
	2.8	4.0	2.6	4.0	3.2	4.1	5.4	2.6	2.6	4.6	3.5	3.6
Santa Rosa	315	0	202	315	202	159	180	158	180	180	180	0
	3.1	1.9	1.6	1.6	1.4	1.4	2.7	2.1	1.6	1.7	1.5	4.8
Stockton	235	316	315	315	270	270	270	270	292	315	135	135
	4.0	2.2	2.5	2.0	1.7	2.4	3.1	2.9	2.5	3.7	2.1	2.3
Thermal	337	157	337	337	337	337	337	157	337	337	337	315
	2.4	2.6	2.3	4.3	3.0	4.4	3.7	4.3	4.1	2.6	2.8	3.4
Victorville	193	270	251	270	270	225	180	180	180	248	248	248
	1.1	1.2	1.3	1.4	1.3	1.1	2.0	1.8	1.4	1.2	1.3	1.4
Colorado												
Denver	315	315	293	338	338	159	338	0	158	0	270	0
	1.2	1.1	1.0	1.2	1.2	1.2	1.8	1.3	1.6	1.1	1.3	1.3
La Junta	248	292	247	0	45	225	67	203	46	248	0	270
	1.5	1.1	1.3	1.1	1.1	1.5	1.0	1.4	1.3	1.3	1.3	1.3
Pueblo	293	337	270	202	225	203	0	337	23	293	292	293
	1.6	1.2	1.6	1.3	1.4	1.3	1.2	1.1	1.2	1.1	1.4	1.5

[a] Clockwise from the north through 360°.

Table 3 (continued)
PREVAILING WIND EROSION DIRECTION[a] AND PREPONDERANCE OF PREVAILING WIND EROSION FORCES IN THE PREVAILING WIND EROSION DIRECTION[58]

Location	Jan	Feb	Mar	Apr	May	Jun	Jul	Aug	Sep	Oct	Nov	Dec
Connecticut												
Bridgeport	293	315	293	293	90	225	270	247	45	248	337	293
	1.2	1.3	1.2	1.5	1.2	1.6	1.3	1.7	1.5	1.3	1.6	1.4
Hartford	315	315	315	338	0	180	180	180	180	0	315	315
	1.8	1.8	1.5	2.0	1.5	1.6	1.4	1.5	1.8	1.4	1.3	1.6
Stratford	315	293	293	225	225	202	—	204	225	247	292	270
	1.3	2.3	1.4	1.4	1.6	1.6	—	1.3	2.5	2.5	1.5	1.6
Windsor Locks	315	315	337	315	180	180	180	180	0	315	337	315
	1.8	4.6	1.3	1.9	2.1	2.0	1.6	1.7	1.7	1.5	1.9	2.0
Delaware												
Wilmington	315	293	315	315	338	315	158	180	180	68	315	293
	1.7	2.0	1.8	1.3	1.1	1.2	1.4	1.3	1.4	1.1	1.7	1.4
Florida												
Avon Park	338	22	180	45	67	270	337	68	68	45	0	0
	1.5	1.6	1.2	1.6	2.0	1.2	1.2	1.7	1.8	2.1	1.5	2.4
Cape Canaveral	337	315	157	135	112	135	135	135	90	45	337	335
	1.5	1.4	1.4	1.4	1.3	1.3	1.6	1.3	1.1	1.4	1.3	1.4
Daytona Beach	202	225	225	68	68	68	90	67	67	45	0	0
	1.2	1.2	1.1	1.3	1.6	1.7	1.6	1.7	1.7	2.0	1.4	1.3
Fort Myers	23	225	45	67	68	90	281	68	67	45	45	22
	1.4	1.3	1.3	1.4	2.0	1.2	1.2	1.6	1.7	1.9	1.8	2.1
Homestead	112	112	90	90	90	90	112	90	90	65	67	112
	1.2	1.4	1.3	2.0	2.4	2.5	2.5	2.5	1.7	2.2	1.6	1.3
Jacksonville	225	247	270	270	270	68	90	45	45	45	225	225
	1.3	1.5	1.5	1.6	1.8	1.7	1.5	1.5	1.9	1.9	1.2	1.3
Key West	0	135	135	113	112	113	135	112	90	45	45	23
	1.1	1.3	1.6	1.3	2.1	1.8	1.7	2.2	1.2	1.9	1.5	1.5
Orlando	315	315	315	114	270	270	158	270	69	67	0	338
	1.2	1.1	1.1	1.1	1.4	1.2	1.1	1.3	1.3	1.3	1.4	1.2
Pensacola	180	202	180	135	158	180	180	203	22	157	0	337
	1.6	1.5	1.8	1.6	1.6	1.8	1.5	1.3	1.1	1.6	2.0	1.5
Perry Field	180	203	225	202	225	225	225	225	225	315	338	180
	1.8	3.8	1.6	3.0	3.2	4.0	2.5	2.2	2.2	1.3	2.3	1.3

Sarasota	0	0	22	202	249	225	337	203	45	22	22	0
	1.6	1.6	1.6	1.3	1.4	1.3	1.3	1.3	2.3	1.7	1.6	1.6
Tallahassee	338	180	180	180	180	180	180	247	45	0	338	338
	1.8	1.5	1.7	1.7	1.4	1.2	1.5	1.2	1.5	1.2	1.9	2.1
Tampa	0	338	202	248	270	270	292	248	45	45	23	22
	1.3	1.2	1.0	1.1	1.3	1.3	1.2	1.3	1.4	2.0	1.7	1.6
Vero Beach	113	112	135	68	67	68	270	47	67	67	67	292
	1.5	1.2	1.3	1.3	1.6	1.6	1.6	1.1	1.8	2.4	1.1	1.2
W. Palm Beach	113	113	135	113	90	112	113	112	68	67	68	90
	1.3	1.5	1.6	1.6	1.6	1.5	2.3	1.8	1.8	2.1	1.4	1.3
Georgia												
Albany	178	180	158	135	222	247	225	225	45	45	293	296
	1.3	1.2	1.3	1.2	1.1	1.4	1.4	1.2	1.6	1.5	1.3	1.1
Athens	270	90	112	248	248	248	225	247	67	67	90	270
	1.4	1.4	1.5	1.4	1.6	1.3	1.5	1.5	2.0	1.8	1.5	1.4
Bainbridge	225	160	158	225	67	157	225	225	67	45	315	270
	1.1	1.0	1.4	1.2	1.1	1.1	1.5	1.3	1.9	1.3	1.5	1.2
Marietta	315	293	315	293	295	292	270	293	90	293	315	293
	1.5	1.5	1.5	1.5	1.3	1.4	1.2	1.4	1.5	1.5	1.9	1.7
Savannah	292	270	270	112	270	270	203	113	68	45	270	270
	1.5	1.6	1.3	1.2	1.2	1.2	1.2	1.1	1.2	1.9	1.3	1.4
Valdosta	270	225	180	203	247	68	225	45	67	67	270	247
	1.2	1.0	1.1	1.2	1.2	1.4	1.4	1.7	2.3	1.6	1.3	1.3
Waycross	293	293	90	90	113	90	0	67	67	67	0	23
	1.3	1.7	1.2	1.2	1.7	1.7	1.2	3.0	2.6	2.1	2.1	1.2
Idaho												
Boise	135	315	315	315	315	315	315	293	315	113	135	135
	3.4	4.9	2.9	3.1	2.6	2.7	2.2	1.7	2.7	2.3	3.5	3.0
Mountain Home	113	135	293	113	113	113	113	135	113	135	113	113
	2.2	2.6	2.4	2.5	2.3	2.0	2.1	2.1	2.3	3.0	2.4	2.3
Pocatello	180	203	247	247	247	225	225	225	247	247	225	203
	2.0	1.5	2.2	2.6	1.8	1.6	1.3	1.3	2.1	1.5	1.3	1.6
Illinois												
Belleville	315	315	293	315	315	90	248	338	2	338	315	315
	1.9	1.9	1.6	1.3	1.4	1.0	1.1	1.0	1.1	1.3	1.4	1.6
Glenview	225	246	23	203	23	203	23	23	203	203	203	225

Table 3 (continued)
PREVAILING WIND EROSION DIRECTION[a] AND PREPONDERANCE OF PREVAILING WIND EROSION FORCES IN THE PREVAILING WIND EROSION DIRECTION[58]

Location	Jan	Feb	Mar	Apr	May	Jun	Jul	Aug	Sep	Oct	Nov	Dec
	1.3	1.1	1.7	1.7	2.3	2.0	2.0	2.5	2.0	1.8	1.4	1.3
Lawrenceville	315	338	225	315	203	203	225	35	180	22	157	292
	1.3	1.2	1.6	1.1	1.8	1.7	1.8	1.6	1.8	1.3	1.2	1.2
Park Ridge	90	292	270	248	225	225	247	225	225	247	270	248
	1.2	1.3	1.4	1.2	1.4	1.2	1.3	1.5	1.2	1.4	1.2	1.2
Peoria	337	293	270	293	180	202	180	180	180	338	315	315
	1.3	1.4	1.2	1.3	1.2	1.3	1.3	1.5	1.7	1.5	1.3	1.2
Rantoul	203	315	248	203	202	224	225	202	202	180	180	158
	1.1	1.1	1.1	1.2	1.4	1.3	1.5	1.6	1.5	1.4	1.2	1.2
Rockford	247	270	248	248	225	247	248	203	203	203	244	292
	1.2	1.4	1.6	1.3	1.2	1.9	1.3	1.6	1.4	1.1	1.1	1.2
Indiana												
Bunker Hill	248	270	225	247	225	224	225	203	203	238	247	226
	1.3	1.7	1.5	1.3	1.5	1.4	1.6	1.6	1.4	1.3	1.5	1.4
Columbus	0	202	248	180	202	203	225	23	180	180	315	180
	1.3	1.1	1.0	1.1	1.5	1.4	1.2	1.4	1.6	1.6	1.1	1.1
Madison	249	270	292	248	158	202	248	202	338	338	247	270
	1.5	1.4	1.4	1.1	1.1	1.2	1.4	1.3	1.5	1.2	1.2	1.6
South Bend	225	270	90	315	338	338	338	0	180	180	225	225
	1.2	1.2	1.3	1.1	1.1	1.3	1.2	1.3	1.3	1.2	1.4	1.2
Iowa												
Burlington	293	293	270	293	248	202	225	0	315	202	315	292
	1.8	1.3	1.2	1.4	1.3	1.3	1.2	1.3	1.2	1.6	1.3	1.1
Des Moines	315	315	315	315	338	203	315	338	315	337	315	315
	1.5	1.1	1.4	1.6	1.3	1.1	1.3	1.2	1.1	1.5	1.7	1.3
Kansas												
Dodge City	0	0	0	0	180	180	180	180	180	180	0	0
	2.1	2.0	2.4	1.7	2.2	2.4	2.1	2.5	2.8	2.5	2.2	2.3
Goodland	338	338	338	338	180	180	180	180	180	158	337	338
	2.0	2.5	2.5	2.1	1.9	1.9	2.0	2.2	2.7	2.4	2.3	2.4

Olathe	180	180	180	180	180	202	202	202	180	180	180	180
	1.7	1.6	1.6	1.3	1.6	1.8	2.2	1.9	2.0	2.0	1.6	1.5
Salina	0	0	0	180	180	180	180	180	180	180	338	0
	2.0	1.9	1.8	1.6	1.8	2.0	2.1	2.0	2.7	2.1	1.8	1.7
Topeka	180	338	180	338	225	0	0	338	180	180	338	180
	1.5	1.5	1.6	1.4	1.2	1.7	1.5	1.2	2.0	2.2	1.5	1.6
Wichita	0	180	180	180	180	180	180	180	180	180	180	180
	2.5	2.1	1.6	2.1	2.1	2.3	2.1	1.9	2.3	2.9	2.1	1.8
Maine												
Bangor	337	337	337	338	338	0	338	337	338	338	338	315
	1.9	1.7	1.7	1.5	1.6	1.6	1.5	1.4	1.6	1.6	1.7	1.9
Brunswick	0	338	338	0	180	202	180	202	202	0	0	0
	1.9	1.6	1.5	1.6	1.9	1.8	1.5	1.8	2.0	1.5	1.4	1.5
Caribou	293	292	293	293	315	315	293	293	292	293	292	292
	1.8	1.6	1.9	1.7	1.5	1.2	1.3	1.4	1.3	1.6	1.6	1.6
Maryland												
Aberdeen	315	315	315	315	0	202	202	22	22	0	315	315
	1.3	1.9	1.6	1.3	1.3	1.3	1.6	1.7	1.4	1.4	1.5	1.5
Annapolis	315	315	315	338	180	180	180	180	180	180	315	315
	1.8	1.9	1.7	1.5	1.5	1.8	2.0	2.0	2.0	1.2	1.4	1.7
Frederick	293	315	315	315	315	315	293	315	315	315	293	293
	2.6	2.9	4.0	2.1	1.9	1.9	2.3	1.3	1.8	1.8	2.6	2.8
Massachusetts												
Bedford	293	293	292	292	270	247	225	225	23	248	293	292
	1.4	1.5	1.3	1.2	1.1	1.4	1.4	1.3	1.2	1.3	1.1	1.5
Chicopee Falls	337	315	338	338	0	180	180	180	0	0	0	315
	1.5	2.0	1.6	1.6	1.6	1.9	2.1	2.1	1.9	1.7	1.5	1.6
Nantucket	315	293	292	23	225	225	225	204	23	45	315	292
	1.4	1.4	1.3	1.3	1.6	1.9	1.9	1.6	1.5	1.3	1.2	1.3
Worcester	248	248	67	270	248	225	248	45	45	67	248	248
	1.9	2.1	1.5	1.7	1.7	1.8	1.8	2.2	3.2	2.2	1.4	1.6
Michigan												
Battle Creek	248	248	248	270	247	248	248	270	270	225	225	225
	1.4	1.4	1.6	1.4	1.5	1.5	1.5	1.3	1.3	1.2	1.3	1.5
Cadillac	248	248	292	292	225	225	247	225	246	203	203	247
	1.4	1.3	1.2	1.5	1.4	1.2	1.3	1.4	1.2	1.4	1.2	1.5

Table 3 (continued)
PREVAILING WIND EROSION DIRECTION[a] AND PREPONDERANCE OF PREVAILING WIND EROSION FORCES IN THE PREVAILING WIND EROSION DIRECTION[58]

Location	Jan	Feb	Mar	Apr	May	Jun	Jul	Aug	Sep	Oct	Nov	Dec
Flint	225	270	248	248	247	248	248	225	225	225	225	225
	1.4	1.4	1.6	1.4	1.3	1.6	1.4	1.8	1.2	1.2	1.6	1.5
Marquette	0	338	0	0	0	180	202	0	180	180	180	180
	1.8	1.8	1.9	1.7	2.1	1.8	1.9	2.9	1.7	2.0	1.8	1.8
Mt. Clemens	225	225	225	203	180	201	202	180	180	202	203	225
	1.5	1.2	1.2	1.3	1.3	1.3	1.4	1.5	1.6	1.5	1.4	1.4
Muskegon	248	270	248	225	205	225	225	203	203	203	225	270
	1.7	1.6	1.4	1.2	1.5	1.7	1.4	2.3	1.4	1.4	1.5	1.1
Oscoda	338	315	270	239	227	270	202	225	248	224	226	315
	1.2	1.2	1.2	1.3	1.1	1.1	1.1	1.3	1.2	1.2	1.1	1.0
Pellston	270	270	270	270	270	248	248	248	248	248	292	270
	1.4	1.7	1.6	1.5	1.5	1.8	2.0	1.7	1.5	1.3	1.4	1.4
Sault Ste. Marie	292	293	293	293	293	293	293	293	293	293	293	292
	1.8	2.3	2.5	2.9	2.5	2.6	3.1	2.2	2.3	2.3	1.9	2.1
Traverse City	203	202	202	202	203	225	203	203	202	202	180	225
	1.3	1.3	1.4	1.4	1.6	1.7	1.6	1.6	1.5	1.7	1.7	1.3
Ypsilanti	248	270	270	270	270	270	270	270	270	248	248	248
	1.5	1.6	1.9	1.8	1.5	1.6	1.6	1.3	1.5	1.3	1.6	1.5
Minnesota												
Duluth	292	270	293	90	90	248	270	68	270	248	293	293
	1.9	1.6	1.3	1.7	1.7	2.0	1.7	1.5	1.8	1.3	1.6	1.7
Int. Falls	292	270	293	292	292	293	292	315	315	293	293	292
	1.4	1.5	1.3	1.4	1.3	1.4	1.5	1.3	1.3	2.3	1.5	1.9
Minneapolis	315	293	292	293	292	293	180	180	315	315	315	293
	1.5	1.7	1.4	1.5	1.3	1.3	1.0	1.2	1.4	1.6	1.7	1.4
Rochester	337	315	315	315	157	158	157	180	180	157	315	337
	2.0	1.9	1.7	1.8	1.6	1.8	1.7	2.1	2.0	2.1	1.6	1.6

Montana												
Billings	203	202	315	0	315	0	0	338	338	338	316	291
	1.4	1.2	1.3	1.3	1.3	1.1	1.4	1.4	1.5	1.2	1.0	1.2
Cut Bank	270	270	270	270	270	247	270	270	270	270	248	270
	4.5	3.3	2.2	2.3	1.9	1.3	2.5	2.5	2.1	1.9	2.6	4.4
Glasgow	293	294	293	293	293	112	292	113	293	293	293	293
	2.9	3.0	2.8	2.1	2.2	2.2	2.3	2.3	2.6	3.9	3.6	3.6
Great Falls	225	225	225	247	248	248	247	248	247	225	225	225
	3.2	3.3	1.9	1.6	1.8	1.6	1.6	1.5	1.9	2.1	3.5	3.6
Helena	180	270	292	293	292	292	293	293	293	270	248	293
	1.4	1.8	1.6	1.5	1.6	1.8	1.4	1.4	1.7	1.2	1.3	1.2
Lewistown	270	270	270	292	293	293	315	292	293	293	225	226
	1.7	1.4	2.1	1.9	2.7	2.0	1.5	2.1	1.6	1.5	1.5	1.6
Miles City	315	293	293	315	293	292	315	315	315	293	293	293
	1.8	1.7	2.1	2.6	2.0	1.8	1.6	1.6	1.9	2.3	3.1	2.5
Missoula	112	292	292	248	270	270	293	315	315	338	293	113
	1.5	1.4	1.5	1.4	1.2	1.3	1.5	1.2	1.2	1.5	1.6	1.6
Nebraska												
Grand Island	338	338	338	0	180	180	180	180	180	158	338	338
	2.2	2.1	1.6	1.6	1.7	2.0	1.6	1.9	2.2	2.4	1.9	2.1
Lincoln	338	337	338	158	337	158	158	158	158	338	337	338
	2.4	1.5	1.5	1.6	1.6	1.6	1.8	1.5	1.8	1.7	2.1	2.1
North Platte	338	338	338	338	338	158	158	158	180	338	338	338
	2.3	2.6	2.1	2.6	2.2	2.4	2.0	2.6	2.4	3.1	3.0	3.4
Omaha	337	337	337	315	180	158	169	158	158	338	337	338
	2.9	1.7	1.7	1.9	1.7	1.6	1.7	1.3	2.0	1.9	2.1	2.1
Scottsbluff	292	293	315	315	337	294	135	315	0	315	337	292
	2.1	2.0	1.9	1.6	1.7	1.6	1.4	1.8	1.3	2.3	2.9	2.7
Nevada												
Fallon	225	203	247	270	270	292	292	248	315	247	159	180
	1.6	1.5	1.2	1.4	1.2	1.8	1.5	1.5	1.1	1.5	1.2	1.5
Las Vegas	180	0	203	225	203	225	225	203	225	203	22	0
	1.8	1.4	1.7	1.9	2.2	2.6	2.0	2.0	2.6	1.9	1.8	1.8
Mercury	180	0	202	180	202	180	181	180	180	180	180	180
	2.8	2.1	2.2	2.2	2.3	2.9	1.3	3.2	2.8	2.8	3.2	3.7

Table 3 (continued)
PREVAILING WIND EROSION DIRECTION[a] AND PREPONDERANCE OF PREVAILING WIND EROSION FORCES IN THE PREVAILING WIND EROSION DIRECTION[58]

Location	Jan	Feb	Mar	Apr	May	Jun	Jul	Aug	Sep	Oct	Nov	Dec
Reno	180	180	158	157	293	293	293	292	315	158	180	180
	2.2	2.1	1.3	1.6	1.6	2.4	2.4	2.7	1.2	1.5	2.1	2.2
Tonopah	157	315	315	315	315	157	158	180	158	315	315	337
	3.0	3.0	3.2	2.7	2.3	2.3	2.1	2.6	2.6	2.5	3.1	2.6
New Hampshire												
Concord	315	315	315	315	315	315	315	315	315	315	315	315
	2.8	2.8	2.2	2.0	2.2	1.8	2.1	1.8	1.6	1.9	2.0	3.0
Manchester	315	315	294	293	315	315	158	315	337	337	315	292
	1.7	1.6	1.4	1.6	1.3	1.2	1.3	1.5	1.4	1.5	1.5	1.8
Portsmouth	315	293	293	293	293	292	315	337	158	315	315	315
	1.4	1.9	1.6	1.4	1.4	1.2	1.4	1.2	1.1	1.4	1.4	1.4
New Jersey												
Atlantic City	292	292	292	292	293	304	180	202	315	270	293	293
	2.2	2.0	1.9	1.6	1.3	1.2	1.2	1.3	1.2	1.1	1.7	1.6
Lakehurst	270	270	270	248	248	247	225	225	225	247	248	270
	1.8	2.0	1.9	1.5	1.7	1.8	2.1	1.7	1.7	1.8	1.6	1.8
Trenton	292	293	292	292	292	249	225	225	22	67	293	292
	1.2	1.6	1.4	1.2	1.2	1.2	1.4	1.1	1.4	1.2	1.5	1.4
New Mexico												
Albuquerque	112	293	293	113	112	135	135	113	113	135	112	113
	1.8	1.6	1.5	1.4	1.3	1.5	1.4	1.9	1.4	1.3	1.4	1.5
Hobbs	248	270	247	270	247	180	180	202	203	202	225	247
	1.8	1.5	2.1	1.3	1.2	1.3	1.6	1.8	2.7	1.8	1.7	1.3
Roswell	202	190	203	202	180	180	180	158	180	180	202	180
	1.6	1.5	1.3	1.7	1.6	2.0	2.2	2.0	2.6	2.2	1.4	1.5
New York												
Geneva	315	293	157	0	337	338	202	180	180	180	180	180
	2.0	1.9	1.8	2.4	2.4	1.8	2.8	2.1	2.3	2.1	1.8	1.4
New York	292	292	292	292	292	157	158	180	158	270	270	292
	1.6	1.9	1.7	1.5	1.1	1.4	1.4	1.2	1.1	1.1	1.5	1.7

Niagara	247	247	226	225	225	225	225	225	225	225	225	247
	1.7	1.5	1.6	2.0	1.6	1.8	2.1	2.2	1.7	1.7	1.8	2.0
Plattsburgh	293	315	315	315	135	135	315	135	157	135	157	157
	1.3	1.3	1.6	1.6	2.0	1.8	1.6	2.1	1.8	2.1	1.3	1.5
Rome	292	293	293	293	293	315	293	293	293	315	293	292
	2.7	3.1	2.6	2.5	2.5	2.1	1.7	2.1	1.9	1.7	2.2	2.5
Schenectady	293	293	293	293	293	292	292	293	293	293	292	293
	5.6	2.8	2.4	1.6	2.6	2.2	2.5	1.9	1.5	2.1	2.0	2.7
West Hampton	315	293	292	248	225	225	225	225	225	67	0	292
	1.2	1.5	1.2	1.2	1.2	1.4	1.8	1.6	1.6	1.3	1.1	1.2
North Carolina												
Cherry Point	23	203	203	203	203	203	203	23	23	23	23	23
	1.7	1.7	2.0	2.6	2.9	2.5	2.8	2.2	2.5	2.3	2.0	1.7
Hatteras	0	22	202	202	203	203	203	203	23	23	0	22
	1.2	1.5	1.4	1.8	2.8	2.6	2.9	1.6	1.7	1.9	1.6	1.3
Jacksonville	337	315	225	207	203	180	203	45	26	23	1	0
	1.3	1.3	1.2	1.6	1.9	1.5	2.1	2.0	1.7	2.4	1.2	1.3
Weeksville	22	202	225	203	225	23	45	45	45	22	338	23
	1.9	2.0	1.5	2.7	1.7	1.6	3.1	2.4	3.8	2.0	3.1	1.6
Wilmington	180	180	338	202	202	180	202	202	0	0	0	0
	1.4	1.5	1.1	1.3	2.0	2.1	2.5	1.6	1.4	2.0	2.0	1.4
Winston-Salem	225	247	225	225	225	45	225	45	45	45	45	225
	1.2	1.2	1.2	1.4	2.3	2.2	2.2	2.3	2.8	2.2	1.5	1.3
North Dakota												
Bismarck	315	293	325	315	315	293	315	157	315	315	315	315
	1.7	2.0	1.3	1.7	1.8	1.7	1.5	1.4	2.1	2.0	2.4	2.4
Fargo	337	337	337	315	337	315	157	158	315	315	337	337
	2.5	2.0	1.9	1.7	1.5	1.2	1.7	1.6	1.6	1.9	2.4	2.2
Grand Forks	338	337	338	337	337	270	337	158	315	337	337	337
	3.2	2.5	2.1	2.0	1.5	1.2	1.4	1.7	1.3	1.8	2.5	2.5
Minot	315	315	315	315	315	293	315	337	293	315	315	315
	1.7	1.8	1.7	1.5	1.5	1.3	1.5	1.4	1.7	1.8	2.2	1.6
Ohio												
Cincinnati	203	225	247	225	225	225	225	225	225	225	225	225
	1.5	1.3	1.3	1.6	1.6	1.8	1.7	1.8	1.8	1.7	1.5	1.6
Cleveland	248	270	247	247	203	225	203	293	202	203	247	225
	1.3	1.2	1.3	1.3	1.3	1.2	1.4	1.1	1.2	1.4	1.3	1.2

Table 3 (continued)
PREVAILING WIND EROSION DIRECTION[a] AND PREPONDERANCE OF PREVAILING WIND EROSION FORCES IN THE PREVAILING WIND EROSION DIRECTION[58]

Location	Jan	Feb	Mar	Apr	May	Jun	Jul	Aug	Sep	Oct	Nov	Dec
Columbus	315	270	247	225	203	202	202	0	180	180	225	203
	1.1	1.1	1.2	1.3	1.4	1.4	1.4	1.4	1.4	1.4	1.3	1.2
Dayton	292	270	270	248	247	225	246	225	203	224	247	225
	1.1	1.3	1.4	1.3	1.3	1.3	1.4	1.2	1.2	1.2	1.2	1.1
Toledo	247	247	248	247	247	225	204	225	248	225	220	225
	1.4	1.5	1.4	1.6	1.7	1.4	1.1	1.3	1.4	1.5	1.6	2.0
Youngstown	248	270	248	248	248	247	225	202	225	225	247	247
	1.4	1.5	1.6	1.4	1.2	1.1	1.3	1.3	1.3	1.3	2.0	1.7
Oklahoma												
Oklahoma City	0	180	180	180	180	180	180	180	180	180	180	180
	2.6	2.4	1.7	2.4	2.1	2.1	2.1	2.0	2.1	3.1	2.6	2.5
Tulsa	0	180	0	0	180	180	180	180	180	0	0	180
	2.6	2.8	1.8	2.2	2.2	2.7	2.0	1.9	2.8	2.7	2.7	2.1
Oregon												
Astoria	203	202	203	202	315	315	315	315	180	203	202	203
	1.6	1.6	1.5	1.2	1.3	1.7	3.1	2.1	1.4	2.3	2.1	1.8
Klamath Falls	180	158	157	157	337	320	337	337	158	159	158	158
	1.8	1.8	1.3	1.5	1.6	1.3	2.0	1.8	1.4	1.9	2.0	1.4
Medford	157	136	157	315	315	293	315	315	315	157	152	158
	1.5	1.5	1.5	1.6	2.3	1.7	1.7	2.2	1.7	1.7	1.9	2.6
Pendleton	248	270	270	270	270	270	270	270	270	270	270	270
	1.2	2.0	3.1	4.1	4.8	4.2	3.2	3.9	3.7	3.1	2.7	1.7
Portland	180	202	202	180	180	158	315	315	180	180	180	180
	1.1	1.5	1.2	1.5	1.4	1.3	1.8	1.8	1.2	1.6	1.1	1.2
Redmond	180	158	315	315	315	315	315	315	337	180	180	180
	1.6	1.8	1.6	1.7	2.6	1.9	2.7	2.7	3.2	1.7	1.5	2.4
Salem	180	180	180	180	180	202	0	0	180	180	180	180
	6.6	5.4	3.8	2.1	1.6	1.4	1.7	1.8	2.6	4.5	5.6	5.5
Rhode Island												
Quonset Point	0	338	0	0	202	202	202	180	0	22	0	338
	1.6	1.3	1.5	1.6	2.0	2.2	2.2	2.1	2.2	2.0	1.5	1.3
South Carolina												
Beaufort	247	270	292	180	270	23	202	225	45	45	202	225

	1.2	1.7	1.6	1.3	1.4	1.5	1.9	1.3	1.6	2.3	1.2	1.3
Florence	203	225	225	225	225	203	203	203	45	23	23	225
	2.1	2.3	1.6	1.5	1.4	2.3	1.8	2.2	1.8	2.1	1.9	1.8
Greenville	45	45	225	225	45	45	45	45	45	45	45	45
	2.0	2.4	2.0	1.9	2.4	2.1	2.0	2.3	3.3	2.7	2.3	2.3
Myrtle Beach	225	225	225	203	203	202	203	202	225	45	247	247
	1.3	1.2	1.2	1.3	1.8	1.7	2.3	2.0	1.4	1.3	1.2	1.4
Sumter	225	225	225	225	225	203	225	45	45	45	225	225
	1.7	1.8	1.5	1.6	1.6	1.4	1.9	1.7	3.1	2.3	1.9	1.7
South Dakota												
Huron	337	337	337	337	157	158	158	158	158	315	315	315
	2.9	2.6	2.6	1.8	1.8	1.9	2.3	2.6	2.0	2.4	2.5	2.7
Rapid City	337	337	337	337	337	337	337	337	337	337	337	315
	2.7	2.9	3.0	2.4	2.0	2.4	2.4	2.3	2.7	2.6	3.0	2.9
Sioux Falls	337	338	315	337	315	315	158	158	337	338	315	337
	2.2	1.5	1.5	1.9	1.4	1.4	1.6	1.6	1.6	1.7	2.3	2.0
Texas												
Amarillo	203	0	225	203	203	202	180	202	202	202	202	23
	1.9	1.7	1.3	1.4	1.3	1.9	1.4	1.9	2.1	2.0	1.7	1.8
Austin	0	0	346	0	180	180	180	0	0	0	0	0
	2.3	1.8	1.7	1.8	2.1	1.9	1.7	1.9	1.9	2.1	2.4	1.8
Brownsville	157	157	157	135	135	135	135	135	157	135	337	158
	3.6	2.6	2.1	2.1	3.1	3.0	3.7	2.5	1.4	1.7	3.2	2.9
Corpus Christi	158	157	158	157	157	157	157	135	135	157	158	338
	3.1	3.0	2.4	2.3	2.3	3.4	2.7	1.8	1.3	1.7	2.5	2.3
Dalhart	225	225	225	203	225	203	202	202	203	203	203	23
	2.4	2.2	2.3	1.8	2.0	2.2	2.3	2.3	2.8	2.6	2.0	2.2
Dallas	180	158	158	180	180	180	180	180	180	180	180	178
	2.9	2.2	1.9	2.3	2.2	2.6	2.1	1.9	1.9	2.5	1.8	2.1
Galveston	337	337	337	135	157	158	180	180	41	135	0	338
	1.9	1.5	1.4	1.3	1.9	2.2	2.3	1.5	1.4	1.2	2.0	1.3
Houston	158	158	158	157	157	180	180	158	45	68	0	338
	1.9	1.6	1.6	1.8	1.7	1.9	1.8	1.6	1.2	1.4	1.6	1.6
Laredo	155	135	135	135	135	135	135	135	135	135	157	157
	1.9	1.8	1.7	2.4	2.9	3.5	3.8	3.2	1.7	2.2	1.8	2.0
Lubbock	227	247	248	225	202	180	180	180	202	203	214	248
	1.3	1.2	1.5	1.1	1.3	1.9	2.0	1.5	2.0	1.6	1.2	1.2

Table 3 (continued)
PREVAILING WIND EROSION DIRECTION[a] AND PREPONDERANCE OF PREVAILING WIND EROSION FORCES IN THE PREVAILING WIND EROSION DIRECTION[58]

Location	Jan	Feb	Mar	Apr	May	Jun	Jul	Aug	Sep	Oct	Nov	Dec
Midland	247	270	225	293	180	180	180	180	180	180	225	0
	1.6	1.1	1.2	1.1	1.5	1.8	1.7	1.5	1.8	1.9	1.4	1.1
Port Arthur	315	180	180	158	158	180	180	22	23	338	158	158
	1.5	1.6	2.2	2.1	2.3	1.6	1.2	1.3	1.3	1.3	1.4	1.4
San Angelo	225	225	204	203	180	180	180	180	202	202	23	225
	1.5	1.3	1.1	1.5	1.8	2.0	2.8	1.6	1.8	1.7	2.1	2.7
San Antonio	23	337	338	21	135	135	135	135	45	22	0	0
	1.6	1.3	1.2	1.1	1.7	1.6	1.9	1.7	1.2	1.5	2.1	1.6
Waco	0	0	180	180	180	180	202	202	202	22	0	0
	2.5	2.4	2.5	2.4	2.5	3.0	2.4	2.3	2.0	2.4	2.5	2.6
Wichita Falls	1	0	338	180	158	180	180	180	180	180	338	0
	1.9	1.6	1.5	1.8	1.9	2.3	1.5	1.4	1.8	2.1	1.6	1.6
Utah												
Dugway	180	0	0	0	0	180	180	180	0	180	0	180
	4.0	3.2	2.8	1.9	2.0	2.1	1.8	2.2	2.0	2.2	2.6	2.6
Ogden	180	45	180	23	67	67	68	67	68	68	67	22
	1.4	1.3	1.3	1.2	1.3	1.2	1.9	1.6	2.1	2.2	1.3	1.1
Salt Lake City	158	157	158	180	337	157	180	157	158	157	158	158
	2.6	2.5	1.7	1.6	1.7	1.7	1.6	1.9	2.4	1.5	2.4	2.4
Wendover	338	315	315	315	315	337	338	157	315	315	315	315
	1.4	1.9	1.9	2.1	1.6	1.5	1.4	1.3	1.6	1.9	1.8	1.9
Virginia												
Blackstone	221	225	203	225	225	225	221	203	203	23	203	225
	1.8	1.2	1.2	1.6	1.9	1.8	2.3	1.6	1.8	1.7	1.5	1.3
Chincoteague	315	293	292	182	225	228	203	203	45	45	315	301
	1.4	1.3	1.3	1.1	1.2	1.1	1.5	1.3	1.7	1.4	1.3	1.3
Hampton	0	0	22	0	202	203	203	22	22	0	0	0
	1.5	1.3	1.4	1.5	1.5	1.7	1.9	1.7	2.1	2.1	1.6	1.5
Oceana	0	353	225	203	225	45	225	67	45	22	338	0
	1.7	1.1	1.1	1.4	1.2	1.5	2.2	1.4	1.5	1.4	1.5	1.3

Quantico	337	337	315	180	180	180	180	180	180	349	338	337
	1.7	1.4	1.3	1.2	1.4	1.2	1.6	1.5	1.7	1.3	1.5	1.5
Washington												
Everett	158	158	158	158	158	338	0	338	338	158	158	158
	2.5	2.4	2.7	2.3	2.7	3.5	2.9	3.1	4.0	2.8	2.5	2.6
Kelso	158	158	158	158	338	338	338	338	0	158	158	158
	7.7	6.3	3.8	2.3	1.8	1.6	1.9	1.8	3.3	4.5	9.2	9.2
Moses Lake	0	180	225	227	270	225	270	0	0	180	0	180
	2.5	1.9	1.3	1.1	1.1	1.2	1.1	1.2	1.3	1.6	1.6	1.9
Olympia	202	203	203	225	247	225	225	225	225	202	203	203
	3.1	2.9	2.3	2.3	2.2	2.3	2.3	2.3	2.3	2.6	2.8	3.1
Spokane	203	203	225	225	225	203	203	203	225	202	203	203
	4.1	2.8	2.5	3.1	3.4	2.1	1.9	1.6	2.6	3.4	3.9	3.5
Tacoma	203	203	203	225	225	225	225	225	203	203	202	202
	2.4	2.6	2.1	2.0	1.9	1.8	1.4	1.4	2.2	2.3	2.2	2.4
Walla Walla	202	180	202	225	225	225	225	225	225	180	180	180
	2.6	2.4	1.8	2.8	2.7	2.9	2.7	2.4	2.7	2.4	2.3	5.5
Whidbey Island	135	135	135	135	270	270	225	270	135	135	135	135
	2.7	2.5	1.9	1.7	1.4	1.7	1.6	1.2	2.2	2.7	2.8	2.8
Yakima	203	223	221	293	293	315	337	315	293	180	180	203
	2.3	1.0	1.1	1.2	1.7	1.8	1.7	2.0	1.5	1.1	1.1	1.4
Wisconsin												
Green Bay	292	228	225	247	225	225	225	225	225	225	270	227
	1.2	1.3	1.4	1.2	1.4	1.5	1.3	1.9	1.7	1.3	1.3	1.2
La Crosse	315	315	315	315	329	157	336	225	315	315	315	338
	1.6	1.9	1.8	1.8	1.1	1.2	1.4	1.2	1.5	1.5	1.9	1.0
Madison	315	270	250	270	248	247	248	225	247	247	270	248
	1.2	1.3	1.4	1.5	1.3	1.3	1.3	1.3	1.3	1.1	1.4	1.2
Milwaukee	225	225	23	225	225	203	202	225	203	180	225	247
	1.2	1.2	1.3	1.1	1.2	1.2	1.7	1.3	1.2	1.2	1.4	1.1
Wyoming												
Cheyenne	270	270	270	315	292	293	337	315	292	293	292	270
	2.4	1.7	1.2	1.5	1.4	1.3	1.5	1.2	1.4	1.7	1.9	2.3
Sheridan	315	315	315	315	315	315	315	315	315	246	315	292
	3.4	2.5	3.0	3.1	2.7	3.1	2.4	2.5	3.0	3.2	3.4	2.3

Table 4
WIND EROSION DIRECTION FACTOR

Preponderance	Angle of deviation (°)										
	0	5	10	15	20	25	30	35	40	45	50
1.0	1.90	1.90	1.90	1.90	1.90	1.90	1.90	1.90	1.90	1.90	1.90
1.2	1.55	1.57	1.60	1.65	1.70	1.74	1.77	1.81	1.84	1.88	1.92
1.4	1.40	1.43	1.46	1.51	1.55	1.60	1.65	1.71	1.78	1.86	1.95
1.6	1.30	1.34	1.38	1.42	1.46	1.51	1.55	1.64	1.73	1.85	1.97
1.8	1.23	2.26	3.30	2.35	1.40	1.44	1.48	1.59	1.70	1.85	2.00
2.0	1.19	1.24	1.30	1.32	1.35	1.39	1.44	1.55	1.67	1.85	2.04
2.2	1.17	1.22	1.27	1.30	1.33	1.37	1.41	1.53	1.66	1.87	2.08
2.4	1.15	1.20	1.25	1.28	1.31	1.35	1.40	1.52	1.65	1.88	2.12
2.6	1.13	1.18	1.23	1.27	1.31	1.35	1.40	1.53	1.67	1.91	2.16
2.8	1.11	1.16	1.22	1.26	1.30	1.36	1.41	1.56	1.71	1.96	2.22
3.0	1.09	1.14	1.20	1.25	1.30	1.36	1.43	1.59	1.75	2.02	2.29
3.2	1.08	1.14	1.19	1.25	1.31	1.38	1.45	1.64	1.82	2.09	2.37
3.4	1.07	1.13	1.19	1.25	1.31	1.40	1.48	1.68	1.89	2.18	2.47
3.6	1.06	1.12	1.18	1.25	1.32	1.42	1.52	1.76	2.00	2.31	2.62
3.8	1.05	1.11	1.18	1.25	1.33	1.45	1.57	1.84	2.11	2.45	2.78
4.0	1.04	1.10	1.17	1.26	1.35	1.48	1.61	1.91	2.22	2.59	2.95

Table 5
DEVIATIONS OF CURVE OF E_2 vs L FROM REFERENCE FOR USE IN DETERMINING E_4 (50)

SOIL LOSS, E_2 TONS/A./YR. and TONS/HA./YR.		MEDIAN UNSHELTERED DISTANCE ACROSS FIELD, L (METERS AND FEET)																		
		3,048	2,438	1,829	1,219	914	610	305	244	183	122	91	61	30	24	18	12	9	6	3
METRIC	ENGLISH	10,000	8,000	6,000	4,000	3,000	2,000	1,000	800	600	400	300	200	100	80	60	40	30	20	10
672	300	0.00	0.00	0.00	0.00	0.00	0.00	0.00	0.00	0.00	0.00	0.25	1.00	2.75	3.50	4.90	6.90	8.50	11.00	15.50
650	290	0.00	0.00	0.00	0.00	0.00	0.00	0.00	0.00	0.00	0.25	0.75	1.25	3.00	3.75	5.10	7.00	8.75	11.50	16.00
628	280	0.00	0.00	0.00	0.00	0.00	0.00	0.00	0.00	0.00	0.25	0.75	1.50	3.50	4.25	5.80	7.90	9.50	12.00	16.50
605	270	0.00	0.00	0.00	0.00	0.00	0.00	0.00	0.00	0.00	0.50	0.80	1.80	3.90	4.70	6.00	8.10	10.00	12.50	17.00
583	260	0.00	0.00	0.00	0.00	0.00	0.00	0.00	0.00	0.00	0.50	1.00	1.80	3.70	4.50	6.00	8.20	10.10	12.80	17.50
560	250	0.00	0.00	0.00	0.00	0.00	0.00	0.00	0.00	0.00	0.50	1.00	2.00	4.00	5.00	6.50	8.75	10.60	13.00	17.60
538	240	0.00	0.00	0.00	0.00	0.00	0.00	0.00	0.00	0.20	0.75	1.10	2.10	4.20	5.20	7.00	9.00	11.00	13.50	18.20
516	230	0.00	0.00	0.00	0.00	0.00	0.00	0.00	0.00	0.25	0.95	1.50	2.50	4.80	5.90	7.50	9.75	11.50	14.00	18.60
493	220	0.00	0.00	0.00	0.00	0.00	0.00	0.00	0.00	0.50	1.00	1.60	2.80	5.00	6.10	7.90	10.00	11.80	14.30	18.80
471	210	0.00	0.00	0.00	0.00	0.00	0.00	0.00	0.20	0.75	1.50	2.00	3.30	5.90	6.90	8.20	10.40	12.20	14.50	18.90
448	200	0.00	0.00	0.00	0.00	0.00	0.00	0.00	0.25	0.80	1.50	2.10	3.40	6.00	7.00	8.75	10.90	12.50	15.00	19.40
426	190	0.00	0.00	0.00	0.00	0.00	0.00	0.50	0.80	1.00	2.00	2.80	4.00	6.10	7.00	8.80	11.00	12.60	15.10	19.75
403	180	0.00	0.00	0.00	0.00	0.00	0.00	0.20	0.50	1.00	1.90	2.50	3.90	6.50	7.50	9.20	11.40	12.70	15.80	20.00
381	170	0.00	0.00	0.00	0.00	0.00	0.00	0.25	0.75	1.10	2.00	3.00	4.25	7.00	8.00	9.80	11.80	13.00	16.00	20.00
359	160	0.00	0.00	0.00	0.00	0.00	0.00	0.50	0.90	1.40	2.40	3.10	4.80	7.40	8.50	10.10	12.30	14.00	16.50	20.60
336	150	0.00	0.00	0.00	0.00	0.00	0.00	0.50	0.90	1.50	2.50	3.50	5.00	7.75	8.90	10.50	12.60	14.50	17.00	21.00
314	140	0.00	0.00	0.00	0.00	0.00	0.00	0.75	1.00	1.80	2.90	3.60	5.20	8.00	9.00	10.60	12.90	14.50	17.00	21.00
291	130	0.00	0.00	0.00	0.00	0.00	0.00	0.80	1.00	1.95	3.00	3.90	5.40	8.00	9.20	11.00	13.00	14.90	17.30	21.10
269	120	0.00	0.00	0.00	0.00	0.00	0.25	1.10	1.80	2.60	3.80	4.75	6.20	9.00	10.00	11.80	13.80	15.50	17.75	21.50
247	110	0.00	0.00	0.00	0.00	0.00	0.30	1.50	1.95	2.90	4.00	5.00	6.75	9.50	10.60	12.30	14.50	16.20	18.60	22.50
224	100	0.00	0.00	0.00	0.00	0.00	0.30	1.50	2.00	3.00	4.25	5.40	7.00	10.00	11.00	13.00	15.00	16.50	19.20	23.00
202	90	0.00	0.00	0.00	0.00	0.00	0.50	1.80	2.25	3.30	4.80	5.90	7.60	10.60	11.50	13.45	15.75	17.50	19.80	24.00
179	80	0.00	0.00	0.00	0.00	0.20	0.75	2.00	2.50	3.60	4.90	6.00	7.60	10.70	11.90	13.60	16.00	17.50	20.00	24.00
157	70	0.00	0.00	0.00	0.50	0.80	1.00	2.75	3.20	4.00	5.25	6.20	8.00	10.80	12.00	13.75	16.00	18.00	20.50	26.10
134	60	0.00	0.00	0.00	0.25	0.60	1.20	2.90	3.30	4.40	5.70	6.80	8.45	11.50	12.70	14.75	17.00	19.00	21.70	26.50
112	50	0.00	0.00	0.00	0.50	0.90	1.50	3.00	3.75	4.80	6.00	7.10	9.00	12.40	13.55	15.60	18.00	20.10	23.00	28.00
90	40	0.00	0.00	0.00	0.60	1.00	1.90	3.50	4.20	5.20	6.80	7.90	9.80	13.10	14.50	16.50	19.20	21.00	-1.00	-1.00
78	35	0.00	0.00	0.25	0.80	1.40	2.10	3.95	4.50	5.60	7.00	8.10	10.10	13.90	15.25	17.40	20.50	-1.00	-1.00	-1.00
67	30	0.00	0.00	0.25	0.90	1.30	2.20	4.00	4.80	6.00	7.50	8.90	11.00	15.00	16.40	19.00	-1.00	-1.00	-1.00	-1.00
56	25	0.00	0.00	0.40	1.00	1.80	2.50	4.50	5.30	6.75	8.25	9.90	12.00	16.00	-1.00	-1.00	-1.00	-1.00	-1.00	-1.00
45	20	0.00	0.00	0.50	1.00	1.80	2.50	5.00	5.80	7.10	9.10	10.80	13.80	-1.00	-1.00	-1.00	-1.00	-1.00	-1.00	-1.00
34	15	0.00	0.25	0.75	1.20	2.00	3.25	5.50	6.50	8.00	10.10	12.00	-1.00	-1.00	-1.00	-1.00	-1.00	-1.00	-1.00	-1.00
22	10	0.00	0.25	0.50	1.50	2.00	3.10	5.75	6.80	8.60	-1.00	-1.00	-1.00	-1.00	-1.00	-1.00	-1.00	-1.00	-1.00	-1.00
11	5	0.00	0.60	1.00	2.00	2.75	3.75	6.00	-1.00	-1.00	-1.00	-1.00	-1.00	-1.00	-1.00	-1.00	-1.00	-1.00	-1.00	-1.00

SCALE USED WITH TABLE ABOVE FOR DETERMINING SOIL LOSS E_4 — **SCALE DIVISIONS** — SOIL LOSS E_3 = I'K'C' OR E_4 = I'K'C' f(L') (TONS/A./YR. AND TONS/HA./YR.)

ENGLISH: 420 380 340 300 280 260 240 220 200 180 160 140 120 100 90 80 70 60 50 40 30 25 20 15 10 5 2 1 .05

METRIC: 941 852 762 672 628 583 538 493 448 403 359 314 269 224 202 179 157 134 112 90 67 56 45 34 22 11 4

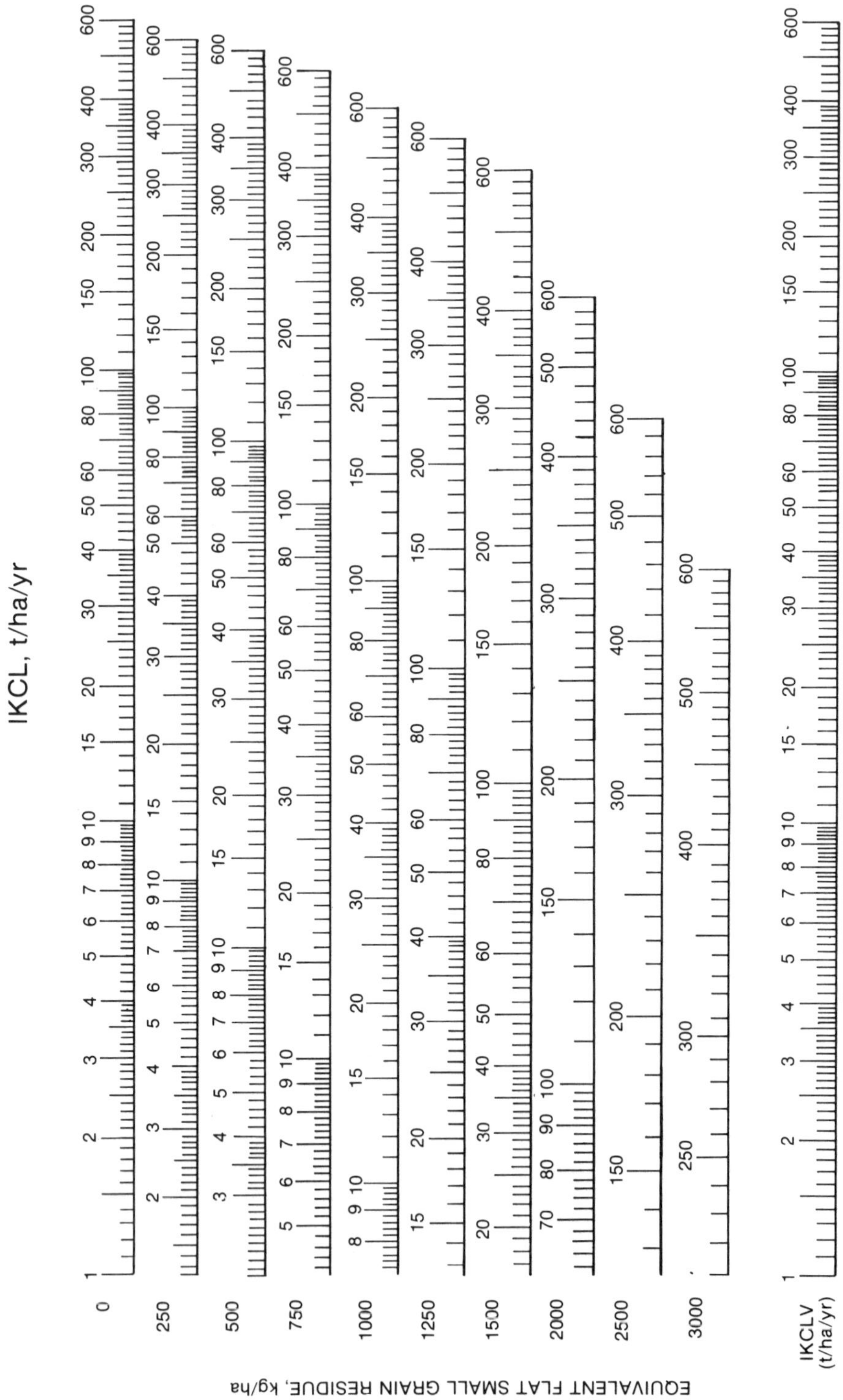

FIGURE 2. Chart for determining soil loss E = IKCLV from soil loss E_4 = IKCL and from equivalent flat small grain residue.

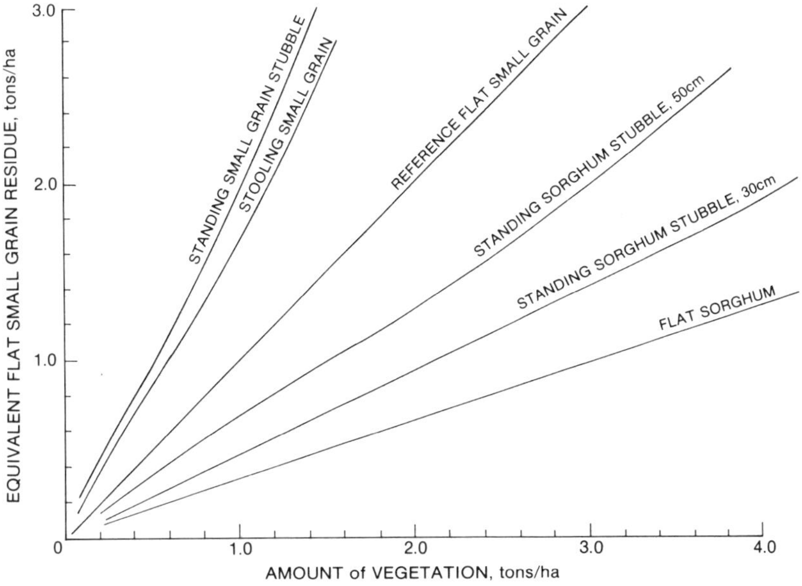

FIGURE 3. Wind erosion protection of small grain and sorghum stubble compared with flat small grain residue. (Adapted from Woodruff, N. P. and Siddoway, F. H.[22])

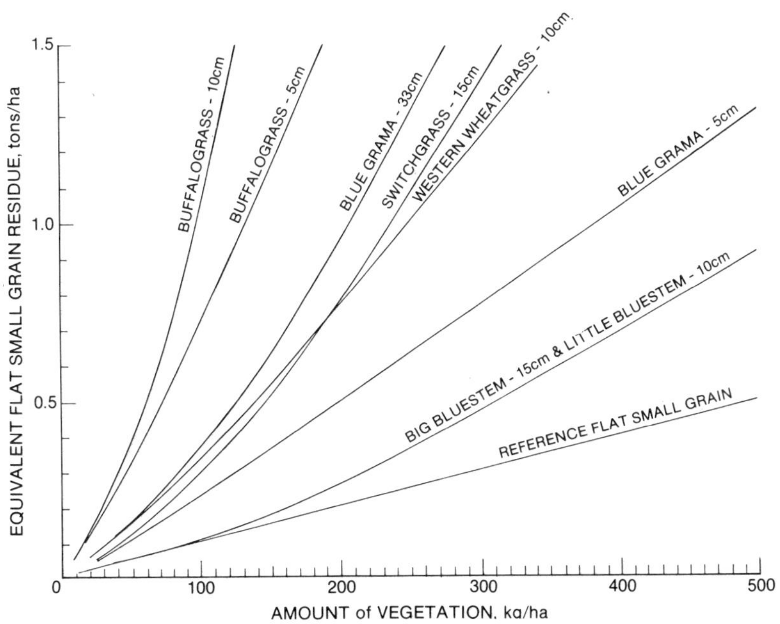

FIGURE 4. Wind erosion protection of various range grasses compared with flat small grain residue. (Adapted from Lyles, L. and Allison, B. E.[41])

Table 6
POTENTIAL AVERAGE ANNUAL SOIL LOSS AS INDICATED BY SOLVING THE WIND EROSION EQUATION FOR INDICATED LEVELS OF THE FACTORS INFLUENCING WIND EROSION

Clods > 0.84 mm (%)	Soil erodibility I (metric tons/ha)	Ridge roughness K	Climatic factor C	Unsheltered distance along prevail. erod. dir. L (m)	Equivalent flat small grain residue V (kg/ha)	Potential avg. ann. soil loss E (tons/ha/year)
10	300	0.75	100	wide[a]	0	225
10	300	0.75	100	wide	1000	98
10	300	0.75	100	wide	2000	15
10	300	0.75	100	200	0	186
10	300	0.75	100	200	1000	75
10	300	0.75	100	200	2000	9.9
10	300	0.75	50	wide	0	113
10	300	0.75	50	wide	1000	39
10	300	0.75	50	wide	2000	3.5
10	300	0.75	50	200	0	87
10	300	0.75	50	200	1000	27
10	300	0.75	50	200	2000	1.9
10	300	0.75	25	wide	0	56
10	300	0.75	25	wide	1000	15
10	300	0.75	25	wide	2000	<1.0
10	300	0.75	25	200	0	40
10	300	0.75	25	200	1000	9.5
10	300	0.75	25	200	2000	<1.0
25	193	0.75	100	wide	0	145
25	193	0.75	100	wide	1000	54
25	193	0.75	100	wide	2000	5.8
25	193	0.75	100	200	0	110
25	193	0.75	100	200	1000	37
25	193	0.75	100	200	2000	3.0
25	193	0.75	50	wide	0	73
25	193	0.75	50	wide	1000	21

25	193	0.75	50	wide	2000	1.3
25	193	0.75	50	200	0	56
25	193	0.75	50	200	1000	15
25	193	0.75	50	200	2000	<1.0
25	193	0.75	25	wide	0	36
25	193	0.75	25	wide	1000	8.2
25	193	0.75	25	wide	2000	<1.0
25	193	0.75	25	200	0	22.4
25	193	0.75	25	200	1000	4.4
25	193	0.75	25	200	2000	<1.0
40	126	0.75	100	wide	0	95
40	126	0.75	100	wide	1000	30
40	126	0.75	100	wide	2000	2.4
40	126	0.75	100	200	0	63
40	126	0.75	100	200	1000	17
40	126	0.75	100	200	2000	<1.0
40	126	0.75	50	wide	0	47
40	126	0.75	50	wide	1000	12
40	126	0.75	50	wide	2000	<1.0
40	126	0.75	50	200	0	27
40	126	0.75	50	200	1000	5.6
40	126	0.75	50	200	2000	<1.0
40	126	0.75	25	wide	0	24
40	126	0.75	25	wide	1000	8
40	126	0.75	25	wide	2000	<1.0
40	126	0.75	25	200	0	9.6
40	126	0.75	25	200	1000	1.4
40	126	0.75	25	200	2000	<1.0

[a] Any further increase in field width would not increase the wind erosion hazard. This condition usually occurs in a field between 500 and 1000 m wide.

REFERENCES

1. **Idso, S. B.,** Dust storms, *Sci. Am.,* 235(4), 108, 1976.
2. Soil erosion by wind and measures for its control on agricultural lands, Agricultural Development Paper No. 71, Food and Agriculture Organization, Genera, 1960.
3. Soil and Water Conservation Needs — a National Inventory, Misc. Publ. 971, U.S. Department of Agriculture, Washington, D.C., 1965.
4. **Bagnold, R. A.,** *The Physics of Blown Sand and Desert Dunes,* William Morrow, New York, 1943, 265.
5. **Bisal, F. and Nielson, K. F.,** Movement of soil particles in saltation, *Can. J. Soil Sci.,* 42, 81, 1962.
6. **Lyles, L. and Krauss, R. K.,** Threshold velocities and initial particle motion as influenced by air turbulence, *Trans. Am. Soc. Agric. Eng.,* 14, 563, 1971.
7. **Chepil, W. S. and Woodruff, N. P.,** The physics of wind erosion and its control, *Adv. Agron.,* 15, 211, 1963.
8. **Vanoni, V. A.,** Sedimentation engineering, *ASCE Man. Rep. Eng. Pract.,* 54, 230, 1975.
9. **Zingg, A. W.,** Some characteristics of aeolian sand movement by saltation process, Editions du Centre National de la Recherche Scientifique 13, Quai Anatole France, Paris, 1953, 197.
10. **Bennett, H. H.,** *Soil Conservation,* McGraw-Hill, New York, 1939, 933.
11. **Hagen, L. J. and Woodruff, N. P.,** Air pollution from duststorms in the Great Plains, *Atmos. Environ.,* 7, 323, 1973.
12. **Chepil, W. S.,** Sedimentary characteristics of dust storms. I. Sorting of wind-eroded soil material, *Am. J. Sci.,* 255, 12, 1957.
13. **Daniel, H. A.,** Physical changes in soils of the Southern High Plains due to cropping and wind erosion and the relation between the sand + silt/clay ratios in these soils, *J. Am. Soc. Agron.,* 28, 570, 1936.
14. **Moss, H. C.,** Some field and laboratory studies of soil drifting in Saskatchewan, *Sci. Agric.,* 15, 665, 1935.
15. **Malina, Frank J.,** Recent developments in the dynamics of wind erosion, *Trans. Am. Geophys. Union,* 262, 1941.
16. **Chepil, W. S.,** Sedimentary characteristics of dust storms. III. Composition of suspended dust, *Am. J. Sci.,* 255, 206, 1957.
17. **Zingg, A. W.,** The wind erosion problem in the Great Plains, *Trans. Am. Geophys. Union,* 35, 252, 1954.
18. **Daniel, H. A. and Langham, W. H.,** The effect of wind erosion and cultivation on the total nitrogen and organic matter of soils in the Southern High Plains, *J. Am. Soc. Agron.,* 28, 587, 1936.
19. **Claflin, L. E., Stuteville, D. L., and Armbrust, D. V.,** Windblown soil in the epidemiology of bacterial leaf spot of alfalfa and common blight of beans, *Phytopathology,* 63, 1417, 1973.
20. **Hayes, W. A.,** Wind erosion equation useful in designing northeastern crop protection, *J. Soil Water Conserv.,* 20, 153, 1965.
21. **Hayes, W. A.,** Guide for wind erosion control in the northeastern states, Soil Conservation Service, U.S. Department of Agriculture, Washington, D.C.
22. **Woodruff, N. P. and Siddoway, F. H.,** A wind erosion equation, *Soil Sci. Soc. Am. Proc.,* 29, 602, 1965.
23. **Chepil, W. S.,** Properties of soil which influence wind erosion. II. Dry aggregate structure as an index of erodibility, *Soil Sci.,* 69, 403, 1950.
24. **Chepil, W. S. and Woodruff, N. P.,** Estimations of wind erodibility of farm fields, Prod. Res. Rep. No. 25, United States Department of Agriculture, Washington, D.C., 1959, 21.
25. **Chepil, W. S.,** Seasonal fluctuations in soil structure and erodibility of soil by wind, *Soil Sci. Soc. Am. Proc.,* 18, 13, 1954.
26. **Chepil, W. S., and Milne, R. A.,** Wind erosion of soil in relation to roughness of surface, *Soil Sci.,* 52, 417, 1941.
27. **Hayes, W. A.,** Designing wind erosion control systems in the Midwest Region, Regional Technical Service Center, Soil Conservation Service, Lincoln, Neb., Technical Note — Agronomy LI-9, 1972.
28. **Chepil, W. S.,** Dynamics of wind erosion. I. Nature of movement of soil by wind, *Soil Sci.,* 60, 305, 1945.
29. **Chepil, W. S.,** Dynamics of wind erosion. II. Initiations of soil movement, *Soil Sci.,* 60, 397, 1945.
30. **Chepil, W. S.,** Dynamics of wind erosion. III. The transport capacity of wind, *Soil Sci.,* 60, 475, 1945.
31. **Zingg, A. W.,** Wind-tunnel studies of the movement of sedimentary materials, in *Proceedings of 5th Hydraulic Conference,* John Wiley & Sons, New York, 1953, 111.
32. **Chepil, W. S.,** Influence of moisture on erodibility of soil by wind, *Soil Sci. Soc. Am. Proc.,* 20, 288, 1956.

33. **Chepil, W. S., Siddoway, F. H., and Armbrust, D. V.,** Climatic factor for estimating wind erodibility of farm fields, *J. Soil Water Conserv.*, 17, 162, 1962.
34. **Chepil, W. S. and Milne, R. A.,** Wind erosion of soils in relation to size and nature of the exposed area, *Sci. Agr.*, 21, 479, 1941.
35. **Chepil, W. S.,** Dynamics of wind erosion. V. Cumulative intensity of soil drifting across eroding fields, *Soil Sci.*, 61, 257, 1946.
36. **Chepil, W. S.,** Width of field strips to control wind erosion, *Kans. Agric. Exp. Stn. Tech. Bull.*, 92, 1957.
37. **Chepil, W. S., Woodruff, N. P., and Zingg, A. W.,** Field study of wind erosion in western Texas, SCS-TP-125, U.S. Department of Agriculture, Washington, D.C., 1955.
38. **Chepil, W. S.,** Utilization of crop residues for wind erosion control, *Sci. Agr.*, 24, 307, 1944.
39. **Siddoway, F. H., Chepil, W. S., and Armbrust, D. V.,** Effect of kind, amount, and placement of residue on wind erosion control, *Trans. Am. Soc. Agric. Eng.*, 8, 327, 1965.
40. **Craig, D. C., and Turelle, J. W.,** Guide for Wind Erosion Control on Cropland in the Great Plains States, Soil Conservation Service, U.S. Department of Agriculture, Washington, D.C., 1964, 104.
41. **Lyles, L. and Allison, B. E.,** Range grasses and their small grain equivalents for wind erosion control, *J. Range Manage.*, 33, 143, 1980.
42. **Lyles, L. and Allison, B. E.,** Equivalent wind erosion protection of selected crop residues, *Trans. Am. Soc. Agr. Eng.*, 24, 405, 1981.
43. **Woodruff, N. P., Lyles, L., Dickerson, J. D., and Armbrust, D. V.,** Using cattle feedlot manure to control wind erosion, *J. Soil Water Conserv.*, 29, 127, 1974.
44. **Skidmore, E. L. and Siddoway, F. H.,** Crop residue requirements to control wind erosion, in *Crop Residue Management Systems*, Oschwald, W. R., Ed., Special Publ. No. 31, American Society of Agronomy, Madison, Wis., 1978, 17.
45. **Skidmore, E. L., Kumar, M., and Larson, W. E.,** Crop residue management for wind erosion control in the Great Plains, *J. Soil Water Conserv.*, 34, 90, 1979.
46. **Hayes, W. A.,** Estimating wind erosion in the field, in *Land Use: Food and Living*, Soil Conservation Society America, San Antonio, 1975, 139.
47. **Carreker, J. R.,** Wind erosion in the southeast, *J. Soil Water Conserv.*, 21, 86, 1966.
48. **Moldenhauer, W. C. and Duncan, E. R.,** Principles and methods of wind erosion control in Iowa, Special Rep. No. 62, Iowa State University, Ames, 1969.
49. **Woodruff, N. P., Lyles, L., Siddoway, F. H., and Fryrear, D. W.,** How to control wind erosion, U. S. Department of Agriculture, Washington, D.C., Agric. Inf. Bull. No. 354, 1972, 22.
50. **Hagen, L. J., Skidmore, E. L., and Dickerson, J. D.,** Designing narrow strip barrier systems to control wind erosion, *J. Soil Water Conserv.*, 27, 269, 1972.
51. Investigations of fugitive dust-sources, emissions, and control, Report prepared under contract No. 68-02-0044, Task Order No. 9, PEDCO-Environmental Specialists, Inc., Office of Air Quality Planning and Standards, U.S. Environmental Protection Agency, Research Triangle Park, N.C., 1973.
52. **Wilson, L.,** Application of the wind erosion equation to predict fugitive dust emissions, *J. Soil Water Conserv.*, 30, 215, 1975.
53. **Gillette, D. A., Blifford, I. H., Jr., and Fenster, C. R.,** Measurements of aerosol size distribution and vertical fluxes of aerosols on land subject to wind erosion, *J. Appl. Meteorol.*, 11, 977, 1972.
54. **Lyles, L.,** Speculation on the Effect of Wind Erosion on Productivity, Special Report to U.S. Department of Agriculture, Task Force on Wind Erosion Damage Estimates, 1974.
55. **Lyles, L.,** Possible effects of wind erosion on soil productivity, *J. Soil Water Conserv.*, 30, 279, 1975.
56. **Lyles, L., Schmeidler, N. F., and Woodruff, N. P.,** Stubble requirements in field strips to trap windblown soil, *Kans. Agric. Exp. Stn. Res. Publ.*, 164, 22, 1973.
57. **Chepil, W. S.,** Soil conditions that influence wind erosion, USDA Tech. Bull. No. 1185, U.S. Department of Agriculture, Washington, D.C., 1958, 40.
58. **Skidmore, E. L. and Woodruff, N. P.,** Wind erosion forces in the United States and their use in predicting soil loss, Handbook No. 346, Agriculture Research Service, U.S. Department of Agriculture, Washington, D.C., 1968, 42.
59. **Skidmore, E. L., Fisher, P. S., and Woodruff, N. P.,** Wind erosion equation: computer solution and application, *Soil Sci. Soc. Am. Proc.*, 34, 931, 1970.
60. **Fisher, P. S. and Skidmore, E. L.,** WEROS: a Fortran IV Program to Solve the Wind Erosion Equation, 41-174, Agricultural Research Service, U.S. Department of Agriculture, Washington, D.C., 1970, 13.
61. **Skidmore, E. L.,** Wind Erosion Calculator: Examples of Use, Wind Erosion Research Unit, Manhattan, Kan., unpublished report, 1978.

GLOSSARY

A

ABC soil — A soil with a distinctly developed profile, including A, B, and C horizons.

absorption, active — Movement of ions and water into the plant root as a result of metabolic processes by the root, frequently against an activity gradient.

absorption, passive — Movement of ions and water into the plant root as a result of diffusion along an activity gradient.

AC soil — A soil having a profile containing only A and C horizons with no clearly developed B horizon.

acid-forming fertilizers — A fertilizer capable of increasing residual soil acidity.

acidity, residual — Soil acidity which is neutralized by lime or other alkaline materials, but which cannot be replaced by an unbuffered salt solution.

acidity, salt-replaceable — The aluminum and hydrogen which can be replaced from an acid soil by an unbuffered salt solution such as KCl or NaCl.

acidity, total — The total acidity in a soil or clay. Usually it is estimated by a buffered salt determination of (cation-exchange capacity − exchangeable bases) = total acidity. It also is approximated by the sum of salt replaceable acidity + residual acidity.

acidulation — Process of treating a material with an acid. The most common acidulation process is the treatment of phosphate rock with an acid or mixtures of acids.

Actinomycetes — A nontaxonomic term applied to a group of organisms with characteristics intermediate between the simple bacteria and the true fungi. Most soil actinomycetes are unicellular microorganisms that produce a slender branched mycelium and spotulate by segmentation of the entire mycelium, or, more commonly, by segmentation of special hyphae. Includes many but not all organisms belonging to the order Actinomycetales.

aerobic — (1) Having molecular oxygen as a part of the environment. (2) Growing only in the presence of molecular oxygen, as aerobic organisms. (3) Occurring only in the presence of molecular oxygen (said of certain chemical or biochemical processes such as aerobic decomposition).

agronomy — A specialization of agriculture concerned with the theory and practice of field-crop production and soil management. The scientific management of land.

air porosity — The proportion of the bulk volume of soil that is filled with air at any given time or under a given condition such as a specified soil water content or soil water matrix potential.

alkaline soil — Any soil having a pH > 7.0. See reaction soil.

alkalinity, soil — The degree or intensity of alkalinity in a soil, expressed by a value > 7.0 for the soil pH.

allophane — A series of naturally occurring minerals which are hydrous aluminum silicates of widely varying chemical compositions, characterized by short range order, by the presence of Si-O-A1 bonds, and by a differential thermal analysis curve displaying a low temperature endotherm and a high temperature exotherm with no intermediate endotherm.

Alluvial soil — A soil from recently deposited alluvium and exhibiting essentially no horizon development or modification of the recently deposited materials.

Alpine Meadow soils — Dark soils of grassy meadows at altitudes above the timberline.

amendment, soil — Any substance added to the soil which alters soil properties. Examples are gypsum, lime, fertilizers, sawdust, etc.

ammoniation — Process of introducing ammonia into superphosphate, forming ammoniated superphosphate, or into a mixture of superphosphate and other fertilizer ingredients to make fertilizer mixtures.

ammonification — The biochemical process whereby ammoniacal nitrogen is released from nitrogen-containing organic compounds.

ammonium fixation — Adsorption of ammonium ions (NH_4^+) by the mineral fraction of the soil in forms that cannot be replaced by a neutral potassium salt solution (e.g., 1*N*KCl).

anaerobic — (1) The absence of molecular oxygen. (2) Growing in the absence of molecular oxygen (such as anaerobic bacteria). (3) Occurring in the absence of molecular oxygen (as a biochemical process).

anion-exchange capacity — The sum total of exchangeable anions that a soil can adsorb. Expressed as milliequivalents per 100 g of soil (or of other adsorbing material such as clay).

anion exclusion — The exclusion or repulsion of anions from the vicinity of soil particle surfaces because of the negative potential associated with most soil particles.

attapulgite clay — A fuller's earth. The main constituent is a hydrous magnesium aluminum silicate having high adsorptive and swelling properties. Used in fertilizer production, including conditioning of fertilizer products and as a suspending agent in suspension fertilizers.

autochthonous flora — That portion of the microflora presumed to subsist on the more resistant soil organic matter and little affected by the addition of fresh organic materials. Contrast with zymogenous flora.

autotrophic — Capable of utilizing carbon dioxide and/or carbonates as a sole or major source of carbon and of obtaining energy for carbon reduction and biosynthetic processes from radiant energy (photoautotroph) or oxidation of inorganic substances (chemoautotroph).

autotrophic nitrification — Oxidation of ammonium to nitrate through the combined action of two chemoautotrophic organisms, one forming nitrite from ammonium and the other oxidizing nitrite to nitrate.

available nutrients — (1) Nutrient ions or compounds in forms which plants can adsorb and utilize in growth, and (2) contents of legally designated "available" nutrients in fertilizers determined by specified laboratory procedures which in most states constitute the legal basis for guarantees.

available water — The portion of water in a soil that can be absorbed by plant roots. Considered by most workers as the amount of water released by the soil when the equilibrium soil water matrix potential is decreased from field capacity to −15 bar.

B

bar — A unit of pressure equal to one million dynes per square centimeter.

base-saturation percentage — The extent to which the adsorption complex of a soil is saturated with exchangeable cations other than hydrogen. It is expressed as a percentage of the total cation-exchange capacity.

BC soil — A soil profile with B and C horizons but with little or no A horizon.

bedding — Preparing a series of flat-topped parallel ridges usually no wider than two crop rows, separated by shallow trenches usually less than the width between crop rows. The resultant structures are beds.

bentonite — Clays, largely composed of smectite clay minerals, produced by the alteration of volcanic ash in situ.

biosequence — A sequence of related soils that differ, one from the other, primarily because of differences in kinds and numbers of *soil organisms* as a soil-forming factor.

Black Earth — A term used by some as synonymous with "Chernozem"; by others (in Australia) to describe self-mulching black clays.

Black Soils — A term used in Canada to describe soils with dark-colored surface horizons of the black (Chernozem) zone; includes Black Earth or Chernozem, Wiesenboden, Solonetz, etc.

BOD (biochemical oxygen demand) — The quantity of oxygen used in the biochemical oxidation of organic matter in a specified time, at a specified temperature, and under specified conditions. An indirect measure of the concentration of biologically degradable material present in organic wastes.

buffer compounds, soil — The clay, organic matter, and compounds such as carbonates and phosphates which enable the soil to resist appreciable change in pH.

bulk density, soil — The mass of dry soil per unit bulk volume. The bulk volume is determined before drying to constant weight at 105°C.

bulk volume — The volume, including the solids and the pores, of an arbitrary soil mass.

buried soil — Soil covered by an alluvial, loessal, or other deposit, usually to a depth greater than the thickness of the solum.

C

calcareous soil — Soil containing sufficient free calcium carbonate or calcium-magnesium carbonate to effervesce visibly when treated with cold 0.1 *N* hydrochloric acid.

calciphytes — Plants that require or tolerate considerable amounts of calcium, or, are associated with soils rich in calcium.

caliche — (1) A layer near the surface, more or less cemented by secondary carbonates of calcium or magnesium precipitated from the soil solution. It may occur as a soft thin soil horizon, as a hard thick bed just beneath the solum, or as a surface layer exposed by erosion. Not a geologic deposit. (2) Alluvium cemented with sodium nitrate, chloride, and/or other soluble salts in the nitrate deposits of Chile and Peru.

carbon cycle — The sequence of transformations whereby carbon dioxide is fixed in living organisms by photosynthesis or by chemosynthesis, liberated by respiration and by the death and decomposition of the fixing organisms, used by heterotrophic species, and ultimately returned to its original state.

carbon-organic nitrogen ratio — The ratio of the weight of organic carbon to the weight of organic nitrogen in soil, organic material, plants, or the cells of microorganisms.

catena — A sequence of soils of about the same age, derived from similar parent material, and occurring under similar climatic conditions, but having different characteristics due to variation in *relief* and in *drainage.* See clinosequence and toposequence.

cation exchange — The interchange between a cation in solution and another cation on the surface of any surface-active material such as clay colloid or organic colloid.

cation-exchange capacity (CEC) — The sum total of exchangeable cations that a soil can adsorb. Expressed in meg/100 g or per gram of soil (or of other exchangers such as clay).

channery — In Scotland and Ireland, gravel; in the U.S. thin, flat fragments of limestone, sandstone, or schist up to 6 in. in major diameter.

chemodenitrification — Nonbiological processes leading to the production of gaseous forms of nitrogen (molecular nitrogen or an oxide of nitrogen).

chisel — To break up soil using narrow tools. It may be performed at other than the normal plowing depth. Chiseling at depths greater than 40 cm is usually termed subsoiling.

chronosequence — A sequence of related soils that differ, one from the other, in certain properties primarily as a result of *time* as a soil-forming factor.

Table 1
CLASSIFICATION SCHEME FOR PHYLLOSILICATES RELATED TO CLAY MINERALS[1]

Type	Group (x = charge per formula unit)	Subgroup	Species[a]
1:1	Kaolinite-serpentine	Kaolinites	Kaolinite, Halloysite (7 A), Halloysite (10 A)
	x ∼ 0	Serpentines	Chrysotile, lizardite, antigorite
2:1	Pyrophyllite-talc	Pyrophyllites	Pyrophyllite
	x ∼ 0	Talcs	Talc
	Smectite	Dioctahedral smectites	Montmorillonite, beidellite, nontronite
	x ∼ 0.25—0.6	Trioctahedral smectites	Saponite, hectorite, sauconite
	Vermiculite	Dioctahedral vermiculite	Dioctahedral vermiculite
	x ∼ 0.6—0.9	Trioctahedral vermiculite	Trioctahedral vermiculite
	Mica	Dioctahedral micas	Muscovite, paragonite
	x ∼ 1	Trioctahedral micas	Biotite, phlogopite
	Brittle mica	Dioctahedral brittle micas	Margarite
	x ∼ 2	Trioctahedral brittle micas	Clintonite
2:1:1	Chlorite	Dioctahedral chlorites (4—5 octahedral cations per formula unit)	
	x variable	Trioctahedral chlorites (5—6 octahedral cations per formula unit)	Pennine, chlinochlore, prochlorite

[a] Only a few examples are given.

class, soil — A group of soils having a definite range in a particular property such as acidity, degree of slope, texture, structure, land-use capability, degree of erosion, or drainage. See soil texture and soil structure.

clay — (1) A soil separate consisting of particles <0.002 mm in equivalent diameter. See soil separates. (2) A textural class. See soil texture.

clay loam — A textural class. See soil texture.

clay mineral — A naturally occurring inorganic crystalline phyllosilicate material found in soils and other earthy deposits. Not limited by particle size.

claypan — A dense, compact layer in the subsoil having a much higher clay content than the overlying material, from which it is separated by a sharply defined boundary; formed by downward movement of clay or by synthesis of clay in place during soil formation. Claypans are usually hard when dry, and plastic and sticky when wet. Also, they usually impede the movement of water and air, and the growth of plant roots.

climosequence — A sequence of related soils that differ, one from the other, in certain properties primarily as a result of the effect of *climate* as a soil-forming factor.

clinosequence — A group of related soils that differ, one from the other, in certain properties primarily as a result of the effect of the *degree of slope* on which they were formed. See toposequence.

coarse sand — See soil separates and soil texture.

colloidal phosphate — A finely divided raw material phosphate or phosphatic clay.

colluvium — A deposit of rock fragments and soil material accumulated at the base of steep slopes as a result of gravitational action. See creep.

compound fertilizer — A mixed fertilizer containing at least two of the primary plant nutrients (N, P_2O_5, and K_2O) formed by the intimately mixing or granulating two or more fertilizer materials, usually by the processes that involve chemical reactions of the materials with each other.

concentrated flow — The flowing of a rather large accumulated body of water over a relatively narrow course. It often causes serious erosion and gullying.

conservation tillage — Any tillage sequence which reduces loss of soil or water relative to conventional tillage.

controlled-release fertilizers — Fertilizers in which one or more of the nutrients have limited solubility in the soil, so that they become available to the growing plants over a controlled period.

creep — Slow mass movement of soil and soil material down relatively steep slopes primarily under the influence of gravity, but facilitated by saturation with water and by alternate freezing and thawing.

critical nutrient concentration — The nutrient concentration in the plant or its organs at the time the nutrient becomes deficient for growth.

crop residue management — Disposition of stubble, stalks, and other crop residues by tillage operations. (1) To remove residues from the soil surface, (2) to anchor residues partially in the surface soil while leaving the residues partially exposed at the surface (mulch tillage), (3) to leave residues entirely at the soil surface intact or cut into smaller pieces. (Residues may be removed by nontillage methods, i.e., harvesting, burning, grazing, etc.)

cumulative infiltration — Total volume of water infiltrated per unit area during a specified time period.

cyclic salt — Salt deposited on soils near the sea or inland salt lakes from the atmosphere.

D

Darcy's law — (1) A law describing the rate of flow of water through porous media. (Named for Henry Darcy of Paris who formulated it in 1856 from extensive work on the flow of water through sand filter beds.) As formulated by Darcy the law is:

$$Q = kS\frac{H + e}{e}$$

where Q = the volume of water passed in unit time, S = the area of the bed, e = the thickness of the bed, H = the height of the water on top of the bed, and "k is a coefficient depending on the nature of the sand" and for cases where the pressure "under the filter is equal to the weight of the atmosphere." (2) Generalization for three dimensions: the rate of viscous flow of water in isotropic porous media is proportional to, and in the direction of, the hydraulic gradient. (3) Generalization for other fluids: the rate of viscous flow of homogeneous fluids through isotropic porous media is proportional to, and in the direction of, the driving force.

deflocculate — (1) To separate the individual components of compound particles by chemical and/or physical means. (2) To cause the particles of the *disperse phase* of a colloidal system to become suspended in the *dispersion medium.*

denitrification — The biochemical reduction of nitrate or nitrite to gaseous nitrogen either as molecular nitrogen or as an oxide of nitrogen.

desert crust — A hard layer, containing calcium carbonate, gypsum, or other binding material, exposed at the surface in desert regions.

desert pavement — The layer of gravel or stones left on the land surface in desert regions after the removal of the fine material by wind erosion.

desert varnish — A glossy sheen or coating on stones and gravel in arid regions.

diffuse layer — A system, when referred to soils, which consists of a charged (negative) particle surface and an equal amount of counterions (positive) accumulated in the liquid near the particle surface.

disperse — (1) To break up compound particles, such as aggregates, into the individual component particles. (2) To distribute or suspend fine particles, such as clay, in or throughout a dispersion medium, such as water.

dispersion medium — The portion of a colloidal system in which the *disperse phase* is distributed.

diversion dam — A structure or barrier build to divert part or all of the water of a stream to a different course.

drainage, excessive — Too much or too rapid loss of water from soils, either by percolation or by surface flow. Loss greater than that necessary to prevent the development of an anaerobic condition for any appreciable length of time.

dry aggregate — A compound or secondary soil particle that is not destroyed by dry sieving.

dryland farming — The practice of crop production in low-rainfall areas without irrigation.

dry-weight percentage — The ratio of the weight of any constituent (of a soil) to the oven-dry weight of the soil. See oven-dry soil.

dust mulch — A loose, finely granular, or powdery condition on the surface of the soil, usually produced by shallow cultivation.

dy — Finely divided, partially decomposed organic material accumulated in peat soils in the transition zone between the peat and the underlying mineral material.

dynamometer — An instrument for measuring draft of tillage implements and for measuring resistance of soil to penetration by tillage implements.

E

ecology — The science that deals with the interrelations of organisms and their environment.

ecosystem — A community of organisms and the surroundings in which they live.

ectodynamomorphic soils — Soils with properties that have been produced or influenced mainly by factors other than parent material. See endodynamomorphic soils.

edaphology — The science that deals with the influence of soils on living things, particularly plants, including man's use of land for plant growth.

effective precipitation — That portion of the total precipitation which becomes available for plant growth.

eluviation — The removal of soil material in suspension (or in solution) from a layer or layers of a soil. (Usually, the loss of material in *solution* is described by the term "leaching". See illuviation and leaching).

endodynamomorphic soils — Soils with properties that have been influenced primarily by parent material.

endotrophic — Nourished or receiving nourishment from within, as fungi or their hyphae receiving nourishment from plant roots in a mycorrihizal association.

equivalent diameter — In sedimentation analysis the diameter assigned to a nonspherical particle, it being numerically equal to the diameter of a spherical particle of the same density and velocity of fall.

equivalent weight of a soil colloid — The weight of clay or organic colloid that has a combining power equivalent to 1 gram atomic weight of hydrogen.

erodible — Susceptible to erosion. (Expressed by terms such as highly erodible, slightly erodible, etc.)

erosion — (1) The wearing away of the land surface by running water, wind, ice, or other geological agents, including such processes as gravitational creep. (2) Detachment and movement of soil or rock by water, wind, ice, or gravity. The following terms are used to describe different types of water erosion:

- **accelerated erosion** — Erosion much more rapid than normal, natural, geological erosion, primarily as a result of the influence of the activities of man or, in some cases, of animals.
- **geological erosion** — The normal or natural erosion caused by geological processes acting over long geological periods. Synonymous with **natural erosion.**
- **gully erosion** — The erosion process whereby water accumulates in narrow channels and, over short periods, removes the soil from this narrow area to considerable depths, ranging from 0.5 meter to as much as 25 to 30 meters.
- **interrill erosion** — The removal of a fairly uniform layer of soil on a multitude of relatively small areas by splash due to raindrop impact and by film flow.
- **natural erosion** — Wearing away of the earth's surface by water, ice, or other natural agents under natural environmental conditions of climate, vegetation, etc., undisturbed by man. See **geological erosion**.
- **normal erosion** — The gradual erosion of land used by man which does not greatly exceed natural erosion. See **natural erosion.**
- **rill erosion** — An erosion process in which numerous small channels of only several centimeters in depth are formed; occurs mainly on recently cultivated soils. See rill.
- **sheet erosion** — The removal of soil from the land surface by rainfall and surface runoff. Often interpreted to include rill and interrill erosion.
- **splash erosion** — The detachment and airborne movement of small soil particles caused by the impact of raindrops on soils.

erosion classes — A grouping of erosion conditions based on the degree of erosion or on characteristic patterns. (Applied to accelerated erosion; not to normal, natural, or geological erosion.) Four erosion classes are recognized for water erosion and three for wind erosion. Specific definitions for each vary somewhat from one climatic zone, or major soil group, to another. (For details see Soil Survey Staff, SCS, USDA, 1951. Soil survey manual, USDA Handbook 18 Washington, D.C.)

erosion index — A measure of the erosive potential of a specific rainfall event. In the Universal Soil Loss Equation it is defined as the product of two rainstorm characteristics: total kinetic energy of the storm times its maximum 30-minute intensity (EI).

erosion pavement — A layer of coarse fragments, such as sand or gravel, remaining on the surface of the ground after the removal of fine particles by erosion.

eutrophic — Having concentrations of nutrients optimal, or nearly so, for plant or animal growth. (Said of nutrient or soil solutions and bodies of water.)

evapotranspiration — The combined loss of water from a given area, and during a specific period of time, by evaporation from the soil surface and by transpiration from plants.

exchange capacity — The total ionic charge of the adsorption complex active in the adsorption of ions. See anion-exchange capacity and cation-exchange capacity.

exchangeable-cation percentage — The extent to which the adsorption complex of a soil is occupied by a particular cation. It is expressed as follows:

$$ECP = \frac{\text{exchangeable cation (meq/100 g soil)}}{\text{cation-exchange capacity (meq/100 g soil)}} \times 100$$

exchangeable phosphate — The phosphate anion reversibly attached to the surface of the solid phase of the soil in such form that it may go into solution by anionic equilibrium reactions with isotopes of phosphorus or with other anions of the liquid phase without solution of the colloid phase to which it was attached.

exchangeable potassium — The potassium that is held by the adsorption complex of the soil and is easily exchanged with the cation of neutral nonpotassium salt solutions.

exchangeable-sodium percentage — The percentage of the cation-exchange capacity of a soil occupied by sodium. It is expressed as follows:

$$ESP = \frac{\text{exchangeable sodium (meq/100 g soil)}}{\text{cation-exchange capacity (meq/100 g soil)}} \times 100$$

F

F layer — A layer of partially decomposed litter with portions of plant structures still recognizable. Occurs below the L layer (O11 horizon) on the forest floor in forest soils. It is the fermentation layer or the 012 layer. See L layer and soil horizon.

fertility, soil — The status of a soil with respect to its ability to supply the nutrients essential to plant growth.

fertilizer, complete — Used formerly to denote a fertilizer containing appreciable contents of N, P, and K but now also to include secondary and micronutrients essential for plant growth.

fertilizer, compound — A fertilizer formulated with two or more fertilizer compounds or ingredients.

fertilizer grade — The grade of a fertilizer indicates the nutrient content expressed in terms of percentages of N, P_20_5, and K_20, in that order.

fertilizer salt index — The ratio of the increase in osmotic pressure by a fertilizer compound or mixture to that produced by the same weight of $NaNO_3$ ($NaNO_3$ = 100).

fertilizer, starter — A relatively small application of fertilizer applied with or near the seed for accelerating early growth of the crop.

fine sand — (1) A soil separate. See soil separates. (2) A soil textural class. See soil texture.

fine sandy loam — See soil texture.

fine texture — Consisting of or containing large quantities of the fine fractions, particularly of silt and clay. (Includes all clay loams and clays; that is, clay loam, sandy clay loam, silty clay loam, sandy clay, silty clay, and clay textural classes. Sometimes subdivided into clayey texture and moderately fine texture.) See soil texture.

fixation — The process of conversion of an element in the soil essential to plants from a readily available to a less available form.

fixed ammonium — The ammonium in soil that cannot be replaced by a neutral potassium salt solution (e.g., 1*N*KC1). See ammonium fixation.

flagstone — A relatively thin fragment, 6 to 15 in. long, of sandstone, limestone, slate, shale, or rarely, of schist. See coarse fragments.

flow velocity (of water in soil) — The volume of water transferred per unit of time and per unit of area normal to the direction of the net flow.

fragipan — A natural subsurface horizon with high bulk density relative to the solum above, seemingly cemented when dry, but when moist showing a moderate to weak brittleness. The layer is low in organic matter, mottled, slowly or very slowly permeable to water, and usually shows occasional or frequent bleached cracks forming polygons. It may be found in profiles of either cultivated or virgin soils but not in calcareous material.

flux — Rate of transfer of a quantity (water, heat, etc.) across a surface.

flux density — Flux across a unit cross-sectional area.

fritted trace elements — Sintered silicates having total guaranteed analyses of micronutrients with controlled (relatively slow) release characteristics.

fulvic acid — A term of varied usage but usually referring to the mixture of organic substances remaining in solution upon acidification of a dilute alkali extract from the soil.

G

gilgai — The microrelief of soils produced by expansion and contraction with changes in moisture. Found in soils that contain large amounts of clay which swells and shrinks considerably with wetting and drying. Usually a succession of microbasins and microknolls in nearly level areas or of microvalleys and microridges parallel to the direction of the slope. See microrelief.

glacial drift — Rock debris that has been transported by glaciers and deposited, either directly from the ice or from the meltwater. The debris may or may not be heterogeneous.

glaciofluvial deposits — Material moved by glaciers and subsequently sorted and deposited by streams flowing from the melting ice. The deposits are stratified and may occur in the form of outwash plains, deltas, kames, eskers, and kame terraces. See glacial drift and till, (1).

granule — A natural soil aggregate or ped which is relatively nonporous. See soil structure and soil structure types.

Gray Desert soil — A term used in Russia, and frequently in the U.S., synonymously with desert soil.

green manure — Plant material incorporated with the soil while green, or soon after maturity, for improving the soil.

gully — A channel resulting from erosion and caused by the concentrated but intermittent flow of water usually during and immediately following heavy rains. Deep enough to interfere with, and not to be obliterated by, normal tillage operations.

gyttja — Sedimentary peat consisting mainly of plant and animal residues precipitated from standing water.

gypsum requirements — The quantity of gypsum or its equivalent required to reduce the exchangeable-sodium percentage of a given increment of soil to an acceptable level.

H

H layer — A layer occurring in mor humus consisting of well-decomposed organic matter of unrecognizable origin.

halomorphic soil — Saline and alkali soils formed under imperfect drainage in arid regions.

halophytic vegetation — Vegetation requiring or tolerating a saline environment.

hardpan — A hardened soil layer, in the lower A or in the B horizon, caused by cementation of soil particles with organic matter or with materials such as silica, sesquioxides, or calcium carbonate. The hardness does not change appreciably with changes in moisture content and pieces of the hard layer do not slake in water. See caliche and claypan.

heavy metals — Those metals which have densities > 5.0.

heterotrophic — An organism capable of deriving energy for life processes from the oxidation of organic compounds.

heterotrophic nitrification — Biochemical oxidation of reduced forms of nitrogen (e.g., ammonium) to nitrate by heterotrophic microorganisms.

humic acid — A mixture of dark-colored substances of indefinite composition extracted from soil with dilute alkali and precipitated by acidification.

humification — The process involved in the decomposition of organic matter and leading to the formation of humus.

humin — The fraction of the soil organic matter that is not dissolved upon extraction of the soil with dilute alkali.

humus — (1) That more or less stable fraction of the soil organic matter remaining after the major portion of added plant and animal residues have decomposed. Usually it is dark colored. (2) Includes the F and H layers in undisturbed forest soils. See soil organic matter, mor, mull, and soil horizon.

hydrogenic soil — Soil developed under the influence of water standing within the profile for considerable periods; formed mainly in cold, humid regions.

hydrologic cycle — The fate of water from the time of precipitation until the water has been returned to the atmosphere by evaporation and is again ready to be precipitated.

hydrology — The science dealing with the distribution and movement of water.

agrohydrology — The science dealing with the distribution and movement of water and soil solution to and from the root zone in agricultural land, and with the distribution and movement of irrigation and surface water in conveyance systems on agricultural land.

surface hydrology — The science dealing with the distribution and conveyance of water on the soil surface.

soil water hydrology — The science dealing with the distribution and movement of the soil solution in the soil profile.

ground-water hydrology — The science dealing with the movement of the soil solution in the saturated zone of the soil profile.

hydromorphic soils — Soils formed under conditions of poor drainage in marshes, swamps, seepage areas, or flats.

hydroxy-aluminum interlayers — Polymers of general composition $[Al(OH)_{3-x}]m^{+nx}$ which are adsorbed on interlayer cation exchange sites. Although not exchangeable by unbuffered salt solutions, they are responsible for a considerable portion of the titratable acidity (and pH-dependent charge) in soils.

hymatomelanic acid — The alcohol-soluble fraction of humic acid.

hysteresis — Nonuniqueness in the relationship between two quantities, in particular in the relationship between the soil water content and the soil-water matric potential.

I

igneous rock — Rock formed from the cooling and solidification of magma, and that has not been changed appreciably since its formation.

illuvial horizon — A soil layer at horizon in which material carried from an overlying layer has been precipitated from solution or deposited from suspension. The layer of accumulation.

illuviation — The process of deposition of soil material removed from one horizon to another in the soil; usually from an upper to a lower horizon in the soil profile.

immature soil — A soil with indistinct or only slightly developed horizons because of the relatively short time it has been subjected to the various soil-forming processes. A soil that has not reached equilibrium with its environment.

immobilization — The conversion of an element from the inorganic to the organic form in microbial tissues or in plant tissues.

impervious — Resistant to penetration by fluids or by roots.

indicator plants — Plants characteristic of specific soil or site conditions.

infiltration — The downward entry of water into the soil.

infiltration rate — The rate at which water enters the soil. It has the dimensions of velocity.

infiltrometer — A device for measuring the rate of entry of fluid into a porous body, e.g., water into soil.

insoluble — As applied to phosphoric oxide in fertilizer, insoluble refers to that portion of the total that is neither soluble in water nor in neutral ammonium citrate solution. As applied to potash and nitrogen, it refers to insolubility in water. As applied to phosphate rock, it refers to the portion insoluble in sulfuric acid, e.g., SiO_2.

intergrade — A soil that possesses moderately well-developed distinguishing characteristics of two or more genetically related great soil groups.

intrazonal soils — A soil with more or less well-developed soil characteristics that reflect the dominating influence of some local factor of relief, parent material, or age, over the normal effect of climate and vegetation.

intrinsic permeability — The property of a porous material that relates to the ease with which gases or liquids can pass through it. The Darcy "k" multiplied by η/pg, where η = the viscosity of the fluid in poises, p = the density of the fluid in g/cm^3, and g = the acceleration of gravity in cm/sec^2.

ions — Atoms, groups of atoms, or compounds, which are electrically charged as a result of the loss of electrons (cations) or the gain of electrons (anions).

iron-pan — An indurated soil horizon in which iron oxide is the principal cementing agent.

irrigation — The artificial application of water to the soil for the benefit of growing crops.

irrigation efficiency — The ratio of the water actually consumed by crops on an irrigated area to the amount of water diverted from the source onto the area.

irrigation methods — The manner in which water is artifically applied to an area. The methods and the manner of applying the water are as follows:

- **border-strip** — The water is applied at the upper end of a strip with Earth borders to confine the water to the strip.
- **check-basin** — The water is applied rapidly to relatively level plots surrounded by levees. The basin is a small check.
- **corrugation** — The water is applied to small, closely-spaced furrows, frequently in grain and forage crops, to confine the flow of irrigation water to one direction.
- **flooding** — The water is released from field ditches and allowed to flood over the land.
- **furrow** — The water is applied to row crops in ditches made by tillage implements.
- **sprinkler** — The water is sprayed over the soil surface through nozzles from a pressure system.
- **subirrigation** — The water is applied in open ditches or tile lines until the water table is raised sufficiently to wet the soil.
- **wild-flooding** — The water is released at high points in the field and distribution is uncontrolled.

isodyne — Points of equal dynamometer pull of a cultivating implement; a line on a map of a cultivated field connecting points of equal dynamometer pull.

isomorphous substitution — The replacement of one atom by another of similar size in a crystal without disrupting or changing the crystal structure of the mineral.

K

kame — An irregular ridge or hill of stratified glacial drift.

kaolin — (1) A group name for aluminum silicates of approximately the composition $2H_20A1_20_3\ 2SiO_2$, and (2) kaolinite is the clay mineral that characterizes most kaolins.

L

L layer (litter) — The surface layer of the forest floor consisting of freshly fallen leaves, needles, twigs, stems, bark, and fruits. This layer may be very thin or absent during the growing season. The 01 horizon.

labile — A substance in soil which readily undergoes transformation or is readily available to plants.

lacustrine deposit — Material deposited in lake water and later exposed either by lowering of the water level or by the elevation of the land.

lagg — The depressed margin of a raised bog.

land classification — The arrangement of land units into various categories based upon the properties of the land or its suitability for some particular purpose.

land forming — Tillage operations which move soil to create desired soil configurations. Forming may be done on a large scale such as gully filling or terracing, or on a small scale such as contouring, ridging, or bedding.

landslide — (1) A mass of material which has slipped down hill under the influence of gravity, frequently assisted by water (that is, when the material is saturated). (2) Rapid movement down slope of a mass of soil, rock, or debris.

Lateritic soil — Soils formed in warm, temperate, and tropical regions.

Latosol — Soils formed under forested, tropical, humid conditions and characterized by low silica-sesquioxide ratios of the clay fractions, low base-exchange capacity, low activity of the clay, low content of most primary minerals, low content of soluble constituents, a high degree of aggregate stability, and usually having a red color.

lattice — A three-dimensional grid of lines connecting the points representing the centers of atoms or ions in a crystal.

leaching — The removal of materials in solution from the soil.

leaching fraction — The fraction of applied irrigation water passing through the root zone under a given water management regime.

leaching requirement — The fraction of applied irrigation water that should be passed through the root zone to control soil salinity at a specified level.

lime, agricultural — A soil amendment consisting principally of calcium carbonate but including magnesium carbonate and perhaps other materials used to neutralize soil acidity and furnish calcium and magnesium for the growth of plants.

lime concretion — An aggregate of precipitated calcium carbonate, or of other material cemented by precipitated calcium carbonate.

lime-pan — A hardened layer cemented by calcium carbonate.

lime requirement — The mass of agricultural lime, or the equivalent of other specified liming material, required to raise the pH of the soil to a specified value under field conditions.

liquid fertilizers — This term applies to anhydrous and aqua ammonia, nitrogen solutions, and liquid mixed fertilizers including clear liquids, suspensions of solids in liquids, and slurry-type mixtures requiring constant stirring to keep the solids suspended in the liquid.

liquid limit — The minimum percentage (by weight) of moisture at which a small sample of soil will barely flow under a standard treatment. Synonymous with "upper plastic limit."

listing (middlebreaking) — A tillage and land-forming operation using a tool which turns two furrows laterally in opposite directions, thereby producing beds or ridges.

lister planting — A method of planting in which the seed is planted in the bottom of lister furrows, usually simultaneously with the opening of these furrows.

lithosequence — A group of related soils that differ, one from the other, in certain properties primarily as a result of differences in the parent rock as a soil-forming factor.

Lithosols — Soils characterized by an incomplete solum or no clearly expressed soil morphology and consisting of freshly and imperfectly weathered rock or rock fragments.

loam — A soil textural class. See soil texture.

loamy — Intermediate in texture and properties between fine-textured and coarse-textured soils. Includes all textural classes with the words "loam" or "loamy" as a part of the class name, such as clay loam or loamy sand. See loam and soil texture.

loamy coarse sand — See soil texture.

loamy sand — See soil texture.

loamy fine sand — See soil texture.

loamy very fine sand — See soil texture.

loess — Material transported and deposited by wind and consisting of predominantly silt-sized particles.

luxury uptake — The absorption by plants of nutrients in excess of their need for growth. Luxury concentrations during early growth may be utilized in later growth.

lysimeter — (1) A device for measuring percolation and leaching losses from a column of soil under controlled conditions. (2) A device for measuring gains (irrigation, precipitation, and condensation) and losses (evapotranspiration) by a column of soil.

M

macronutrients — Includes the primary and secondary nutrients nitrogen, phosphorus, potassium, calcium, magnesium, and sulfur.

made land — Areas filled with earth, or with earth and trash mixed, usually by or under the control of man. A miscellaneous land type.

maintenance application — Application of fertilizer materials in amounts and at intervals to insure a certain minimum level of an available nutrient.

manure — (1) The excreta of animals, with or without the admixture of bedding or litter, in varying stages of decomposition. Also referred to as "barnyard manure" or "stable manure." (The usual meaning of the term as used in the U.S.; in many other countries "manure" refers to any fertilizer.)

marl — Soft and unconsolidated calcium carbonate, usually mixed with varying amounts of clay or other impurities.

marsh — Periodically wet or continually flooded areas with the surface not deeply submerged. Covered dominantly with sedges, cattails, rushes, or other hydrophytic plants. Subclasses include fresh-water and salt-water marshes. See swamp.

mass flow (nutrient) — The movement of solutes associated with net movement of water.

mature soil — A soil with well-developed soil horizons produced by the natural processes of soil formation and essentially in equilibrium with its present environment.

meander land — Unsurveyed land along a lake shore or stream border that has developed by the receding of the shore line or of the stream since the last cadastral survey of the area. A miscellaneous land type.

medium-texture — Intermediate between fine-textured and coarse-textured (soils). (It includes the following textural classes: very fine sandy loam, loam, silt loam, and silt.)

mellow soil — A very soft, very friable, porous soil without any tendency toward hardness or harshness.

mesophilic bacteria — Bacteria whose optimum temperature for growth falls in an intermediate range of approximately 15 to 45C.

metamorphic rock — Rock derived from pre-existing rocks but that differ from them in physical, chemical, and mineralogical properties as a result of natural geological processes, principally heat and pressure, originating within the earth. The preexisting rocks may have been igneous, sedimentary, or another form of metamorphic rock.

microclimate — (1) The climatic condition of a small area resulting from the modification of the general climatic conditions by local differences in elevation or exposure. (2) The sequence of atmospheric changes within a very small region.

microflora — Bacteria, including actinomycetes, viruses, and fungi.

micronutrients — Also referred to as trace or minor elements, fertilizer micronutrients include boron, chlorine, cobalt, copper, iron, manganese, molybdenum, and zinc. While their functions as plant nutrients are not fully understood, most of them are necessary for the activation of certain enzymes and for plant growth.

microrelief — Small-scale, local differences in topography, including mounds, swales, or pits that are only a few feet in diameter and with elevation differences of up to 6 ft.

mine wash — Water-deposited accumulations of sandy, silty, or clayey material recently eroded in mining operations. It may clog streams and channels, and damage land on which it is deposited. A miscellaneous land type.

mineralization — The conversion of an element from an organic form to an inorganic state as a result of microbial decomposition.

mineralogical analysis — The estimation or determination of the kinds or amounts of minerals present in a rock or in a soil.

mineral soil — A soil consisting predominantly of, and having its properties determined predominantly by, mineral matter. Usually contains < 20% organic matter, but may contain an organic surface layer up to 30 cm thick.

minimum tillage — The minimum soil manipulation necessary for crop production or meeting tillage requirements under the existing soil and climatic conditions.

miscellaneous land type — A mapping unit for areas of land that have little or no natural soil, that are too nearly inaccessible for orderly examination, or that for any reason it is not feasible to classify the soil.

moderately-coarse texture — Consisting predominantly of coarse particles. (In soil textural classification, it includes all the sandy loams except the very fine sandy loam.)

moderately-fine texture — Consisting predominantly of intermediate-size (soil) particles or with relatively small amounts of fine or coarse particles. (In soil textural classification, it includes clay loam, sandy clay loam, and silty clay loam.)

moisture retention curve — A graph showing the soil moisture percentage (by weight or by volume) vs. applied tension (or pressure). Points on the graph are usually obtained by increasing (or decreasing) the applied tension or pressure over a specified range.

moisture tension (or pressure) — The equivalent negative pressure in the soil water. It is equal to the equivalent pressure that must be applied to the soil water to bring it to hydraulic equilibrium, through a porous permeable wall or membrane, with a pool of water of the same composition. The pressures used and the corresponding percentages most commonly determined are:

fifteen-atmosphere percentage — The percentage of water contained in a soil that has been saturated, subjected to, and is in equilibrium with, an applied pressure of 15 atm. (Pressure applied in a pressure membrane or ceramic pressure plant apparatus. Usually expressed as a weight percentage but may be expressed as a volume percentage. Approximately the same as 15-bar percentage.)

fifteen-bar percentage — The percentage of water contained in a soil that has been saturated, subjected to, and is in equilibrium with, an applied pressure of 15 bars. (Pressure applied in a pressure membrane or ceramic pressure plant apparatus. Usually expressed as a weight percentage but may be expressed as a volume percentage. Approxmately the same as the 15-atm percentage.)

one-third-atmosphere percentage — The percentage of water contained in a soil that has been saturated, subjected to, and is in equilibrium with, an applied pressure of 1/3 atm. (Pressure applied in a ceramic plate apparatus. Usually expressed as a weight percentage, but may be expressed as a volume percentage. Approximately the same as one-third bar percentage. Also, for medium to coarse-textured soils approximately numerically equal to moisture equivalent.)

one-third bar percentage — The percentage of water contained in a soil that has been saturated, subjected to, and is in equilibrium with, an applied pressure of 1/3 bar. (Pressure applied in a ceramic plate apparatus. Usually expressed as a weight percentage. Approximately the same as one-third atmosphere percentage. Also, for medium- to coarse-textured soils approximately numerically equal to moisture equivalent.)

sixty-centimeter percentage — The percentage of water contained in a soil that has been saturated, subjected to, and is in equilibrium with, an applied pressure (or tension) equivalent to a column of water 60 cm high. (The pressure may be applied in a pressure plate apparatus or, as a tension, on a tension table. May be expressed on a weight or volume basis. Considered by many to approximate the "field-moisture capacity," especially in medium- to coarse-textured soils.)

moisture volume percentage — The ratio of the volume of water in a soil to the total bulk volume of the soil.

moisture weight percentage — The moisture content expressed as a percentage of the oven-dry weight of soil.

montmorillonite — An aluminosilicate clay mineral with a 2:1 expansible layer structure; that is, with two silicon tetrahedral sheets enclosing an aluminum octahedral sheet. Considerable expansion may be caused along the C-axis by water moving between silica layers of contiguous units.

mor — A type of forest humus in which the H layer is present and in which there is practically no mixing of surface organic matter with mineral soil; that is, the transition from the H layer to the A1 horizon is abrupt. (Sometimes differentiated into thick mor, thin mor, granular mor, greasy mor, or felty mor.)

mottled zone — A layer that is marked with spots or blotches of different color or shades of color. The pattern of mottling and the size, abundance, and color contrast of the mottles may vary considerably and should be specified in soil description.

mottling — Spots or blotches of different color or shades of color interspersed with the dominant color.

muck — Highly decomposed organic material in which the original plant parts are not recognizable. Contains more mineral matter and is usually darker in color than peat.

muck soil — (1) A soil containing between 20 and 50% of organic matter. (2) An organic soil in which the organic matter is well decomposed (U.S. usage).

mulch — (1) Any material such as straw, sawdust, leaves, plastic film, loose soil, etc., that is spread upon the surface of the soil to protect the soil and plant roots from the effects of raindrops, soil crusting, freezing, evaporation, etc. (2) To apply mulch to the soil surface.

mulch farming — A system of farming in which the organic residues are not plowed into or otherwise mixed with the soil but are left on the surface as a mulch.

mulch tillage — Tillage or preparation of the soil in such a way that plant residues or other materials are left to cover the surface; also, mulch farming, trash farming, stubble mulch tillage, plowless farming.

mull — A type of forest humus in which the F layer may or may not be present and in which there is no H layer. The A1 horizon consists of an intimate mixture of organic matter and mineral soil with gradual transition between the A1 and the horizon beneath. (Sometimes differentiated into firm mull, sand mull, coarse mull, medium mull, and fine mull.)

mycorrhiza — Literally "fungus root". The association, usually symbiotic, of specific fungi with the roots of higher plants.

N

neutral soil — A soil in which the surface layer, at least to normal plow depth, is neither acid nor alkaline in reaction.

nitrate assimilation — Uptake of nitrate by living organisms and conversion to cell substances (e.g., protein).

nitrification — Process of formation of nitrates in the soil, from other compounds of nitrogen, by various microorganisms or bacteria.

nitrogen assimilation — The incorporation of nitrogen into organic cell substances by living organisms.

nitrogen cycle — The sequence of biochemical changes undergone by nitrogen wherein it is used by a living organism, liberated upon the death and decomposition of the organism, and converted to its original state of oxidation.

nitrogen fixation — Biological conversion of molecular dinitrogen (N_2) to organic combinations or to forms utilizable in biological processes.

nitrogen solutions — Solutions of nitrogenous fertilizers in water. Solutions are used in manufacturing liquid or dry mixed fertilizers and/or are applied to the soil either with special applicators or in irrigation water.

nitrogenous material — Materials which contain nitrogen, whether in organic or inorganic form.

nutrient antagonism — A reciprocal relationship among concentrations of two or more nutrients in plants (K and Mg, Fe and Mn, etc.).

nutrient balance — An as-yet undefined ratio among concentrations of nutrients essential for plant growth which permits maximum growth rate and yield. An imbalance results when one or more nutrients are present in either deficient or excess supply.

nutrient concentration vs. content — Terms often used interchangeably in regard to nutrients in plants. Concentration is specific, usually expressed in percent; content is synonymous with uptake unless used with percent.

nutrient, diffusion — The movement of nutrients in soil that results from a concentration gradient.

nutrient interaction — A statistical term used to denote departure from additive responses to two or more nutrients applied separately, and together.

O

oil wasteland — Land on which oily waters have accumulated. Includes slush pits and adjacent areas affected by oil waste. A miscellaneous land type.

organic phosphorus — Phosphorus present as a constituent of an organic compound, or a group of organic compounds such as glycerophosphoric acid, inositol phosphoric acid, cytidlyic acid, etc.

organic soil — A soil which contains a high percentage (> 15 or 20%) of organic matter throughout the solum.

orstein — An indurated layer in the B horizon of Podzols in which the cementing material consists of illuviated sesquioxides (mostly iron) and organic matter.

oven-dry soil — Soil which has been dried at 105C until it reaches constant weight.

oxidation ditch — An artificial open channel for partial digestion of liquid organic wastes in which the wastes are circulated and aerated by a mechanical device.

P

pans — Horizons or layers, in soils, that are strongly compacted, indurated, or very high in clay content.

pan, genetic — A natural subsurface soil layer of low or very low permeability, with a high concentration of small particles, and differing in certain physical and chemical properties from the soil immediately above or below the pan.

pan, pressure or induced — A subsurface horizon or soil layer having a higher bulk density and a lower total porosity than the soil directly above or below it, as a result of pressure that has been applied by normal tillage operations or by other artifical means. Frequently referred to as plowpan, plowsole, or traffic pan.

parent material — The unconsolidated and more or less chemically weathered mineral or organic matter from which the solum of soils is developed by pedogenic processes.

particle density — The mass per unit volume of the soil particles. In technical work, usually expressed as grams per cubic centimeter. See bulk density, soil.

particle size — The effective diameter of a particle measured by sedimentation, sieving, or micrometric methods.

particle-size analysis — Determination of the various amounts of the different separates in a soil sample, usually by sedimentation, sieving, micrometry, or combinations of these methods.

particle-size distribution — The amounts of the various soil separates in a soil sample, usually expressed as weight percentages.

parts per million (ppm) — Weight units of any given substance per 1 million weight units (of oven-dry soil); or, in the case of (soil) solution, the weight units of solute per million weight units of (soil) solution.

peat — Unconsolidated soil material consisting largely of undecomposed, or only slightly decomposed, organic matter accumulated under conditions of excessive moisture.

peat soil — An organic soil containing more than 50% organic matter. Used in the U.S. to refer to the stage of decomposition of the organic matter, "peat" referring to the slightly decomposed or undecomposed deposits and "muck" to the highly decomposed materials. See peat, muck, and muck soil.

ped — A unit of soil structure such as an aggregate, crumb, prism, block, or granule, formed by natural processes (in contrast with a clod, which is formed artificially).

peneplain — A once high, rugged area which has been reduced by erosion to a low, gently rolling surface resembling a plain.

penetrability — The ease with which a probe can be pushed into the soil. (May be expressed in units of distance, speed, force, or work depending on the type of penetrometer used.)

percolation, soil water — The downward movement of water through soil. Especially, the downward flow of water in saturated, or nearly saturated soil at hydraulic gradients of the order of 1.0 or less.

permafrost — (1) Permanently frozen material underlying the solum. (2) A perennially frozen soil horizon.

permanent wilting percentage — The water content of a soil when indicator plants growing in that soil wilt and fail to recover when placed in a humid chamber. Often estimated by the 15-bar percentage.

permeability, soil — The ease with which gases, liquids, or plant roots penetrate or pass through a bulk mass of soil or a layer of soil. Since different soil horizons vary in permeability, the particular horizon under question should be designated.

pH soil — The negative logarithm of the hydrogen-ion activity of a soil. The degree of acidity (or alkalinity) of a soil as determined by means of a glass, quinhydrone, or other suitable electrode or indicator at a specified moisture content or soil-water ratio, and expressed in terms of the pH scale.

pH-dependent charge — The portion of the cation or anion exchange capacity which varies with pH. See acidity, residual.

phase, soil — A subdivision of a soil type or other unit of classification having characteristics that affect the use and management of the soil but which do not vary sufficiently to differentiate it as a separate type. A variation in a property of characteristic such as degree of slope, degree of erosion, content of stones, etc.

phosphate rock — Naturally occurring ore containing one or more calcium phosphate minerals of sufficient purity and quantity as to permit its use in commercial product manufacture.

phosphates — Salts of phosphoric acid (H_3PO_4) may be formed by combining phosphoric acid with an alkali or metal. Phosphates of sodium, potassium, ammonium, and calcium, especially the latter two, are used in fertilizers.

phosphoric acids — Usually a solution of P_2O_5 in water but frequently referred to merely as P_2O_5. More accurate terms for P_2O_5 are phosphoric anhydride, phosphorus pentoxide, or phosphoric oxide.

phosphorus efficient plant — Plants able to absorb, translocate, and utilize P effectively in growth.

physical weathering — The breakdown of rock and mineral particles into smaller particles by physical forces such as frost action.

phytomorphic soils — (Canada) Well-drained soils of an association which have developed under the dominant influence of the natural vegetation characteristic of a region.

plant food — Any substance that contains in readily available form significant amounts of the nutrients required for the proper growth and development of plants.

plant nutrient — An element which is absorbed by plants and is necessary for completion of the life cycle.

plastic soil — A soil capable of being molded or deformed continuously and permanently, by relatively moderate pressure, into various shapes.

plastic limit — The minimum moisture percentage by weight at which a small sample of soil material can be deformed without rupture. Synonymous with "lower plastic limit." See liquid limit and plasticity number.

plasticity number — The numerical difference between the liquid and the plastic limit or, synonymously, between the lower plastic limit and the upper plastic limit. Sometimes called "plasticity index."

plasticity range — The range of moisture weight percentage within which a small sample of soil exhibits plastic properties.

platy — Consisting of soil aggregates that are developed predominately along the horizontal axes; laminated; flaky.

pocosin — A swamp, usually containing organic soil, and partly or completely enclosed by a sandy rim. The Carolina Bays of the Southeastern U.S.

Podzol — Soils formed in cool-temperate to temperate, humid climates, under coniferous or mixed coniferous and deciduous forest, and characterized particularly by a highly-leached, whitish-gray (Podzol) A2 horizon.

podzolization — A process of soil formation resulting in the genesis of Podzols and Podzolic soils.

polyphosphates — Salts of polyphosphoric acids such as ammonium polyphosphates and calcium polyphosphates.

polyphosphoric acid — Any of a series of phosphoric acids whose molecular structure contains more than one atom of phosphorus.

pore-size distribution — The volume of the various sizes of pores in a soil. Expressed as percentages of the bulk volume (soil plus pore space).

pore space — Total space not occupied by soil particles in a bulk volume of soil.

porosity — The volume percentage of the total bulk not occupied by solid particles.

potash (K_2O) potassium oxide — The trade term "potash" is used interchangeably with the word "potassium" and expresses percentage of potassium oxide (K_2O) in potash salts and mixtures.

potassium-supplying power of soils — The capacity of the soil to supply potassium to growing plants from both the exchangeable and the moderately available forms.

Prairie soils — Soils formed under temperate to cool-temperate, humid regions under tall grass vegetation.

precipitation interception — The stopping, interrupting and temporary holding of precipitation in any form by a vegetative canopy or vegetation residue.

pressure membrane — A membrane, permeable to water and only very slightly permeable to gas when wet, through which water can escape from a soil sample in response to a pressure gradient.

primary mineral — A mineral that has not been altered chemically since deposition and crystallization from molten lava.

prismatic soil structure — A soil structure type with prismlike aggregates that have a vertical axis much longer than the horizontal axes.

productivity, soil — The capacity of a soil, in its normal environment, for producing a specified plant or sequence of plants under a specified system of management. The "specified" limitations are necessary since no soil can produce all crops with equal success nor can a single system management produce the same effect on all soils. Productivity emphasizes the capacity of soil to produce crops and should be expressed in terms of yields.

profile, soil — A vertical section of the soil through all its horizons and extending into the parent material.

psammophytes — Plants which prefer or tolerate sand, particularly fine to medium sand, as a habitat.

R

R horizon — See soil horizon.

rainfall interception — See precipitation interception.

reaction, soil — The degree of acidity or alkalinity of a soil, usually expressed as a pH value. Descriptive terms commonly associated with certain ranges in pH are: extremely acid, <4.5; very strongly acid, 4.5—5.0; strongly acid, 5.1—5.5; moderately acid, 5.6—6.0; slightly acid, 6.1—6.5; neutral, 6.6—7.3; slightly alkaline, 7.4—7.8; moderately alkaline, 7.9—8.4; strongly alkaline, 8.5—9.0; and very strongly alkaline, >9.1.

Red Desert soil — Soils formed under warm-temperate to hot, dry regions under desert-type vegetation, mostly shrubs.

red earth — Highly leached, red clayey soils of the humid tropics, usually with very deep profiles that are low in silica and high in sesquioxides.

Red-Yellow Podzolic soils — Soils formed under warm-temperate to tropical, humid climates, under deciduous or coniferous forest vegetation and usually, except for a few members of the Yellow Podzolic Group, under conditions of good drainage.

regolith — The unconsolidated mantle of weathered rock and soil material on the earth's surface; loose earth materials above solid rock. (Approximately equivalent to the term "soil" as used by many engineers.)

Regosol — Any soil without definite genetic horizons and developing from or on deep, unconsolidated, soft mineral deposits such as sand, loess, or glacial drift.

Rendzina — Soils with brown or black friable surface horizons underlain by light gray to pale yellow calcareous material; developed from soft, highly calcareous parent material under grass vegetation or mixed grasses and forest in humid and semiarid climates.

residual material — Unconsolidated and partly weathered mineral materials accumulated by disintegration of consolidated rock in place.

residual shrinkage — The decrease in the bulk volume of soil in addition to that caused by the loss of water.

retentivity profile, soil — A graph showing the retaining capacity of a soil as a function of depth. The retaining capacity may be for water, for water at any given tension, for cations, or for any other substances held by soils.

reticulate mottling — A network of streaks of different color; most commonly found in the deeper profiles of Lateritic soils.

rhizobia — Bacteria capable of living symbiotically in roots of legumes, from which they receive energy and often utilize molecular nitrogen. Collective common name for the genus *Rhizobium*.

rhizoplane — The external surface of plant roots. (As used by some, it includes the surface of the adhering soil particles.)

rhizosphere — The zone of soil where the microbial population is altered both quantitatively and qualitatively by the presence of plant roots.

river wash — Barren alluvial land, usually coarse-textured, exposed along streams at low water and subject to shifting during normal high water.

rock land — Areas containing frequent rock outcrops and shallow soils. Rock outcrops usually occupy from 25 to 90% of the area.

rough broken land — Land with very steep topography and numerous intermittent drainage channels but usually covered with vegetation.

rubble land — Land areas with 90% or more of the surface covered with stones and boulders.

runoff — That portion of the precipitation on an area which is discharged from the area through stream channels. That which is lost without entering the soil is called surface runoff and that which enters the soil before reaching the stream is called ground water runoff or seepage flow from ground water. (In soil science "runoff" usually refers to the water lost by surface flow; in geology and hydraulics "runoff" usually includes both surface and subsurface flow.)

S

saline-sodic soil — A soil containing a combination of soluble salts and exchangeable sodium sufficient to interfere with the growth of most crop plants. The electrical conductivity and sodium-adsorption ratio of the saturation extract are > 2 mmol/cm at 25°C and > 15, respectively. The pH is usually 8.5 or less in the saturated soil paste. (Formerly called saline-alkali soil.)

saline soil — A nonsodic soil containing sufficient soluble salt to impair its productivity. The electrical conductivity of the saturation extract is > 2 mmho/cm at 25°C.
salinization — The process whereby soluble salts accumulate in soil.
salt balance — The relation between the quantity of dissolved salts carried to an area in irrigation water and the quantity of dissolved salts removed by drainage water.
salt-affected soil — Soil that has been adversely modified for the growth of most crop plants by the presence of soluble salts, exchangeable sodium, or both.
salt index — Salt index is a measure of the relative tendency of a fertilizer to increase the osmotic pressure of the soil solution as compared to the increase caused by an equal weight of sodium nitrate as a reference material.
salt tolerance — (1) The relation between average soil salinity and plant yield. (2) The average soil salinity required to produce a specified decrease in plant yield.
sand — (1) A soil particle between 0.05 and 2.0 mm in diameter. (2) Any one of five soil separates, namely, very coarse sand, coarse sand, medium sand, fine sand, and very fine sand. See soil separates. (3) A soil textural class.
sandy — Containing a large amount of sand. (Applied to any one of the soil classes that contains a large percentage of sand.)
sand clay loam — A soil textural class.
sandy clay — A soil textural class.
sandy loam — A soil textural class. See class, soil, and soil texture.
saturate — (1) To fill all the voids between soil particles with a liquid. (2) To form the most concentrated solution possible under a given set of physical conditions in the presence of an excess of the solute. (3) To fill to capacity, as the adsorption complex with a cation species, e.g., H-saturated, etc.
saturation extract — An increment of solution obtained from a saturated soil paste.
scoria land — Areas of slaglike clinkers, burned shale, and fine-grained sandstone; characteristic of burned-out coal beds. Such areas commonly support a sparse cover of grasses, but are of low agricultural value.
secondary mineral — A mineral resulting from the decomposition of a primary mineral or from the reprecipitation of the products of decomposition of a primary mineral. See primary mineral.
secondary nutrient elements — Calcium, magnesium, and sulfur are the secondary macronutrient elements, so-called because they are essential to plant growth in lesser quantity than the primary macronutrients.
sedimentary rock — A rock formed from materials deposited from suspension or precipitated from solution and usually being more or less consolidated. The principal sedimentary rocks are sandstone, shales, limestones, and conglomerates.
self-mulching soil — A soil in which the surface layer becomes so well aggregated that it does not crust and seal under the impact of rain but instead serves as a surface mulch upon drying.
shear — Force, as of a tillage implement, acting at right angles to the direction of movement.
Sierozem — Soils with pale grayish A horizons grading into calcareous material at a depth of 1 ft or less, and formed in temperate to cool, arid climates under a vegetation of desert plants, short grass, and scattered brush.
silica-aluminum ratio — The molecules of silicon dioxide (SiO_2) per molecule of aluminum oxide (Al_2O_3) in clay minerals or in soils.
silica-sesquioxide ratio — The molecules of silicon dioxide (SiO_2) per molecule of aluminum oxide (Al_2O_3) plus ferric oxide (Fe_2O_3) in clay minerals or in soils.
silt — (1) A soil separate consisting of particles between 0.05 and 0.002 mm in equivalent diameter. See soil separates. (2) A soil textural class. See soil texture.

silt loam — A soil textural class containing a large amount of silt and small quantities of sand and clay. See soil texture and class, soil.

silty clay — A soil textural class containing a relatively large amount of silt and clay and a small amount of sand. See soil texture and class, soil.

silty clay loam — A soil textural class containing a relatively large amount of silt, a lesser quantity of clay, and a still smaller quantity of sand. See class, soil, and soil texture.

site index — (1) A quantitative evaluation of the productivity of a soil for forest growth under the existing or specified environment. (2) The height in feet of the dominant forest vegetation taken at or calculated to an index age, usually 50 or 100 years.

slickens — Fine-textured materials separated in placer mining and in ore-mill operations; may be detrimental to plant growth unless confined in specially constructed basins.

slit planting — A method of mulch tillage. Two to four weeks prior to seeding time an area 30 to 50 cm wide is thoroughly tilled by plowing, middle busting, or disk tilling to form a conventional seedbed for each row. At planting time, or at first cultivation, the remaining living mulch material in the row middle is cut loose and killed or retarded.

slow release — A fertilizer term used interchangeably with delayed release, controlled release, controlled availability, slow acting, and metered release to designate a rate of dissolution (usually in water) much less than is obtained for completely water-soluble compounds. Slow release may involve either compounds which dissolve slowly or soluble compounds coated with substances impermeable to water.

sodic soil — (1) A soil containing sufficient exchangeable-sodium to interfere with the growth of most crop plants. (2) A soil in which the sodium-adsorption ratio of the saturation extract is 15 or more.

sodication — The process whereby the exchangeable sodium content of a soil is increased.

sodium-adsorption ratio (SAR) — A relation between soluble sodium and soluble divalent cations which can be used to predict the exchangeable-sodium percentage of soil equilibrated with a given solution. It is defined as follows:

$$\text{SAR} = \frac{\text{sodium, mmoles}/\ell}{(\text{calcium} + \text{magnesium})^{1/2}\ (\text{mmoles}/\ell)^{1/2}}$$

soil amendment — Any material, such as lime, gypsum, sawdust, or synthetic conditioner, that is worked into the soil to make it more amenable to plant growth. Amendments may contain important fertilizer elements, but the term commonly refers to added materials other than fertilizer.

soil complex — A mapping unit used in detailed soil surveys where two or more defined taxonomic units are so intimately intermixed geographically that it is undesirable or impractical, because of the scale being used, to separate them. A more intimate mixing of smaller areas of individual taxonomic units than that described under soil association.

soil-formation factors — The variable, usually interrelated natural agencies that are active in and responsible for the formation of soil. The factors are usually grouped into five major categories: parent rock, climate, organisms, topography, and time.

soil genesis — (1) The mode of origin of the soil with special reference to the processes or soil-forming factors responsible for the development of the solum, or true soil, from the unconsolidated parent material. (2) A division of soil science concerned with soil genesis (1).

soil horizon — A layer of soil or soil material approximately parallel to the land surface and differing from adjacent genetically related layers in physical, chemical, and biological properties or characteristics such as color, structure, texture, consistency, kinds, and numbers of organisms present, degree of acidity or alkalinity.

soil management — The combination of all tillage operations, cropping practices, fertilizer, lime, and other treatments conducted on or applied to the soil for the production of plants.

soil map — A map showing the distribution of soil types or other soil mapping units in relation to the prominent physical and cultural features of the Earth's surface. The following kinds of soil maps are recognized in the U.S.

soil map, detailed — A soil map on which the boundaries are shown between all soil types that are significant to potential use as field-management systems. The scale of the map will depend upon the purpose to be served, the intensity of land use, the pattern of soils, and the scale of the other cartographic materials available. Traverses are usually made at 1/4-mile, or more frequent, intervals. Commonly a scale of 4 in. = 1 mile (1:15,840) is now used for field mapping in the U.S.

soil map, detailed reconnaissance — A reconnaissance map on which some areas or features are shown in greater detail than usual, or than others.

soil map, generalized — A small-scale map which shows the general distribution of soils within a large area and thus in less detail than on a detailed soil map. Generalized soil maps may vary from soil association maps of a county, on a scale of 1 in. = 1 mile (1:63,360), to maps of larger regions showing associations dominated by one or more great soil groups.

soil map, reconnaissance — A map showing the distribution of soils over a large area as determined by traversing the area at intervals varying from about 1/2 mile to several miles. The units shown are soil associations. Such a map is usually made only for exploratory purposes to outline areas of soil suitable for more intensive development. The scale is usually much smaller than for detailed soil maps.

soil map, schematic — A soil map compiled from scant knowledge of the soils of new and undeveloped regions by the application of available information about the soil-formation factors of the area. Usually on a small scale (1:1,000,000 or smaller). See soil-formation factors.

soil mechanics and engineering — A subspecialization of soil science concerned with the effect of forces on the soil and application of engineering principles to problems involving the soil.

soil microbiology — A subspecialization of soil science concerned with soil-inhabiting microorganisms and with their relation to agriculture, including both plant and animal growth.

soil mineral — (1) Any mineral that occurs as a part of or in the soil. (2) A natural inorganic compound with definite physical, chemical, and crystalline properties (within the limits of isomorphism), that occurs in the soil. See clay mineral.

soil separates — Mineral particles, < 2.0 mm in equivalent diameter, ranging between specified size limits. The names and size limits of separates recognized in the U.S. are very coarse sand,* 2.0 to 1.0 mm; coarse sand, 1.0 to 0.5 mm; medium sand, 0.5 to 0.25 mm; fine sand, 0.25 to 0.10 mm; very fine sand, 0.10 to 0.05 mm; silt, 0.05 to 0.002 mm; and clay,** < 0.0002 mm.

* Prior to 1947 this separate was called "fine gravel;" now fine gravel includes particles between 2.0 mm and about 12.5 mm in diameter.

** Prior to 1937, "clay" included particles < 0.005 mm in diameter, and "silt," those particles from 0.05 to 0.005 mm.

The separates recognized by the International Society of Soil Science are (1) coarse sand, 2.0 to 0.2 mm; (2) fine sand, 0.2 to 0.02 mm; (3) silt, 0.02 to 0.002 mm; and (4) clay, < 0.002 mm.

soil structure — The combination or arrangement of primary soil particles into secondary particles, units, or peds. These secondary units may be, but usually are not, arranged in the profile in such a manner as to give a distinctive characteristic pattern. The secondary units are characterized and classified on the basis of size, shape, and degree of distinctness into classes, types, and grades, respectively.

soil structure classes — A grouping of soil structural units or peds on the basis of size.

soil structure grades — A grouping or classification of soil structure on the basis of inter- and intra-aggregate adhesion, cohesion, or stability within the profile. Four grades of structure designated from 0 to 3 are recognized:

0 **structureless** — no observable aggregation or no definite and orderly arrangement of natural lines of weakness. Massive, if coherent; single-grain, in noncoherent.

1 **weak** — poorly formed indistinct peds, barely observable in place.

2 **moderate** — well-formed distinct peds, moderately durable and evident, but not distinct in undisturbed soil.

3 **strong** — durable peds that are quite evident in undisturbed soil, adhere weakly to one another, withstand displacement, and become separated when the soil is disturbed.

soil survey — The systematic examination, description, classification, and mapping of soils in an area. Soil surveys are classified according to the kind and intensity of field examination.

soil test — A chemical, physical, or microbiological operation which estimates a property of the soil pertinent to the suitability of the soil to support plant growth. (Sometimes used as an adjective to define fractions of soil components, e.g., "soil test phosphorus").

soil texture — The relative proportions of the various soil separates in a soil as described by the classes of soil texture. The limits of the various classes and subclasses are as follows:

sand — Soil material that contains 85% or more of sand; percentage of silt, plus 1.5 times the percentage of clay, shall not exceed 15.

coarse sand — 25% or more very coarse and coarse sand, and < 50% of any other one grade of sand.

sand — 25% or more very coarse, coarse, and medium sand, and < 50% fine or very fine sand.

fine sand — 50% or more fine sand (or) < 25% very coarse, coarse, and medium sand and < 50% very fine sand.

very fine sand — 50% or more very fine sand.

loamy sand — Soil material that contains at the upper limit 85 to 90% sand, and the percentage of silt plus 1.5 times the percentage of clay is not less than 15; at the lower limit it contains not less than 70 to 85% sand, and the percentage of silt plus twice the percentage of clay does not exceed 30.

loamy coarse sand — 25% or more very coarse and coarse sand, and < 50% any other one grade of sand.

loamy sand — 25% or more very coarse, coarse, and medium sand, and < 50% fine or very fine sand.

loamy fine sand — 50% or more fine sand (or) < 25% very coarse, coarse, and medium sand and < 50% very fine sand.

loamy very fine sand — 50% or more very fine sand.

sandy loam — Soil material that contains either 20% clay or less, and the percentage of silt plus twice the percentage of clay exceeds 30, and 52% or more sand; or < 7% clay, < 50% silt, and between43% and 52% sand.

coarse sandy loam — 25% or more very coarse and coarse sand and < 50% any other one grade of sand.

sandy loam — 30% or more very coarse, coarse, and medium sand, but < 25% very coarse sand, and < 30% very fine or fine sand.

fine sandy loam — 30% or more fine sand and < 30% very fine sand (or) between 15 and 30% very coarse, coarse, and medium sand.

very fine sandy loam — 30% or more very fine (or) > 40% fine and very fine sand, at least half of which is very fine sand < 15% very coarse, coarse, and medium sand.

loam — Soil material that contains 7 to 27% clay, 28 to 50% silt, and < 52% sand.

silt loam — Soil material that contains 50% or more silt and 12 to 27% clay (or) 50 to 80% silt and < 12% clay.

silt — Soil material that contains 80% or more silt and < 12% clay.

sandy clay loam — Soil material that contains 20 to 35% clay, < 28% silt, and 45% or more sand.

clay loam — Soil material that contains 27 to 40% clay and 20 to 45% sand.

silty clay loam — Soil material that contains 27 to 40% clay and < 20% sand.

sandy clay — Soil material that contains 35% or more clay and 45% or more sand.

silty clay — Soil material that contains 40% or more clay and 40% or more silt.

clay — Soil material that contains 40% or more clay, < 45% sand, and < 40% silt.

soil water — Water in soil is subject to several force fields originating from: the presence of the soil solid phase; the dissolved salts; the action of external gas pressure; and, the gravitational field. These effects may be quantitatively expressed by assigning an individual component potential to each. The sum of these potentials is designated the total potential of soil water and may be identified with the partial specific Gibbs free energy of the soil water relative to free pure water at the same temperature. Pure water refers to the chemically pure compound H_2O.

total potential (of soil water) — The amount of work that must be done per unit quantity of pure water in order to transport reversibly and isothermally an infinitesimal quantity of water from a pool of pure water, at a specified elevation and at atmospheric pressure, to the soil water (at the point under consideration). The total potential (of soil water) consists of the following:

osmotic potential — The amount of work that must be done per unit quantity of pure water in order to transport reversibly and isothermally an infinitesimal quantity of water from a pool of pure water, at a specified elevation and at atmospheric pressure, to a pool of water identical in composition to the equilibrium soil solution (at the point under consideration), but in all other respects being identical to the reference pool.

gravitational potential — The amount of work that must be done per unit quantity of pure water in order to transport reversibly and isothermally an infinitesimal quantity of water, identical in composition to the soil water, from a pool at a specified elevation and at atmospheric pressure, to a similar pool at the elevation of the point under consideration.

matric potential — The amount of work that must be done per unit quantity of pure water in order to transport reversibly and isothermally an infinitesimal quantity of water, identical in composition to the soil water, from a pool at the elevation and the external gas pressure of the point under consideration, to the soil water.

gas pressure potential — This potential component is to be considered only when external gas pressure differs from atmospheric pressure as, e.g., in a pressure membrane apparatus. A specific term and definition is not given.

soil water pressure — The pressure (positive or negative), relative to the external gas pressure on the soil water, to which a solution identical in composition to the soil water must be subjected in order to be in equilibrium through a porous permeable wall with the soil water. May be identified with the **matric potential** defined above.

osmotic pressure — The pressure to which a pool of water identical in composition to the soil water, must be subjected in order to be in equilibrium, through a semipermeable membrane, with a pool of pure water (semipermeable means permeable only to water.) May be identified with the **osmotic potential** defined above.

total pressure — The pressure (positive or negative) relative to the external gas pressure on the soil water, to which a pool of pure water must be subjected in order to be in equilibrium through a semipermeable membrane with the soil water. Total pressure is thus equal to the sum of soil water pressure and negative value of the osmotic pressure. Total pressure may also be derived from the measurement of the partial pressure of the water vapor in equilibrium with the soil water. May be identified with the *total potential* defined above when gravitational and external gas pressure potentials can be neglected.

hydraulic head — The elevation with respect to a specified reference (level), usually the soil surface, at which water stands in a piezometer connected to the point in question in the soil. Its definition can be extended to soil above the water table if the piezometer is replaced by a tensiometer. The hydraulic head in systems under atmospheric pressure may be identified with a potential expressed in terms of the height of a water column. More specifically, it can be identified with the sum of gravitational and capillary potentials, and may be termed the *hydraulic potential.* For unsaturated soil, the hydraulic head, referenced to the soil surface is equal to the soil suction (— cm) at a given point plus the depth (— cm) to that point, with positive taken as upwards.

water content — The amount of water lost from the soil upon drying to constant weight at 105°C; expressed either as the weight of water per unit weight of dry soil or as the volume of water per unit bulk volume of soil.

differential water capacity — The absolute value of the rate of change of water content with soil water pressure. The water capacity at a given water content will depend on the particular desorption or adsorption curve employed. Distinction should be made between volumetric and specific water capacity.

Terms Relating to the Movement of Water in Soil

Experimentally it has been established that generally the flow of a fluid in a porous medium can be described by Darcy's law which states that the flux of fluid is proportional to the driving force. In viscous flow of water in soils, the driving force equals the negative gradient of the hydraulic potential.

hydraulic conductivity — The proportionality factor in Darcy's law as applied to the viscous flow of water in soil, i.e., the flux of water per unit gradient of hydraulic potential. If conditions require that the viscosity of the fluid be divorced from the conductivity of the medium, it is convenient to define the permeability (intrinsic permeability has been used in some publications) of the soil as the conductivity, expressed in $g^{-1}cm^2$ sec multiplied by the viscosity in poise. For the purpose of solving the partial differential equation of the nonsteady-state flow in unsaturated soil it is often convenient to introduce a variable termed the soil water diffusivity.

soil water diffusivity — The hydraulic conductivity divided by the differential water capacity (care being taken to be consistent with units), or the flux of water per unit gradient of moisture content in the absence of other force fields.

Symbol, Dimension, and Unit for the Above Given Terms When One Gram Mass is Used as Unit Quantity of Water

Term	Symbol[a]	Dimension	Unit
Total potential	↓	L^2T^{-2}	erg g^{-1}, joule kg^{-1}
Osmotic potential	O	L^2T^{-2}	erg g^{-1}, joule kg^{-1}
Gravitational potential	Z	L^2T^{-2}	erg g^{-1}, joule kg^{-1}
Capillary potential	M	L^2T^{-2}	erg g^{-1}, joule kg^{-1}
Soil water pressure		$ML^{-1}T^{-2}$	dyne cm^{-2}, bar, cm water, cm Hg
Osmotic pressure		$ML^{-1}T^{-2}$	dyne cm^{-2}, bar, cm water, cm Hg
Total pressure		$ML^{-1}T^{-2}$	dyne cm^{-2}, bar, cm water, cm Hg
Hydraulic head	H	L	cm, m
Hydraulic potential	ϕ	L^2T^{-2}	erg g^{-1}, joule kg^{-1}
Water content	w		cm^3cm^{-3}, g g^{-1}
Differential water capacity	C	$M^{-1}LT^2$	cm^2 $dyne^{-1}$, bar^{-1}
Hydraulic conductivity	K	b	b
Permeability	k	L^2	cm^2, Darcy
Soil water diffusivity	D	L^2T^{-1}	cm^2 sec^{-1}

[a] Symbols as C, K, and D may have w or h as a subscript if they in the same paper are used for water as well as for heat.

[b] The dimension would depend on the units used to describe the driving force as shown in the table below:

		Hydraulic conductivity	
Driving force	Dimension	Dimension	Unit
Hydraulic potential gradient	LT^{-2}	T	sec
Hydraulic head gradient	LL^{-1}	LT^{-1}	cm sec^{-1}
Pressure gradient	$ML^{-2}T^{-2}$	$M^{-1}L^3T$	g^{-1} cm^3 sec

solclime — The temperature and moisture conditions of the soil; the soil climate.

solodized soil — A soil that has been subjected to the processes responsible for the development of a Soloth and having at least some of the characteristics of a Soloth.

Solonchak — Soils with gray, thin, salty crust on the surface, and with fine granular mulch immediately below being underlain with grayish, friable, salty soil; formed under subhumid to arid, hot or cool climate, under conditions of poor drainage, and under a sparse growth of halophytic grasses, shrubs, and some trees.

Solonetz — Soils with a very thin, friable, surface soil underlain by a dark, hard columnar layer usually highly alkaline; formed under subhumid to arid, hot to cool climates, under better drainage than Solonchaks, and under a native vegetation of halophytic plants.

solubility — To be available to plants a nutrient must be at least slightly soluble in the soil solution. The amount of substance that will dissolve at a given temperature in 100 parts of water is known as its solubility.

soluble-sodium percentage (SSP) — The proportion of sodium ions in solution in relation to the total cation concentration, defined as follows:

$$\text{SSP} = \frac{\text{soluble-sodium concentration (meq/}\ell\text{)}}{\text{total cation concentration (meq/}\ell\text{)}} \times 100$$

solum (plural: sola) — The upper and most weathered part of the soil profile; the A and B horizons.

spatial variability — The variation in soil properties both vertically downward through the soil profile and laterally across the soil surface at a given depth or with a given horizon.

speciality fertilizers — Fertilizers recommended or used principally for growing flowers, lawns, home gardens, house plants, or for any purpose other than growing agricultural crops.

starter fertilizer — Liquid or solid fertilizer, placed near or in contact with the seed or seedling roots, is commonly considered a starter fertilizer.

sticky point — (1) A condition of consistency at which the soil barely fails to stick to a foreigh object. (2) Specifically and numerically, the weight moisture percentage of a well-mixed, kneaded soil that barely fails to adhere to a polished nickel or stainless steel surface when the shearing speed is 5 cm/sec.

Stokes' law — An equation relating the terminal settling velocity of a smooth, rigid sphere in a viscous fluid of known density and viscosity to the diameter of the sphere when subjected to a known force field. Used in particle-size analysis of soils by the pipette, hydrometer, or centrifuge methods. The equation is

$$V = \frac{2gr^2 (d_1 - d_2)}{9\eta}$$

where

V = velocity of fall (cm sec^{-1})
g = acceleration of gravity (cm sec^{-2})
t = "equivalent" radium of particle (cm)
d_1 = density of particle (g cm^{-3})
d_2 = density of medium (g cm^{-3})
η = viscosity of medium (dyne sec cm^{-2})

structural charge — The negative charge on a clay mineral which is caused by ismorphous substitution within the layer. (Expressed as equivalents per formula weight of a clay or meg/100 g or per gram of clay.)

structure index — Any measurement of a soil physical property, such as aggregation, porosity, permeability to air or water, or bulk density, that denotes or indicates the structural condition of a soil.

stubble mulch — The stubble of crops or crop residue left essentially in place on the land as a surface cover before and during the preparation of the seedbed at least partly during the growing of a succeeding crop.

subsoiling — Breaking of compact subsoils, without inverting them, with a special knife-like instrument (chisel) which is pulled through the soil at depths usually of 30 to 60 cm and at spacings usually of 60 to 150 cm.

subsurface tillage — Tillage with a special sweeplike plow or blade which is drawn beneath the surface at depths of several inches and cuts plant roots and loosens the soil without inverting it or without incorporating the surface cover.

sulfur cycle — The sequence of transformation whereby sulfur is oxidized or reduced through both organic or inorganic products.

summation curve, particle size — A curve showing the accumulative percentage by weight of particles within increasing (or decreasing) size limits as a function of diameter; the percent by weight of each size fraction is plotted accumulatively on the ordinate as a function of the total range of diameters represented in the sample plotted on the abscissa.

surface-charge density — The excess of negative or positive charge per unit of surface area of soil or soil mineral.

surface soil — The uppermost part of the soil, ordinarily moved in tillage, or its equivalent in uncultivated soils and ranging in depth from 3 or 4 in. to 8 or 10. Frequently designated as the "plow layer", the "Ap layer", or the "Ap horizon".

specific water capacity — The rate of change of soil water content with change in soil water matric potential.

suspension fertilizer — Liquid fertilizer in which the solids are held in suspension (prevented from settling) by the use of a suspending agent, usually a swelling type clay. It must be fluid enough to be mixed, pumped, and applied to the soil and yet remain homogeneous during application.

symbiosis — The living together in intimate association of two dissimilar organisms, the cohabitation being mutually beneficial.

symmetry concentration — That quantity of cations (or anions) equivalent to the exchange capacity of a soil. For example, if the cation-exchange capacity of a soil is 10 meq/100 g of soil, then 1 symmetry concentration is 10 meq of any cation.

symmetry value — The percentage of the adsorbed ion released when one symmetry concentration of another ion is added.

T

talus — Fragments of rock and other soil material accumulated by gravity at the foot of cliffs or steep slopes.

talud — A short, steep slope formed gradually at the downslope margin of a field by deposition against a hedge, a stone wall, or other similar barrier.

tensiometer — A device for measuring the soil-water matric potential (or tension) of water in soil *in situ;* a porous, permeable ceramic cup connected through a water-filled tube to a manometer, vacuum gauge, pressure transducer, or other pressure measuring device.

terrace — (1) A level, usually narrow, plain bordering a river, lake, or the sea. Rivers sometimes are bordered by terraces at different levels. (2) A raised, more or less level or horizontal strip of earth usually constructed on or nearly on a contour and supported on the downslope side by rocks or other similar barrier and designed to make the land suitable for tillage and to prevent accelerated erosion. For example, the ancient terraces built by the Incas in the Andes. (3) An embankment with the uphill side sloping toward and into a channel for conducting water, and the downhill side having a relatively sharp decline; constructed across the direction of the slope for the purpose of conducting water from the area above the terrace at a regulated rate of flow and to prevent the accumulation of large volumes of water on the downslope side of cultivated fields. The depth of the channel, the width of the terrace ridge, and the spacings of the terraces on a field are varied with soil types, cropping systems, climatic conditions, and other factors.

thermal analysis — Measurement of changes in physical or chemical properties of materials as a function of temperature, usually heating or cooling at a uniform rate. (1) DTA, measures temperature difference (ΔT) between a sample and reference material. (2) DSC, measures the differential heat flow between a sample and reference material. (3) TG, measures weight loss or gain.

thermogenic soils — Soils with properties that have been influenced primarily by high temperature as the dominant soil-formation factor; developed in subtropical and equatorial regions.

thermophilic organisms — Organisms which grow readily at temperatures above 45°C.

thermosequence — A sequence of related soils that differ, one from the other, primarily as a result of temperature as a soil-formation factor.

till — (1) Unstratified glacial drift deposited directly by the ice and consisting of clay, sand, gravel, and boulders intermingled in any proportion. (2) To plow and prepare for seeding; to seed or cultivate the soil.

topdressing — An application of fertilizer to a soil after the crop stand has been established.

topsequence — A sequence of related soils that differ, one from the other, primarily because of topography as a soil-formation factor. See clinosequence.

tortuosity factor — The reciprocal of the increase in diffusion path that an ion must take in diffusing through the water present in the soil when it moves along a concentration gradient as compared to the path in water.

truncated — Having lost all or part of the upper soil horizon of horizons.

tuff — Volcanic ash usually more or less stratified and in various states of consolidation.

tundra — A level or undulating treeless plain characteristic of artic regions.

tundra soils — (1) Soils characteristic of tundra regions. (2) Soils with dark-brown peaty layers over grayish horizons mottled with rust and having continually frozen substrata; formed under frigid, humid climates, with poor drainage, and native vegetation of lichens, moss, flowering plants, and shrubs.

U

underground runoff (seepage) — Water flowing toward stream channels after infiltration into the ground.

undifferentiated soil groups — Soil mapping units in which two or more similar taxonomic soil units occur, but not in a regular geographic association. For example, the steep phases of two or more similar soils might be shown as a unit on a map because topography dominates the properties.

unit — The term "unit" represents percent or 20 pounds of nutrient (N, P_2O_5, or K_20) in a 2000-lb ton of fertilizer. Thus, a 5-20-10 fertilizer contains five units of N, 20 units of P_20_5, and 10 units of K_20 per ton.

unsaturated flow — The movement of water in a soil which is not completely filled with water.

V

varve — A distinct band representing the annual deposit in sedimentary materials regardless of origin and usually consisting of two layers, one a thick, light-covered layer of silt and fine sand and the other a thin, dark-colored layer of clay.

vescular arbuscular (VA) mycorrhiza — A common endomycorrhizal association produced by phycomycetous fungi of the genus *Endogone.* Host range includes many agricultural and horticultural crops.

vertical mulching — A subsoiling operation in which a vertical band of mulching material is placed into a vertical slit in the soil, for example, immediately behind the soil-opening implement.

volcanic-ash land — Areas of volcanic ash so recently deposited that the ash is essentially unmodified and shows little or no evidence of soil development. The areas have almost no vegetation on them. A miscellaneous land type.

W

water-stable aggregate — A soil aggregate which is stable to the action of water such as falling drops, or agitation as in wet-sieving analysis.

water table — The upper surface of ground water or that level in the ground where the water is at atmospheric pressure.
water table, perched — The water table of a saturated layer of soil which is separated from an underlying saturated layer by an unsaturated layer.
water use efficiency — Dry matter or harvested portion of crop produced per unit of water consumed.

X

xerophytes — Plants that grow in or on extremely dry soils or soil materials.

Y

yield, sustained — A continual annual, or periodic, yield of plants or plant material from an area; implies management practices which will maintain the productive capacity of the land.

Z

zonal soil — A soil characteristic of a large area, or zone.
zymogenous flora — Organisms found in soils in large numbers immediately following the addition of readily decomposable organic materials.

REFERENCES

1. *Glossary of Soil Science Terms*, Soil Science Society of America, Madison, Wis., 1978.
2. **Kilmer, V. J.**, Soil fertility — key to abundant food, *Our Land and Its Care*, The Fertilizer Institute, Washington, D.C. 1977.

INDEX

A

D

E

F

G

H

N

O

P

Q

T

U

V

W

X

Y

Z